T0235194

Much of the world's surface, even under the oceans, is covered in thick deposits of sedimentary particles – gravel, sand, silt, and clay. The nature of the deposits and their formation is very much dependent on the distribution of particles of different sizes. However, different instruments measure different attributes of a particle's size, based on how fast a particle settles in water, or the surface area of a particle, or its length. This book provides information on the how and why of particle size analysis in terms of understanding these sediment deposits.

Sponsored and encouraged by the International Union of Geological Sciences, this book presents a synthesis of the state of the art in particle size characterization. Thirty-three authors have combined their expertise to provide information on the latest theoretical principles, laboratory and field techniques of particle size analysis, and the manipulation and application of particle size data. The theory, procedure, calibration, and accuracy of various techniques are discussed, including settling tubes, sieves, image analysis, light scattering analysis, and thin-section analysis of rocks. Results of a world calibration (interinstrument and interlaboratory) experiment of particle size analyzers are highlighted, with recommendations for reporting size information in the scientific literature.

The needs both of research professionals and of students in earth, planetary, marine, and environmental sciences are addressed in this comprehensive and balanced work. Geographers, geologists, geophysicists, sedimentologists, stratigraphers, geotechnical engineers, geochemists, hydrologists, and oceanographers will all find this an important and useful resource.

Principles, methods, and application of particle size analysis

Principles, methods, and application of particle size analysis

Edited by

JAMES P. M. SYVITSKI

Atlantic Geoscience Center,
Geological Survey of Canada,
Bedford Institute of Oceanography

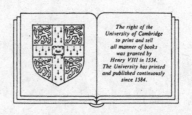

The right of the
University of Cambridge
to print and sell
all manner of books
was granted by
Henry VIII in 1534.
The University has printed
and published continuously
since 1584.

CAMBRIDGE UNIVERSITY PRESS

Cambridge
New York Port Chester Melbourne Sydney

CAMBRIDGE UNIVERSITY PRESS
Cambridge, New York, Melbourne, Madrid, Cape Town, Singapore, São Paulo

Cambridge University Press
The Edinburgh Building, Cambridge CB2 8RU, UK

Published in the United States of America by Cambridge University Press, New York

www.cambridge.org
Information on this title: www.cambridge.org/9780521364720

© Cambridge University Press 1991

This publication is in copyright. Subject to statutory exception
and to the provisions of relevant collective licensing agreements,
no reproduction of any part may take place without the written
permission of Cambridge University Press.

First published 1991
Reprinted 1997
This digitally printed version 2007

A catalogue record for this publication is available from the British Library

Library of Congress Cataloguing in Publication data

Principles, methods, and application of particle size analysis /
James P. M. Syvitski, editor.

   p.   cm.

Includes bibliographical references.

ISBN 0-521-36472-8

1. Particle size determination.   2. Sediments (Geology) – Analysis.
I. Syvitski, James P. M.
QE471.2.P75   1991
551.3′04 – dc20   90–25625

ISBN 978-0-521-36472-0 hardback
ISBN 978-0-521-04461-5 paperback

*Thanks to my new and
wonderful family:*

Dianne L. Syvitski and
Eric W. H. Hutton

"The voyage of discovery lies not in seeking new horizons, but in seeing with new eyes."

                              – Marcel Proust

Contents

Contributors

Y. C. AGRAWAL, Flow Research Inc., 21414-68th Street S., Kent, Washington 98032, USA.

KENNETH W. ASPREY, Geological Survey of Canada, Atlantic Geoscience Centre, Bedford Institute of Oceanography, Dartmouth, N.S., Canada B2Y 4A2.

A. J. BALE, Plymouth Marine Laboratory, Prospect Place, The Hoe, Plymouth PL1 2BP, UK.

DALE E. BUCKLEY, Geological Survey of Canada, Atlantic Geoscience Centre, Bedford Institute of Oceanography, Dartmouth, N.S., Canada B2Y 4A2.

CHRISTIAN CHRISTIANSEN, Department of Earth Sciences, University of Aarhus, Ny Munkegade Build. 520, 8000 Aarhus C, Denmark.

D. A. CLATTENBURG, Geological Survey of Canada, Atlantic Geoscience Centre, Bedford Institute of Oceanography, Dartmouth, N.S., Canada B2Y 4A2.

JOHN P. COAKLEY, Water Survey of Canada, National Water Research Institute, Burlington, Ontario, Canada L7R 4A6.

RAY E. CRANSTON, Geological Survey of Canada, Atlantic Geoscience Centre, Bedford Institute of Oceanography, Dartmouth, N.S., Canada B2Y 4A2.

KRISTIAN DALSGAARD, Department of Earth Science, University of Aarhus, Ny Munkegade, Build. 520, DK-8000, Aarhus C, Denmark.

J. K. GHOSH, Statistics-Mathematics Division, Indian Statistical Institute, Calcutta 700 035, India.

DANIEL HARTMANN, Geology & Mineralogy Department, Ben Gurion University of the Negev, Beer Sheva, Israel.

DAVID E. HEFFLER, Geological Survey of Canada, Atlantic Geoscience Centre, Bedford Institute of Oceanography, Dartmouth, N.S., Canada B2Y 4A2.

FRANCES J. HEIN, Department of Geology & Geophysics, University of Calgary, Calgary, Alberta, Canada T2N 1N4.

JENS LEDET JENSEN, Department of Theoretical Statistics, Institute of Mathematics, University of Aarhus, Ny Munkegade, Build. 530, DK-8000, Aarhus C, Denmark.

MIROSLAW JONASZ, Technicon Instruments Corp., Tarrytown, New York 10591, USA.

STEPHEN K. KENNEDY, Faculty of Arts and Sciences, Department of Geology and Planetary Science, University of Pittsburgh, Pittsburgh, Pennsylvania 15260, USA.

KATE KRANCK, Physical & Chemical Sciences Branch, Department of Fisheries & Oceans, Bedford Institute of Oceanography, Dartmouth, N.S., Canada B2Y 4A2.

K. WILLIAM G. LEBLANC, Geological Survey of Canada, Atlantic Geoscience Centre, Bedford Institute of Oceanography, Dartmouth, N.S., Canada B2Y 4A2.

MARTIN D. MATTHEWS, Texaco E&P Technology Division, 3901 Briarpark, Houston, Texas 77042, USA.

B. S. MAZUMDER, Illinois State Water Survey, University of Illinois, Champaign-Urbana, Illinois 61801 USA.

JIM MAZZULLO, Department of Geology, Texas A&M University, College Station, Texas 77843, USA.

I. N. MCCAVE, Department of Earth Sciences, University of Cambridge, Downing Street, Cambridge CB2 3EQ, UK.

T. G. MILLIGAN, Physical & Chemical Sciences Branch, Department of Fisheries & Oceans, Bedford Institute of Oceanography, Dartmouth, N.S., Canada B2Y 4A2.

A. W. MORRIS, Plymouth Marine Laboratory, Prospect Place, The Hoe, Plymouth PL1 2BP, UK.

JULIAN D. ORFORD, School of Geosciences, The Queen's University of Belfast, Belfast, BT7 1NN, Northern Ireland.

DAN B. PRAEG, Department of Geology, Northern Illinois University, DeKalb, Illinois 60115, USA.

J. B. RILEY, MIT Lincoln Laboratories, MS KB350, Box 73, Lexington, Massachusetts 02173, USA.

SUPRIYA SENGUPTA, Department of Geology and
    Geophysics, Indian Institute of Technology,
    Kharagpur 721 302, India.

SONG TIANRUI, Institute of Geology, Chinese
    Academy of Geological Sciences, 100037,
    Beijing, China.

MICHAEL SØRENSEN, Department of Theoretical
    Statistics, Institute of Mathematics, University
    of Aarhus, Ny Munkegade, Build. 530, DK-
    8000, Aarhus C, Denmark.

JAY A. STRAVERS, Department of Geology,
    Northern Illinois University, DeKalb, Illinois
    60115, USA.

JAMES P. M. SYVITSKI, Geological Survey of
    Canada, Atlantic Geoscience Centre, Bedford
    Institute of Oceanography, Dartmouth, N.S.,
    Canada B2Y 4A2.

WILLIAM F. TANNER, Regents Professor,
    Geology Department, Florida State University,
    Tallahassee, Florida 32306-3026, USA.

W. BRIAN WHALLEY, School of Geosciences,
    The Queen's University of Belfast, Belfast,
    BT7 1NN, Northern Ireland.

Preface

In 1983, members of the International Union of Geological Sciences – Committee on Sedimentology (IUGS–COS) expressed concerns on the state of automated particle size instruments used in the earth sciences and their apparent lack of calibration. The concerns of Professor E. Seibold (past president, IUGS) centered on the proliferation of user-built instruments (such as settling towers), that were essentially one of a kind and uncalibrated in the traditional analytical sense. Particle size information was typically reported in the international literature without concern for the analytical errors associated with such results. Were some of the presented data overinterpreted, considering the sampling, subsampling, and analytical techniques employed?

The IUGS–COS struck a working group on "modern methods of grain size analysis" in 1984, and I was asked by Dr. K. A. W. Crook (AUN, chairman IUGS–COS) to be its convener. Our objective was to determine the precision and accuracy of modern automated methods of particle size analysis and the role of these methods in investigation of earth processes and properties. Two meetings were convened (Dartmouth, Canada, in 1985, and Heidelberg, Germany, in 1987) with participants representing twenty countries. The working group agreed to initiate and participate in a world calibration experiment for particle size analyzers (see Chapter 13), and to publish these results together with overviews on automated instruments and the theory of particle size characterization.

Our result is this textbook formulated to address the needs of both students and research professionals in the earth science and oceanographic community. This community comprises:
1. sedimentologists and physical geographers, who use particle size data to understand the erosion, transport, and deposition of sediment;
2. geologists, who examine trends and patterns in the solid earth in response to surface processes of the past (stratigraphy, mapping, mineral deposits);

3. geotechnical and geological engineers, who determine the stability of sedimentary deposits under load;
4. geochemists and environmental scientists, who ascertain the kinetic reactions that occur between liquids and particles, either while particles are in motion or part of the solid earth; and
5. hydrogeologists, who study the flow of fluids through the solid earth, particularly sedimentary deposits that host the world's reserves of hydrocarbon deposits.

The twenty-four chapters that comprise this text are loosely organized into five parts. The first three introductory chapters (Part I) discuss the basic principles behind particle size analysis, including discussion on the nature of geological samples, the effect of grain shape and density on size measurement, the effect of pretreatment on geological samples, and the theory used in particle size analysis.

The ten chapters of Part II present the theory, methods, and calibration of the principal methods employed in particle size analyzers. These include settling tubes (Chapter 4), sieves (Chapter 5), image analysis (Chapter 6, size; Chapter 7, shape), electroresistance particle size analysis (Chapter 8), laser diffraction size analysis (Chapter 9), x-ray size analysis (Chapter 10), light scattering analysis (Chapter 11), and thin-section analysis of sedimentary rocks (Chapter 12). Chapter 13 presents the results of the IUGS–COS-sponsored interinstrument and interlaboratory experiment, where the precision and accuracy of these methods is ascertained using a variety of known standards.

Part III takes us out of the laboratory and into the geological environment, where particle size and concentration are being determined with minimal interference while the particles remain in motion. The science of in situ techniques remains immature, and we highlight just two techniques: laser diffraction (Chapter 14) and stereo photography (Chapter 15).

How particle size data are interpreted and

manipulated, the subject of Part IV of the book, is a most controversial subject in the earth science community. Hundreds of scientists have tried to discern patterns in particle size information so as to understand the modern geological environments and provide a means to reconstruct the environmental conditions of previous geological periods. Some of these attempts have been successful, but most are based on regional trends that could not be considered universal. The chapters presented in Part IV provide examples of advances in this long history of sedimentological research. Chapter 16 introduces the concept of suite statistics, wherein statistical parameters of a group of samples (rather than a single sample) are investigated. Chapter 17 describes size frequency distributions of sediment samples in terms of the hyperbolic distribution, a distribution type that includes both the exponential and Gaussian distributions as end members. Chapter 18 provides a theoretical and numerical examination of multivariate analysis of large grain size data sets using Q-mode factor analysis. Chapter 19 uses the theory of fluid dynamics, supported by laboratory experiments, to understand the nature of size frequency distributions.

Finally, Part V provides examples of how grain size data can be applied in the earth sciences. They include applications to stratigraphy (Chapter 20), glacial geology (Chapter 21), marine geochemistry (Chapter 22), oceanography (Chapter 23), and marine geotechniques (Chapter 24). Together these chapters provide overviews or case histories that demonstrate the need

for precise and accurate grain size data from sedimentary samples.

A reader, upon an initial glance at the table of contents, may wonder why the classical techniques of pipette and hydrometer, among other techniques developed over the past hundred years, were not included in a text on particle size analysis. These techniques are discussed throughout the book: In Chapter 1 they are described in terms of theory, and in many chapters they are considered in terms of interinstrument calibration. However, as clearly documented in Chapter 13 on instrument calibration, these manual techniques have had their day. They are imprecise and time consuming, and we encourage laboratories to consider their demise during new equipment acquisitions. They have been well described in at least a dozen sedimentological textbooks, most still available from earth science publishers.

Finally, I must add a disclaimer common to textbooks that describe commercial instruments and their performance. I, as a representative of the IUGS, in no way endorse or discredit any of the commercial instruments mentioned. Our contributors have presented facts that stand on their own merit. Readers should reach their own conclusions as to the appropriateness of a particular instrument or technique. Commercial instruments constantly change, mostly through improvements based on comments from scientists and design engineers.

James P. M. Syvitski
February 1991

Acknowledgments

First I thank Dr. Jiri Brezina (Granulometry, Germany) for cochairing the two IUGS–COS sponsored meetings dealing with automated size analyzers and their calibration. This text is a direct outgrowth of and response to those meetings. Dr. Keith Crook (ANU) is acknowledged for his leadership at the helm of the IUGS–COS and for recognizing the need for illumination in the field of geological (as opposed to engineering) particle size characterization.

My role as editor was helped considerably by experienced colleagues at the Geological Survey of Canada, Bedford Institute of Oceanography, whom I deeply thank. On behalf of all the contributors, I applaud K. W. G. LeBlanc, who cheerfully dealt with fifteen word processing languages, permission-to-publish communication, and technical aspects of the book format. Thank you, Bill.

W. Gregory (DEMR, Communications Branch) reviewed many of the manuscripts for style. K. W. Asprey and his technical team in the AGC Soft Sediment Laboratory provided the IUGS calibration standards and their distribution to participants. A. Cosgrove and his graphics team provided drafting assistance on a number of chapters. I thank my management within the Geological Survey of Canada for their support of my seemingly never-ending work on this book. This volume comprises GSC Contribution No. 26590.

Each chapter has been subjected to the normal review process pertaining to international scientific journals. Submissions from countries where English is not the dominant language received initial editing before external review. Manuscripts were each reviewed by two or three experts in the field. I have been fortunate to have the following distinguished list of reviewers offer their time and expertise:

S. Ackleson (BLOS)	M. Matthews (TRC)
C. Amos (DEMR)	J. Mazzullo (TA&M)
T. Bale (PML)	I.N. McCave (CU)
C. Christiansen (AU)	G. V. Middleton (MU)
M. Church (UBC)	T. Milligan (DFO)
J. Coakley (CCIW)	K. Moran (DEMR)
W. Coulbourn (HIG)	J. Orford (QUB)
R. Dalrymple (QU)	D. Piper (DEMR)
T. Day (SSC)	C. Schafer (GSC)
D. Forbes (GSC)	N. Silverberg (DFO)
R. Gilbert (QU)	D. Sego (UA)
F. Hein (CU)	R. Sheldon (DFO)
L. Jansa (GSC)	J. Stravers (UNI)
M. Jonasz (TI)	P. Stoffyn (BIO)
K. Kranck (DFO)	W. Tanner (FSU)
J. Kravitz (ONR)	B. Topliss (DFO)
K.W. LeBlanc (BIO)	G. Vilks (GSC)
S. LeRoy	G. Winters (DEMR)
D. Loring (DFO)	

I *Introduction*

There are two scientific disciplines that work with powders and their particle characterization. One is earth science, involving the study of natural deposits of gravels, sands, silts, and clays. The interest is not just in describing these varied deposits of sediment, but on ascertaining the origin of such deposit. Textbooks on lab methods have included Krumbein and Pettijohn (1938), Griffiths (1967), Carver (1971), and Folk (1974). The field is closely linked to the petroleum industry, mining industry, agriculture, forestry, fishing industry, and space programs.

The second discipline is that of powder technology, principally involving the chemical industry, where the properties of manmade powders and their quality control is of prime importance. Biomedical and military research, the paint industry, ceramic industry, and industrial incineration are but a few of the many sides to this research. Textbooks within this field include Allen (1981), Kaye (1981), and Barth (1984). Although a much younger science compared to geology, the field of powder technology has led the way in recent years, developing automated methods of particle size analysis. With close links to industry, automation means increased speed of analysis and precision of results, and thus increased profits.

Earth scientists have borrowed heavily from the powder technology industry, adapting many of the automated methods for use in their laboratories. In the earth sciences, the number of analyses required every year grows. Grossly approximated, the number of sediment samples analyzed worldwide is 800,000 per annum. At least 70% of these analyses involve automated size analyzers. Thus the field of powder technology has a significant impact on earth scientists, helping them cope with this ever-increasing deluge of samples to be analyzed.

In Part I, three authors, representing the triad of research (university, government, and industry), have provided their ideas on the nature of geological samples and the principles of size analysis. Although the theory of size-analytical methods based on a particle's settling rate should be familiar to many earth scientists, other fields of particle size theory, outlined in Chapter 1, may not be. They include optical attenuation and Mie theory, resistance pulse counting, laser diffraction spectroscopy, and photon correlation spectroscopy. Although this text does not address acoustic techniques for the determination of particle size, this field has become more of a science and less of an art with the passing of each year.

Chapter 2 reminds us that particle size may inherently be the most important fundamental property of a sediment sample. Secondary properties of a sediment, porosity and permeability, are very dependent on particle size. However, particle size is itself dependent on the user's definition and thus on two other fundamental properties of a sediment sample – grain density and grain shape.

Chapter 3 provides readers with a review on the nature of sediment samples and the effect of sample preparation (pretreatment). Are we interested in the size of natural particles, including pellets, aggregates, floccules, and agglomerates? Or are we interested in the nature of the size of individual grains that comprise these natural bundles of grains?

REFERENCES

Allen, T. (1981). *Particle Size Measurement.* London: Chapman & Hall.
Barth, H. G. (1984). *Modern Methods of Particle Size Analysis.* New York: Wiley.
Carver, R. E. (1971). *Procedures in Sedimentary Petrology.* New York: Wiley.
Folk, R. L. (1974). *Petrology of Sedimentary Rocks.* Austin, Texas: Hemphills Publishing.
Griffiths, J. C. (1967). *Scientific Analysis of Sediments.* New York: McGraw–Hill, 508 pp.
Kaye, B. H. (1981). *Direct Characterization of Fine Particles.* New York: Wiley.
Krumbein, W. C., & Pettijohn, F. J. (1938). *Manual of Sedimentary Petrography.* New York: Appleton-Century-Crofts.

1 Principles and methods of geological particle size analysis*

I. N. MCCAVE AND J. P. M. SYVITSKI

Introduction

Particle size is a fundamental property of sedimentary materials that may tell us much about their origins and history. In particular the dynamical conditions of transport and deposition of the constituent particles of rocks is usually inferred from their size. The size distribution is also an essential property for assessing the likely behaviour of granular material under applied fluid or gravitational forces, and gauging the economic utility of bulk materials ranging from foundry sands to china clay.

Among solid bodies only a sphere has a single characteristic linear dimension. Irregular sedimentary particles possess many properties from which several characteristic linear dimensions may be obtained. These include a particle's projected area, settling velocity, volume, lengths, and the size of a hole through which it will pass. These dimensions are, of course, not equivalent, save in special circumstances (e.g., for a sphere), a fact which is generally appreciated but usually overlooked. Krumbein & Pettijohn (1938) give a detailed analysis of the properties used as measures of particle size. Their book is *the* reference for all that we shall refer to as "classical" in this chapter. There is a huge variety of commercially available instruments, but we have not attempted to list them all (but see the appendix for names and addresses). A concise survey of the state of the market in 1987 was provided by Stanley-Wood (1987a) and a little earlier by Bunville (1984).

The nature of geological sediment samples

At the outset one must remember that we do deal with samples and determine sample, not

*Geological Survey of Canada Contribution No. 49889.

population, properties. The strictures bearing on the statistical inferences that may be drawn from samples are lucidly set out by Griffiths (1967) and will not be rehearsed here.

Geological materials commonly contain a wide range of particle sizes from tens of millimetres down to clay of colloidal (<1 μm) size. Both the wide range and the fact that there is no lower size cutoff present analytical problems. The chemical industry also requires size analysis of powders and other materials (Barth, 1984). These are usually intended to contain a well characterized, usually narrow size (and shape) range with little material in the colloidal range, not wide spectrum size distributions. Many of the modern instruments currently being adapted for geological use were designed with these narrow-spectrum analytical requirements in mind (e.g., the Coulter Counter for blood cells, the laser particle sizer for fuel spray droplet size). Geologists should note that these instruments do not sense the whole clay range. It is as though they give a detailed description of the size of tails without saying whether they are attached to elephants or fish!

Under special circumstances we encounter materials that are well sorted (i.e., their grain size distributions have low standard deviations) and unimodal – for example, dune sands, gravels in some fluvial bars, and loess. However, many samples, even when moderately well sorted, are polymodal, and methods of analysis need to be able to resolve modal structure.

The degree of lithification of geological materials is very variable, but much apparently indurated material can be disaggregated by prolonged agitation together with some ultrasonic treatment. Rock that cannot be disaggregated can be analyzed only by measurement and counting in thin section for sand and silt, or equivalent outcrop methods for conglomerates. Only with complex back-calculations involving particle size, shape, and density values, can thin-section data be used for an understanding of the hydraulic properties of geological samples. Because of the small area of thin sections, the small sample volume of poorly sorted samples means that the coarser end of the size spectrum is underrepresented. Even if a sample may be disaggregated, this does not guarantee unaltered material. The fine fraction may have suffered diagenetic change, carbonate sand grains may have been removed in solution, and other grains may have overgrowths.

The most familiar samples are those obtained from deposits, whether rock or modern sediment. Material in transit can also be sampled, and this may either yield a flux-related sample (e.g., from a sediment trap at sea or an aeolian saltation trap) or simply a sample of the content of a given volume of water in the sea or a river obtained by filtration. Such samples are typically rather small, but modern methods are particularly good at dealing with them.

There is an important difference between flux samples and inventory samples because the former contain the dimension of time (dimensions $ML^{-2}T^{-1}$) whereas the latter do not (dimensions ML^{-3}). Geological deposits are clearly flux related, but our ability to know the time value involved with a thin layer of sediment is usually rather poor. In varved sediments, analyses of size can be converted into annual fluxes of different grades of material, whereas analyses of a bulk sample from a debris flow cannot.

Finally we may distinguish material whose size is sufficiently controlled by the dynamics of the depositional process that analysis of size allows some inference of the nature and intensity of that process (e.g., see Middleton [1976] for inference of fluvial shear velocities from size distributions) from those where the process of transport and deposition does not have much affect on the size distribution (e.g., ice rafting or debris flow), though it may well yield a characteristic fabric.

Field versus laboratory techniques

Most size analysis is carried out in the laboratory. Approximate estimation using a grain size comparator is commonly undertaken for sand sizes in the field, but any instrumental work is laboratory based. The exception is measurement of the size of gravels and conglomerates. For coarse stream gravels it is usually impracticable to transport sufficiently large samples back to the lab. Large-diameter steel plate sieves are used in the field with a simple weighing balance (e.g., kitchen scales). For conglomerates a net strewn over the outcrop is used to select the grains for measurement by ruler and point counting of grains and matrix.

Boundaries and methods

There are three principal categories of material in terms of mode of transport. Gravel is nearly always bedload and mud is nearly always carried in suspension, whereas sand may be either. Mud is further divided into silt and clay, a distinction of mineralogical but not dynamical significance. There are, of course, exceptions to this, in that gravel may be carried in suspension in very fast flows such as turbidity currents or giant floods, and a small amount of silt is carried as bedload, saltating in the viscous sublayer of the turbulent boundary layer. The boundaries between these grades are arbitrary but most branches of science – engineering, soil science, clay mineralogy, sedimentology – agree on them to within a factor of two. We believe the most useful schemes to be modifications of the Udden–Wentworth and Atterberg boundaries as shown in Table 1.1.

Classical versus modern techniques

The classical techniques of particle counting, sieving, and settling are still in use though often using automated instruments. New ways of sensing old variables, such as discriminating and counting grains by image analyzers, have also been developed. In addition there are some completely new principles in use – laser diffraction and photon correlation spectroscopy. Table 1.2 sets out the principles and size ranges to which they are applicable with brief notes on methods and instruments. Subsequent sections of this chapter, as well as other chapters in this book, will examine these in depth. An overview is provided here, highlighting recent advances.

Counting

Modern counting of particles uses either image analyzers or several varieties of particle counter. Images are obtained traditionally with transmitted light microscopy or, for smaller particles, via scanning electron microscopy. Images of suspended particles in situ may be obtained from plankton or marine snow cameras (Honjo et al., 1984; Netherlands Institute for Sea Research, 1989; Syvitski et al., 1989). In ordinary

Table 1.1. *Sizes, names, and principles*

Altered Atterberg[a]	Modified Udden-Wentworth[b]	Classical analysis	Modern analysis
Gravel	Gravel	Count	Field sieve
------- 2 mm --			
			Settle
Sand (3 grades)	Sand (5 grades)	Sieve	Count Scatter light
------- 63 μm --			
			Count
Silt (3 grades)	Silt (5 grades)	Settle	Settle Scatter light
------- 2 μm --			
Clay	Clay	Settle	Settle Centrifuge
------- molecular --			

[a]Atterberg originally used major boundaries on a decade variation of 2×10^n. Thus fine sand was 20–200 μm, coarse sand 200–2,000 μm. He also used subgrades divided at 6×10^n as an approximation to the geometric midpoint 6.32×10^n. Reverting to his usage without approximation and moving the sand–silt boundary to 63 μm permits three grades in silt and sand in 2–6.3–20–63–200–630–2,000-μm intervals.

[b]This is changed only by putting the silt–clay boundary at 2 μm (e.g., as by Friedman & Sanders, 1978).

light microscopy and SEM work the image analysis system may operate directly on images without the intermediary of photographs. Image analyzers are designed to sense the boundaries of particles, usually through highly sensitive gray levels (64–120 in number). Once the outlines of all the particles are discriminated, the particle size and shape parameters may be obtained either through line scanning, pixel counting, or outline tracing. These parameters may include minimum or maximum grain diameters, or the circular area-equivalent diameter. In the case of thin-section data, these particle diameter distributions must be corrected for sectioning bias (Friedman, 1958).

Instrumental methods of particle counting are dominated by the electrical sensing zone particle counters, of which the best known makes are from Coulter and Elzone. Light-blockage counters are less widely used but cover a similar range, mainly silt but straying a little way into sand and down into clay in ordinary usage.

Table 1.2. *Size ranges and applicable principles*

Range	Count	Sieve	Settle from top	Settle suspension	Laser scatter	Photon correlation spectrosc.
Gravel	√[a]	√	X	X	X	X
Sand	√[b]	√	√[f]	X	√[i]	X
Silt	√[c]	√[e]	√[g]	√[h]	√[i]	X
Clay	X[d]	X	√[g]	√[h]	X[i]	√[j]

Symbols: √ = technique applicable; X = technique of limited applicability or impossible.
[a]Counting and measuring on outcrop or from photographs.
[b]Thin section or grain mount under the microscope analyzed directly or by image analyzer; electrical resistive pulse counter (particles suspended in saline glycerol) or light-blockage counter.
[c]Image analysis of SEM pictures; electrical resistive pulse and light-blockage counters.
[d]The electrical resistive pulse method will cover some of the clay range, down to ~0.5 μm with some difficulty.
[e]Only part of the range is routinely accessible (down to 20–25 μm), but special sieves down to 5 μm can be obtained (Fritsch). The latter are more for separation than for precise size distributions.
[f]Settling tubes with either a visual accumulation, weight, differential pressure, or light attenuation method of sensing are used.
[g]The Joyce-Loebl disk centrifuge performs sedimentation size analysis of fine sediment introduced via a density gradient and sensed by light attenuation. Discrimination down to 0.01 μm is claimed by the manufacturers.
[h]Homogeneous settling suspensions may be analyzed by bottom withdrawal or sedimentation balance, or via the pipette method with evaporation and weighing (or its modern analogues of light attenuation or x-ray attenuation).
[i]Many makes of laser forward scattering devices are now available (Malvern, Cilas, Microtrac, Fritsch, Coulter, Horiba).
[j]Coulter, Brookhaven, Langley-Ford, and Nicomp market versions of this instrument.

Sieving

There is no new principle in sieving, but automated systems for wet sieving, sonic, electromagnetic, and air-jet particle agitation have been added to the standard mechanical dry sieve shaking method. Electroformed sieves are now available that take the range of analysis below the 31 μm permitted by woven-wire sieves. This has extended the range down to 5 μm, but not for routine analysis. This chapter will not deal further with sieving as the principles are well covered in older texts (Krumbein & Pettijohn, 1938; Carver, 1971; Folk, 1974); the manufacturers of modern sieves and sieve shakers are given by Stanley-Wood (1977, 1987b).

Settling of suspensions

In sedimentation analysis one must distinguish between the settling of a well-mixed homogeneous dispersion and that of particles introduced at the top of a column. The only significant new departure in this area is the use of x-ray attenuation to sense the time-varying density of a settling suspension; otherwise all the methods are at least thirty years old. There are some new tricks in centrifugal sedimentation in which the basic sedimentation method can be extended down well into the clay range. One such novel idea is to analyze fine-grained suspensions introduced at the top of a clear column. Normally, if this were done simply in water overlain by clay suspension, the clay suspension would sink in plumes rather than as single particles. This has been overcome in the Joyce–Loebl disk centrifuge by introducing the particles via a density gradient.

Laser diffraction

Methods based on the scattering of light by suspensions began to be implemented in the late 1970s. The original three manufacturers – Cilas Granulometer, Malvern Particle Sizer, and Leeds & Northrup Microtrac – have recently been joined by Fritsch, Coulter, and Horiba.

Photon correlation spectroscopy

There are very few techniques that will yield data well into the submicron range. Photon correlation spectroscopy, which depends on the Brownian motion of suspended colloidal particles, is one. The method is not appropriate for larger particles, so there could be problems of data overlap.

Accuracy and precision

It is extremely difficult to specify the accuracy of a measurement of size distribution when the particles are of variable irregular shape and density. To a great extent the accuracy ("approach to the true value") is dependent upon the definition of the size being determined – projected area size by image analyzer, intermediate diameter size by sieve, volume size by Coulter or Elzone counter, or quartz-equivalent spherical

sedimentation diameter by settling tube. At one step removed from these more philosophical considerations is the performance of an instrument in determination of the size of a standard. It is axiomatic that all instruments will perform well on their manufacturers' standards. Earth scientists need to develop standards for grain size analysis to compare different instruments. The method adopted by the IUGS–COS subcommittee for automated size analytical methods (Syvitski, 1986, 1987) for a recent intercalibration exercise was image analysis of the standards to determine their size properties (cf. Chapter 13). So I'm afraid we have to side-step the problem of true accuracy because it is too difficult, and substitute accuracy as the degree to which results approach the values for a known standard. Just how that standard is "known" is a matter requiring agreement, preferably of an international sedimentological standards body. Precision, meaning the reproducibility of results, may be assessed by systematic study of a few samples. Some authors now report their precision, and all should. A noteworthy early study of precision (though it was entitled "accuracy") was that of Poole et al. (1951), who both examined reproducibility of sand-settling analyses and tried to assess the contribution of sample splitting to the overall variation. Their precision for the Emery settling tube was 0.6–2% and for sieving 0.4–0.7%. Among modern methods McCave and Jarvis (1973) report $\pm 0.02\,\phi$ for sand size by Coulter Counter and Jones et al. (1988) report a standard deviation of $0.11\,\phi$ for silt sizes and $\pm 1.9\%$ error in modal peak height by SediGraph. McCave et al. (1986) report a standard deviation of the median size of 0.17 μm for a laser diffraction instrument.

For the future it is important that precisions be reported. The frequency curves (rather than the cumulative curve) should be given because these are much more revealing of errors. The precision both of the size on the abscissa and of modal peak heights is required. Only such reporting will give a satisfactory basis for comparison of samples.

Speed and processing of analyses

Most modern instrumental methods are more rapid than classical techniques. SediGraph samples can be processed at twenty per day, laser sizer samples at ten per hour, and electrical particle counter analyses at about the same rate. However these figures, beloved of manufacturers, disguise the fact that, for geological materials, analysis time is not the rate-determining step. Disaggregation, splitting, and often wet sieving to remove the <63-μm fraction are the operations that control the rate at which samples can be processed. Do not be deceived!

Caution should also be exercised in assessing the resolution of presented results. The widespread availability of curve-fitting programs for PCs, the usual output device for modern size analyses, means that interpolated points can be presented at much finer resolution than the actual determination. For example, the Horiba laser sizer has an eighteen-element detector, but can present the results as a fifty-six-division histogram. Also some programs fit the data to a prescribed size distribution model (e.g., lognormal or Rosin–Rammler), and the parameters are derived from that model. A goodness-of-fit is also reported.

When data were acquired by hand, the simple graphic statistics of Inman (1952) or Folk and Ward (1957) were used. With digital data acquired directly by the machine it is now more straightforward to use the moment statistical or other numerical measures to which the graphic ones were just an approximation. However, it becomes all too easy to report moment measures from distributions for which they are quite invalid, being polymodal or heavily skewed. For these a modal decomposition of the frequency curve may be the geologically most revealing route of analysis (Curray, 1961; van Andel, 1973).

Although we applaud the heroic role of Krumbein and Pettijohn's book in promoting sound sedimentological technique, we disagree with their promotion of the reversed-scale plotting of size data. The great majority of physical data, as well as grain size data obtained by engineers and powder technologists, is plotted with quantities increasing to the right on the abscissa. Many sedimentologists also plot their data that way now. We believe that all should, and that sedimentologists should join the rest of the physical world. Krumbein and Pettijohn's

(1938, p. 185) justification – that the results of most mechanical analyses "are obtained in order from coarsest to finest" and that sedimentation curves are obtained in the reversed sense – does not appear compelling at the present time with rapid data manipulation and plotting machines available. We advocate the plotting of log size in μm or mm increasing to the right on the lower side of a diagram and the phi parameter increasing to the left along the top.

Sedimentation of a homogeneous suspension
Theory
The sedimentation of a homogeneous polydisperse (mixture of sizes) suspension was formulated by Oden (1915). He reasoned that in such a system, after time t, the material settled to the bottom of a chamber of height h comprised two fractions:

1. a portion of the sediment with settling velocity $w_s(t) < h/t$, and
2. all of the sediment with $w_s \geq h/t$.

The continuous settling velocity distribution function $f(w_s)$ is defined as $dm = f(w_s) \, dw_s$, where dm is the mass of particles per unit volume of suspension having settling velocity between w_s and $(w_s + dw_s)$. For $w_s < w_s(t)$ the fraction of dm that has settled is $f(w_s) \, dw_s \cdot (w_s t/h)$. For the whole suspension the mass sedimented at time t is

$$P(t) = \int_0^{w_s(t)} (w_s t/h) \, f(w_s) \, dw_s + \int_{w_s(t)}^{\infty} f(w_s) \, dw_s$$

$$(1.1)$$

Differentiating yields:

$$\frac{dP(t)}{dt} = \int_0^{w_s(t)} (w_s/h) \, f(w_s) \, dw_s \qquad (1.2)$$

Thus the partially sedimented fraction, the first term on the right of equation (1.1) is $t \, dP(t)/dt$. If a curve is drawn of the cumulative mass sedimented $P(t)$ as a function of time t (Fig. 1.1), it will be seen that the intercept on the ordinate of the tangent to the curve (y') gives the portion that has completely settled out of suspension. Expressing y' as $f(t)$ (or $f(w_s)$ for given h) then gives the cumulative settling velocity curve. A

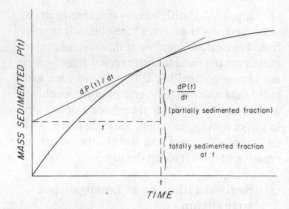

Figure 1.1. The Oden curve. The ordinate is the total mass sedimented at time t from an initially homogeneous dispersion.

second differentiation yields the settling velocity weight frequency curve. For specified density the size frequency curve can be obtained from this curve using Stokes's Law.

The bottom withdrawal tube

The tube is a laboratory instrument designed to generate the sedimentation curve of Figure 1.1, and Owen's (1971) tube was its field equivalent for estuaries. The method, detailed by Vanoni (1975) and Owen (1976), involves removing the lowest 10 cm of an initially homogeneous 1-m-long column of sedimenting suspension at ten successive logarithmically spaced time intervals. This is accomplished via a tap at the bottom of the tube. The withdrawn samples are then filtered or taken to dryness to determine the sediment content. The weights are corrected for the fact that successive samples come from a progressively smaller volume of the whole suspension, and the times are corrected for the fact that the later the sample the shorter the distance settled. The corrected cumulative weights are then plotted against corrected time as in Figure 1.1.

The sedimentation balance

The provision of a balance pan in the bottom of the tube allows the accumulating weight of settled material to be determined continuously. Thus the curve of Figure 1.1 can be determined continuously without the need to make corrections for changing suspension volume

or height because no fluid is withdrawn. This method yields the Oden curve directly, and differentiating it twice yields the settling velocity weight frequency. The procedure is rather slow, however, because the sedimentation process is sensed at a fixed point at the bottom of the column. In a 25-cm column the analysis to 9 ϕ (2 μm) at 20 °C takes about twenty hours. In this time the electronic balance can suffer baseline drift and the requirement for a uniform temperature is most stringent, since the slower-settling fractions are most susceptible to convection. The environment must thus be insulated from diurnal temperature changes. Procedures and programs for making the sedimentation balance analyses and processing the results are given by Thiede et al. (1976).

The method also depends on differentiating the sedimentation-time curve twice to obtain a frequency curve. With noisy data even differentiation once can result in large uncertainties in the derivative; thus this method is not widely used. The exceptions are natural low-concentration suspensions, where a large volume must be processed in order to yield reliable weights for the sedimentation curve (Owen, 1971, 1976; Bartz et al., 1985). In these cases the objective is generally to obtain a settling velocity distribution. This is not converted to size because the particles are natural aggregates of variable, nonquartz, density.

Methods based on sampling of a thin layer: Theory

The theory of sedimentation of a homogeneous suspension is dealt with particularly elegantly by Fisher & Oden (1924). They define the settling velocity distribution function $f(w_s)$ in terms of the log of w_s such that

$$f(w_s) = w_s \frac{dm}{dw_s} \tag{1.3}$$

thus

$$f(w_s) = \frac{dm}{d(\log w_s)} \tag{1.4}$$

where m is the fraction of the sediment having a settling velocity less than w_s. At depth z and time t, then, $w_s(t) = z/t$.

The bulk density of the suspension of concentration C at time zero is

$$\rho_0 = 1 + \frac{C}{\rho_s}(p_s - 1) \qquad (1.5)$$

and at time t, depth z

$$\rho_{tz} = 1 + \frac{C}{\rho_s}(p_s - 1)\,m = 1 + km \qquad (1.6)$$

Noting that

$$\frac{\partial w_s}{\partial t} = \frac{-z}{t^2} \text{ and that } \frac{\partial w_s}{\partial z} = \frac{1}{t} \qquad (1.7)$$

at constant depth, ρ varies with time as

$$\frac{\partial \rho}{\partial t} = k\,\frac{dm}{dw_s}\cdot\frac{dw_s}{dt} = k\,\frac{f(w_s)}{w_s}\cdot\frac{(-z)}{t^2} \qquad (1.8)$$

$$f(w_s) = -t/k \cdot d\rho/dt \qquad (1.9)$$

Thus the settling velocity distribution function (i.e., the frequency curve) may be obtained from the first derivative of the variation of density with time at a point. Contrast this with the preceding analysis of the bottom withdrawal method, where the frequency curve is obtained from the second derivative of the sediment accumulation-time curve.

Similarly, the variation of density with depth at constant time is

$$\frac{\partial \rho}{\partial z} = k\,\frac{dw_s}{dz}\cdot\frac{dm}{dw_s} = \frac{k}{z}f(w_s) \qquad (1.10)$$

$$f(w_s) = (z/k)\,(\partial\rho/\partial z) \qquad (1.11)$$

Direct sensing of density: The hydrometer

The most direct application of the principle in equation (1.9) is determination of changing density using a hydrometer. Once used widely, it is not used much now, save in some areas of soil science, because it is not very accurate. Several designs of hydrometer were produced; these are detailed by Krumbein & Pettijohn (1938).

The pipette method

Sedimentation methods dominated the analysis of fine particles, and among them the pipette method was most commonly used. It too is based on the variation of density at a point as a function of time. In practice, the principle used is that in a settling suspension at time t, depth z, a sample taken from a thin layer gives the origi-

nal concentration of all particles of $w_s \leq z/t$. It is assumed that particles settle independently with no hindered settling effects, no flocculation, at constant temperature, and that a very thin layer is taken from a suspension that was perfectly homogeneous at time zero. In practice, a pipette is inserted to depth z, and a small volume of suspension is withdrawn. The solid content is determined by evaporation and weighing, with correction for weight of dissolved salts. A cumulative curve may then be drawn of the weights of the successively slower-settling fractions. The time axis at constant z (i.e., w_s) is converted to size using Stokes's Law,

$$w_s = \Delta\rho\,g\,d^2/18\mu \qquad (1.12)$$

where $\Delta\rho$ is the density contrast between fluid and grains, d is grain diameter, μ is molecular viscosity, and g is the acceleration due to gravity. Most commonly, $\Delta\rho$ is assumed to be for quartz in water.

Many books give details of how to conduct a pipette analysis; examples are Krumbein & Pettijohn (1938), Galehouse (1971), Folk (1974, or most recent edition), and McManus (1988).

Optical attenuation: Photosedimentation ± centrifuge

The so-called photoextinction method actually depends on the degree of attenuation (c) of a light beam by a suspension of particles. The attenuation is caused by absorption (a) and scattering (b); thus $a + b = c$, with units m^{-1}. Old versions of this method, and regrettably most commercial instruments, use white light and do not treat the optics of light scattering by particles rigorously (see Jonasz, Chap. 11, this volume). The scattering function is related to size, and so a solution to the inference of size from light attenuation by a time- and size-varying suspension is required. A rigorous optical approach to the problem is outlined by Zaneveld et al. (1982) using Mie theory. The attenuation coefficient c_p for mixed-size particles is given by

$$c_p = \pi/4 \int_{d_1}^{d_2} N(d)\,d^2\,K(d)\,dd \qquad (1.13)$$

For a size d_* the value of c_p can be obtained from measurements of c at closely spaced times

t_1, t_2 corresponding to sizes d_1, d_2, which bracket d_*.

$$c_p(d_1, d_2) \approx (\pi/4) \, N(d_1, d_2) \, d_*^2 K(d_*) \qquad (1.14)$$

Thus the number of particles between d_1 and d_2 is

$$N(d_1, d_2) = (4/\pi) \cdot [c_p(d_1, d_2) \,]/[d_*^2 K(d_*)] \qquad (1.15)$$

where $c_p(d_1, d_2)$ is measured, d_* is obtained from Stokes's Law, and the attenuation efficiency $K(d_*)$ is obtained from Mie theory. In its simplest form for nonabsorbing particles,

$$K = 2 - (4/\alpha)(\sin \alpha) + (4/\alpha^2)(1 - \cos \alpha) \qquad (1.16)$$

where

$$\alpha = (2\pi d) \, \lambda^{-1} | \, n - 1 | \qquad (1.17)$$

in which λ is the wavelength of light and n is the real part of the complex index of refraction.

With a good optical system using a parallel beam of monochromatic light, and accounting for near-forward-scattered light, a good system can be constructed. Zaneveld et al. (1982) describe one and discuss its performance.

There are several commercially available systems employing this principle, and some claim to involve Mie theory in data reduction. If this is properly done, the results should be satisfactory. We are not aware of any recent published comparisons using natural fine sediments. A test comparing the Fritsch Scanning Photo-Sedimentograph with pipette analysis showed 1% clay by the former and 14% by the latter method (P. Balson, pers. commun.).

Centrifugal sedimentation is usually sensed by optical means. Horiba and Joyce–Loebl make instruments of this type. The time T to deposit particles at radius r_2, having started at radius r, is

$$T = (18\mu) \cdot \ln(r_2/r_1) \, / \, (\Delta\rho \, \omega^2 \, d^2) \qquad (1.18)$$

where ω is the angular velocity, radians s^{-1}. Normally one thinks of fine-particle systems being analyzed from a starting condition as a homogeneous dispersion. However, the Joyce–Loebl instrument can deal with introduction of the particles at the top (i.e., r_1) by use of a density gradient to prevent bulk suspension streaming. Analysis down to 0.01 μm is claimed in advertising, and Coll and Searles (1987) confirm good results to less than 0.1 μm. This is one of the few ways of obtaining results in the colloidal range. Again, we have not seen results from analysis of natural sediments.

X-ray attenuation: The SediGraph

The SediGraph is a particle sizer that determines the concentration of particles remaining at decreasing sedimentation depths in a suspension-filled cell. The cell is continuously lowered relative to a finely collimated x-ray beam, whose attenuation by the settling suspension is measured as a function of time and height. The ratio of the x-ray transmission of the cell when filled with suspension to its transmission when filled with pure sedimentation liquid is measured and transformed into concentration values. These are indicated as the cumulative mass percent on the y axis of an x–y chart. The data are provided as equivalent spherical diameter (ESD) corresponding to particular sedimentation depths, assuming Stokes's Law of settling. The principle is similar for the pipette method (Welch et al., 1979).

It is claimed that the SediGraph can complete analyses over the size range 0.1–100 μm (Micromeritics, 1984). However, more realistic limits are 1–70 μm, the lower limit for acceptably accurate results being 1 μm. (See the discussion in Hendrix & Orr [1972], where the developers of the instrument were forced to admit that the reality of submicron-size distributions might owe more to temperature fluctuations than to true size features!) A known quantity of sample is not required for an analysis. To avoid hindered settling effects, however, and thus recording of excessive amounts of fines, sediment suspensions should be of the order of 1–2-vol % concentration (Stein, 1985). These low concentrations can lead to analyses being initiated with a starting percentage below 100%. Results then have to be scaled up to yield the true cumulative mass percentages. The lowest starting percentage advisable is 50% (Stein, 1985).

The SediGraph's graphic output is a plot of cumulative mass percent finer than a corresponding equivalent spherical diameter. For modal analysis the raw electronic output of the machine needs to be smoothed and differentiated. This is usually done by digitizing the electronic signals and processing the result, as demonstrated by Jones et al. (1988).

Further details of theory and instrumental operation are described by Hendrix & Orr (1972), Vitturi & Rabitti (1980), Micromeritics (1984), Stein (1985), Jones et al. (1988), and Coakley & Syvitski (Chap. 10, this volume).

Sedimentation of particles through a clear water column

Principles

Similar to the above size-analytical techniques based on the sedimentation of a homogeneous suspension, another series of techniques also utilize the settling of grains through a liquid but from the same starting point at the top. In this case, a sediment sample is introduced into a clear column of water. This principle is almost exclusively used for analysis of sands, the only exception known to us being the disk centrifuge described earlier.

Settling tubes are the devices used to determine the fall velocity distribution of particles assumed to settle out individually neither hindered by other settling particles, nor involved in convective plumes of high concentration, nor retarded by upflow of displaced fluid. This is accomplished by introducing only coarse (>50-μm) sediment particles, typically in the sand size range, in low concentration, at a common level near the top of a sufficiently wide settling tube. At the lower end of the tube, the arrival of the falling particles is sensed, and thus the frequency distribution of the sample as a function of settling duration is obtained. It can be a distribution in terms of number of particles, particle volume, particle weight, or projected area, depending upon the detection method used to determine particle arrival (Slot & Geldof, 1986) (Fig. 1.2). Weight is sensed most commonly. The frequency distribution based on settling duration is converted into the fall velocity distribution using the particles' settling distance.

In the interpretation of settling tube analyses, it is generally assumed that the settling velocity of a particle is a more fundamental dynamic property than any geometrically defined measure of size with reference to its behaviour in a hydrodynamic environment. Since particle

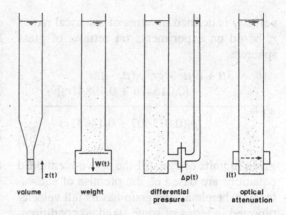

Figure 1.2. The four principal methods of sensing the sedimentation of particles in a column. These may be applied both to sedimentation of a homogeneous dispersion (particularly weight and attenuation) and to particles all starting from the top of the column.

mobility is dependent on the ratio between shear velocity and settling velocity (Francis, 1973), the settling velocity distribution and therefore the sedimentation diameter (ESSD) distribution are more appropriate than sieve-determined size distribution (Middleton, 1976; Bridge, 1981). Settling tube results are also compatible with normal geologic methods for the determination of the mud fraction (i.e., pipette and SediGraph methods).

Settling velocities

For settling of sand and gravel-sized particles, a general equation (cf. Kennedy and Koh, 1961) may be written:

$$w_s = [4/3 \, g \, d(\rho_s - \rho_f)/C_d \rho_f]^{0.5} \qquad (1.19)$$

where C_d is the particle's drag coefficient. In this form, tables or nomograms of C_d versus w_s must be used, with particle density or fluid density held constant. More typically least-squares polynomial regression equations are used. For instance, to relate spherical grain diameter to settling velocity, Bridge (1981) used the form

$$\ln (Re) = -2.96 + 9.04 \ln(Ya) - 0.013 \ln^2(Ya) \qquad (1.20)$$

where $Ya = (\rho_s - \rho_f) \, g \, d^3/(\rho v^2)$, v is the kinematic viscosity, and $Re = w_s d/v$. The regression equation most commonly used is from the work of Gibbs et al. (1971), in which settling velocity is defined in terms of spherical radius r, based on experiments on settling of glass spheres:

$$w_s = \frac{-3\mu + [9\mu^2 + gr^2\rho_f(\rho_s - \rho_f) \cdot (0.015476 + 0.19841r)]^{0.5}}{\rho_f(0.011607 + 0.14881r)}$$

$$(1.21)$$

given in units of cgs. All the above-mentioned equations are based on the premise of equivalency, wherein a sand grain has its fall velocity reported in terms of some standard conditions. These may include reporting measured velocities with either theoretical or empirical results of spheres of constant density (typically $\rho_s = 2.65$ g cm^{-3}) settling in quiescent, distilled water of infinite extent at a temperature of 20° or 24°C.

The conversion of settling velocity to particle diameter is philosophically not simple, and unfortunately there is no agreed standard in the geologic and engineering community. There are those who feel that the conversion may not be warranted, and that the particle velocity distribution should be reported directly either as a base-2 logarithmic distribution (e.g., Psi = $\log_2 w_s$ in cm s^{-1}, after Middleton [1967]), or simply as a base-10 logarithmic velocity distribution (Taira & Scholle, 1979). However, most researchers do convert their velocity data to some equivalent grain diameter, normally reported as the equivalent spherical sedimentation diameter (ESSD) or sedimentation diameter. Natural sand grains, however, are never spheres, and they invariably settle at a velocity lower than would be expected for a sphere of equivalent size (defined as a sphere whose diameter equals the intermediate diameter of the particle) (Baba & Komar, 1981). This relates to irregularities in grain shape (nonsphericity, asymmetry) rather than to grain roundness (angularity). Komar and Cui (1984) suggest that the relationship

$$w_s = 1.026w_{sm}^{1.095} \qquad (1.22)$$

should also be used to convert the measured settling velocity w_{sm} into intermediate grain diameters d_i, after substitution of w_s into Gibbs's equation. Brezina (1979) introduced the hydraulic (Corey) Shape Factor ($= c/(ab)^{1/2}$, where a,b,c are the long, intermediate, and short

diameters), following the work of Komar & Reimers (1978), such that

$$Kw_s^{-2} + Lw_s^{-1} + Mw_s^{-0.5} + C = 0 \qquad (1.23)$$

where w_s is in units of cm/s, $K = -d\,g\,\Delta\rho/\rho_f$, and L, M, and C are multiple-regression coefficients involving the shape factor, phi size, and fluid viscosity.

For a full list of similar regression equations, the reader is referred to Warg (1973) and Brezina (1979).

Settling tube technology, accuracy, and precision

Multigrain settling, as in the case of settling tubes, must deal with a variety of problems inherent in the method. These include:

1. *hindered settling,* which is caused by a counterflow of fluid induced by the falling sediment, thus retarding the settling velocity of individual particles;

2. *settling convection,* whereby vortices and pressure gradients created in the sedimentation fluid by the movement of particles can interfere with the condition of water "quiescence" and the fall path of particles;

3. *thermal convection,* a result of temperature differential along the length of the fall path; and

4. *mass settling,* where smaller grains are dragged down in the wakes of larger grains.

Specifically, settling tube errors include those associated with:

1. the initial position and initial velocity of the particles (i.e., the sample introduction method);

2. particle behaviour in the settling tube (see above, but also influenced by the wall effect and concentration effect); and

3. the quality of the detection method for particle arrival at the lower end of the settling tube.

Kranenburg and Geldof (1974) derived a procedure for estimating the magnitude of convection and hindered settling. For a dimensionless concentration (sample volume per tube volume) of less than 10^{-4}, relative error is <5% for small samples of large (>1-mm) particles or as high as 60% for large samples (>50 g) of small particles. In general, the error associated with concentration effect was shown to increase both with decreasing particle size and with increasing sediment concentration. Gibbs (1972) presents a nomogram to be used for estimating the degree of error to be expected for a multigrain sample depending on particle diameter, sample weight, and tube size. He suggests that minimum dimensions of 13–16-cm inside diameter and 140 cm of effective length be observed to ensure that particles fall at their terminal velocity over as long a distance as possible, while the large diameter reduces wall effects. Vanoni (1975) proposed a formula relating the settling velocity w_d of a particle measured in a settling tube to its unconfined settling velocity w_s, as

$$w_s/w_d = 1 + (9/4 \ d/D) + (9/4 \ d/D)^2 \quad (1.24)$$

where D is the tube diameter. Chapman (1981) also developed an equation that compensates for concentration effects and yields the hydraulic mean diameter of a multigrain sample.

Detection errors include those associated with reaction time of the measuring device, instrument drift, instrument noise, and nonlinearity effects. A full error analysis to be used in settling tube evaluation is given in Slot & Geldof (1979). Detection methods can be divided into four categories (Fig. 1.2), measuring: (1) sand volume, (2) differential pressure in the fluid, (3) light attenuation, and (4) particle weight.

In the first category, the rate of increase of the volume of sand at the base of the tube is measured. Bulk density value(s), preferably variable as a function of time (size), must be used to convert the volume of accumulated sediment to weight. However, the rather imprecise choice of bulk density values, wall-effect problems in the lower portion of the tube, operator dependence, and low resolution of results have made this class of tubes obsolete, although their cost is very low. This method was first proposed by van Veen (1936) and Emery (1938).

The second category includes settling tubes where the suspension of particles in a settling tube causes a piezometric head at a certain level that differs from the head in clear water, the difference being proportional to the submerged weight of the particles that are present above the level of measurement. However, with conventional transducers this method of detection re-

quires too large a sample and thus suffers from increased concentration effects. This method was proposed by Mason (1949) and popularized by Zeigler et al. (1960) and Schlee (1966).

The third category includes settling tubes that measure the arrival of particles near the bottom of the tube by means of an interruption of a beam of light. Although a sensitive and precise technique, it is labour intensive as it requires frequent changing of the fluid (Taira & Scholle, 1977). The treatment of the optics neglects several features, such as forward scattering, dealt with in depth by Zaneveld et al. (1982).

The most common and, in many cases, the most precise class of settling tubes involves the recording of the weight of sediment as a function of time. The weight can be measured by means of a strain gauge situated on a cantilever from which the sediment accumulation pan is suspended (van Andel, 1964), or through a pan suspended by wire connected to a balance at the top of the settling tube (Doeglas, 1946). The most sophisticated in terms of cost and complexity, albeit with increased precision and accuracy, uses an underwater balance supported by leaf springs, with the movement of the pan measured by inductive displacement transducers (Brezina, 1972; Slot & Geldof, 1986). This settling tube at the Delft University of Technology has an overall accuracy of fall velocity better than 3% (which takes into consideration concentration effects), and a precision of <4% (Slot & Geldof, 1986).

Particle counting
Resistance pulse counters

There are two principal manufacturers of these instruments, Coulter and Elzone. They employ the principle that if an electric field is maintained in an electrolyte, a particle passing through this field will cause a change in voltage, providing the particle resistivity differs from that of the electrolyte. The change in resistance ΔR due to a particle of diameter d is given by:

$$\Delta R = [(8r_f d)/(3\pi D^4)]$$
$$\cdot [1 + 4/5 \ (d/D)^2 + 24/35 \ (d/D)^4 \ ...]$$
$$(1.25)$$

where r is the resistivity of the fluid and D is the aperture diameter.

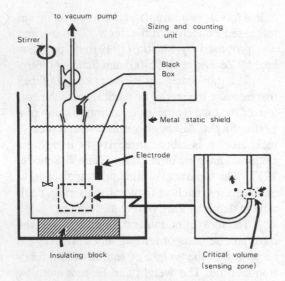

Figure 1.3. The fundamental setup of an electrical sensing zone particle size analyzer. The principal component of the "black box" is a multichannel analyzer (from McCave & Jarvis, 1973).

In practice, a tube containing an electrode and having a small aperture in the side is immersed in an electrolyte containing suspended particles and another electrode (Fig. 1.3). An electric current thus flows by way of the small aperture. A vacuum is then applied to the top of the tube and electrolyte plus particles sucked through the small aperture. The passage of each particle generates a voltage pulse whose magnitude is proportional to the volume of the particle. The machine counts and classifies the pulses according to size at a rate of up to 5,000 per second. The concentration of particles is adjusted so that the probability of two particles being in the aperture at the same time is low. Originally particles whose diameter was 2%–40% of the aperture diameter could be sized; modern claims now suggest a 30-fold range for each aperture. A number of tubes with overlapping ranges are thus required for material with a wide range in size (e.g., apertures of 30 μm for 0.6–12-μm particles, 200 μm for 8–80 μm, and 2,000 μm for 40–800 μm). Presentation of the information is as either differential or cumulative number or volume of particles.

The voltage pulses are related to particle volumes through a series of calibration experiments. The old Model T Coulter and Elzone counter record both number and magnitude of the pulses generated by the passage of particles through the aperture. Pulses are assigned to one of up to 256 size channels, permitting fine resolution in both the Elzone and Coulter Multisizer. These instruments use very low concentrations because there should be only one particle at a time in the sensing zone. This poses splitting problems as only a few milligrams are used. This is a simplified account of the basic principles of the instrument. More detailed treatments are given by Allen (1981), Muerdter et al. (1981), and Milligan and Kranck (Chap. 8, this volume).

The counter will size material 2%–40% of the aperture diameter. Below 2% particles tend to be swamped by electronic noise; particles much above 40% tend to block the aperture. The more recent Coulter Counter model TA II does not count particles but sums the peak heights in each channel. In order to obtain particle number size distributions back-calculation is necessary, assuming the average particle volume for each channel.

Instrumental effects. Because some particle populations counted contained a substantial proportion of aggregates, the interpretation of results depends on knowing what the instrument senses and how it may have affected the aggregates. In chemical engineering studies of coagulating pure mineral suspensions, it is generally accepted that the Coulter Counter senses the solid part of an aggregate, not the total aggregate volume (Camp, 1969; Treweek and Morgan, 1977) and expresses the result as a volume-equivalent sphere. The electrolyte in the aggregate structure conducts current unless it is completely occluded from its surroundings, and this does not happen in pure mineral aggregates. However, there are some data suggesting pure mineral aggregates contain occluded water (Gartner & Carder, 1979). Though in theory the recorded size of such an aggregate is that of a coalesced sphere of its mineral components, Treweek & Morgan (1977) show that the distribution of material within an aggregate, its porosity, and its stretching on passage through the aperture can cause pulse height to decrease and pulse width to increase. As the instrument analyzes pulse heights, aggregates may be recorded at less than their coalesced solid content. The result is that the actual coalesced solid diameter may be up to 25% greater than that recorded. The recorded diameter will in general be smaller than the aggregate diameter that might be seen photographically in the suspension or under a microscope.

Particle rupture in the aperture is another significant instrumental effect. Mean flow velocity through the aperture is 5.4 m s^{-1}, with a residence time of about 50 μs in the sensing zone and only 25 μs in the highest shear of the aperture itself. Hunt (1980) notes aggregates of kaolinite breaking up in a region of higher shear before they enter the sensing zone. Particles that break up while in the sensing zone contribute to the pulse broadening effect noted by Treweek & Morgan (1977). Thus, there will be some undercount of the larger aggregates in a spectrum.

Limitations of the method. The principal limitations of these counters is that the instrument can size only a fairly narrow range on a single pass through the aperture (e.g., 2–40 μm. It is possible to run analyses with multiple overlapping apertures, but this is by no means straightforward as it may involve resetting calibration values and changing the vacuum (Mc-Cave, 1985). Allied to this is the often unrecognized fact that the counter sees only a portion of the distribution and, in particular, does not sense all the clay. Thus these machines would give the same result for the size distribution for the 2–40-μm silt fraction of a sediment sample whether that fraction was 20% or 80% of the total. Shideler (1976) showed the obvious result that a Coulter Counter distribution, even when taken down to 0.6 μm, is coarser than that produced by a pipette. There is a perfectly simple way around this: Make a single-point pipette analysis to determine the amount <2 μm and thus, for a <63-μm wet sieved sample, to obtain the silt: clay ratio. The particle counter then provides details of the size distribution in the silt to coarse clay range.

Light-blockage counters

The primary principal instrument using this principle is the HIAC counter from Pacific Scientific. As with the Coulter and Elzone counters a dilute suspension is sucked through a sensing zone. In this case it is illuminated by a collimated light beam. The particles, which again must pass singly through the beam, decrease the amount of the beam falling on a photodiode detector. Calibration experiments with particles of known size relate the output voltage pulse height (depth) to particle diameter. The particle size sensed is related to a projected particle area containing the long axis.

Experiments conducted by the first author show that the HIAC counter undercounts relative to a Coulter Counter for sizes $\lesssim 7$ μm or ten times the wavelength of light. This is probably because the particles are no longer perfectly blocking the light, and diffracted light reaches the sensor. This finding is confirmed by an ASTM comparison (ASTM, 1983). In this study the mean sizes of particles analyzed by image analyzer (d_l, longest diameter), Coulter Counter (d_v), HIAC counter (d_0), and sedimentation analysis (d_s) were compared, yielding: $d_s = 0.625 d_l$, $d_0 = 0.77 d_l$, $d_v = 0.58 d_l$, from which other ratios may be obtained.

Laser pulse counter

There is an instrument, the Spectrex laser particle counter, that sizes and counts particles on the basis of the intensity and number of forward-reflected flashes of light, respectively. It counts and sizes particles according to scattered light intensity at ~15° from the axis of the beam. The intensity of the forward reflection is proportional to particle area, and the voltages produced by a photodiode are analyzed and classified according to peak height. The author has encountered some poor results (E. T. Baker, pers. commun.) and has seen only the paper by Chung (1982) in the literature.

Image analyzers

As previously mentioned, the first step in the size analysis of an image is to define the edges of individual particle grains, usually in some x–y coordinate system. Edge detection is not a trivial matter to automate, and many times human intervention is needed (Schäfer, 1982). For example, cracks through grains sensed from thin sections of sedimentary rocks may result in two smaller particles being counted where only one exists. Also the proper setting of an instrument "gray-level" discriminator may need to vary widely between samples of mostly transparent mineral grains (such as quartz and feldspar) and those with significant amounts of dark minerals (e.g., ilmenite, pyroxene). Edge detection remains problematic in automated analysis of grains that are in contact with each other or that contain pseudoboundaries, such as rock fragment grains that are composed of a variety of mineral phases.

The second step in image analysis is the determination of one or more measures of grain diameter (e.g., maximum or circular-area equivalent). Though mathematically trivial, problems remain in the accuracy of these measurements. Where the user may wish a low magnification to count a "statistically reliable" population of grains, the resolution and precision of the instrument-calculated diameter decreases rapidly for the finer-sized particles. However, at a higher magnification, the number of particles in a given frame or cell of the image is greatly reduced, and thus more cells must be analyzed. Automated image sizers, such as those by Cambridge (Quantimet), Leitz, AMS, Kontron, Leco, Joyce–Loebl (Magiscan), Quantel, and Nucleopore deal with these problems to varying degrees (also see Kennedy & Mazzullo, Chap. 6, this volume).

Laser diffraction spectroscopy

Laser diffraction analysis is based on the principle that particles of a given size diffract light through a given angle, which increases with decreasing particle size. A parallel beam of monochromatic light is passed through a suspension, and the diffracted light is focused onto a multielement ring detector. The detector senses the angular distribution of scattered light intensity.

A lens, placed outside the illuminated sample with the detector at its focal point, focuses the undiffracted light to a point at the centre of the detector. This leaves only the surrounding diffraction pattern, which does not vary with particle movement (Fig. 1.4). Thus, a stream of particles can be passed through the beam to generate a stable diffraction pattern. This angular distribution of light intensity $I(\theta)$ is given by:

$$I(\theta) = \frac{1}{\theta} \int_0^\infty r^2 n(r) \, J_1^2 \, (kr\theta) \, dr \qquad (1.26)$$

where θ is the scattering angle, r is the particle radius, $n(r)$ is the size distribution function, $k = 2\pi/\lambda$ (with λ being the wavelength of the light), and J_1 is the Bessel function of the first kind. Having measured $I(\theta)$, this expression must be inverted to obtain the size distribution. For details see Swithenbank et al. (1977), Weiner (1979), and Agrawal & Riley (1984).

Several instruments have been made to perform size analysis by this method. Malvern Instruments' Laser Sizer (McCave et al., 1986), CILAS Granulometer (Cornillault, 1972), and the Leeds and Northrup Microtrac (Cooper et al., 1984) are well established. New instruments of this type have also recently been produced by Fritsch, Horiba, and Coulter. These instruments differ in the precise nature of the detector system and the inversion method employed. For the Malvern, three lenses are available with focal lengths of 62, 100, and 300 mm, the length determining the size range that is analyzed (1.2–118 μm, 1.9–188 μm, and 5.6–564 μm, respectively). The instrument does not give a good indication of the amount of material below the analytical range. The amount indicated between 0.5 μm (wavelength of light) and 2 μm was only 16%–20% of the amount <2 μm actually present (McCave et al., 1986). This aspect has not been checked on other instruments but is probably true to some extent of all. Below sizes of a few micrometers, the particles do not diffract light in the manner required for valid application of Fraunhofer diffraction theory because their diameter approaches that of the wavelength of light (de Boer et al., 1987). In fact, the approximation becomes increasingly poor below ~7 μm or $d \approx 10\lambda$ (Bayvel & Jones, 1981), and the full Mie theory should be used.

The advantages of this method are that it is rapid and precise. Its disadvantages are that it is suspect in the indication of the amount of submicron material, that some versions may give spurious modes in the frequency curve, and that it does not resolve polymodal distributions very accurately (McCave et al., 1986; Singer et al., 1988; see also Agrawal et al., Chap. 9, this volume).

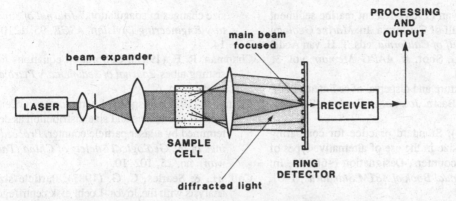

Figure 1.4. Schematic diagram of a laser particle size analyzer showing the primary components of light source, sample, focusing lens, detector, and processing system (from McCave et al., 1986).

Photon correlation spectroscopy

A suspension of particles diffusing by Brownian motion scatters light. The amount of scattered light fluctuates as particles move randomly in the beam. The diffusion of the particles is related to the autocorrelation function for the scattered light signal. The autocorrelation function shows an exponential decay with time, and $1/e$ of its maximum value is equal to $1/DK^2$. K is simply a geometrical optics parameter involving wavelength, refractive index, and scattering angle, but D is the Brownian diffusivity of the particles,

$$D_i = KT / 3\pi\mu d_i \qquad (1.27)$$

where K is Boltzmann's constant and T is absolute temperature. Thus the fluctuations of light scattering contain information about particle size d. Because many sizes are present, the signal has to be analyzed to optimize the best-fit size distribution. In simple cases a unimodal form is assumed, but more sophisticated programs and systems with multiple-angle scattering are available (Stanley-Wood, 1987b). This is the principal method for obtaining size data well into the colloidal range, with a lower limit of about 1 nm (Weiner, 1984).

In the geologically usual case of a wide-spectrum size distribution, there is also the problem that the light scattering function is also size dependent. Intensity weighting functions for each size are required. These may be optimized via the first derivative of the decay coefficient (equal to $2K^2D$) at time zero. An assumption also has to be built into the processing about the particle shape, spheres giving different optical scattering than rods. Most of the commercially available instruments use 90° scattering and give a mean and standard deviation for the size distribution (i.e., an assumed Gaussian form [instruments from Brookhaven and Coulter], or fit a bimodal distribution [Nicomp]). Other manufacturers are Malvern and Langley Ford. For further details Bunville (1984) and other authors in Barth (1984) outline principles.

Conclusion

The pace of development of modern size instrumentation is extremely rapid. Most electronic instruments are obsolescent when purchased. It is most important that geologists realize the limitations of modern instruments. These often involve assumption of spherical particle geometry, fitting of data to a specified form of distribution, reporting of data in size classes far more numerous than the number of points sensed on the curve, and assumption of uniform particle mineralogy and density. However, there is much to be gained from the new instruments in speed and precision if you understand what is being offered. As the estate agents say, *caveat emptor*.

REFERENCES

Agrawal, Y. C., & Riley, J. B. (1984). Optical particle sizing for hydrodynamics based on near-forward scattering. *Society of Photo-optical Instrumentation Engineers*, 489: 68–76.

Allen, T. (1981). *Particle Size Measurement*, 3rd ed., London: Chapman & Hall, 399 pp.

Andel, T. H. van (1964). Recent marine sediment of the Gulf of California. In: *Marine Geology of the Gulf of California*, eds. T. H. van Andel and G. G. Shor, Jr. *AAPG Memoir*, vol. 3, 216–310.

(1973). Texture and dispersal of sediments in the Panama Basin. *Journal of Geology*, 81: 434–57.

ASTM (1983). Standard practice for comparing particle size in the use of alternative types of particle counters, Designation F-660-83. In: *1983 Annual Book of ASTM Standards*, 1401, 771–9.

Baba, J., & Komar, P. D. (1981). Measurements and analysis of settling velocities of natural quartz sand grains. *Journal of Sedimentary Petrology*, 51: 631–40.

Barth, H. G. (ed.) (1984). *Modern Methods of Particle Size Analysis (Chemical Analysis*, vol. 73). New York: Wiley, 309 pp.

Bartz, R., Zaneveld, J. R. V., McCave, I. N., Hess, F., & Nowell, A. R. M. (1985). ROST and BEAST: devices for in-situ measurement of particle settling velocity. *Marine Geology*, 66: 381–95.

Bayvel, L. P., & Jones, A. R. (1981). *Electromagnetic Scattering and Its Applications*. London: Applied Science, 289 pp.

Boer, G. B. J. de, Weerd, C. de, Thoenes, D., & Goossens, H. W. J. (1987). Laser diffraction spectrometry: Fraunhofer diffraction versus Mie scattering. *Particle Characterisation*, 4: 14–19.

Brezina, J. (1972). Stratified sedimentation above the Stokes' range and its use for particle size analysis. In: *Particle Size Analysis 1970*, eds. M. J. Groves & J. L. Wyatt-Sargent. London: Society of Analytical Chemistry, pp. 255–66.

(1979). Particle size and settling rate distributions of sand-sized materials. *Proceedings of the 2nd European Symposium on Particle Characterisation*, Nürnberg, Federal Republic of Germany, 21 pp. + 23 pp. of tables.

Bridge, J. S. (1981). Hydraulic interpretation of grain size distributions using a physical model for bedload transport. *Journal of Sedimentary Petrology*, 51: 1109–24.

Bunville, L. G. (1984). Commercial instrumentation for particle size analysis. In: *Modern Methods of Particle Size Analysis*, ed. H. G. Barth. New York: Wiley, pp. 1–42.

Camp, T. R. (1969). Discussion of "agglomerate size changes in coagulation." *Journal of Sanitary Engineering Division, ASCE*, 95: 1210–14.

Chapman, R. E. (1981). Calibration equations for settling tubes. *Journal of Sedimentary Petrology*, 51: 644–6.

Chung, Y. (1982). Suspended particulates in ocean waters: abundance and size distribution as determined by a laser particle counter. *Proceedings of the Geological Society of China (Taiwan)*, no. 25, 102–10.

Coll, H., & Searles, C. G. (1987). Particle size analysis with the Joyce–Loebl disk centrifuge: A comparison of the line-start with the homogeneous-start method. *Journal of Colloid and Interface Science*, 115: 121–9.

Cooper, L. R., Haverland, R. L., Hendricks, D. M., & Knisel, W. G. (1984). Microtrac particle-size analysis: an alternative particle-size determination method for sediment and soils. *Soil Science*, 138: 138–46.

Cornillault, J. (1972). Particle size analyser. *Applied Optics*, 11: 265–8.

Curray, J. R. (1961). Tracing sediment masses by grain size modes. *Proceedings 21st International Geological Congress, Copenhagen*, 23, 119–30.

Doeglas, D. J. (1946). Interpretation of the results of mechanical analysis. *Journal of Sedimentary Petrology*, 16: 19–40.

Emery, K. O. (1938). Rapid method of mechanical analysis of sands. *Journal of Sedimentary Petrology*, 8: 105–11.

Fisher, R. A., & Oden, S. (1924). The theory of the mechanical analysis of sediments by means of the automatic balance. *Proceedings of the Royal Society of Edinburgh*, 44: 98–115.

Folk, R. L. (1974). *Petrology of Sedimentary Rocks*. Austin, Texas: Hemphills Publishing, 170 pp.

Folk, R. L., & Ward, W. C. (1957). Brazos River bar, a study in the significance of grain-size parameters. *Journal of Sedimentary Petrology*, 27: 3–27.

Francis, J. R. D. (1973). Experiments on the motion of solitary grains along the bed of a water-stream. *Proceedings of the Royal Society of London*, A332: 443–71.

Friedman, G. M. (1958). Determination of sieve-size distribution from thin-section data for sedimentary petrological studies. *Journal of Geology*, 66: 394–416.

Friedman, G. M., & Sanders, J. E. (1978). *Princi-*

ples of Sedimentology. New York: Wiley, 792 pp.

Galehouse, J. S. (1971). Sedimentation analysis. In: *Procedures in Sedimentary Petrology*, ed. R. E. Carver. New York: Wiley-Interscience, pp. 69–94.

Gartner, J. W., & Carder, K. L. (1979). A method to determine specific gravity of suspended particles using an electronic particle counter. *Journal of Sedimentary Petrology*, 49: 631–3.

Gibbs, R. J. (1972). The accuracy of particle size analysis utilizing settling tubes. *Journal of Sedimentary Petrology*, 42: 141–5.

Gibbs, R. J., Matthews, M. D., & Link, D. A. (1971). Relationship between sphere size and settling velocity. *Journal of Sedimentary Petrology*, 41: 7–18.

Griffiths, J. C. (1967). *Scientific Analysis of Sediments*. New York: McGraw–Hill, 508 pp.

Hendrix, W. P., & Orr, C. (1972). Automatic sedimentation size analysis instrument. In: *Particle Size Analysis 1970*, eds. M. J. Groves and J. L. Wyatt-Sargent. London: Society of Analytical Chemistry, pp. 133–46.

Honjo, S., Doherty, K. W., Agrawal, Y. C., & Asper, V. L. (1984). Direct optical assessment of large amorphous aggregates (marine snow) in the deep ocean. *Deep-Sea Research*, 31: 67–76.

Hunt, J. R. (1980). Prediction of oceanic particle size distributions from coagulation and sedimentation mechanisms. In: *Particulates in Water*, eds. M. C. Kavanaugh & J. O. Leckie, *Advances in Chemistry*, 189: 243–57.

Inman, D. L. (1952). Measures for describing the size distribution of sediments. *Journal of Sedimentary Petrology*, 22: 125–45.

Jones, K. P. N., McCave, I. N., & Patel, P. D. (1988). A computer-interfaced SediGraph for modal analysis of fine-grained sediment. *Sedimentology*, 35: 163–72.

Kennedy, J. F., & Koh, R. C. Y. (1961). The relation between the frequency distributions of sieve diameters and fall velocities of sediment particles. *Journal of Geophysical Research*, 66: 4233–46.

Komar, P. D., & Cui, B. (1984). The analysis of grain-size measurements by sieving and settling tube techniques. *Journal of Sedimentary Petrology*, 54: 603–14.

Komar, P. D., & Reimers, C. E. (1978). Grain shape effects on settling rates. *Journal of Geology*, 86: 193–209.

Kranenburg, C., & Geldof, H. J. (1974). Concentration effects on settling tube analysis. *Journal of Hydraulic Research*, 12: 337–55.

Krumbein, W. C., & Pettijohn, F. J. (1938). *Manual of Sedimentary Petrography*. New York: Appleton-Century-Crofts, 549 pp.

McCave, I. N. (1985). Properties of suspended sediment over the HEBBLE area on the Nova Scotian Continental Rise. *Marine Geology*, 66: 169–88.

McCave, I. N., Bryant, R. J., Cook, H. F., & Coughanowr, C. A. (1986). Evaluation of a laser diffraction size analyser for use with natural sediments. *Journal of Sedimentary Petrology*, 56: 561–4.

McCave, I. N., & Jarvis, J. (1973). Use of the Model T Coulter Counter in size analysis of fine to coarse sand. *Sedimentology*, 20: 305–15.

McManus, J. (1988). Grain size determination and interpretation. In: *Techniques in Sedimentology*, ed. M. E. Tucker. Oxford: Blackwells, pp. 63–85.

Mason, M. A. (1949). A manometric settling velocity tube. *Transactions of the American Geophysical Union*, 30: 533–8.

Micromeritics (1984). *Instruction Manual, Sedi-Graph 5000ET Particle Sizer*.

Middleton, G. V. (1967). Experiments on density and turbidity currents, 3. The deposition of sediment. *Canadian Journal of Earth Sciences*, 4: 475–505.

(1976). Hydraulic interpretation of sand size distributions. *Journal of Geology*, 84: 405–26.

Muerdter, D. R., Dauphin, J. P., & Steele, G. (1981). An interactive computerized system for grain size analysis of silt using electroresistance. *Journal of Sedimentary Petrology*, 51: 647–50.

Netherlands Institute for Sea Research (1989). Flocculation and deflocculation of suspended matter – in situ suspension camera. *Annual Report 1988*, NIOZ. Texel: The Netherlands, p. 15.

Oden, S. (1915). Eine neue Methode zur mechanischen Bodenanalyse. *Internationale Mitteilung für Bodenkunde*, 5: 257–311.

Owen, M. W. (1971). The effect of turbulence on the settling velocities of silt flocs. *Proceedings of the 14th Congress of the International Association for Hydraulic Research, Paris*, paper D-4, 27–32.

(1976). Determination of the settling velocities of cohesive muds. Hydraulics Research Station,

Wallingford, U.K., Report IT 161, 8 pp.

Poole, D. M., Butcher, W. S., & Fisher, R. L. (1951). The use and accuracy of the Emery settling tube for sand analysis. U.S. Army Corps of Engineers, Beach Erosion Board, Technical Memorandum no. 23, 18 pp.

Schäfer, A. (1982).The Kontron videoplan, a new device for determination of grain size distributions from thin sections. *Neues Jahrb. Geol. Paläontol. Monatsch.* 115–28.

Schlee, J. (1966). A modified Woods Hole rapid sediment analyser. *Journal of Sedimentary Petrology*, 36: 403–13.

Shideler, G. L. (1976). A comparison of electronic particle counting and pipette techniques in routine mud analysis. *Journal of Sedimentary Petrology*, 46: 1017–25.

Singer, J. K., Anderson, J. B., Ledbetter, M. T., McCave, I. N., Jones, K. P. N., & Wright, R. (1988). Assessment of analytical techniques for the size of fine-grained sediments. *Journal of Sedimentary Petrology*, 58: 534–43.

Slot, R. E., & Geldof, H. J. (1979). *Design Aspects and Performance of a Settling Tube System*. Laboratory of Fluid Mechanics, Delft University of Technology, Internal Report no. 6-79, 18 pp.

(1986). *An Improved Settling Tube System for Sand*. Communications on Hydraulic and Geotechnical Engineering Report no. 86-4, Delft University of Technology, 45 pp.

Stanley-Wood, N. G. (1977). Survey of some of the current instruments and methods available for particle and powder characterisation, Pts. 1–3. *Powder Metallurgy International*, 9: 27–9, 84–7, 138–42.

(1987a). Trends in particle characterisation. In: *Particle Size Analysis 1985*, ed. P. J. Lloyd. London: Wiley, pp. 3–24.

(1987b). Particle characterisation trends. *Trends in Analytical Chemistry*, 6: 100–5.

Stein, R. (1985). Rapid grain-size analyses of clay and silt fraction by SediGraph 5000D: comparison with Coulter Counter and Atterberg methods. *Journal of Sedimentary Petrology*, 55: 590–3.

Swithenbank, J., Beer, J. M., Taylor, D. S., Abbot, D., & McGreath, G. C. (1977). A laser diagnostic technique for the measurement of droplet and particle size distributions. In: *Experimental Diagnostics in Gas Phase Combustion Systems*, ed. B. J. Zinn, *Progress in Astronautics and Aeronautics*, 53: 421–47.

Syvitski, J. P. M. (1986). Minutes of the first meeting of the IUGS–COS sponsored working group on modern methods of grain size analysis, held at the Bedford Institute of Oceanography, Dartmouth, Canada. Copies may be obtained via Box 1006, Dartmouth, Canada, B2Y 4A2.

(1987). Minutes of the second meeting of the IUGS–COS sponsored working group on modern methods of grain size analysis, held at the University of Heidelberg, West Germany. Copies may be obtained via Box 1006, Dartmouth, Canada, B2Y 4A2.

Syvitski, J. P. M., Asprey, K. W., & Heffler, D. E. (1989). The Floc camera: A 3-D imaging system of suspended particulate matter. In: *The Microstructure of Fine-grained Sediments: From Muds to Shale*, eds. R. H. Bennett, W. R. Bryant, & M. H. Hulbert. *Frontiers in Sedimentary Geology*. New York: Springer-Verlag, pp. 281–9.

Taira, A., & Scholle, P. A. (1977). Design and calibration of a photo-extinction settling tube for grain-size analysis. *Journal of Sedimentary Petrology*, 47: 1347–60.

(1979). Origin of bimodal sands in some modern environments. *Journal of Sedimentary Petrology*, 49: 777–86.

Thiede, J., Chriss, T., Clauson, M., & Swift, S. A. (1976). *Settling Tubes for Size Analysis of Fine and Coarse Fractions of Oceanic Sediment*. Corvallis: Oregon State University School of Oceanography, Report no. 76-8, 87 pp.

Treweek, G. P., & Morgan, J. J. (1977). Size distributions of flocculated particles: Application of electronic particle counters. *Environmental Sciences and Technology*, 11: 707–14.

Vanoni, V. A. (ed.) (1975). *Sedimentation Engineering*. ASCE Manuals and Reports of Engineering Practice no. 54, New York, 745 pp.

Veen, J. van (1936). *Onderzoekingen in de Hoofden in Verband Met de Gesteldheid der Nederlandsche Kust*. Algemeene Landsdrukkerij, s'-Gravenhage, 252 pp.

Vitturi, L. M., & Rabitti, S. (1980). Automatic particle-size analysis of sediment fine fraction by SediGraph 5000D. *Geologia Applicata e Idrologia*, 15: 101–8.

Warg, J. B. (1973). An analysis of methods for calculating constant terminal-settling velocities of spheres in liquids. *Mathematical Geology*, 5: 59–72.

Weiner, B. B. (1979). Particle and spray sizing using laser diffraction. *Society of Photo-optical Instrument Engineers*, 170: 53–6.

(1984). Particle sizing using photon correlation spectroscopy. In: *Modern Methods of Particle Size Analysis*, ed. H. G. Barth. New York: Wiley, pp. 93–116.

Welch, N. H., Allen, P. B., & Galindo, D. J. (1979). Particle size analysis by pipette and SediGraph. *Journal of Environmental Quality*, 8: 543–6.

Zaneveld, J. R. V., Spinrad, R. W., & Bartz, R. (1982). An optical settling tube for the determination of particle size-distribution. *Marine Geology*, 49: 357–76.

Zeigler, J. M., Whitney, G. G., & Hayes, C. R. (1960). Woods Hole rapid sediment analyzer. *Journal of Sedimentary Petrology*, 30: 490–5.

APPENDIX: ADDRESSES OF MANUFACTURERS OF INSTRUMENTS FOR PARTICLE SIZE ANALYSIS

AMS, London Road, Pampisford, Cambridge, UK, CB2 4EF.

Brookhaven Instruments Corporation, 200 Thirteenth Avenue, Ronkonkoma, New York 11779, USA.

Cambridge Instruments Ltd. (Quantimet), Clifton Road, Cambridge, UK, CB1 3QH.

CILAS-Compagni Industrielle des Lasers, Route de Nozay 91460, Marcoussie, France.

Coulter Electronics, Inc., 590 West 20th Street, Hialeah, Florida 33010, USA.

Fritsch GmbH, Hauptstrasse 542, D6580 Idar-Oberstein 1, Federal Republic of Germany.

HIAC/Royco Instruments Division, 141 Jefferson Drive, P.O. Box 2168, Menlo Park, California 94025, USA.

Horiba Instruments Incorporated, 1021 Duryea Avenue, Irvine, California 92714, USA.

Joyce-Loebl Division, Vickers Instruments Inc., #00 Commercial St., Malden, Massachusetts 02148, USA.

Kontron Instruments Ltd., Blackmoore Lane, Croxley Centre, Watford, UK, WD1 8XQ.

Langley-Ford Instruments, 29 Cottage Street, Amherst, Massachusetts 01002, USA.

Leco Instuments (UK) Ltd., Newbury Road, Hazel Grove, Stockport, Cheshire, UK, SK7 5DA.

Leeds and Northrup Company, Microtrac Division, 3000 Old Roosevelt Blvd., St. Petersburg, Florida 33543, USA.

Malvern Instruments Ltd., Spring Lane, Malvern Worcestershire, UK, WR14 1AL.

Micrometrics Instruments Corporation, 5680 Goshen Springs Road, Norcross, Georgia 30093, USA.

Nicomp Instruments, P.O. Box 6463, Santa Barbara, California 93111, USA.

Nuclepore, 7035 Commerce Circle, Pleasantum, California 94566, USA.

Particle Data, Inc., Box 265, 111 Hahn Street, Elmhurst, Illinois 60126, USA.

Quantel, 31 Turnpike Road, Newbury, Berkshire, UK, RG13 2NE.

Spectrex Corporation, 3594 Haven Avenue, Redwood City, California 94063, USA.

Wild-Lietz UK Ltd., Davy Avenue, Knowlhill, Milton Keynes, UK, MK5 8LB.

2 The effect of grain shape and density on size measurement

MARTIN D. MATTHEWS

Introduction

The size of a sedimentary particle is commonly described in terms of a characteristic length. If sedimentary particles were all the same regular shape and composed of the same material, the choice of an operational definition of size (length, area, volume, sieve, settling velocity) would not be a problem because the various definitions could be mathematically related. Sediments, however, are variable in shape and composition, causing the choice of a definition of size to be more critical.

A complex three-dimensional description of populations of irregular particles is currently impractical on a routine basis. For simplicity and ease of analysis, particle size is currently related to an "equivalent" particle such as a circle or a sphere. This permits the use of a single length to be a complete but idealized descriptor. Unfortunately, because there are multiple techniques to arrive at a characteristic length, the reported size of the particle becomes partially dependent on the technique employed.

Commonly a complete size description of many sedimentary units cannot be achieved by one measurement technique because of the great range of sizes in the sample. This necessitates the physical separation of the sample into two or more fractions and often involves the mathematical combination of two or more measurement techniques. Standard analytical procedure for the determination of grain size distributions (Krumbein & Pettijohn, 1938; Folk, 1974) usually consists of separating the population into a coarse and a fine fraction by wet sieving at 62 μm. The coarse fraction has been traditionally analyzed by sieving, although settling tubes are increasingly being utilized. The exact sieve diameter of a particle is defined as the width of a sieve opening through which the particle will just pass. Operationally, a set of sieves is used, and the aperture of the sieve that the particle is retained on is recorded. The fine fraction is commonly analyzed by pipetting or similar settling techniques. Size obtained by hydraulic settling has two common definitions: *hydraulic diameter* [*Hydraulischer werth*] (Schone, 1868), "the diameter of a quartz sphere with a settling velocity equal to that of the particle of interest" and *sedimentation diameter* (Wadell, 1934), "the diameter of a sphere of the same specific gravity and the same settling velocity, in the same fluid, as the particle of interest." In addition, direct optical measurements of length or area are utilized, as well as measurements of volume.

The comparison and mathematical merging of two different techniques and the possible bias that results in the integrated size distribution have been the concern of sedimentologists for several years. Sengupta & Veenstra (1968), Sanford & Swift (1971), and others have discussed the differences in grain size distributions obtained by sieving and settling techniques.

This chapter summarizes the fundamental principles behind several size analysis techniques and describes the theoretical relationships among volume, sieving, and settling techniques as a function of density and deviations from spherical shape. For mathematical simplicity, the irregularly shaped particles will be treated as triaxial ellipsoids. A detailed comparison of sieving and settling points to the danger in combining two different size measurement techniques.

This chapter also treats the measuring techniques as if each particle was individually measured, that is, no particle–particle interactions occurred. (Particle–particle interactions are covered elsewhere in the book; see Chapters 3, 14, and 23). Particles are also considered to be measured as whole grains, not by sectioning techniques that bias the measurements. For a discussion on thin-section bias, the reader is referred to Krumbein & Pettijohn (1938).

Theoretical effects
Density

Size measurements involving only length, area, and volume are unaffected by density variation. The effect of particle density on grain size analysis is, however, important for techniques that directly involve some measure of sample weight, such as sieving or volume estimation by weight. Size determined by settling techniques is also a function of grain density (Gibbs et al., 1971; Komar, 1981). The effect of density will be presented with the simplifying assumption that all grains are spherical. This removes the effect of shape on sieving and settling.

Heavy minerals tend to be found preferentially in the finer sieve classes (Rubey, 1933). Since sieving yields a quasi–three-dimensional size, this causes a density-related bias in sieve size distributions, artificially shifting the population toward the finer sizes, although the effect is generally small.

The effect of density on settling is more complicated. When the density of the particle is known and is entered into the appropriate settling equation, the observed settling velocity may be used to calculate the sedimentation diameter of the particle. This calculated size will be identical to the physical size measures of these spherical particles. A sediment sample, however, is composed of mixtures of particles of differing densities. This variability necessitates the assumption of a particular density for the bulk population. The value of quartz (2.65 g/cm³) is generally chosen, resulting, by definition, in the hydraulic diameter of the grain. The hydraulic diameter will differ from the physical diameter of the sphere when the grain density is not that of quartz. Spheres of a given physical size whose densities are lower than quartz will have lower terminal velocities than a quartz sphere of equal size. This low velocity will then be converted into a hydraulic diameter less than the physical diameter of the spherical particle. Similarly, spherical particles denser than quartz will have hydraulic diameters greater than their physical size. This effect is shown in Figure 2.1, whereas Figure 2.2 shows the relationship of the particle's physical size to its density when its hydraulic diameter is that of a 62-μm quartz sphere.

If settling velocity is measured by weight as a function of time, as in a sedimentation balance, there also will be a bias caused by any preferential occurrence of minerals of different densities in different velocity classes.

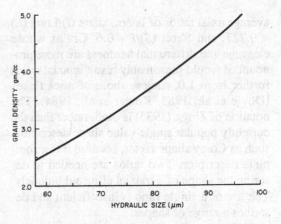

Figure 2.1. Hydraulic size of 62-μm spheres as a function of grain density.

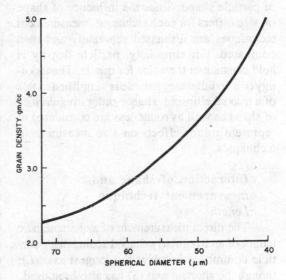

Figure 2.2. Spherical diameter of grains with a settling velocity equal to a 62-μm quartz sphere.

Shape

The assumption of spherical grains in the preceding section is seldom appropriate for natural materials. The grain size, shape, and density of sedimentary particles are causally linked by the action of weathering and transportation on the crystallographic properties of minerals and the conditions of their crystallization. Smalley (1966) described the most probable shape of grains, produced by random fractures in an isotropic material, as consisting of three orthogonal, unequal axes (blades). Ludwick & Henderson (1968), considering quartz sands, also report blade shapes as being dominant, with the average axial ratios of Intermediate (I)/Long (L) = 0.727, and Short (S)/I = 0.6. Grains whose cleavage and differential hardness are more pronounced would presumably result in axial ratios further from 1.0, such as those of mica flakes (Doyle et al., 1983; Komar et al., 1984). The notation of Zingg (1935) is used, rather than the currently popular single-value shape descriptors such as Corey shape factor, because it is a complete descriptor. Two ratios are needed to describe the shape of a triaxial ellipsoid uniquely. The use of a single ratio is insufficient and describes a range of shapes.

The size of an irregular particle, determined by any measurement technique, is a function of particle shape. Since the influence of shape on size differs for each technique, measurement techniques are discussed separately and then compared. For simplicity, particle density is held constant at the value for quartz. The topology of a sedimentary particle is simplified to that of a triaxial ellipsoid. Higher-order irregularities of shape as well as roundness are considered to represent minor effects on size measurement techniques.

Interaction of shape and measurement technique
Length
The direct measurement of a characteristic length of an irregularly shaped sedimentary particle commonly involves the longest axis (L); though the shortest axis (S) has also been used.

In the case of a triaxial ellipsoid or a more irregular particle, three or more axes are needed to describe the particle completely. Thus, the measurement of a single axis is an inadequate definition. It is only satisfactory for a regular body, such as a sphere or cube. This is easily understood by considering two people of equal height, one person weighing 110 kilos and the other 75 kilos. The first person would clearly be considered "larger" than the other, even though their heights are equal.

Area
Area is usually measured by placing a grain in a stable position, thereby defining an area that approaches the maximum projected area of the grain. Thus, information perpendicular to this area is not available. Area is a better measure of size than a single length. Area is commonly converted into a length by using a circle as a reference shape. If the projected area is measured directly, either manually or by using a digital scanning technique and a computer, the diameter of a circle of equivalent area is called the *nominal section diameter* (Wadell, 1935). Other, but less accurate, methods of measuring projected area and converting it to a characteristic length use an inscribed circle, a circumscribed circle, and some sort of average of the long and intermediate (I) axes [arithmetic mean [($L + I$)/2], geometric mean (LI)$^{1/2}$, etc.]. The total area of the three-dimensional grain can also be measured. In this case a spherical form is generally chosen as a reference shape.

In the case of a triaxial ellipsoid or a more irregular particle, any measurement of an area is better than a single length but is still an inadequate definition. The measurement is only satisfactory where the third axis of the ellipsoid is constant or where the grain is a sphere.

Volume
Volume accounts for all three dimensions of a sedimentary particle and as such can be considered an adequate descriptor. Rather than record the volume directly, it is commonly converted into a length by using a sphere as a reference shape. If the volume is measured directly, the diameter of a sphere of equivalent volume is called the *true nominal diameter* (Wadell 1932). Other, but less accurate, methods of measuring volume and comparison to an equivalent sphere include an inscribed sphere, a circumscribed sphere, and some sort of average of the long, intermediate, and short axes [arithmetic mean [($L + I + S$)/3], geometric mean (LIS)$^{1/3}$, harmonic mean [(3 LIS)/((LI)+(IS)+(LS))], etc.]. Volume is generally not practical as a basis for size measurement for particles smaller than gravel because of the tedium involved in obtaining a size distribution. A volume-related property is used for small sizes via electrical sensing zone particle counters (Coulter, Elzone, etc.).

Sieving
Baba & Komar (1981), Sahu (1965), and others have shown that the sieve size of sedi-

mentary grains is conceptually proportional to the intermediate axis of the particle. The shape dependency of sieving was recognized by Rittenhouse (1943) and investigated in detail by Ludwick & Henderson (1968). The absolute ability of a sieve to pass or trap a particular particle depends on the clearance between the particle and the sieve aperture when the particle's smallest cross section is contained in the plane of the sieve. A sieve is capable of passing or trapping a particle depending on the orientation of the particle relative to the sieve aperture. A particle capable of passing through the sieve with its I axis diagonal in the sieve opening and its L axis perpendicular to the sieve may be trapped by the sieve when its I axis is not diagonal. The orientation of the particle's major axis also controls the trapping or passing of particles. A particle that would pass through a sieve when its L axis was at high angles to the plane of the sieve would be trapped at lower angles.

For a given orientation, the probability of a sieve passing a particular grain increases as the end-on clearance of the particle in the sieve aperture increases. Particles whose axial ratios approach 1.0 are also more likely to pass a given sieve than those with lower axial ratios. The probability of a particle being retained on a sieve whose size is larger than the true sieve size of the particle is, therefore, dependent on particle shape and the ability of the particle to achieve the proper orientation for passage during the times allotted for sieving. This results in an apparent coarsening of the population relative to its true sieve size. The shape dependency of sieving was shown by Ludwick & Henderson (1968) to explain the occurrence of identical particles on as many as five sieves.

Sieve size is therefore a combined measure of least cross-sectional area and maximum axial length, further complicated by an orientation effect. For particles of equal minimum cross section, those with the greatest maximum length have a greater probability of being trapped on a larger sieve than those whose maximum lengths are closer to the intermediate length. The combination of an area measurement with a probability-related length measurement leaves us with a $2\frac{1}{2}$-dimensional measure, not quite as good as a true three-dimensional measure, but

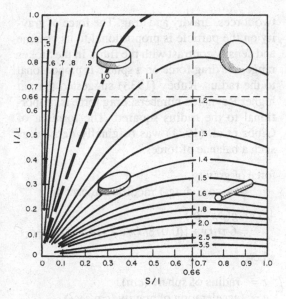

Figure 2.3. Nominal diameter (mass) of triaxial ellipsoids with constant sieve size.

better than a simple one- or two-dimensional measure.

The nominal diameter of triaxial ellipsoids having identical sieve sizes is shown in Figure 2.3. Note that most grains have nominal diameters (mass) greater than that of a sphere passing through the same sieve. This is because distortion of a particle from a spherical shape generally enables it to pass through a finer mesh than it would undistorted. Thus sieving is expected to underestimate the nominal diameter of a sedimentary grain as observed by Baba & Komar (1981). Exceptions occur in the case of extreme disk-shaped particles.

Settling

Size analysis by settling techniques depends on knowing the relationship between fall velocity and size that is appropriate for the particle of interest. For mineral particles of the same density it is intuitively understood that larger particles will fall faster than smaller ones. In actuality, however, the shape of a particle also affects its settling velocity. The effect of shape on size analysis by settling techniques has been discussed by McNown & Malaika (1950), Janke (1966), Komar & Reimers (1978), and others.

As seen below, the settling of a particle in a fluid is traditionally considered as a balance of two forces, gravity and drag. The force of gravity on the particle is proportional to its volume and density contrast with the fluid. In the Stokes range, the drag force on a sphere is proportional to the radius. Rubey (1933) suggested that, at higher Reynolds numbers, drag force is proportional to the radius squared. The formula of Gibbs et al. (1971) was originally derived as such a balance of forces:

force of gravity
     = viscous drag + modified impact drag

$$4/3 \, \pi r^3 g(\rho_p - \rho_f)$$
$$= 6\pi\eta r v + [(0.078/ \, (r/6.73) + 1)] \, \pi r^2 v^2 \rho_f$$

where

$r =$ radius of sphere (cm),
$g =$ acceleration of gravity (cm/sec^2),
$\rho_p =$ density of particle (g/cm^3),
$\rho_f =$ density of fluid (g/cm^3),
$\eta =$ dynamic viscosity of fluid (poises [P]), and
$v =$ velocity of sphere (cm/s).

The similarity to Rubey's formulation is readily apparent, the principal difference being the addition of a function of r to Rubey's impact term. This function is the formula for a hyperbola that was used to modify the impact drag term and shape the transition from Stokes's Law to the impact region.

     Theoretical or empirical equations such as those above or Slot's equation (1984) are used to convert measured fall velocity to an equivalent length, such as hydraulic diameter or sedimentation diameter.

     The hydraulic or sedimentation diameter is dependent on volume and some mixture of length (r) and/or area. It is therefore similar to sieving in being a $2\frac{1}{2}$-dimensional measure.

     The conversion of a settling velocity into a hydraulic diameter tempts the investigator to think in terms of physical size rather than conceptual size. This is especially misleading when the shape of the particle deviates from that of a sphere. For the purpose of discussion and comparison, the physical size of the particle is represented by its true nominal diameter, and the hydraulic diameter will denote its hypothetical size. Shapes more streamlined than a sphere are

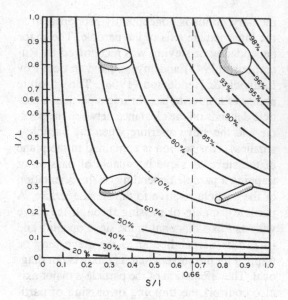

Figure 2.4. Settling velocity of triaxial ellipsoids expressed as a percentage of the velocity of an equivalent nominal sphere (drag).

rarely attained by sediment grains in nature and are therefore not considered.

     The stable position of settling grains is one in which maximum cross-sectional areas are perpendicular to the direction of motion (Krumbein, 1942), ignoring instabilities introduced by Karman vortex interactions. Therefore, at constant volume (fixed true nominal diameter), nonstreamlined distortions of a sphere cause increased drag; this decreases the settling velocity, resulting in a hydraulic (or sedimentation) diameter smaller than the particle's true nominal diameter, as discussed by Komar & Reimers (1978). Figure 2.4 shows the terminal velocities of triaxial ellipsoids, expressed as a percentage of the terminal velocity of its nominal equivalent, predicted by combining the formulas of McNown & Malaika (1950), reported to be applicable near and below a Reynolds number of 10, with that of Gibbs et al. (1971). Settling, therefore, is expected to underestimate the nominal diameter of a sedimentary grain. This relationship has been observed by Lane (1938).

     The calculation of hydraulic diameters rather than sedimentation diameters is recommended because the former are more closely related to the settling velocity distribution of the sample and do not require the separation of the sample into subsamples of known density ranges.

Comparison of shape effects on sieving and settling

Previously the effect of shape on sieve and hydraulic diameter was compared to the particle's true nominal diameter. In this section the hydraulic diameters of various triaxial ellipsoids are calculated theoretically and compared to their equivalent sieve diameters.

A grain's shape may cause it to have different sieve and hydraulic diameters. Similarly, two physically different grains may have identical sieve and hydraulic sizes. Triaxial ellipsoids may, therefore, be divided into three groups based on the relationship of their hydraulic and sieve sizes:

1. hydraulic diameter > sieve diameter,
2. hydraulic diameter < sieve diameter, and
3. hydraulic diameter = sieve diameter.

The first group is characterized by the particle's mass increasing more rapidly than its drag, with increasing major diameter and constant least cross section (rod). This increased mass causes the particle to fall more rapidly than a spherical particle of the same sieve size, resulting in a hydraulic diameter greater than the particle's sieve diameter. The second group of particle shapes behaves in an opposite manner. The drag of these particles increases more rapidly than their mass, with decreasing major diameter and constant least cross section (disks). These particles fall less rapidly than a sphere of the same sieve diameter and therefore have hydraulic diameters smaller than their sieve diameters. The third group is composed of particles whose mass and drag alter at the same rate, resulting in their hydraulic and sieve diameter being the same. Figure 2.5 presents the hydraulic diameters of triaxial ellipsoids that can pass diagonally through a 62-μm sieve. Relationships near and below Reynolds numbers of 10, similar to Figure 2.5 but for other sieve sizes, may be constructed using McNown & Malaika's (1950) formulas or Figure 2.3 and an appropriate hydraulic diameter–fall velocity relationship for spherical particles. Figure 2.6 shows the relationships for several other sizes from 16 to 140 μm.

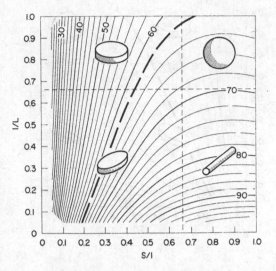

Figure 2.5. Hydraulic sizes in microns of triaxial ellipsoids with a sieve size of 62 μm.

Measured differences in sieving and settling

Procedure

Several sediment samples from beaches, rivers, and dunes were selected as representative of typical samples. These samples were carefully sieved into half-phi-size intervals. Three-inch-diameter sieves were used, each sieve being shaken until the sediment passing through it became negligible (>15 min). The mean particle shape of each sieved interval, expressed as axial ratios (Zingg, 1935), was estimated from 100 grains observed under a microscope using a graduated reticule and a mirror at 45°. Splits of these intervals, not exceeding one gram, were obtained using a microsplitter. The hydraulic size distribution of these splits was then obtained using a settling tube (Gibbs, 1974). The settling column was 108 cm long and 16.9 cm in diameter with a sensitivity on the order of 0.05% and an accuracy to within 2% (Gibbs, 1972). The formula of Gibbs et al. (1971) was used to convert the observed settling velocities to their equivalent hydraulic diameters.

Results

Approximately 40% of the area of Figures 2.5 and 2.6 represents particles whose sedimentation diameters are smaller than their sieve size; the remaining 60% represents particles that are coarser.

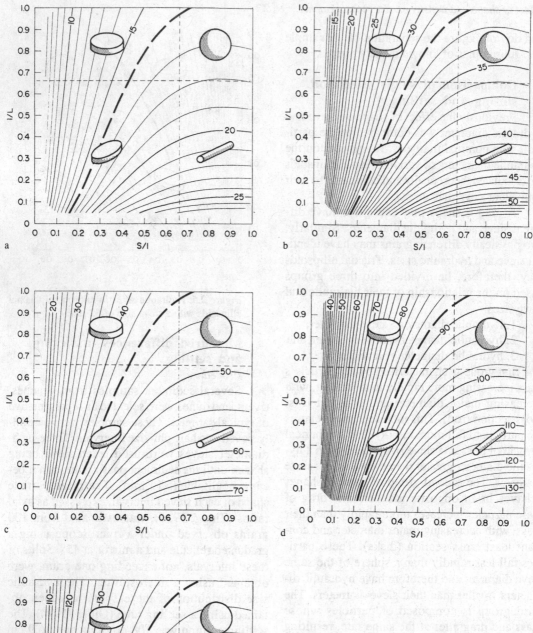

Figure 2.6. Hydraulic sizes in microns of triaxial ellipsoids with a sieve size of: (a) 16 μm, (b) 32 μm, (c) 44 μm, (d) 88 μm, (e) 140 μm.

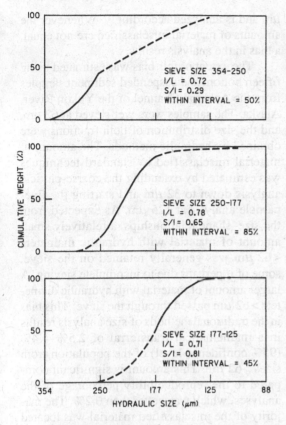

Figure 2.7. Cumulative frequency of hydraulic diameters of selected sieve intervals. Dashed lines represent portions of the distribution not contained within the sieve interval.

In the absence of data on average particle shape, one would expect that the average hydraulic size of particles would be slightly larger than their sieve size. However, particle shape and the relationship between the sieve and hydraulic diameters of particles are functions of grain size, as recognized by Sengupta & Veenstra (1968). The larger sieve sizes tend to have settling diameters smaller than their sieve size, whereas the finer sieve sizes show a settling diameter larger than their sieve size (Sanford & Swift, 1971; Baba & Komar, 1981; Bergman, 1983).

My analysis showed that sieve sizes ≥250 μm generally contain a high proportion of particles whose hydraulic diameters are finer than the sieve interval in which they occur (Fig. 2.7). These grains had a high proportion of disk shapes, although other shapes do occur. The observed size relationships are in agreement with that theoretically predicted from the grain shapes present. The majority of particles whose sieve sizes were near the 250–177-μm interval tended to have comparable hydraulic diameters, though both coarser and finer hydraulic diameters also occurred (Fig. 2.7). Shape analysis of these grains showed them to be predominantly equant, as was expected from the similarity of their sieve and hydraulic diameters.

Sieve sizes finer than 177 μm tended to have a high proportion of grains whose density is greater than quartz and whose shape approaches that of a rod. These grains would, therefore, be expected to have a high proportion of hydraulic diameters greater than the coarse limit of the sieve interval. This predicted relationship was demonstrated by all samples analyzed and has been observed by Wang & Komar (1985). In a detailed analysis of two-dimensional shape, Ehrlich et al. (1980) found a discrete change in the frequency distribution of shape at about 125 μm. The proportion of grains whose hydraulic diameters are smaller than the lower bound of the sieve interval in which they occur increases with increasing sieve size, due to their preferential disk shapes. Similarly, the proportion of grains with hydraulic diameters greater than the upper limit of the sieve interval increases as the sieve size decreases, due to their higher density and preferred rod shapes (Fig. 2.8). This relationship was studied in sieve sizes of 44 μm and larger. In finer sieves the relationship would be expected to reverse because of a change in the particle mineralogy from a high proportion of elongated heavy minerals to less dense, platy clay minerals.

Effect of sieve and settling analyses on size distributions

The systematic relationship between sieve and hydraulic diameters causes a difference in the overall size distribution obtained by the two methods of analysis. The standard deviation of sieve diameters is theoretically greater than that of the hydraulic diameters measured on the same sample due to shape and density effects. This relationship is demonstrated by Sengupta & Veenstra (1968) and by Sanford & Swift (1971). The hydraulically measured population would

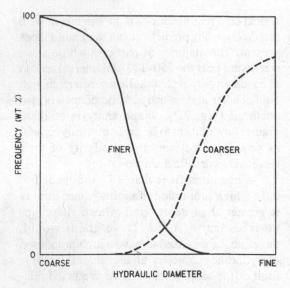

Figure 2.8. Weight frequency of settling sizes larger or smaller than their sieve size. Settling diameters greater than the upper boundary of a sieve interval, due to high grain density and the preferential occurrence of rod shapes, is shown by the dashed line.

be expected to be more leptokurtic than the sieve size distribution for similar reasons. The concentration of heavy minerals in the finer sieve sizes in contrast to their hydraulic equivalent size would cause differences in the mean and skewness of settling and sieve size distributions. Similarly, an imbalance in the shape effect on the coarse and fine fraction will cause differences in the mean and skewness of the settling and sieve size distributions.

The difference in size obtained by sieving and settling is particularly significant when the sample is sieved at 62 μm and the fraction passing through the sieve is analyzed separately from that retained on the sieve. In these analyses an assumption is that the fraction that passed through the sieve contains no material greater than 62 μm. Any material with hydraulic size <62 μm passing the sieve is not analyzed by standard methods (Krumbein & Pettijohn, 1938; Folk, 1974). The standard procedures, therefore, cause some material that is smaller than 62 μm to be mathematically considered as belonging to the size class immediately *above* 62 μm. Similarly, some >62-μm-material is considered to belong to the size class immediately *below* 62

μm and is analyzed accordingly. Whenever the amounts of material misclassified are not equal, a bias in the analysis results.

The extent of this bias was estimated using fifteen bottom and suspended sediment samples from the Acharon Channel of the Yukon River, Alaska. The samples were wet-sieved at 62 μm, and the size distribution of both fractions were obtained by hydraulic methods. The amount of material misclassified by standard techniques was estimated by extending the coarse-particle analysis down to 32 μm and starting the fine-particle analysis at 125 μm. As expected from the size–shape relationships, a relatively small amount of material with hydraulic diameters <62 μm was generally retained on the sieve, some of it perhaps due to incomplete sieving. A larger amount of material with hydraulic diameters >62 μm passed through the sieve. This bias in the traditional methods of size analysis results in a mathematical transferral of 2.5%–4.4% (95% confidence level) of the population from sizes >62 μm. This amount is significant compared to the reproducibility percentage of the analyses, which was better than 0.2%. The majority of the misclassified material was located in size classes adjacent to the wet sieve size. Significant amounts did, however, extend up to 125 μm and down to 32 μm.

Conclusions

Ideally the size of an object should be independent of such attributes as grain shape and density. This is not true for assemblages of sedimentary particles that are of mixed shapes and densities. Three values are required to describe completely the size of a triaxial ellipsoid. It is unreasonable to expect a single length value to describe completely the size of an irregularly shaped sediment grain. Realizing this, we continue to use a single value because of historic precedent and convenience. The reduction of the multitude of information needed to describe completely the size of an irregular grain to a single value is accomplished through the use of a reference shape such as a sphere or circle. The effect of density on size is comparatively easily handled.

Of the techniques examined (direct measurement of length, area, or volume, sieving, and settling), length and area are considered an inadequate description of size because fewer than three dimensions of the grain are utilized to arrive at the single size value. Sieving is considered to provide a $2\frac{1}{2}$-dimensional measure and has been successfully used for a long time; however, it is not considered to be a good size descriptor because of the probability of one particle being found on multiple sieves. When this effect is minimized, sieving approaches a pure two-dimensional technique and would be considered inadequate. Settling is also a $2\frac{1}{2}$-dimensional measure and is considered adequate. Volume-based techniques are three dimensional and are considered to be an adequate representation of the three-dimensional size of the grain.

The inability to describe the size of a sedimentary grain uniquely by a single value causes us to choose the size analysis method that befits the purpose of the study. If the purpose is to describe the hydraulic conditions of the depositional environment, a settling technique may be chosen, believing deposition from suspension is important. Alternatively, traction might be considered important and an area-based technique chosen, emphasizing drag at the water–sediment interface; or volume, believing the inertia of the particle to be important. Similarly, if the study was concerned with diagenesis, surface area would be the method of choice if reaction rate was emphasized; or volume would be chosen to emphasize bulk changes in the solid phases. Perhaps the most complete description of size would be to use several techniques in combination and look at the interrelationships between the measures.

Grain shape and density variations interact with size measurement techniques to create a consistent bias in grain size estimates. If one consistent technique is used for all sizes, and if comparisons are confined to similar techniques, no difficulty is expected. If, however, different techniques are used, apparent differences in size can arise because of variations in mineralogy and shape. In order to relate volume, sieving, and settling estimates of size to each other, it is necessary to know the relationship among grain size, shape, and density. Theoretically the size of an object should be independent of such attributes as the shape and density of the object.

This is not true for average properties of assemblages of sedimentary particles. If particle size is simplified to a triaxial ellipsoid, there is an observed general tendency for large particles ($>250\ \mu m$) to be disk shaped and finer particles ($<177\ \mu m$) to be rod shaped. Grain density generally increases discontinuously with decreasing sediment size. These size-related differences in shape and density may have a beneficial or detrimental effect on the ability of a particular size technique to be used in a genetic interpretation.

Direct, detailed comparison of two different techniques is, however, difficult to impossible. In general, sieving will underestimate the true nominal size of a grain. Settling similarly underestimates the true nominal size of a grain. The relationship between sieving and settling is more complex. Settling size distributions are expected to be more leptokurtic and have a smaller standard deviation than the sieve size distribution measured on the same sample. The mean and skewness of the two measured populations will generally also differ. The exact relationship between these two measures of grain size depends on particle shape and heavy mineral content, controlled by the availability of mineral types in the source area and the extent to which the deposition environment concentrates heavy minerals and is shape selective. The differences in any two analyses cannot be eliminated by calibrating a settling tube against a sieve standard or by standardizing to a true nominal diameter reference. The difference depends on the unique relationships of grain size to particle shape and density inherent in each sample.

The joining of size distributions measured by two different techniques into a single size distribution is particularly dangerous. This practice is commonly followed, however, when the sample is wet sieved and the coarse size analysis done by sieving and fine analysis by settling. The assumption implicit in joining these two analyses is that no material $>62\ \mu m$ passed through the sieve, but this is not well founded when the size of the fine particles is determined hydraulically. As a result, the distribution is physically depleted of material in the size classes immediately $>62\ \mu m$. This material is then considered to belong in the size class just below 62 μm. This artificial depletion caused by the analytical method is great enough to explain the dearth of material in the 125–62-μm size interval, which was noted by Pettijohn (1957).

A relatively minor alteration of the standard methods will greatly reduce but not eliminate this bias. The pipette sample taken to estimate the total weight of the material in fine-grained analysis should be taken as soon as possible after cessation of mixing and as deep as practical: 5 sec and 30 cm are suggested. Samples representing the amount of material finer than a given particle size are then taken, starting with 125 μm rather than 44 μm. This procedure will allow calculation of the percentage of material coarser than 125 μm and the amount in each smaller size class. The material hydraulically coarser than the sieve aperture it passed through can then be added to the appropriate sieve intervals of the coarse-particle analyses.

Because none of the current techniques are really adequate to describe the size of an irregularly shaped grain, it is appropriate to describe a potentially adequate technique that could be developed using existing technology. The proposed method is based on an extrapolation of the technique used by Ehrlich & Weinberg (1970) to describe two-dimensional shapes. Combined laser and computer techniques are capable of scanning and storing a digital representation of the external form of a three-dimensional object. This array of data could then be rotated into a standard frame of reference and reduced to a smaller number of descriptors by Fourier or other techniques. This would yield a complete description of size suitable for a complex particle, rather than a single value that is suggestive and convenient but inadequate. In order to be practical, this technique, or any other involving examining individual particles, must be automated.

REFERENCES

Baba, J., & Komar, P. D. (1981). Measurements and analysis of settling velocities of natural quartz sand grains. *Journal of Sedimentary Petrology*, 51: 631–40.

Bergman, P. C. (1983). Comparison of sieving, settling and microscope determination of sand grain size, in near shore sedimentology. In: *Proceedings, Sixth Symposium on Coastal Sedimentology, Geology Department, Florida State University*, ed. W. F. Tanner. Tallahassee, Florida, pp. 35–6.

Doyle, L. J., Kendall, L. G., & Steward, R. G. (1983). The hydraulic equivalence of mica. *Journal of Sedimentary Petrology*, 53: 643–8.

Ehrlich, R., Brown, P. S., Yarns, J. M., & Przysocki, R. S. (1980). The origin of shape frequency distributions and the relationship between size and shape. *Journal of Sedimentary Petrology*, 50: 475–84.

Ehrlich, R., & Weinberg, B. (1970). An exact method for the characterization of grain shape. *Journal of Sedimentary Petrology*, 40: 205–12.

Folk, R. L. (1974). *Petrology of Sedimentary Rocks*. Austin, Texas: Hemphills, 182 pp.

Gibbs, R. J. (1972). The accuracy of particle size analyses utilizing settling tubes. *Journal of Sedimentary Petrology*, 42: 141–5.

(1974). A settling tube system for sand-size analysis. *Journal of Sedimentary Petrology*, 44: 583–8.

Gibbs, R. J., Matthews, M. D., & Link, D. A. (1971). The relationship between sphere site and settling velocity. *Journal of Sedimentary Petrology*, 41: 7–18.

Janke, N. C. (1966). Effect of shape upon the settling velocity of regular convex geometric particles. *Journal of Sedimentary Petrology*, 36: 370–6.

Komar, P. D. (1981). The applicability of the Gibbs equation for grain settling velocities to condition other than quartz grains in water. *Journal of Sedimentary Petrology*, 51: 1125–32.

Komar, P. D., Baba, J., & Cui, B. (1984). Grain-size analysis of mica within sediments and the hydraulic equivalence of mica and quartz. *Journal of Sedimentary Petrology*, 54: 1379–91.

Komar, P. D., & Reimers, C. E. (1978). Grain shape effects on settling rates. *Journal of Geology*, 86: 193–209.

Krumbein, W. C. (1942). Settling velocities and flume behaviour of non-spherical particles. *Transactions of the American Geophysical Union*, 41: 621–33.

Krumbein, W. C., & Pettijohn, F. J. (1938). *Manual of Sedimentary Petrography*. New York: Appleton-Century-Crofts, 549 pp.

Lane, E. W. (1938). Notes on the formation of sand. *Transactions of the American Geophysical Union*, 19: 505–8.

Ludwick, J. C., & Henderson, P. L. (1968). Particle shape and influence of size from sieving. *Sedimentology*, 11: 197–235.

McNown, J. S., & Malaika, J. (1950). Effects of particle shape on settling velocity at low Reynolds numbers. *Transactions of the American Geophysical Union*, 31: 74–82.

Pettijohn, F. J. (1957). *Sedimentary Rocks*. New York: Harper Brothers, 718 pp.

Rittenhouse, G. (1943). Relationship of shape to the passage of grains through sieves. *Industrial Engineering Chemistry (Analytical Edition)*, 15: 153–5.

Rubey, W. W. (1933). Settling velocities of gravel, sand, and silt particles. *American Journal of Science*, 25: 325–38.

Sahu, B. K. (1965). Theory of sieving. *Journal of Sedimentary Petrology*, 35: 750–3.

Sanford, R. B., & Swift, D. J. P. (1971). Comparison of sieving and settling techniques for size analysis, using a Benthos Rapid Sediment Analyzer. *Sedimentology*, 17: 257–64.

Schone, E. (1868). Ueber einen neuen apparat für die Schlammanalyze. *Zeitschrift für Analytische Chemie*, 7: 29–47.

Sengupta, S., & Veenstra, H J. (1968). On sieving and settling techniques for sand analysis. *Sedimentology*, 11: 83–98.

Slot, R. E. (1984). Terminal velocity formula for objects in a viscous fluid. *Journal of Hydraulic Research*, 22: 235–43.

Smalley, I. J. (1966). The expected shapes of blocks and grains. *Journal of Sedimentary Petrology*, 36: 626–9.

Wadell, H. (1932). Volume, shape, and roundness of rock particles. *Journal of Geology*, 40: 443–51.

(1934). Some new sedimentation formulas. *Physics*, 5: 281–91.

(1935). Volume, shape, and roundness of quartz particles. *Journal of Geology*, 43: 250–80.

Wang, C., & Komar, P. D. (1985). The sieving of heavy mineral sands. *Journal of Sedimentary Petrology*, 55: 479–82.

Zingg, T. (1935). Beitrag zur schofferanalyze. *Schweizerische Mineralogische und Petrographische Mitteilungen*, 15: 39–140.

3 The effect of pretreatment on size analysis

MARTIN D. MATTHEWS

Introduction

A commonly held assumption in particle size analysis is that the sample consists of discrete separate grains. The validity of this assumption for naturally occurring earth materials is questionable, and under certain circumstances it is not even desirable to consider them as separate grains. A significant mode of sediment transport is in the form of aggregates. Aggregates develop as part of the soil-forming processes; however, organic binding by fungi, filter feeders, and burrowing organisms are locally important during sediment transportation and deposition.

An aggregate particle, or "ped," is defined here as consisting of two or more primary particles (produced by the weathering of rock) bound together by strong cohesive forces. Aggregates are stable under "normal" dispersive techniques such as stirring and dispersion with sodium hexametaphosphate. This definition eliminates such composite particles as floccules (unstable in "normal" dispersion) at one end of the spectrum and rock fragments (primary particle) at the other end.

Particle size analysis is generally utilized to interpret the conditions controlling the sample's occurrence (such as the provenance area or the physical/chemical conditions of erosion/transport/deposition), or for description/comparison. Sample handling, and the application of a particular pretreatment technique, may lead to modification of the grain size distribution in either a beneficial or a deleterious manner. Almost every sample collected for size analysis is pretreated. In some cases the treatment may be relatively gentle, such as stirring the sample in water, or even mild agitation/rubbing to ensure dispersion and destruction of any lumps formed during sample storage. In other instances the treatment may be more vigorous, as with the use of ultrasonics and/or chemical reagents to remove binding materials (organics, carbonates, and/or iron oxides). It is important that careful consideration be given to pretreatment procedures to ensure that the resultant size analysis is meaningful to the goal of the investigation.

Conditions of aggregate formation in soil horizons and during transport and deposition are described first, including some estimates of the ratios of aggregates to individual particles. This is followed by a discussion of the expected range of hydraulic particle sizes that constitute an aggregate and, finally, an example from Acheron Channel of the Yukon River in Alaska.

Aggregate formation

Aggregate particles are initially formed in soil horizons. The process of weathering and soil formation is often thought of as the process of breaking down rocks into separate, quasi-stable mineral grains. The process of soil formation, however, also binds these individual particles into a dynamic soil structure. The processes of aggregation and disaggregation reach a steady state in the soil system with continuous formation and destruction of multiparticle grains. Soil scientists have identified a wide range of agents that are responsible for binding soil particles into aggregates: precipitation of soluble and insoluble salts, chemical and physical action of microorganisms and mesofauna, and clay mineral bonding.

Bonding mechanisms involving the precipitation of soluble salts such as halites are summarized by Price (1963) and Bowler (1973). The processes of silica and carbonate cement formation in soil horizons as nodules or crusts, and their microscaled precursors, is summarized by Flach et al. (1969). The cementation of soil particles by sesquioxides, iron, manganese, and alumina (Brewer & Blackmore, 1976) is less well understood and remains controversial. Lutz (1936) reported a correlation between ferric iron content and aggregate stability. McIntyre (1956) and Turchene & Oades (1978) suggest that the effectiveness of iron oxides as an aggregating agent may be confined to the finest fraction of the clay population. Greenland et al. (1968) showed by electron microscopy that iron oxides often occur as discrete microcrystals rather than coatings and therefore would not be expected to be efficient bonding agents. Deshpande et al. (1968) suggested that the allophane (aluminum oxides/hydroxides) associated with the iron oxides is the important bonding agent. However, the role of allophane as a binding agent is unclear because it usually occurs with an admixture of organic material (Riezebos & Lustenhouwer, 1983).

A wide variety of humic organic compounds occur in soils. These compounds range in molecular weight up to several million atomic mass units (amu) and often act as both antiwetting agents and binders (Russell, 1971). Organisms are also directly responsible for aggregate formation through the binding action of fungi (Swaby, 1949; Griffiths, 1965), the mechanical action of termites (Fitzpatrick, 1983), or the formation of fecal casts by beetle larvae or earthworms.

Clay minerals are probably the most common material found in aggregates. Their role as binding agents (Brewer & Blackmore, 1976) is, however, poorly understood. Electrochemical bonding may be the dominant force but the importance of electrochemical bonding is confounded with the action of organic binders and inorganic cements. Although these cements and binding agents are capable of aggregating materials that do not include clay particles, the majority of aggregates have an appreciable clay content, suggesting that clays themselves may be an important factor. Aggregates commonly extend from clay to sand size particles. Volume changes in clay soil due to moisture loss or gain are two of the most important processes responsible for the disintegration of soil structure into clods, crumbs, aggregates, and primary particles. Towner (1988) showed that during moisture loss the larger primary grains in a soil tend to act as centres for the shrinkage process, and therefore aggregates tend to consist of a large grain to which a variety of smaller grains are attached.

Young (1980) has summarized the work of Gabriels & Moldenhauer (1978) and others who studied the erosion of soils by simulated rainfall and rill formation through field and laboratory experiments. The proportions of aggre-

gate and primary grains eroded from a soil are controlled primarily by the texture and aggregate proportion of the original soil, suggesting that many soil aggregates are stable under the forces of erosion. Some experiments showed a concentration of aggregate particles within the fine fractions, whereas others were more uniformly distributed with size. Young (1980) concluded that the dominant mechanism of erosion of clay-sized primary particles occurs as silt-sized aggregates.

A large proportion of the aggregates created in soil horizons persist during transport in both aqueous and aeolian phases. Rust & Nanson (1989) reported that sand-sized mud aggregates formed by pedogenic processes occur in the bed load of modern and ancient rivers. These aggregates are similar to those noted by Fitzpatrick (1984) in clay-rich soils, supporting the contention that some soil aggregates resist destruction during erosion and transportation. Butler (1956) reports a similar relationship between the formation of calcareous clay-rich aggregates in soil horizons, and their erosion/transportation by aeolian processes and deposition as a material he called Parna.

Primary and aggregate particles produced in the soil horizons within a given provenance area are often modified during transport. Some aggregates are broken into either smaller aggregates or primary particles by constant immersion in water accompanied by oxidation and physical impact processes, while other particles are modified by organisms to form new organo-mineralic aggregates. Suspended microflora have been observed to attach themselves to suspended particles and bind them into aggregates. Experiments have shown that organisms can aggregate suspended matter in as little as three days (Paerl, 1973). The interaction of zooplankton with sediment is occasionally strong enough to alter mineralogy as well as bind the particles within fecal pellets (Lewis & Syvitski, 1983). Filter feeders also play a major role in the aggregation of fine particles through the formation of feces and pseudofeces. A majority of the primary particles are in the 1–5-μm range, whereas the fecal pellets are usually in the 500–3,000-μm range (Haven & Morales-Alamo, 1972). In some cases, fecal pellet size controls the behavior of the primary particles in the environment; in others the pellets are broken down into smaller aggregate fragments and/or primary grains. Sediment transported down a river system is often redeposited on floodplains where it is again subjected to soil formation processes. There is also good potential for reworking fine-grained concentrates into fine clasts. Buller & Green (1976) suggested that the reworking of thin peaty layers into peaty sand clasts cause their deposition in higher-energy settings than would be assumed based on their disaggregated size distribution. Within the depositional environment, the production of aggregates by pelagic organisms and suspension filter feeders is supplemented by the ability of bottom-feeding infauna to rework deposited sediment into coarser grain sizes. Rhoads (1963) estimated that the infauna in Buzzards Bay is capable of recycling the annual sediment accumulation several times over. Thus bioaggregation can effectively decrease the high-energy winnowing of fines deposited during low-energy conditions.

Estimates of aggregate frequency of occurrence

The few estimates of the proportion of aggregate grains in modern transport systems show that their presence is the rule rather than the exception. Gabriels & Moldenhauer (1978) found that aggregate particles present in soils, and those eroded from them, ranged from 2% or 3% to as much as 90% of the total particles in a size interval. Schubel (1971) estimated that <2% of the suspended sediment in Chesapeake Bay is composed of individual grains and that 97% of the aggregates are composed of more than three individual grains. Traction transport, due to its higher energy of grain–grain contact, is expected to contain a smaller proportion of aggregate grains. Meglis (1987) showed that near-bottom suspended sediment in a coastal lagoon contained a slightly lower proportion of aggregate grains (81%) during conditions of high turbulence than under low-turbulence conditions (87%). Meglis (1987) also reported that the proportion of pellets in suspension increased from 19% to 54% under the conditions of increased turbulence, and that despite the tendency of pellets to break up under the higher-energy

conditions, the resuspension of bottom sediment diluted the proportion of aggregates (broken pellets and soil particles) from 68% to 27%.

Size of aggregates and their dispersed equivalent

Riezebos & Lustenhouwer (1983) and others have shown that aggregates from the soil horizon are sufficiently strong to resist "normal" laboratory pretreatment techniques. Butler (1956), however, points out that the origin of clay material in silts and sands is often misinterpreted since the clay's particle size is measured in the dispersed state rather than the aggregated state in which it was transported, demonstrating the possibility of overdispersion in the laboratory.

Johnson (1974) has reported that destruction of aggregates occasionally releases particles whose hydraulic size is larger than that of the original aggregate. This is in agreement with my findings for disaggregation of Yukon River particles discussed later in this chapter, and may be explained by Towner's (1988) observation that the dominant soil aggregate formation mechanism is the aggregation of less-dense grains around a denser nuclear grain. Figure 3.1 shows the theoretical density of a spherical aggregate formed about a spherical nucleus of density 2.65 g/cc by adding material of density 1.10 g/cc. The density rapidly decreases as the aggregate diameter increases. Given a constant shape, the hydraulic behaviour of an aggregate grain is therefore governed by the joint variation of the size of the aggregate and its average density. Figure 3.2 shows the result of this interaction with respect to the settling velocity of the nuclear grain. Notice that line 1.0 represents a spherical grain whose settling velocity is constant. This is achieved by balancing the increase in diameter by a decrease in average density. Above the line we have the expected situation where the aggregate grain falls faster than the nucleus. Below the line, however, we find that the aggregate falls more slowly than the nucleus since the decreasing density function outstrips the increasing physical diameter parameter. Although the effect of this interaction is probably slight, its occurrence in nature is supported by the observations on densities of aggregates. Ag-

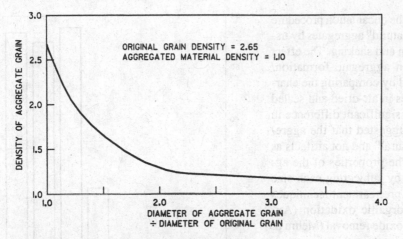

Figure 3.1. Change in aggregate density with the ratio of aggregate diameter to nucleus diameter as low-density (1.1 g/cc) material is attached to a dense (2.65 g/cc) nuclear particle.

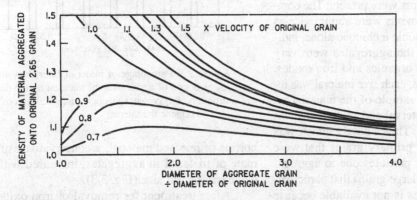

Figure 3.2. Change in fall velocity of an aggregate particle as a function of the ratio of the diameter of the nuclear grain to the diameter of the aggregate and the density of material aggregated around a 2.65-g/cc nucleus.

gregates densities are usually less than those of their primary particles (Rhoton et al. 1983). Fecal pellet densities have been reported to vary from 1.2 to 2.45 g/cc, with large pellets usually denser than the smaller ones (Meglis, 1987). Young (1980) suggests that soil aggregate densities range from 2.65 down to 1.5, with the larger particles generally being less dense.

Characteristics of aggregate particles in the Yukon River
Method of analysis and results
Samples of sediment from Kwiklauk Pass near the Alukanuk Slew of Acheron Pass in the delta of the Yukon River, Alaska were studied to determine the characteristics of the material in transport. Bottom samples were obtained using a pipe dredge. Suspended samples were obtained by pressure filtration using a 0.45-μm nominal pore diameter filter. Samples were sealed in plastic bags in the field to minimize drying. In the laboratory the suspended sediment samples were removed from the filter by a stream of distilled water from a wash bottle with the aid of a rubber spatula. Each bottom sample was wet-split and separated into several size intervals by decantation. Decantations were repeated more than fifteen times until the supernatant fluid achieved a low and relatively constant turbidity. Extrapolation of the degree of incompleteness of separation of the fine boundary suggested that <0.3% by weight of the sample was below the lower size boundary.

Care was taken during the decantation procedure to minimize breaking "natural" aggregates by using only gentle agitation and shaking. The effect of sample handling on aggregate formation/breakage was estimated by comparing the characteristics of aggregates in air-dried and sealed samples. The lack of a significant difference in the size distribution suggested that the aggregates studied were "natural" and not artifacts as a result of handling. The properties of the aggregates were studied by subjecting each size fraction to a variety of pretreatment techniques including ultrasonic, organic oxidation (Anderson, 1963) and iron oxide removal (Mehra & Jackson, 1960). In order to achieve a more complete separation it was necessary to alternate the two chemical treatments several times on each sample, suggesting that perhaps multiple layers of binding agents were present. The combined chemical treatments were about twice as effective as the ultrasonic technique alone, indicating that many of the aggregates were very tightly bound by the organics and iron oxides. After the pretreatment, each size interval was re-decanted to obtain a sample of the material remaining within the interval and another that was finer than the interval. Thus a good estimate was obtained of the fine primary grains that were transported as larger particles due to aggregation. The estimate of large grains that participated in the aggregations is not available because they remained within the interval. The mineralogy and size distribution of the primary grains released from the aggregated state were evaluated using x-ray diffraction and pipette analysis. Further details are given in Matthews (1973). Despite differences in grain size distribution between the suspended and bottom samples, the proportion, size distribution, and mineralogy of the aggregates were similar, suggesting a common origin. Particles within any selected size interval prior to and after treatment accounted for a total of 86% by weight of the total sample. Particles transported within a coarse interval because of their aggregation accounted for the remaining 14% with organics and iron oxides being equally effective as binding agents. The amount of fine material released from each size interval was related directly to the total amount of material within that size interval, and the pro-

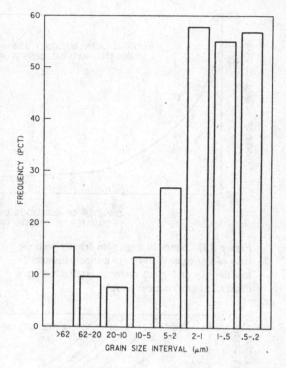

Figure 3.3. Percentage of material finer than the lower bound of a size interval released by the destruction of aggregates through combined iron oxide and organic treatments.

portion of released material, a conservative estimate of material in aggregates, increased with decreasing grain size (Fig. 3.3).

After treatment for removal of iron oxide and organic matter, the distribution of fine particles released from a size interval was similar for the various other size intervals, regardless of treatment (Fig. 3.4). In each case, the distribution was distinctly bimodal, the majority of the discrete particulate material occurring in both the very fine sizes and the size interval immediately finer than that from which it was produced, as expected from Towner (1988).

An estimate of the effect of pretreatment on size distribution statistics is demonstrated by the analysis of the Kwiklauk Pass samples, summarized in Table 3.1. Note that, as expected, the disaggregated sediments have a finer mean and a greater standard deviation than that measured in the aggregated state.

The mineralogy of the fine material released by the removal of binding material was identical at the 95% confidence level and thus is

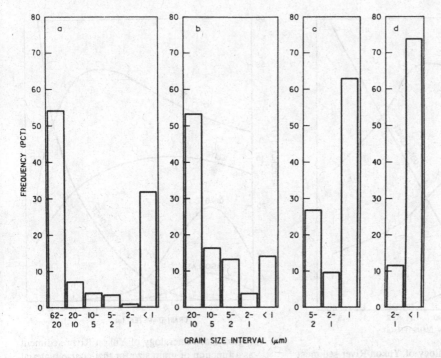

Figure 3.4. Representative distributions of the size of material released by the destruction of aggregates by iron oxide or organic treatment techniques. (a) <62-μm material produced from sand-sized material after organic treatment. (b) <20-μm material produced from 62–20-μm material after organic treatment. (c) <5-μm material produced from 10–5-μm material after iron oxide treatment. (d) <2-μm material produced from 5–2-μm material after iron oxide treatment.

Table 3.1. *Analysis of Kwiklauk Pass sediment samples*

	Suspended		Bottom	
	Mean	s.d.	Mean	s.d.
State				
Aggregated	8.2 φ	2.0 φ	5.4 φ	1.5 φ
Disaggregated	8.6 φ	2.25 φ	6.4 φ	2.6 φ

treated collectively. The mineralogy of this released material is shown in Figure 3.5 and that of the material remaining in the interval in Figure 3.6. The similarity of these distributions is demonstrated by the decrease in quartz and plagioclase in the finer sizes, the increase in montmorillonite in the finer sizes, and the modes of illite and chlorite. If the released population (Fig. 3.5) is offset by two size intervals finer from that from which it was produced (Fig. 3.6), the distribution of minerals of the released population is correlated at the 95% confidence level with the distribution of minerals of the material remaining in the size interval.

Discussion

Aggregate particles in the Yukon River are formed primarily by iron oxide and organic materials binding many fine particles onto a larger particle. This is in agreement with the observations of Bellivich (1962) and Biddle & Milles (1972) who showed that transported aggregates are often polymineralic and commonly consist of large feldspars or quartz grains surrounded by a mass of finer clay particles. These observations are also in agreement with those made for soil aggregates (Towner, 1988). The formation of an aggregate about a large central grain is supported by the bimodal frequency distribution of primary particles composing the aggregates. The lagged correlation of the mineralogy of the aggregated material to that of the primary grains also supports a probabilistic mechanism of aggregate formation.

The increase in the proportion of aggregates in the finer size fractions may be partially

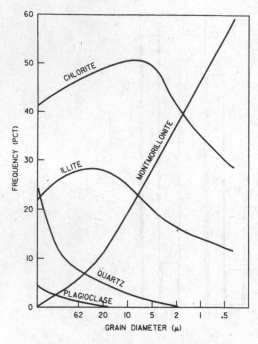

Figure 3.5. Mineralogy of Yukon River sediment as a function of grain size for the fine material released from an aggregated state by organic and iron oxide treatments.

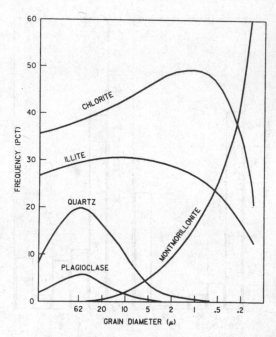

Figure 3.6. Mineralogy of Yukon River sediment as a function of grain size for that coarse material that was either an individual particle or a nuclear particle of an aggregate.

explained by the geometric decrease in size interval with decreasing particle size. The mode of formation and destruction of aggregates within the soil horizon also favors an increase in aggregate proportion with decreasing size. Although the majority of aggregates appear to be composed of both large and small particles, there are undoubtedly aggregates composed entirely of small particles. It is also reasonable that large aggregates are preferentially destroyed during transport. The larger the particle, the more energy is involved during impact, and therefore the greater the chance of fracturing and/or abrasion. Small particles are also cushioned to a greater degree by surrounding water molecules, which would further decrease the destructive effect of their impact.

It is interesting that the characteristics of the suspended and bottom sediment aggregates are generally similar. This relationship indicates that the majority of aggregates were not formed locally during temporary deposition within the river bed – otherwise there would be a higher percentage of aggregates in the bottom samples.

The high proportion of aggregates in the fine fractions suggests that these aggregates are not fecal in origin. Most aggregates therefore, are believed to have formed by soil processes, either during transport or in the source area. The alternation of wet and dry conditions in the soil horizon along with the availability of a considerable quantity of organic and iron oxide material should provide ideal conditions for the formation of tightly bound aggregates. The effect of aggregates on grain size statistics was greater for bottom sediment with respect to both the mean and standard deviation. This indicates that the larger traction particles contained a significant proportion of aggregates. These may have been formed in the soil horizon or perhaps locally by filter feeders or infauna.

Summary

Aggregate particles may comprise a significant part of any sediment sample. In order to measure a meaningful grain size distribution, sample pretreatment must be properly matched to the aim of the study. A proper choice of the degree of chemical and physical treatment to which the sample is exposed must be made to allow either retention of the aggregates in their natural sizes, or disaggregation into individual grains. If the wrong choice is made, the extent of bias on the overall population can be in excess of 1ϕ in both the mean and the standard deviation. More subtle grain size methodologies, such as breaking the size distribution profile into individual components for interpretation, may be completely invalidated by improper sample handling and pretreatment.

The choice of an appropriate pretreatment technique is easily understood by considering a study of the grain size distribution of the surface of a talus slope where the particles can be seen using a hand lens. If we study the processes of transport on the talus slope, we would not disaggregate them. It would be ludicrous to study their size distribution by smashing them with a hammer as a pretreatment technique. If, however, we wished to infer something about the grain size of the parent rock from which the talus was derived, it would clearly then be desirable to break the particles at their grain boundaries and look at the size distribution of these individual, monomineralic grains. The same sort of choice must be made for finer particles although, because we generally do not physically look at them, the choices are not so obvious. However, the results of an inappropriate pretreatment choice may be just as dramatic.

REFERENCES

Anderson, J. U. (1963). An improved pretreatment for mineralogic analysis of samples containing organic matter. *Clays and Clay Minerals*, 10: 380–8.

Bellivich, E. F. (1962). Particle structure of suspended river sediment. *Izvestiya Akademiya Nauk SSSR, Seriya Geograficheskaya*, 2: 71–3.

Biddle, P., & Milles, J. H. (1972). The nature of contemporary silts in British estuaries. *Sedimentary Geology*, 7: 23–3.

Bowler, J. M. (1973). Clay dunes: Their occurrence, formation and environmental significance. *Earth-Science Reviews*, 9: 315–38.

Brewer, R., & Blackmore, A. V. (1976). Subplasticity in Australian soils. II: Relationships between subplasticity rating, optically oriented clay, cementation, and aggregate stability. *Australian Journal Soil Research*, 14: 237–48.

Buller, A. T., & Green, C. D. (1976). The role of organic detritus in the formation of distinctive sandy tidal flat sedimentary structures, Tay estuary, Scotland. *Estuarine and Coastal Marine Science*, 4: 115–18.

Butler, B. E. (1956). Parna – an aeolian clay. *The Australian Journal of Science*, 18: 145–51.

Deshpande T. L., Greenland, D. J., & Quirk, J. P. (1968). Changes in soil properties associated with the removal of iron and aluminum oxides. *Journal Soil Science*, 19: 108–22.

Fitzpatrick, E. A. (1983). *Soils, Their Formation, Classification and Distribution*. London: Longman, 353 pp.

Fitzpatrick, E. A. (1984). *Micromorphology of Soils*. London: Chapman and Hall, 433 pp.

Flach, K. W., Nettleton, W. D., Gile, L. H., & Cady, J. G. (1969). Pedocementation: induration by silica, carbonates, and sesquioxides in the quaternary. *Soil Science*, 107: 442–53.

Gabriels, D., & Moldenhauer, W. C. (1978). Size distribution of eroded material from simulated rainfall: Effect over a range of texture. *Soil Science Society America Journal*, 42: 953–8.

Greenland, D. J., Oades, J. M., & Sherwin, T. W. (1968). Electron-microscope observations of iron oxides in some red soils. *Journal Soil Science*, 19: 123–6.

Griffiths, E. (1965). Micro-organisms and soil structure. *Biological Review*, 40: 129–42.

Haven, D. S., & Morales-Alamo, R. (1972). Biodeposition as a factor in sedimentation of fine suspended solids in estuaries. *Geological Society America Memoir*, 133: 121–30.

Johnson, R. G. (1974). Particulate matter at the sediment–water interface in coastal environments. *Journal Marine Research*, 32: 313–22.

Lewis, A. G., & Syvitski, J. P. M. (1983). The interaction of plankton and suspended sediment in fjords. *Sedimentary Geology*, 36: 81–92.

Lutz, J. F. (1936). The relation of free iron oxide in the soil to aggregation. *Soil Science Society of America Proceedings*, 1: 43–5.

McIntyre, D. S. (1956). The effect of free ferric oxide on the structure of some Terra Rosa and Rendzina soils. *Journal Soil Science*, 7: 302–6.

Matthews, M. D. (1973). Flocculation as exemplified in the turbidity maximum of Acheron Channel, Yukon River Delta, Alaska. Ph.D. dissertation, Northwestern University, 88 pp.

Meglis, A. J. (1987). Bioaggregates and their role in inorganic sediment transport in suspended sediments of a coastal lagoon complex near stone harbour, New Jersey. M.S. thesis, Lehigh University, 72 pp.

Mehra, D. P., & Jackson, M. L. (1960). Iron oxide removal from soils and clays by a dithionite–citrate system buffered with sodium bicarbonate. In: *Clays and Clay Mineral 7th National Conference,* ed. A. Swinford. Washington, D.C., pp. 319–27.

Paerl, H. W. (1973). Detritus in lake Tahoe: structural modification by attached microflora. *Science,* 180: 496–8.

Price, W. A. (1963). Physicochemical and environmental factors in clay dune genesis. *Journal of Sedimentary Petrology,* 33: 766–78.

Rhoads, D. C. (1963). Rates of sediment reworking by Yoldia Limatula in Buzzards Bay, Massachusetts and Long Island Sound. *Journal of Sedimentary Petrology,* 33: 723–7.

Rhoton, F. E., Meyer, L. D., & Whisler, F. D. (1983). Densities of wet aggregated sediment from different textured soils. *Soil Science Society America Journal,* 47: 576–8.

Riezebos, P. A., & Lustenhouwer, W. J. (1983). Characteristics and significance of composite particles derived from a Colombian andosol profile. *Geoderma,* 30: 195–217.

Russell, E. W. (1971). Soil structure: its maintenance and improvement. *Journal of Soil Science,* 22: 137–51.

Rust, R. R., & Nanson, G. C. (1989). Bedload transport of mud as pedogenic aggregates in modern and ancient rivers. *Sedimentology,* 36: 291–306.

Schubel, R. J. (1971). A few notes on the agglomeration of suspended sediment in estuaries. In: *The Estuarine Environment, American Geological Institute Short Course Notes,* ed. J. Schubel. Washington, D.C.: AGI, pp. 1–29.

Swaby, R. J. (1949). The relationship between micro-organisms and soil aggregation. *Journal of General Microbiology,* 3: 235–54.

Towner, G. D. (1988). The influence of sand- and silt-size particles on the cracking during drying of small clay-dominated aggregates. *Journal of Soil Science,* 39: 347–56.

Turchenek, L. W., & Oades, J. M. (1978). Organo-mineral particles in soils. In: *Modification of Soil Structure,* ed. W. W. Emerson, R. D. Bond, and A. R. Dexter. New York: Wiley–Interscience, pp. 137–44.

Young, R. A. (1980). Characteristics of eroded sediment. *Transactions of American Society of Agricultural Engineers,* 23: 1139–42, 1146.

II *Theory and methods*

There have been many techniques developed for the size characterization of geological samples. The earliest was the Oden balance (Oden, 1915), which utilized tangents to the Oden curve (weight of accumulated sediment vs. settling velocity). Some were developed for specific purposes: For instance, the falling-drop method (Moum, 1965) and the volume size analysis (VSA – Syvitski & Swinbanks, 1980) were separately developed for the analysis of small volumes of mud samples. Many of the early techniques have been superseded by faster and more precise automated instruments. In Part II we describe the principal techniques used in modern geological particle size analysis.

Settling tubes provide an ever-popular means for the size characterization of sands and coarse silt samples. Results are in the form of particle settling velocity or sedimentation diameter. Both are considered more fundamental than other geometrically defined measures of size, with reference to a particle's behaviour in the hydrodynamic environment. Chapter 4 documents the complexities and varieties of settling columns, including an analysis of errors, the principles of tube calibration, and a simplified description on how to construct a settling tube.

Sieving of sand is a classical size technique, being at least 50 years old. Since those early days, the variety of sieves in use has increased, including sieves constructed of plastics, brass, and stainless steel. Some are particularly useful for wet sieving of biological matter. Chapter 5 documents the more classical methodology of dry sieving and calibration of wirecloth sieves, but with particular reference to small sieves useful in the analysis of small sample weights.

Image analysis is considered the standard technique in the characterization of particle shape. It is also a useful and straightforward technique for providing a size frequency distribution of a sediment sample. Chapter 6 discusses the concept of edge-point acquisition using the ARTHUR image analysis system. Prob-

lems specifically inherent in image analysis are further discussed in Chapter 13, including the problems in microsplitting and bias in particle counting.

Chapter 7 provides a detailed discussion on quantitative grain form analysis through the use of image analysis techniques. Numerical shape techniques described include axial : length ratios, angular decomposition (Fourier analysis), and chord/length decomposition (fractal analysis).

One of the best established automated particle size techniques, at least within the oceanographic community, is the electroresistance particle size analyzer (e.g., Coulter Counter, Elzone). Chapter 8 discusses its operational principles, sample preparation, and analytical procedure. Its long-standing popularity in the ocean sciences is a direct result of its ability to analyze low concentrations of suspended sediment samples directly, as sea water is an electrolytic solution.

Laser diffraction size analysis is a new method of analyzing both dry powders and suspended sediment samples. It is fast and precise and uses only small sample weights. It is capable of analyzing fine sand particles, an improvement on the SediGraph. The instrumentation is still undergoing the growing pains associated with a new technique. Accuracy can be disappointing (see also Chapter 13). Chapter 9 provides a detailed description of the principles, analytical method, and problems with the analysis of geological samples.

Another well-established technique is the SediGraph based on the attenuation of x rays through a column of turbid water. The technique has replaced the classical pipette technique, remains limited to mud samples (cf. laser diffraction instruments), and requires a minimum of 1 g of sample (cf. electroresistance counters). Chapter 10 reviews the operational principles and theory, analytical procedures, and calibration, with an assessment of the method's accuracy and precision.

The theory of light scattering by various sized particles is over thirty years old; yet laboratory techniques remain unpopular. This may reflect the associated theoretical complexities. Chapter 11 eloquently reviews the theoretical principles on which future instruments will be based, detailing how light scattering occurs, with reference to particle size and shape.

All previous chapters have been concerned with unlithified sedimentary samples; however, most sedimentary deposits are lithified. As it is difficult at best to separate the individual grains from a cemented sample, particle size analysis has historically utilized thin-section analysis. Chapter 12 provides a back-to-basics approach to the analysis of particle size and shape using thin sections of rocks, but on a modern image analysis system. The chapter also provides a simple method to determine the maturity of silliclastic rocks.

Chapter 13, the final chapter in Part II, documents the results of a worldwide inter-instrument calibration experiment, wherein the precision and accuracy of many of the methods described above are addressed. The paper recommends:

replacing classical techniques with modern automated and digital techniques wherever possible;

reporting in the published literature the accuracy and precision of instruments used in particle size analysis; and

limiting scientific conclusions in accordance with the errors associated with particle size analyzers.

Chapter 13 also details how geological standards can be prepared in pursuit of these recommendations.

REFERENCES

Moum, J. (1965). Falling drop used for grain-size analysis of fine-grained materials. *Sedimentology*, 5: 343–7.

Oden, S. (1915). Eine neue methode zur mechanischen Bodenanalyze. *Internationale Mitteilungen für Bodenkunde*, 5: 257–311.

Syvitski, J. P. M., & Swinbanks, D. D. (1980). VSA: A new fast size analysis for low sample weight based on Stokes' settling velocity. *Canadian Geotechnical Journal*, 17: 304–12.

4 Principles, design, and calibration of settling tubes

J. P. M. SYVITSKI, K. W. ASPREY,
AND D. A. CLATTENBURG

Introduction

Over the past fifty years, the textural characterization of an unlithified sample of sandy sediment has been accomplished mainly by sieving and settling tube analysis. Both methods provide a mass frequency distribution of particle size, with the effects of particle shape and size combined. Settling tube analysis is additionally affected by particle density. Sieving became the early standard technique, accepted by both civil engineers and sedimentologists. Standard sieves were mass produced at a variety of mesh sizes, and once sieving time and tapping frequency were agreed upon, interlab precision was acceptable ($<\pm 0.25 \phi$, where $\phi = -\log_2 d$ and d is particle diameter in mm). Sedimentary petrologists, in their study of lithified sedimentary deposits, sized and identified sedimentary particles under a microscope using thin sections from rocks. Sieving and thin-section methods dominated sediment laboratories until the mid-1970s. With the proliferation of microcomputers in the 1980s, sieving has given way to settling tube analysis, and manual thin-section analysis has given way to image analysis.

The basic justification for settling tube analysis is that the settling velocity of a particle is a more fundamental dynamic property than any geometrically defined measure of size (i.e., sieving, thin sections, image analysis) with reference to its behaviour in a hydrodynamic environment (Gillespie & Rendell, 1985). Because particle mobility in liquids (air, water) is dependent on the ratio between shear velocity and settling velocity (Francis, 1973), the settling velocity distribution and therefore the "sedimentation diameter" distribution is argued to be more valid for the characterization of sand texture than sieve-determined size distributions (Middleton,

1976; Bridge, 1981). Settling tube results are also compatible with the standard geologic methods for the determination of the mud fraction of a sedimentary sample, i.e., pipette and Sedi-Graph methods.

In addition to these hydrodynamic arguments supporting the use of settling tubes, they are largely controlled by microcomputers, with the concomitant abilities to reduce human errors during data transfer and to statistically manipulate and display the results. Settling tubes can also provide a data set that is 5–10 times greater in resolution than the typical 0.5ϕ and 0.25ϕ sieve interval, and at a much greater speed (15 min compared to 1 hour for sieving). With the elimination of sieve dust and noise, many sediment laboratories have converted, or are converting, to settling tube analysis as the choice for sand characterization.

There are, however, important problems with settling tube analysis: Very few tubes are identical, with variation in tube dimensions, particle sensing apparatus, sample introduction systems, and environmental control. There is no agreement on the formulae used to convert settling velocity to equivalent particle diameter. Additionally, most tubes have not been properly calibrated; that is, the precision and accuracy of each instrument remains unknown. No manufacturer of settling tubes dominates the market, and most laboratories have constructed their own sedimentation towers. This chapter reviews the principles, instrumentation, results, and problems, and provides a detailed description of one of the more simple and widely used tubes.

Principles

The force of resistance of a spherical particle moving through a fluid depends on the diameter and relative velocity of the particle, and on the density and viscosity of the fluid. Settling tubes are based on the principle that a sample's size distribution can be obtained from measurement of the mass–velocity distribution of sand grains settling through an otherwise turbid-free liquid.

The use of settling tubes is based on the assumption that particles settle out individually, neither hindered by other settling particles, nor involved in convective plumes of high concentration, nor retarded by upflow of displaced fluid. This is accomplished by introducing only coarse (>50-μm) sediment particles, typically in the sand size range (2,000–62.5 μm), in low concentration, at the top of a sufficiently wide settling tube. After being released, the particles are stratified in the sedimentation fluid according to their respective settling velocities. At the lower end of the tube, the arrival of the falling particles is detected, and thus a frequency distribution of the sample as a function of settling duration is obtained. The distribution can be in terms of number of particles, particle volume, particle weight, or projected area, depending on the detection method used to determine particle arrival (Slot & Geldof, 1986; see Fig. 4.1B). The frequency distribution based on settling duration is converted into the fall velocity distribution using the particles' settling distance.

Settling velocity and grain size

For the settling of a sphere of any density in any Newtonian fluid and gravitational field, a general equation (cf. Kennedy & Koh, 1961) may be written:

$$w_s = [4/3 \, g \, d_s (\rho_s - \rho_f)/C_d \rho_f]^{0.5} \quad (4.1)$$

where w_s is the particle's fall or settling velocity, g is the acceleration due to gravity, d_s is the particle's sedimentation diameter, ρ_s is the density of the particle, and ρ_f is the density of the fluid. C_d is the particle's drag coefficient defined by

$$C_d = F/[A \, \rho_f (w_s^2/2)] \quad (4.2)$$

where F is the force of resistance exerted by the fluid on the falling particle, and A is the cross-sectional area of the particle. In this form, tables or nomograms of C_d versus w_s must be used, with particle density or fluid density held constant (see Komar, 1981).

A general formula for the velocity w_s of a sphere of diameter d, induced by an acting force F moving in a viscous fluid with kinematic viscosity v, has the form

$$w_s = (v/2df_\beta) \, [-1 + (1 + (4/3\pi) \, (F/\rho v^2) f_\beta)^{0.5}] \quad (4.3)$$

where f_β is a function of dimensionless quantity

($F/\rho v^2$). If $f_\beta \approx 0.0125 + 0.348[F/\rho v^2)]^{-0.33}$, the formula holds for Reynolds numbers up to 2,000 (associated with a settling grain up to 4.2 mm in diameter) with an accuracy of 2% (Slot, 1983a,b). Equation (4.3) may be rewritten to yield sedimentation diameter d_s:

$$d_s = (0.1125/\Delta g) w_s^2 \, [1 + (1 + (1418.4\Delta g/w_s^2) \\ \cdot (v/w_s + 0.348Q))^{0.5}] \quad (4.4)$$

where $\Delta = (\rho_s - \rho_f)/\rho_f$, $Q = [6v^2/(\pi \Delta g)]^{0.33}$.

The particle velocity distribution may be reported directly as:

1. a base-2 logarithmic distribution (e.g., Psi = $\log_2 w_s$ in cm s^{-1}: Middleton, 1967);

2. as a base-10 logarithmic velocity distribution (Taira & Scholle, 1979); or

3. some equivalent grain diameter, normally reported as the equivalent spherical sedimentation diameter (ESSD) or sedimentation diameter (Guy, 1969).

A popular method of converting settling velocities to grain diameters is through the use of least-squares polynomial regression equations based on empirical results of settling particles of known shape, size, and density (see Gibbs et al., 1971). Such equations are based on the premise of equivalency, wherein a sand grain has its fall velocity reported in terms of some standard conditions. These may include reporting measured velocities with either theoretical or empirical results of spheres of constant density (typically $\rho_s = 2.65$ g cm^{-3}) settling in quiescent, distilled water of infinite extent at a temperature of 20° or 24 °C. Equations and data relating settling velocity to grain size can be found in Rubey (1933), Rouse (1937), ICWS (1958), Graf & Acaroglu (1966), Schlee (1966), Cook (1969), Watson (1969), Warg (1973), Brezina (1979), and Riley & Bryant (1979).

Natural sand grains are rarely spherical and invariably settle at a velocity lower than would be expected for a sphere of equivalent size (defined as a sphere whose diameter equals the intermediate diameter of the particle; Baba & Komar, 1981). This relates to irregularities in grain shape (nonsphericity and asymmetry) rather than grain roundness (angularity). Komar & Cui (1984) suggest that the relationship

$$w_s = 1.026\, w_{sm}^{1.095} \qquad (4.5)$$

should be used to convert w_{sm}, the measured settling velocity, into intermediate grain diameters d_i, after substitution of w_s into Gibbs et al.'s (1971) equation. Brezina (1979) modified his measured fall velocities with predetermined triaxial shape information (i.e., the hydraulic [Corey] Shape Factor), following the work of Komar & Reimers (1978).

If a population of particles has a similar particle shape, and if this shape can be approximated by a single shape factor sf, where sf = (short axis) / (long axis × intermediate axis)$^{0.5}$, then Slot (1983a,b) provides the following expression for the nominal diameter d_n:

$$d_n = (K_0 w_s^2)/((\Delta g/9) - (2K_2 w_s^2/Q))$$
$$\cdot\, [1 + (1 + 2((\Delta g/9) - (2K_2 w_s^2/Q)) /$$
$$(K_0 w_s^2)\,(v/w_s + K_1 Q))^{0.5}] \qquad (4.6)$$

If $K_0 \approx 0.109 - 0.1sf$, $K_1 \approx 0.635 - 0.253sf$, $K_2 \approx (8.95 - 6.34sf)\cdot(10^{-5})$, the formula holds for Reynolds numbers up to 8,000 (into the fine gravel-size range of particles) with an accuracy of 5% (Slot, 1983a,b).

Settling tube technology, accuracy, and precision

The goal in designing a settling tube is to have the measured fall velocities of a set of particles correspond closely to their ideal fall velocity. Multigrain settling, as in the case of settling tubes, must deal with a variety of problems inherent in the method. These include:

1. *hindered settling,* which is caused by a counterflow of fluid induced by the falling sediment, and thus retarding the settling velocity of individual particles (Thacker & Lavelle, 1977, 1978);

2. *settling convection,* whereby vortices and pressure gradients created in the sedimentation fluid by the movement of particles can interfere with the condition of water "quiescence" and the fall path of particles (Kuenen, 1968);

3. *thermal convection,* a result of temperature differential along the length of the fall path; and

4. *mass settling,* where smaller grains are dragged down in the wakes of larger grains.

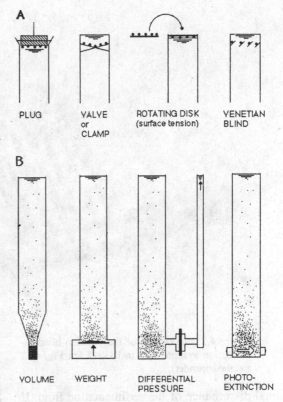

Figure 4.1. (A) Four types of sample introduction systems for settling tubes (after Geldof & Slot, 1979). (B) Four classes of detection methods for settling tubes (after Geldof & Slot, 1979).

Specifically, settling tube errors include those associated with:

1. the initial position and initial velocity of the particles (i.e., the sample introduction method);

2. particle behaviour in the settling tube (see above, but also influenced by the wall effect and concentration effect); and

3. the quality of the detection method for particle arrival at the lower end of the settling tube.

Introduction systems

A number of introduction systems have been designed for settling tubes (Fig. 4.1A). These include plug systems, valve or clamps, rotating disks, and venetian blind (e.g., Fig. 4.2). An introduction system must ensure that all particles are released evenly from a given holding platform, at the same time and with min-

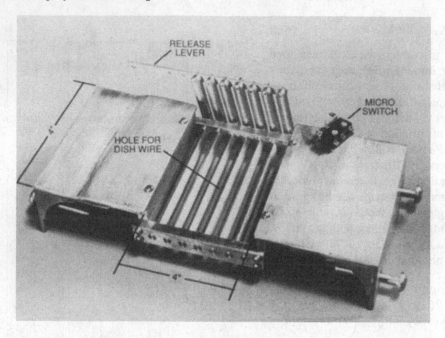

Figure 4.2. Photograph of the venetian-blind introduction system used in BIST-2 (cf. Fig. 4.12 and the appendix).

imal disturbance of the sedimentation fluid. If particles enter the sedimentation media over a time ΔT, then an error will be associated with the determination of each particle size interval. This error, while depending on the magnitude of ΔT, will decrease with a particle's transit time and thus with increasing particle size.

The introduction system must also ensure that all particles are released. If the holding platform is dirty or electrostatically charged, small particles may temporarily stick. Two types of errors could result:

1. The final distribution may be biased to the coarser particles if fine particles remain stuck to the sample holder throughout the analysis.

2. The final distribution may be biased to the finer particles if previously stuck particles are released during the analysis.

The magnitude associated with these errors is highly variable, but typically <1%.

The introduction system should also be designed not to interfere with the detection sensor. For instance, the opening or closing of valves or venetian blinds should not interfere with the cable supporting the collection pan. Poorly constructed systems have mechanically induced

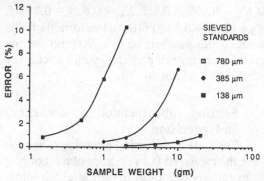

Figure 4.3. The influence of sample weight on three sieved size fractions ($0.25\,\phi$ interval) on the detection error of the Delft University Settling Tube (DUST) (after Slot & Geldof, 1986). For fine sand samples, the initial sample weight should be <1 g.

noise introduced at the start of sample analysis. This provides one of the upper limitations on coarseness of particles that can be analyzed (typically <2 mm). Similarly a pressure wave, set up within the tube from plug or rotating disk introduction systems, can disturb the sedimentation fluid.

Particle behaviour

Kranenburg & Geldof (1974) derived a procedure for estimating the magnitude of convection and hindered settling in settling tubes.

According to their research, for a dimensionless concentration (sample volume per tube volume) of less than 10^{-4}, the relative error is <5% for hindered settling; settling convection may produce a relative error of 5% for small samples of large (>1 mm) particles or as high as 60% for large samples (>50 g) of small particles. In general, the error associated with concentration effect was shown to increase both with decreasing particle size and with increasing sediment concentration (Fig. 4.3). For example, Figure 4.4D demonstrates the increase in the mean grain size (by 0.2ϕ) as the sample weight decreases from 4 g to 0.25 g.

Gibbs (1972) presents a nomogram to be used for estimating the degree of error to be expected for a multigrain sample depending on particle diameter, sample weight, and tube size. He suggests that minimum dimensions of 13–16 cm inside diameter and 140 cm of effective length be observed to ensure that particles fall at their terminal velocity over as long a distance as possible, while the large diameter reduces wall effects and concentration effects.

Figures 4.4A–C, show the effect of increasing fall distance on moment measures of mean and standard deviation for coarse sand. Fall distance primarily affects the estimation of the settling velocity of coarse particles: The estimation of mean size of coarse sand particles may vary by up to 0.3ϕ; the standard deviation by up to 0.15ϕ. For particles 1ϕ or finer, the fall distance should exceed 140 cm. For the measurement of particles coarser than -0.5ϕ, the fall distance should exceed 180 cm. We recommend a settling tube length of ≈ 200 cm.

Vanoni (1975) proposed a formula relating the settling velocity of a particle measured in a settling tube (w_d) to its unconfined settling velocity (w_s), as

$$w_s/w_d = 1 + (9/4 \ d/D) + (9/4 \ d/D)^2 \quad (4.7)$$

where D is the tube diameter. Chapman (1981) also developed an equation that compensates for concentration effects and yields the hydraulic mean diameter of a multigrain sample.

Detection errors

Detection errors include those associated with reaction time of the measuring device, in-strument drift, instrument noise, and nonlinearity effects (i.e., imperfections in the inductive transducers, the Wheatstone-bridge amplifier, and the springs of the balance). A full error analysis to be used in settling tube evaluation is given in Slot & Geldof (1979). Detection methods can be divided into four categories (Fig. 4.1B): those measuring the accumulated volume, particle weight, differential pressure within the fluid, or light attenuation at a given depth.

In the first category, the rate of increase in the volume of sand accumulating at the base of the tube is measured. Bulk density value(s), preferably variable as a function of time (i.e., particle size), must be used to convert the accumulating volume of sediment to weight. However, the rather imprecise choice of bulk density values, wall effect problems in the lower portion of the tube, large sample size (with the above-mentioned problems), and low resolution of results have made this class of tubes obsolete, although their cost is very low. For further information see van Veen (1936), Emery (1938), Poole et al. (1951), Colby & Christiansen (1956), Poole (1957), Guy (1969), Rukavina & Duncan (1970), and Vanoni (1975).

The most common, and in many cases, the most precise class of settling tubes involves the recording of the weight of sediment as a function of time. The weight can be measured (Fig. 4.5) by means of a strain gauge situated on a cantilever from which the sediment-accumulation pan is suspended (van Andel, 1964; Felix, 1969; Thiede et al., 1976; Flemming & Thum, 1978; Anderson & Kurtz, 1979), or through a pan suspended by wire connected to a balance at the top of the settling tube (Doeglas, 1946; Plankeel, 1962; Bienek et al., 1965; Gibbs, 1974; Halka et al., 1980; Amos et al., 1981; Rigler et al., 1981; Sengupta & Veenstra, 1986). More sophisticated systems, in terms of cost and complexity albeit with increased precision and accuracy, use an underwater balance. For instance, the Macrogranometer (Fig. 4.5C; Brezina, 1972) is supported by leaf springs, with the movement of the pan measured by inductive displacement transducers. External vibrations, both local and distant, are dampened by large air shock absorbers (Fig. 4.6B). The settling tube at the Delft University of Technology

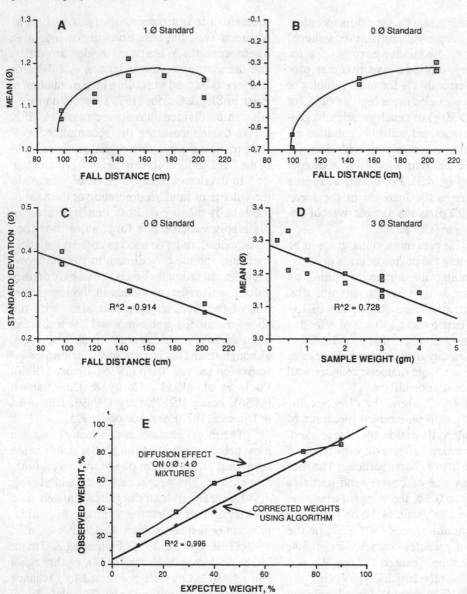

Figure 4.4. Calibration results from an early Bedford Institute Settling Tube (BIST). The influence of particle fall distance on: (A) the mean diameter using a 0.75–1 φ sieved standard; (B) the mean diameter; and (C) the standard deviation of a –0.25–0 φ sieved standard. (D) The influence of sample weight on the mean diameter using a 2.75–3 φ sieved standard. (E) The effect of particle diffusion on pan collection (where the pan diameter is smaller than the inner diameter of the settling tube; see text for details). The observed weights (y axis) refers to the relative contribution of the coarse fraction (–0.25–0 φ) of size frequency distributions from various combinations of 0 φ : 4 φ sieved standards. The expected weights (x axis) refer to the fraction of the 0 φ population used in the preparation of the standards. The solid straight line is a result of an algorithm developed to compensate for the effect of differential particle diffusion.

(DUST: Slot & Geldof, 1986) also includes a feedback loop in which a part of the output signal is subtracted from the input signal by means of a coil and magnet (Fig. 4.5D). External vi-

brations are reduced by means of mounting the settling tube on a large concrete block (1,400 kg), supported by air springs and damped by glycerine (Fig. 4.6A). DUST has an overall ac-

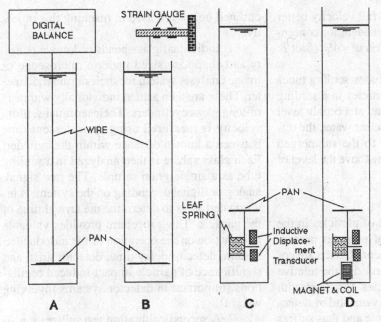

Figure 4.5. Four principal weight detection methods used in settling tubes (after Geldof & Slot, 1979).

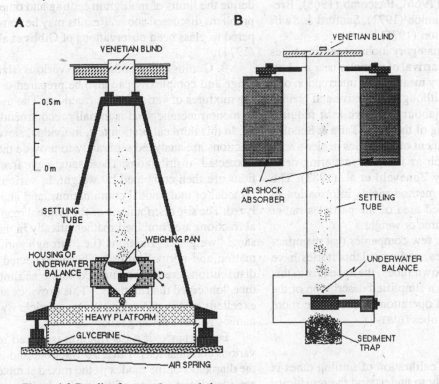

Figure 4.6. Details of two underwater balance settling tubes: (A) the Delft University Settling Tube (after Slot & Geldof, 1986); (B) Macrogranometer (after Brezina, 1979).

curacy of measured particle fall velocity better than 3% (which takes into consideration concentration effects), and a precision of <4% (Slot & Geldof, 1986).

The third category includes settling tubes where the suspension of particles in a settling tube causes a piezometric head at a certain level that differs from the head in clear water, the difference being proportional to the submerged weight of the particles present above the level of measurement – that is,

$$V/A + (V_s/A)\Delta$$

where V is the total volume of particles in the sample, V_s is the volume of particles present above the level of measurement, A is the area of settling tube cross section, and Δ is the relative apparent density of the particles. However, with conventional transducers this method of detection requires too large a sample and thus suffers from increased concentration effects. Readers are referred to Mason (1949), Zeigler et al. (1960), Schlee (1966), Bascomb (1968), Brezina (1969), Channon (1971), Sanford & Swift (1971), and Nelson (1976).

The fourth category includes settling tubes that measure the arrival of particles near the bottom of the tube by means of an interruption of a beam of light. Although a sensitive and precise technique, it is labour intensive as it requires frequent changing of the fluid (Taira & Scholle, 1977). The treatment of the optics neglects several features such as forward scattering dealt with in depth by Zaneveld et al. (1982). The photoextinction method yields information in terms of projected area of the particles rather than particle volume or weight.

As there are few companies that manufacture settling tubes, and most laboratories have constructed their own tubes, the appendix to this chapter provides a simplified description of the construction and operation of one of the more popular settling tubes (BIST-2).

Calibration

The proper calibration of settling tubes is essential, not only to understand the sensitivity of a particular introduction or detection system, but also to ascertain the precision and accuracy of a settling tube. The three calibration tests outlined below are used in reaching this objective.

1. Individual glass beads of known density and shape are sized under a microscope or image analysis system for their nominal diameter. These are then settled individually within 1-m-long glass cyclinders: Their terminal settling velocity is measured using their descent time between a known distance within the cylinder. Each glass sphere is then analyzed in a settling tube as a single grain sample. The raw signal, analog or digital depending on the system, is interrogated so as to determine the arrival time of the particle. This procedure provides valuable information on the operation of the introduction system, detector delay time, detector drift, and significance of particle impact-induced oscillations (important in detector systems involving weight).

2. A second calibration test utilizes multibead samples run at various concentrations and combinations (multimodal mixtures). These tests define the limits of multigrain settling and other problems discussed above. Results may be compared to glass bead observations of Gibbs et al. (1971).

3. Geological standards of various size range and complexity can also be prepared using mixtures of various sieve fractions. The use of monomineralic sand is initially recommended. In this third calibration test, individual sieve fractions are analyzed separately to provide the "expected" distributions. These same sieve fractions are then combined by weight, in various bimodal or multimodal combinations, and analyzed. The size distributions of the original modal fractions are combined mathematically in the same "weighted" fractions (i.e., through summation and normalization). These "expected" distributions are then compared to the settling tube "observed distributions." This provides an excellent means of understanding particle dynamics within a settling tube.

For instance, if a 0ϕ fraction is mixed in various combinations with a 3ϕ fraction, and if the diameters of the modes of the mixed sample are identical to those of the individually run fractions, then the effects of hindered settling and mass settling can be considered negligible at the concentrations used. Below we describe how this third calibration technique can be used to quantify the effect of particle diffusion that is problematic with some settling tube designs.

We note that a large number of settling tubes use a collection pan having a diameter smaller than the inner diameter of the tube. This can lead to problems relating to differential particle diffusion. Coarse sand particles settle with a vertical settling velocity up to 300 mm/s. This is of sufficient magnitude that the advective component due to gravity is much larger than the diffusive component due to the concentration gradient of the initial settling sample mass. Thus all coarse settling particles will be collected by the pan and sensed. However, very fine sand particles settle slowly (≤ 3 mm/s), and consequently gradient diffusion is able to eliminate any across-tube gradient in particle concentration by the time these particles reach the bottom of the tube. The finer particles will spread out and a significant percentage may miss the pan. Thus in a bimodal sample more coarse particles are captured by the collection pan, which can lead to large errors.

The third calibration technique helped us to identify the magnitude of this problem (on an older settling tube). The tube had an inner diameter of 15 cm, the pan had an outer diameter of 9 cm, providing an area ratio of pan to tube of 0.36. Various combinations of a 0ϕ and a 4ϕ fraction were analyzed (1:9, 2.5:7.5, 4:6, 5:5, 6:4, 7.5:2.5, 9:1). Figure 4.4E shows how the contribution (by weight) of the coarse fraction was observed to be invariably greater than what we should have expected from our original modal combinations. An empirical equation was developed to compensate effectively for this diffusion effect (Fig. 4.4E).

Class interval

Depending on the sophistication of the detection system, the raw data generated by settling tubes can easily exceed a size interval of 0.01ϕ for all or portions of the sand size spectrum (i.e., a 5-ϕ range of 2,000–62.5 μm). Rather than manipulating these large data arrays, data reduction programs are commonly employed to provide histograms or frequency tables of grain size versus weight percent at coarser class intervals (typically $<0.1\phi$ interval).

The reasons for this vary from trying to recreate sieve intervals (i.e., whole- or half-phi) to tracking the magnitude of a particular size interval through a series of sediment samples. However, choosing too large a class interval may lead to an often overlooked aspect in manipulating frequency data. As shown in Figure 4.7, the mode of a size frequency distribution can change over a range of 0.6ϕ, depending on the class interval chosen. If the class interval is $\leq 0.2\phi$, the effect becomes negligible, providing a difference in distribution means of $\pm 0.05\phi$.

Sieve diameter versus settling diameter

As discussed in the introduction to this chapter, settling tubes are replacing or augmenting the standard sieving technique. There remains some concern as to the difference in the results of the two methods, although algorithms have been developed to make comparisons more compatible (see Komar & Cui, 1984). Before using these published and very much empirical algorithms, we suggest the many possible operational and design errors associated with the various types of settling tubes discussed above should be considered. It is not easy to separate instrument effects from those relating to particle shape and density in determining particle size.

Figure 4.8 compares the mean diameter and standard deviation of forty deltaic samples. First these were analyzed by sieving into 0.25ϕ intervals; then the reconstituted sieve fractions were analyzed as a sample on a settling tube (see the appendix for tube description) from which the raw data were grouped into 0.2ϕ intervals. Essentially the identical sample material was analyzed using both techniques. The sieve mean was found statistically to be 0.15ϕ finer than the settling tube mean ($r^2 = 0.96$ over the particle range of 0.5–5ϕ; Fig. 4.8A). The size frequency distributions obtained from the sieving data are better sorted (one standard deviation, $\sigma \geq 0.2\phi$) than those obtained from the settling tube ($\sigma \geq 0.4\phi$), although the scatter is large ($r^2 = 0.73$; Fig. 4.8B).

A more telling method of comparing sieve data to settling tube data is with the shape of the resultant size frequency distribution (SFD).

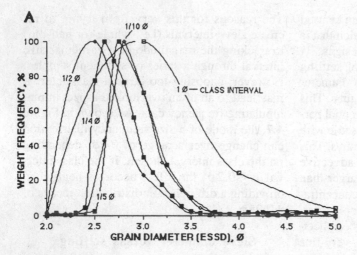

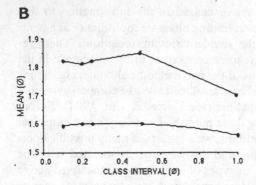

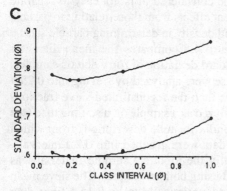

Figure 4.7. The effect of arbitrarily grouping a 0.33 ϕ interval "raw" signal generated from the BIST-2 (cf. Appendix 1) settling tube into 0.1 ϕ, 0.2 ϕ, 0.25 ϕ, 0.5 ϕ, and 1 ϕ intervals. Grain diameters are in terms of equivalent spherical sedimentation diameters (ESSD). (A) Size frequency distributions have been normalized to the maximum frequency class interval of the whole phi distribution for convenience in comparison. (B) and (C) show this interval size effect on the reported mean grain size and standard deviation, respectively, for two sieved standards.

In Figure 4.9, two of the aforementioned forty deltaic samples are compared. As the class intervals were slightly different between the two methods (0.2 ϕ vs. 0.25 ϕ), the settling tube SFDs were renormalized to the frequency of the maximum class interval of the sieve SFDs. In these examples, the settling tube SFDs are better sorted with a single mode of hydraulically equivalent grains; the sieve SFDs were bimodal. Why this difference?

In theory and through observation, both methods produce identical size frequency distri-

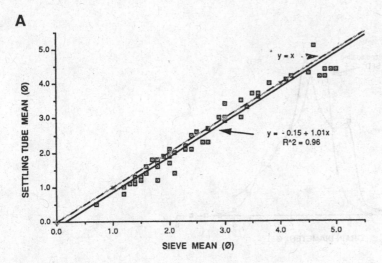

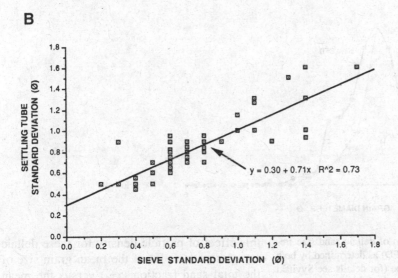

Figure 4.8. Comparison of (A) mean grain size (ϕ), and (B) standard deviation (ϕ), from deltaic sand samples as determined by both sieving and settling tube analysis (for details see Syvitski & Farrow [1983]). Note that statistically the settling tube means are $\approx 0.15\,\phi$ finer than those determined from sieving. There is more scatter in the standard deviation data set, reflecting the nature of the two methods as they are affected by grain shape and density. Settling tube SFDs appear to be more poorly sorted than their sieve counterparts, especially when the samples are coarse grained.

butions of a sample composed of glass spheres of constant density (Fig. 4.10B: Flemming & Thum, 1978). Both methods, however, experi-

ence the influence of particle shape in different ways and thus produce very different SFDs, especially when the sample is composed of skeletal carbonate grains. In such a case (e.g., Fig. 4.10A), the sieve SFD is very poorly sorted, with a mean diameter of 1.8 mm; the settling tube SFD is well sorted, with a mean diameter of 0.6 mm. The effect of particle density is demonstrated in Figure 4.10C, where the hydraulic SFD of a sample of garnet sand is $0.6\,\phi$ coarser than the geometric SFD, as expected based on the concept of equivalency. In other words, the settling tube SFD is presented in terms of equivalent spherical sedimentation diameters (ESSD) of quartz rather than garnet.

A

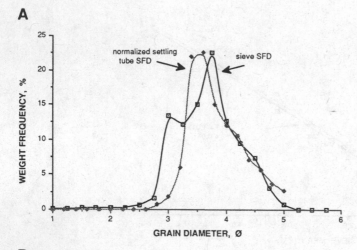

B

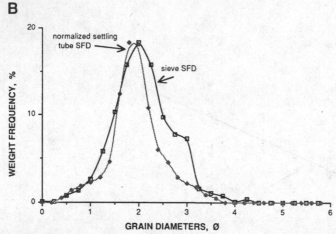

Figure 4.9. Comparison of deltaic sand size frequency distributions (SFD) as determined by both sieving and settling tube (for details see Syvitski & Farrow [1983]). Note that for these two examples, the settling tube SFDs are better sorted and very slightly coarser than those for the sieve data. The settling tube SFDs have been normalized to the maximum peak of the sieve SFDs due to the fact that the latter is from 0.25ϕ class intervals and the former is from 0.2ϕ class intervals.

If we return to the deltaic samples shown in Figure 4.9, the SFD differences between sieving and settling can be discussed in terms of particle shape and particle density effects. As these samples contain no skeletal carbonate grains and have only a 30% range in their Corey shape factor (see equation (4.6)), variations between the two methods must relate principally to particle density. In Figure 4.11 we explore

this effect of particle density for these deltaic samples with a plot of the mean grain size of the total sand fraction (x_{Ts}) versus the mean grain size of the magnetite fraction (x_M). Figure 4.11A shows an increasing deviation between x_{Ts} and x_M in the finer-grained samples: The magnetite fraction becomes increasingly coarser compared to the total sand fraction. Thus as the relative abundance of heavy minerals increases, so will the deviation in SFD between the two methods (Fig. 4.11B). Even when the density differential for magnetite is accounted for, compared to the quartz standard, sample BD-6B remains coarser in the magnetite fraction. Thus the size frequency distribution of a deltaic sample is found to be bimodal when heavy minerals are not removed, and very well sorted when magnetite is removed. The original bimodal nature of the sample reflects sampling across lami-

A

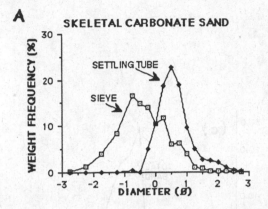

B

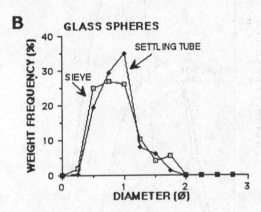

C

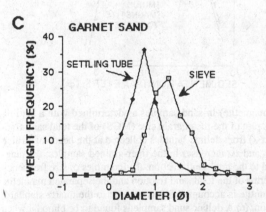

Figure 4.10. The difference in size frequency distributions (SFD) as obtained by sieving and settling tube analysis of the same sample. Note the striking difference in mean diameter and sorting, particularly in the skeletal carbonate fraction (settling tube SFD is finer grained) and the garnet sample (settling tube is coarser grained), with little difference in the glass sphere SFDs (after Flemming & Thum, 1978).

nae (Emery, 1978), whereby a thin layer of heavy mineral sand lies above or below a thin layer of light mineral sands.

Summary

Settling tubes have advanced in many respects over sieving for the characterization of geological sand samples. Settling tubes:

1. provide a data set 5–10 times greater in resolution,

2. provide much faster analysis,

3. produce no dust or noise,

4. provide compatibility with standard geologic methods for the size characterization of muds, and

5. involve digital data transfers, which reduce human error.

Settling tubes have been further justified regarding their ability to provide results in terms of settling velocity (or hydraulic size equivalence). Samples are limited to small (<1 g) sample weights.

The number of potential design problems is rather large, including those relating to the sample introduction system, particle behaviour in the sedimentation column, and the method used to detect particle arrival at the lower end of the tube. We have outlined three calibration techniques and encourage all users and manufacturers of settling tubes to undertake extensive error analysis and become involved in interlab and interinstrument calibration experiments. A well designed settling tube should obtain an overall fall velocity better than 5% over the range of particle sizes being analyzed.

APPENDIX: BIST-2, A MASS-SENSING SETTLING TUBE

In this appendix we outline the design and operational aspect of a simple settling tube that may be easily constructed from "off-the-shelf" components. The tube has been routinely used to analyze thousands of geological samples a year at the Atlantic Geoscience Centre (AGC) sediment laboratory.

The settling tower is comprised of a glass tube 15 cm in diameter and 200 cm in length. The sample is introduced into the top of the tube by a brass venetian-blind-type introduction system. The collection pan is suspended near the bottom of the tube by a thin length of kevlar® cable, which is connected to

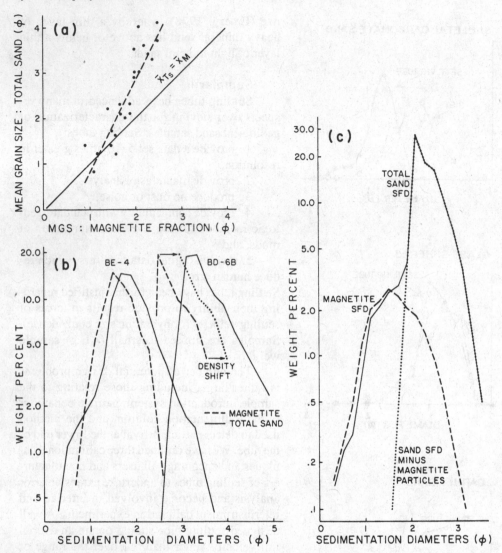

Figure 4.11. Size fractionation of heavy minerals (magnetite) in sand samples as determined with a settling tube (for details see Syvitski & Farrow [1983]). (a) A plot of the mean grain size (MGS) of the total sand fraction (x_{Ts}) versus the MGS of the magnetite fraction (x_M) from deltaic samples collected at the heads of British Columbia fjords. Note that the deviation between x_{Ts} and x_M increases in the finer-grained samples; the magnetite fraction becomes increasingly coarser compared to the total sand fraction. (b) Examples of size frequency distributions (SFDs) of two deltaic sand samples in terms of the total sand fraction and the separated magnetite fraction. Even when the density differential for magnetite is accounted for, compared to the quartz standard, sample BD-6B remains coarser in the magnetite fraction. (c) A deltaic sand sample is found to be bimodal when heavy minerals are not removed. When magnetite is removed, the remaining sands are found to be very well sorted. The original bimodal nature of the sample reflects sampling across laminae.

a standard AE-163 Mettler balance. The balance is interfaced to an Apple II-Plus microcomputer via an RS-232 Super Serial Card. The data logging is controlled by a simple program in BASIC (or TURBO-PASCAL).

Glass tube

The settling tube consists of the following easily purchased components from Corning Industrial Glass Pipe (in Canada – Pegasus, P.O. Box 316, Arincourt, Ontario):

6-in. glass pipe (cat. no. 72-7515; special order to
cut glass pipe to a length of 172 cm)
6-in.–9-in. adapter (cat. no. 86-0525)
9-in. glass cap (cat. no. 86-0502)
6-in. flange plates (cat. no. 72-0739)
9-in. flange plates (cat. no. 86-9006)

Two modifications needing the expertise of a
glass blower are necessary to make operations convenient:

1. shortening the length of the tube so that the
fall distance of a particle is roughly 2 m;

2. installation of a drain spigot in the bottom
cap of the tube.

The glass components are assembled as shown in
Figure 4.12. Two gaskets made from 3.2-mm neo-
prene® act as seals. The collection pan, made from
3.2-mm perspex®, must be obtained separately.
The tube is mounted vertically in a wooden cradle. It
is very important for the tube to be vertical.

Collection pan

The collection pan is simple in design, and
made of lightweight perspex® so as to maximize the
sample weight : pan weight ratio. It is a round plate
machined and glued to a central perspex® post. This
post is drilled with a two-step hole, such that the
pan may be attached to the cable with a concealed
knot so as not to trap sediment.

The pan is suspended by a length of thin kev-
lar® cable (part no. 1948383: Cortland Cable Co.,
Cortland, N.Y.). The cable is fed down through the
centre post of the pan, and the knot is pulled inside
the centre post of the collection pan (Fig. 4.12). The
cable is then fed through a small hole in the introduction system and attached to the pan hook below
the balance.

Introduction system

The introduction system is a venetian-blind
type, modified from the introduction device reported
in Amos et al. (1981). It is made entirely of brass,
chosen both for its ease in machining and its non-
corrosibility in water. The blind consists of seven
rollers, each 9 mm in diameter and 8.6 cm in length.
Each roller is machined so that half the roller is flat.
The fourth roller has a sizable notch at its centre to
allow the cable to pass (see Fig. 4.2). The introduction system is held on the top of the glass tube by
four bolts. The rollers are synchronized by small
connecting rods and one tie rod. The last connecting
rod activates a microswitch that begins the logging
routine when the introduction system is opened. The
switch is connected to the games paddle port of an

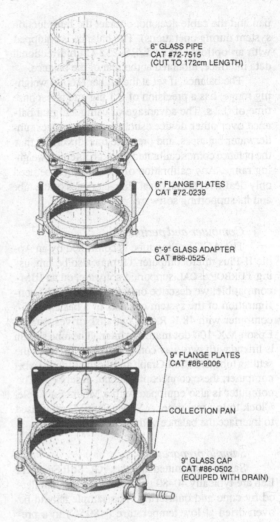

6" GLASS PIPE
CAT #72-7515
(CUT TO 172cm LENGTH)

6" FLANGE PLATES
CAT #72-0239

6"-9" GLASS ADAPTER
CAT #86-0525

9" FLANGE PLATES
CAT #86-9006

COLLECTION PAN

9" GLASS CAP
CAT #86-0502
(EQUIPED WITH DRAIN)

Figure 4.12. Details of a suspended pan settling
tube, BIST-2, located at the Atlantic Geoscience
Centre, Dartmouth, Canada, wherein the collection
pan is greater than the inner diameter of the settling column.

Apple-II computer. The introduction system is submerged ≈1 cm in water at the top of the glass tube.
This allows the sample to be introduced in a wetted
state, eliminating surface tension problems and thus
floating grains. The total surface area of the introduction system is 71.25 cm^2.

Balance

The balance is a standard AE-163 Mettler digital balance, positioned on a shelf above the settling
tube in such a manner that the centre of the pan is
directly above the centre of the introduction system
(i.e., the balance pan is aligned with the collection

pan and the cable does not contact the introduction system during operations). The balance is equipped with an optional 011 CL/RS-232-C unidirectional data interface, mounted at the rear of the balance.

The balance, if set at the higher (160 g) weighing range, has a precision of 0.1 mg and a response time of 0.2 s. The advantage of using a digital balance over other devices such as strain gauges, underwater balances, and pressure transducers, is that the balance comes calibrated through its entire weighing range. Any calibration of the settling tube need only deal with the physical characteristics of the tube and its supporting software.

Computer and peripherals

The AGC settling tube is controlled by an Apple II-Plus microcomputer. (It may also be run, using TURBOPASCAL-supported software, on an IBM-compatible; we describe only the former.) The configuration of the system consists of a basic microcomputer with 48 K RAM, two disk drives, and an Epson MX-100 dot matrix printer. (The tube system is linked with two-way communication between the settling tube and a SediGraph and the mainframe Cyber computer; these connections are not described.) The computer is also equipped with a Mountain Apple Clock to log real time. A Super Serial Card is used to interface the balance with the computer.

Sample preparation

Samples submitted into the AGC Soft Sediment Lab are typically 40–80 g. Subsampling is completed by cone and quartering. The sample should be oven dried at low temperature ($\approx 80\,^\circ$C) in a pre-weighed dish. Once dried, the weight of the sample is measured. If the sample contains < 62.5-μm particles, the sample is dispersed in a 500-ml bottle filled partly with distilled water and shaken for 1 h on an automated shaker. After the sample is dispersed, it is washed through a sieve to separate the mud fraction from the sand fraction. If the fine fraction is also to be analyzed (i.e., using a SediGraph), and the results overlapped with the settling tube data, the sample must be washed through a 53-μm sieve. If the fine fraction is only going to be used for a percent mud value, the sample may be washed through a standard 62.5-μm sieve.

The fine fraction may be retained for future analysis or discarded. The coarser fraction is again oven dried, sieved through a 2-mm sieve, and weighed. The sand fraction is then split down to a 1-g sample using a small microsplitter. The recorded weights are used to calculate percent gravel, sand,

and mud. The small sand fraction is ready to be run on the settling tube.

Running a sample

The balance is turned on and the following parameters set: range at 160, integration at 1, and the stability detector turned off. The balance is allowed to warm for at least an hour. With the computer and monitor turned on, an interactive data logging program is made active. (This and other grain size computer programs are available in Asprey & Syvitski [1990]. Other public programs are available [e.g., Goldberg & Tehori, 1987].) The computer will electronically check the Mountain Apple Clock and determine if enough free space is available on the data disk.

The computer will ask a series of sample identification and startup questions. The introduction system should be fully closed and the sample loaded before these questions are answered. The 1-g sample is placed in a crescent shape around the small hole in the introduction system. Grains found floating due to surface tension may be eliminated by spraying a fine mist of water from a spray bottle. After the sample is loaded and the computer-prompted questions are answered, the introduction system is opened and the computer takes over the operation. When the sample is finished, the program returns and another sample may be run.

REFERENCES

Amos, C. L., Asprey, K. W., & Rodgers, N. A. (1981). Bedford Institute Sedimentation Tube – B.I.S.T., *Bedford Institute of Oceanography, Report Series*, no. Bi-R-81-14, 60 pp.

Andel, T. H. van (1964). Recent marine sediment of the Gulf of California. In: *Marine Geology of the Gulf of California*, eds. T. H. van Andel and G. G. Shor, Jr. *AAPG Memoir*, 3: 216–310.

Anderson, J. B., & Kurtz, D. D. (1979). RUASA: An automated rapid sediment analyzer. *Journal of Sedimentary Petrology*, 49(2): 625–7.

Asprey, K. W., & Syvitski, J. P. M. (1990). Computer programs and code used in the operation of the automated granulometric instruments within the Atlantic Geoscience Centre soft sediment laboratory. *Geological Survey of Canada Open File Report 2292*, 234 pp.

Baba, J., & Komar, P. D. (1981). Measurements and analysis of settling velocities of natural quartz sand grains. *Journal of Sedimentary Petrology*, 51: 631–40.

Bascomb, C. L. (1968). A new apparatus for particle size distribution. *Journal of Sedimentary Petrology*, 38: 878–84.

Bienek, B., Huffmann, H., & Meder, H. (1965). Korngrössenaanalysen mit Hiffe von Sedimentationswaagen. *Erdölund Kohle-Erdgas-Petrochemie*, 18: 509–13.

Brezina, J. (1969). Granulometer – a sediment analyzer directly writing grain size distribution curves. *Journal of Sedimentary Petrology*, 39: 1627–32.

(1972). Stratified sedimentation above the Stokes' range and its use for particle size analysis. In: *Particle Size Analysis 1970*, eds. M. J. Groves & J. L. Wyatt-Sargent. London: Society of Analytical Chemistry, pp. 255–66.

(1979). Particle size and settling rate distributions of sand-sized materials. *Proceedings of the 2nd European Symposium on Particle Characterisation*, Nürnberg, Federal Republic of Germany, 21 pp. + 23 pp. of tables.

Bridge, J. S. (1981). Hydraulic interpretation of grain-size distributions using a physical model for bedload transport. *Journal of Sedimentary Petrology*, 51: 1109–24.

Channon, R. D. (1971). The Bristol fall column for coarse sediment grading. *Journal of Sedimentary Petrology*, 41: 867–70.

Chapman, R. E. (1981). Calibration equations for settling tubes. *Journal of Sedimentary Petrology*, 51: 644–6.

Colby, B. C., & Christiansen, R. P. (1956). Visual accumulation tube for size analysis of sand. *Journal of the Hydrology Division (Proceedings of ASCE)*, 82: 1–17.

Cook, D. O. (1969). Calibration of the University of Southern California automatically recording settling tube. *Journal of Sedimentary Petrology*, 39: 781–6.

Doeglas, D. J. (1946). Interpretation of the results of mechanical analysis. *Journal of Sedimentary Petrology*, 16: 19–40.

Emery, K. O. (1938). Rapid method of mechanical analysis of sands. *Journal of Sedimentary Petrology*, 8: 105–11.

(1978). Grain size in laminae of beach sand. *Journal of Sedimentary Petrology*, 48: 1203–12.

Felix, D. W. (1969). An inexpensive recording settling tube for analysis of sands. *Journal of Sedimentary Petrology*, 39: 777–80.

Flemming, B. W., & Thum, A. B. (1978). The settling tube – a hydraulic method for grain size analysis. *Kieler Meeresforschungen Sonderheft*, 4: 82–95.

Francis, J. R. D. (1973). Experiments on the motion of solitary grains along the bed of a waterstream. *Proceedings of the Royal Society of London*. A332: 443–71.

Geldof, H. J., & Slot, R. E. (1979). Settling tube analysis of sand. *Delft University of Technology, Department of Civil Engineering Internal Report*, no. 4-79, 31 pp.

Gibbs, R. J. (1972). The accuracy of particle size analysis utilizing settling tubes. *Journal of Sedimentary Petrology*, 42: 141–5.

(1974). A settling tube system for sand-size analysis. *Journal of Sedimentary Petrology*, 44: 583–8.

Gibbs, R. J., Matthews, M. D., & Link, D. A. (1971). Relationship between sphere size and settling velocity. *Journal of Sedimentary Petrology*, 41: 7–18.

Gillespie, R. T., & Rendell, C. M. (1985). *Centre for Cold Ocean Resources Sedimentation Tube User Manual*. C-CORE Publ. no. 85-6. Memorial University of Newfoundland, St. John's, Nfld., 41 pp.

Goldberg, R., & Tehori, O. (1987). SEDPAK – A comprehensive operational system and data-processing package in Applesoft Basic for a settling tube, sediment analyzer. *Computers and Geosciences*, 13(6): 565–85.

Graf, W. H., & Acaroglu, E. R. (1966). Settling velocities of natural grains. *International Association of Science and Hydrology, Publ.* 11(4): 27–43.

Guy, H. P. (1969). Laboratory theory and methods for sediment analysis. *Techniques of Water Resources Investigations of the USGS*. Washington, D.C., USGS, bk. 5, chap. C1: 58 pp.

Halka, J. P., Conkwright, R. D., Kerhin, R. T., & Wells, D. V. (1980). The design and calibration of a rapid sediment analyzer and techniques for interfacing to a dedicated computer system. *Maryland Geological Survey Circular* no. 32: 32 pp.

ICWS – Interagency committee on water resources, subcommittee on sedimentation. (1958). Some fundamentals of particle size analysis. Washington D.C.: U.S. GPO, 105 pp.

Kennedy, J. F., & Koh, R. C. Y. (1961). The relation between the frequency distributions of sieve diameters and fall velocities of sediment particles. *Journal of Geophysical Research*, 66: 4233–46.

Komar, P. D. (1981). The applicability of the Gibbs equation for grain-settling velocities to conditions other than quartz grains in water. *Journal of Sedimentary Petrology*, 51: 1125–32.

Komar, P. D., & Cui, B. (1984). The analysis of grain-size measurements by sieving and settling tube techniques. *Journal of Sedimentary Petrology*, 54: 603–14.

Komar, P. D., & Reimers, C. E. (1978). Grain shape effects on settling rates. *Journal of Geology*, 86: 193–209.

Kranenburg, C., & Geldof, H. J. (1974). Concentration effects on settling tube analysis. *Journal of Hydraulic Research*, 12: 337–55.

Kuenen, Ph. H. (1968). Settling convection and grain size analysis. *Journal of Sedimentary Petrology*, 38(3): 817–31.

Mason, M. A. (1949). A manometric settling velocity tube. *Transactions of the American Geophysical Union*, 30: 533–8.

Middleton, G. V. (1967). Experiments on density and turbidity currents, 3. The deposition of sediment. *Canadian Journal of Earth Sciences*, 4: 475–505.

(1976). Hydraulic interpretation of sand size distributions. *Journal of Geology*, 84: 405–26.

Nelson, T. A. (1976). An automated rapid sediment analyzer (ARSA). *Sedimentology*, 23(6): 867–72.

Plankeel, D. M. (1962). An improved sedimentation balance. *Sedimentology*, 1: 158–63.

Poole, D. M. (1957). Size analysis of sand by a sedimentation technique. *Journal of Sedimentary Petrology*, 27: 460–8.

Poole, D. M., Butcher, W. S., & Fisher, R. L. (1951). The use and accuracy of the Emery settling tube for sand analysis. *U.S. Army Corps of Engineers, Beach Erosion Board*, Technical Memorandum no. 23: 18 pp.

Reed, W. E., LeFever, R., & Moir, G. J. (1975). Depositional environment interpretation from settling-velocity (PSI) distributions. *Geological Society of America Bulletin*, 86: 1321–8.

Rigler, J. K., Collins, M. B., & Williams, S. J. (1981). A high precision digital-recording sedimentation tower for sands. *Journal of Sedimentary Petrology*, 51(2): 642–4.

Riley, S. J., & Bryant, T. (1979). The relationship between settling velocity and grain-size values. *Journal of the Geological Society of Australia*, 26: 313–15.

Rouse, H. (1937). Nomogram for the settling velocity of spheres. *Division Geological Geo.*

Exhibit D, NRC, Washington, D.C., 57–64.

Rubey, W. W. (1933). Settling velocities of gravel, sand and silt particles. *American Journal of Science*, 25: 325–38.

Rukavina, N. A., & Duncan, G. A. (1970). F.A.S.T. – fast analysis of sediment textures. *Proceedings of the 13th Conference of the International Association Great Lakes Research*, Ann Arbor, 1970, pp. 274–81.

Sanford, R. B., & Swift, D. J. P. (1971). Comparison of sieving and settling techniques for size analysis, using a Benthos Rapid Sediment Analyser. *Sedimentology*, 17: 257–64.

Schlee, J. (1966). A modified Woods Hole rapid sediment analyser. *Journal of Sedimentary Petrology*, 36: 403–13.

Sengupta, S., & Veenstra, H. J. (1986). On sieving and settling techniques for sand analysis. *Sedimentology*, 11: 83–98.

Singer, J. K., Anderson, J. B., Ledbetter, M. T., McCave, I. N., Jones, K. P. N., & Wright, R. (1988). Assessment of analytical techniques for the size of fine-grained sediments. *Journal of Sedimentary Petrology*, 58: 534–43.

Slot, R. E. (1983a). *An Improved Settling Tube System*. Laboratory of Fluid Mechanics, Delft University of Technology Internal Report no. 7-83: 40 pp.

(1983b). Terminal velocity formula for objects in a viscous fluid. *Journal of Hydraulic Research*, 22, 235–43.

Slot, R. E., & Geldof, H. J. (1979). *Design Aspects and Performance of a Settling Tube System*. Laboratory of Fluid Mechanics, Delft University of Technology Internal Report no. 6-79: 18 pp.

(1986). *An Improved Settling Tube System for Sand*. Communications on Hydraulic and Geotechnical Engineering Report no. 86-4, Delft University of Technology: 45 pp.

Syvitski, J. P. M., & Farrow, G. E. (1983). Structures and processes in bayhead deltas: Knight and Bute Inlet, British Columbia. *Sedimentary Geology*, 36: 217–44.

Taira, A., & Scholle, P. A. (1977). Design and calibration of a photo-extinction settling tube for grain-size analysis. *Journal of Sedimentary Petrology*, 47: 1347–60.

(1979). Origin of bimodal sands in some modern environments. *Journal of Sedimentary Petrology*, 49: 777–86.

Thacker, W. C., & Lavelle, J. W. (1977). Two-phase flow analysis of hindered settling. *Physics of Fluids*, 20: 1577–9.

(1978). Stability of settling of suspended sediments. *Physics of Fluids*, 21: 291–2.

Thiede, J., Chriss, T., Clauson, M., & Swift, S. A. (1976). *Settling Tubes for Size Analysis of Fine and Coarse Fractions of Oceanic Sediment*. Corvallis: Oregon State University School of Oceanography, Report no. 76-8: 87 pp.

Vanoni, V. A. (ed.) (1975). *Sedimentation Engineering*. ASCE Manuals and Reports of Engineering Practice no. 54. New York: ASCE, 745 pp.

Veen, J. van. (1936). Onderzoekingen in de Hoofden in Verband Met de Gesteldheid der Nederlandsche Kust. Algemeene Landsdrukkerij, s'-Gravenhage, 252 pp.

Warg, J. B. (1973). An analysis of methods for calculating constant terminal-settling velocities of spheres in liquids. *Mathematical Geology*, 5: 59–72.

Watson, R. L. (1969). A modified Rubey's law accurately predicts sediment settling velocities. *Water Resources Research*, 5: 1147–50.

Zaneveld, J. R. V., Spinrad, R. W., & Bartz, R. (1982). An optical settling tube for the determination of particle size-distribution. *Marine Geology*, 49: 357–76.

Zeigler, J. M., Whitney, G. G., & Hayes, C. R. (1960). Woods Hole rapid sediment analyzer. *Journal of Sedimentary Petrology*, 30: 490–5.

5 Methodology of sieving small samples and calibration of sieve set

KRISTIAN DALSGAARD,
JENS LEDET JENSEN, AND
MICHAEL SØRENSEN

Introduction

This chapter covers some general comments on sieving, a method of sieve calibration, and some empirical studies on sieving analysis made by the Aarhus sand group. This group comprises geomorphologists, physicists, and statisticians, and the present study is part of a larger research project on the physics of wind-blown sand. The investigations presented here are focused on the effect of sieving in a set, the statistical variation in the sieving process, and calibration of a sieve set. The sieving procedure used by our group is described in the appendix. A general treatment of sieving analysis can be found in Ingram (1971), Allen (1981), and Kaye (1981).

Recent sedimentological studies call for a precise determination of the particle size distribution of small samples. Examples are small sediment samples from cores (McManus, 1965), studies of short-range variations in sand sorting (Barndorff-Nielsen et al., 1982), and sand collected in traps in wind tunnel and field experiments (Jensen et al., 1984; Sørensen, 1985). Furthermore, there is often a great variation in the amount of sand retained on the individual sieves in a set. For a well-sorted sand sample, the sieves representing the flanks of the distribution may well contain less than a thousandth of the weight in the sieves around the modal point. This has implications for the analysis because the sieving efficiency increases with a smaller sediment load on the sieve, whereas the coefficient of variation on the single sieve at repeated sieving grows with decreasing load.

McManus (1965) investigated the effect of sieve loading and showed that the number of finer near-mesh particles (i.e., those just able to pass through the sieve) is important for the sieving efficiency. One near-mesh particle per aperture is the ideal load, whereas with a load of five near-mesh particles per aperture, using 10-min sieving time on an American shaker, 5%–6% of the finer near-mesh particles will be retained on the sieve. McManus suggests that as the maximum load. As it is the maximum and not the minimum load that causes the problem, there seems to be no reason for using small, 4-in. sieves as McManus proposes. The 8-in. sieves have an area four times that of the 4-in. ones, and are therefore more suitable to cover the great variation in the load on the sieves. The increased time to clean the 8-in. sieves is the price to pay for the greater variation in sample size they can handle, and for the greater precision obtained by avoiding splitting of samples, which should be avoided (Emmerling & Tanner, 1974).

Janke (1973) presents a semiempirical sieve load equation to estimate the optimal sample size based on a relation among the number of particles, the number of apertures in a single sieve, particle density, and particle shape. The problem that sieves sort according not only to size but also to form is treated by Sahu (1965), who derives a mathematical relationship for the passage of triaxial ellipsoids through square sieve apertures. Nielsen (1985) considers the passage of superellipsoidal particles through square apertures. Empirical results by Wang & Komar (1985), obtained by sieving heavy mineral sands with varying shape and composition, agree with one of the equations in Sahu (1965), demonstrating that it is the ratio of the smallest particle diameter to its intermediate diameter that controls the passage through the sieve. Reproducibility of the sieving analysis is treated by Rogers (1965), who considers the stability of the empirical mean and variance.

Sieving in sets

Sieving time studies described in the literature have usually been carried out with only one sieve at a time and with large samples (see Whitby, 1959; Kaye, 1981 [wherein further references can be found]). Few have focused on the influence of sieving time when sieving a sample through a set of sieves. Mizutani (1963) makes a theoretical investigation of this problem, based on experimental results for single sieves. Here we report on an experimental investigation of the process of sieving small samples through a set of sieves.

First, we consider the rate at which grains of a particular size pass a sieve and the influence of the presence of larger grains: a supplement to previous investigations using large samples. Next, we investigate the position of a grain in a stack of sieves as a function of the sieving time. This is followed by an experiment where we measure the complete size distribution for three sieving times (10, 20, and 60 min). Finally, for a fixed sieving time, we consider the reproducibility of the sieving process, by sieving the same sample several times. The study of the statistical variation under repeated sieving is also extended to the parameters of the hyperbolic distribution fitted to the size distribution. Hyperbolic distributions are treated in Chapter 17; the discussion here is a summary based on Dalsgaard & Jensen (1985).

The influence of sieving time

The mechanical process of sieving consists of two parts:

1. In the first very short time, grains with a major axis smaller than the aperture size will pass through the sieve.

2. The grains with a size close to the size of the aperture are left on the sieve. These grains must be oriented in a particular way in order to pass the sieve or must find larger holes in the sieve. This means that the sieving time at this stage depends on the shape of the grains and the uniformity of the sieve (Whitby, 1959).

In the first two experiments, a marine sand with a large content of fine material and a sieve with a small aperture are used. One expects that a longer sieving time is needed when much fine material is present, because the fine material must pass the greatest number of sieves before reaching its final position, and it generally takes longer to pass a sieve with a small aperture size (Mizutani, 1963).

A sand sample of 50 g was sieved on a complete column of $\sqrt[4]{2}$ sieves with mesh sizes of 38 μm for the smallest and 1.19 mm for the largest. The sieving time was 1 h, after which

1.5 g were found on the 38-μm sieve (i.e., 1.5 g of the sand were in the size range 38–45 μm). The sand retained on the 38-μm sieve was now placed on the empty 45-μm sieve and sieved once more. It turned out that after 1 h of sieving only 97.2% by weight had passed the sieve; after 100 min this fraction was increased only to 97.4%.

This shows clearly that the larger grains present in the first sieve analysis (i.e., grains in the size range 45–53-μm) have some sort of hammering effect that enables more grains to pass the 45-μm sieve. The causes for the increased passing rate are not understood in every detail; the same effect is, however, obtained when pieces of rubber or brass are placed on the sieves (see Batel, 1971). The finding above also means that there is a group of grains that cannot, in a well-defined way, be said to be either greater or smaller than a certain aperture size. This was also shown by Sengupta & Veenstra (1968) using coloured grains. Plotting the sand that actually passed the 45-μm sieve during the 100 min (Dalsgaard & Jensen, 1985) shows that during the first 2–4 min there is a sharp exponential decrease in the amount retained on the 45-μm sieve (88% by weight has passed after 4 min); but then the rate of decrease becomes smaller and smaller, and after 60 min the above-mentioned 97.2% has passed. The latter behaviour corresponds to the second stage described above. The results show a higher passing rate than used by Mizutani (1963). The reasons for this most likely are the higher load in Mizutani's experiment and a sieving machine that worked with less tapping per minute. It seems preferable to discuss the sieving process in terms of the number of taps instead of the physical sieving time.

We wanted to determine if the results for single sieves have any implications for the sieving in a set, and thus undertook the following experiment. For a stack of sieves we wished to know at what time the sand, which would eventually reach the pan below the sieves, reached the sieve just above the pan. To investigate this we stopped the sieving after 1 min and considered the sand just above the pan separately (i.e., the sand that had to pass the greatest number of sieves before reaching its final position). Siev-

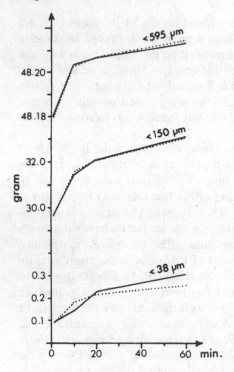

Figure 5.1. The amount of sand on the pan for each of the three columns of sieves is plotted against time. At 10, 20, and 60 min, two different experiments have been made. The dotted line shows the sieving of sand reaching the lowest sieve within the first minute. The full line is the measurement from the complete column of sieves (see the text).

ing this sand on a single sieve, we measured the amount of sand falling down into the pan after 10, 20, and 60 min of total sieving time, including the first minute of sieving. Putting the sand in the pan back onto the sieve just above the pan, the sieving – stopped after 1 min – with the whole stack of sieves was continued, and the weights on the individual sieves after 10, 20, and 60 min were measured.

After having performed the experiment with the first stack of sieves in the size range 1,410–595 μm, the sand found on the pan was placed on top of a new stack of sieves in the size range 500–150 μm and the experiment was repeated. Finally, the experiment was conducted on a third stack of sieves in the size range 125–38 μm.

We first considered the amount of sand falling into the pan. After 1 min we had one

measurement, and at 10, 20, and 60 min we had two, namely the one with only a single sieve on top of the pan, containing the sand that reached the lowest sieve within 1 min, and the one from the size analysis with the complete column of sieves on top of the pan. The data are plotted in Figure 5.1, and it is seen that the two curves are very close. We therefore have the interesting conclusion that after 1 min of sieving almost all the sand that will eventually reach the pan has already reached the sieve just above the pan. There is also a marked difference in the rate of load increase in the pan between 1–20 min and 20–60 min. When plotting the weights found on each sieve at 10, 20, and 60 min (see Dalsgaard & Jensen, 1985), little change occurred after 10 min of sieving, and only with the small group of grains whose size is close to the sieve aperture.

In the last experiment in this section we made three complete size analyses based on 10, 20, and 60 min of sieving. The first of the three stacks of sieves mentioned above was placed in the Ro-Tap machine for 10 min. The content of the pan was then placed on top of the second stack of sieves, which was then sieved for 10 min. Finally, the third stack was sieved for 10 mins. The weight on each sieve was then measured. This procedure was repeated with 20 and 60 min of sieving, respectively.

Three different sand samples were used, one with a narrow distribution (760731, 45 g) and two with a very broad distribution (desert sand, 61 g; Fredbjerg, 44 g), one of which had two modes (desert sand). We have plotted the size distribution in the form of log-histograms in Figure 5.2. For a general discussion of the desirability of using a log scale we refer to Bagnold (1941) and Bagnold & Barndorff-Nielsen (1980). The changes are very small and the distributional form is almost the same (Fig. 5.2). Especially, we note that there is no systematic tendency for the distribution to be shifted toward a distribution with more fine material as the sieving time is increased.

Reproducibility of sieving for different sample sizes

In experiments that involve comparison of several sand samples, it becomes important to

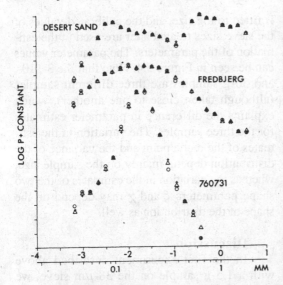

Figure 5.2. For three different sand samples the results of size analyses after (+) 10, (△) 20, and (◦) 60 min of sieving are plotted. The results are plotted as a log-histogram using the natural logarithm. Note that the three log-histograms have been vertically displaced, for which reason there are no numbers on the ordinate axis. The scale of the ordinate is one per division.

know what statistical variation is present when sieving a sand sample. In the experiments reported in this section we assess the errors associated with the sieving process by sieving the same sand sample several times. Three sand samples of approximately 30, 10, and 3 g, respectively, were used, and each sample sieved for 20 min ten times, with remixing in between. For each sample we then have the weights w_{ij}, where $i = 1, \ldots, 10$ is the number among the ten sieve analyses and $j = 1, \ldots, 19$ indicates the size range (1: 841–1,000 $\mu m, \ldots$, 19: 38–45 μm).

The loss associated with the sieving is treated in Dalsgaard & Jensen (1985). For all three samples there is a general decrease corresponding to an average loss per analysis of 0.02%, 0.007%, and 0.06% for the 30-, 10-, and 3-g samples, respectively. This loss is fairly small and can be of consequence only for the very extreme tails, where the amount of sand is comparable with the total loss. However, even though there seems systematically to be a higher loss in the fine-grain tail, this loss is smaller than the total loss. Furthermore, for the comparison of different distributions and for the de-

termination of the distributional form, to be discussed below, the losses were negligible.

Let us turn to the individual weights w_{ij}. For the 30-g sample these are given in Dalsgaard & Jensen (1985). Considering the changes over the ten sieve analyses there is some correlation between two neighbouring sieves, since a reduction on one sieve usually implies an increase on the neighbouring sieves. Generally, a curve fluctuation around a mean value is observed, but in a few cases there is a general increase or decrease. Especially for the very fine grains, there seems to be a decreasing tendency. To assess the importance of these fluctuations for the determination of the size distribution of the sand sample, we first look at the fluctuations on each sieve relative to the average mass on that sieve. Since we are really interested in the changes from one sieve analysis to the next, and not in the cumulated changes associated with ten sieve analyses, we estimate the mean square fluctuation from the differences $w_{ij} - w_{i-1,j}$; that is, we use

$$\sigma_j{}^2 = 1/9 \sum_{i=2}^{10} (w_{ij} - w_{i-1,j})^2 \qquad (5.1)$$

In Figure 5.3 we have plotted the root mean square fluctuations σ_j for the jth sieve divided by the mean weight $w_j = \Sigma\, w_{i,j}/10$ as a function of the mean weight. In the figure we have included the 3-, 10-, and 30-g samples. First, we see that the relative fluctuation is quite small when the mass of sand on the sieve is large. Next, we note that the relative fluctuation seems to be determined mainly by the mass of sand on the sieve and not by the aperture size.

We shall now look at a simple measure of the discrepancy between the distributions obtained by the ten sieve analyses, namely the so-called Kullback–Leibler distance (for information on which see Kotz & Johnson [1982, vol. 4, pp. 421–5]). If any two distributions P and Q are both concentrated on d points with point probabilities $p_1, \ldots, p_d$ and $q_1, \ldots, q_d$, the Kullback–Leibler distance $K_1(P,Q)$ is defined by

$$K_1(P,Q) = \Sigma\, p_i \log(p_i/q_i) \qquad (5.2)$$

This measure is zero if and only if the two distributions are identical, and positive otherwise. In the following we use a symmetric version

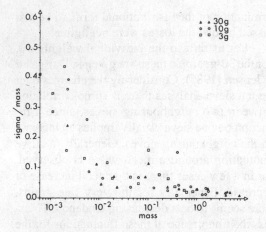

Figure 5.3. The root mean square fluctuation for each sieve divided by the average mass on that sieve is plotted against the average mass. The root mean square fluctuation is based on ten consecutive sieve analyses. Three sand samples, of 3, 10, and 30 g, respectively, have been considered.

Table 5.1. *Kullback–Leibler distances between two consecutive sieve analyses*

	Sample size		
	3 g	10 g	30 g
1–2	0.0011	0.0025	0.00027
2–3	0.0012	0.0004	0.00012
3–4	0.0015	0.0016	0.00020
4–5	0.0013	0.0022	0.00020
5–6	0.0039	0.0015	0.00020
6–7	0.0038	0.0023	0.00022
7–8	0.0047	0.0017	0.00022
8–9	0.0055	0.0012	0.00053
9–10	0.0019	0.0020	0.00062
mean	0.0028	0.0017	0.00029
s.d.	0.0017	0.0006	0.00017

defined by $K(P,Q) = (K_1(P,Q) + K_1(Q,P))/2$. In Table 5.1 we give the Kullback–Leibler distances between all consecutive sieve analyses (i.e., between the first and the second analysis, between the second and the third, and so on). From Table 5.1 there is an almost inverse proportionality between the spread and magnitude of these numbers and the sample size.

To consider changes in the estimates of the parameters when using a parametric model for the size distribution, we shall use the hyperbolic distribution (see Chapter 17). The distribution is fitted to log size, and the calibrated values of the sieve sizes treated later are used in the estimation of the parameters. The parameter values can be seen in Figure 5.4. Note that the 3-, 10-, and 30-g samples are three different samples (although taken close to one another), which explains the difference in parameter estimates for the three samples. The variation in the estimates of the mode point and the variance of the distribution depend mainly on the sample size, whereas the variation in the estimates of the two shape parameters ξ and χ may depend on the shape of the distribution as well.

Discussion

In the first experiment described above with a 1.5-g sample on the 38-μm sieve, we found a sharp exponential decrease in the mass retained on the sieve in the first short part of the sieving. The rate of decrease observed here is much larger than the rates quoted by Mizutani (1963), who worked mostly with >5 g on the sieve. For the comparison of the rates obtained here with those given in Mizutani (1963), we calculate the rates as mass per tapping of the sieving machine, since we believe that the number of tappings, and not the sieving time itself, is the important scale. When sieving a sample of 50 g through a column of sieves, the amount of sand on the individual sieves will, during most of the sieving time, be of the order of 5 g or less. Therefore, the rates used in Mizutani (1963) are inapplicable, the true rates being larger. This implies that the theoretical calculations in Mizutani (1963) on the time needed to sieve a sand through a column of sieves are much too pessimistic. This is in accordance with the findings from the two experiments where sand was sieved through a column of sieves: Only small changes were observed when the sieving time was increased from 10 to 20 or 60 min. Furthermore, the changes occurring at this stage are probably caused by grains finding the greatest apertures. This is also supported by the finding above that during most of the sieving time a grain will either be on its final position or at the sieve just above the final position. If one is interested in the form of the size distribution, then from an operational point of view it seems sufficient to fix a standard number of taps – say,

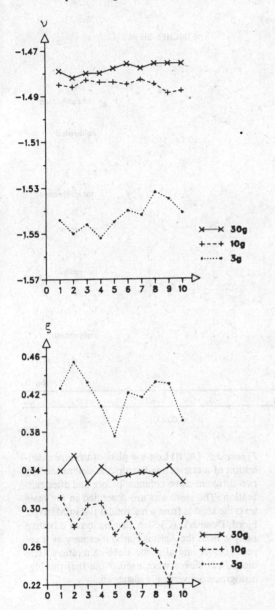

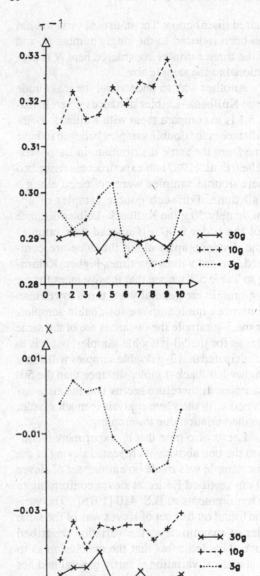

Figure 5.4. Estimates of the four parameters of the hyperbolic distribution, fitted to the logarithm of size, with a 3-, 10-, and 30-g sample, are plotted against the number in ten consecutive size analyses. The parameters are mode point of log size υ, measure of spread τ^{-1}, measure of kurtosis ξ, and measure of skewness χ.

6,000 taps for small sand samples (20 min of sieving with our Ro-Tap machine).

Let us now discuss the statistical variation found above when sieving the same sample several times. A tempting idea would be to describe the variation in terms of a multinominal distribution, that is to consider $(w_{i1}, w_{i2}, \ldots, w_{i19})$ as an observation from a multinominal distribution with unknown number parameter N. The Kullback–Leibler distances, given in Table 5.1, times N would then be observations from a chi-squared distribution with eighteen degrees of freedom. Actually, defining N to be 10^3 times the sample weight in grams (i.e., $N = 3 \cdot 10^3$, 10^4, and $3 \cdot 10^4$ for the 3-, 10-, and 30-g samples, respectively), the numbers in Table 5.1 times N accord reasonably well with a chi-squared distribution. The statistical variation has then been reduced to the single number N, and for the three samples considered here N is proportional to the sample size.

Another way to understand the magnitude of the Kullback–Leibler distances given in Table 5.1 is to compare them with Kullback–Leibler distances for double samples believed to have come from the same distribution. In Barndorff-Nielsen et al. (1982) an experiment is described where double samples were collected along a small dune. For such double samples of approximately 50 g the Kullback–Leibler distance is of the order 0.003. If instead two samples some distance apart along the dune are compared, we get values ten times higher. Returning to Table 5.1 we see that the distances for the 30-g sample are generally ten times lower than the number quoted above for double samples. For the 3-g sample the distances are of the same order as for the 50-g double samples, but it is to be anticipated that 3-g double samples will show a higher Kullback–Leibler distance than the 50-g samples. It therefore seems that the error associated with the sieve analysis is much smaller than the variation due to sampling.

Let us also note that an experiment identical to the one above with repeated sieving of the same sample was made on another set of sieves (20-cm certified Endecott sieves conforming to the requirements of B.S. 410 [1976]). The variation found on this set of sieves was of the same order of magnitude as the variation described above. This indicates that the conclusion as to the statistical variation is fairly general and not entirely related to the actual set of sieves used.

Calibration

The weaving of sieves cannot be perfect. The apertures are not always square, nor are they of exactly uniform size. As a consequence, size distributions determined by sieving samples of particles often show a kinky curve when plotted against the nominal sieve apertures. Examples of this are provided by the log-log plots given in Figure 5.5. Here A and B show the size distribution of a sand sample from a marine deposit, as determined by two different sieve columns, whereas C is a bimodal sample from an estuary.

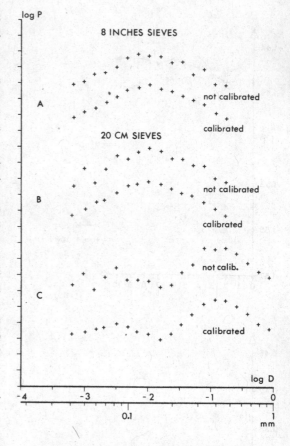

Figure 5.5. (A, B) Log-log plots of the size distribution of a marine sand sample as determined by two different sieve columns before and after calibration. The sieve sets are described in the main text; the sand is from a microtidal flat in Mariager Fjord, Denmark. (C) Size distribution of a bottom sample from the Aarhus Harbour estuary as it appears before and after the sieve calibration. Here the 20-cm sieves were used. Note that the log-histograms are vertically displaced.

A kink in the size distribution curve might arise from a real trait of the particle grading. However, if kinks of the same kind occur at the same size fraction in a series of samples from different localities and with different size distributions (see Figure 5.5B,C), it is very likely that the kinks are caused by one or two inaccurate nominal sieve apertures. Bagnold (1941, p. 124) proposed to adjust the nominal apertures in such a way that these kinks disappear simultaneously in all the size distributions considered. Here we present a computer-based version of

Bagnold's method that uses several samples simultaneously in an objective way.

Using calibrated sieves, Barndorff-Nielsen et al. (1982) were able to detect minor deviations from the hyperbolic distribution, which they explained by the inhomogeneity in specific gravity of the sand grains. When sand samples with a flat or even polymodal size distribution are analyzed, adjustments can be essential for revealing the structure of the distribution. An example of this is given by Figure 5.5C.

The magnitude of the corrections is dependent on the laboratory procedures used. The study reported by Heywood (1946) illustrates this point well. A series of coal dust samples were sieved at seven laboratories, first using the normal procedure of each laboratory, and then using identical procedures at all laboratories. An average size distribution was obtained for each sample by pooling the results from the individual laboratories. This average curve was used to calibrate the sieve sets used. It turned out that the corrections were halved when the same procedure was used at all laboratories.

The calibration method

Our calibration method utilizes the observation that many unimodal size distributions of sand sorted by wind or water under homogeneous conditions have a regular shape that is well described by a log-hyperbolic distribution (see Chapter 17 and the references therein). Sieve calibration could, in a way analogous to ours, be carried out using sand samples with nonhyperbolic types of log-size distributions and another parameterized class of probability distributions relevant to these samples. The main reason for using the log-hyperbolic distributions is that it is easy to find a collection of sand samples with rather different size distributions that are all well fitted by log-hyperbolic distributions. Another reason is that an efficient method has been developed for estimating the parameters in this class of probability distributions (see Chapter 17). A computer program that carries out this estimation procedure is available.

It is perhaps also necessary to state at this point that philosophical considerations concerning the interpretation of size distributions by weight as probability distributions (they do,

incidentally, have such an interpretation) are not relevant here. The probability distributions should simply be thought of as curves describing the size distributions.

A useful and simple measure of dissimilarity between two probability distributions is provided by the Kullback–Leibler information that we used earlier in this chapter. We shall label the sieves in order of increasing aperture by $1, \ldots, n$ and denote the aperture of the jth sieve d_j. Suppose that we have a sand sample for which we have found the fraction r_j on the jth sieve $j = 1, \ldots, n-1$. Only the part of the sand with sieve diameters between d_1 and d_n is considered. Let $h_j(\phi, \gamma, \delta, \mu)$ denote the theoretical probability mass on the jth sieve calculated from the log-hyperbolic distribution with parameters ϕ, γ, δ, and μ. Then the Kullback–Leibler information of the observed size distribution with respect to this log-hyperbolic distribution is defined by

$$K = \sum_{j=1}^{n-1} r_j \log(r_j / h_j(\phi, \gamma, \delta, \mu)) \qquad (5.3)$$

This quantity is nonnegative, and it equals zero if and only if the observed size distribution r_j is equal to the hyperbolic distribution h_j. If a log-hyperbolic distribution is fitted to the observed size distribution by the method of maximum likelihood, this corresponds to choosing the values of ϕ, γ, δ, and μ that minimize K (see Barndorff-Nielsen, 1977).

The Kullback–Leibler information of the observed size distribution with respect to the fitted log-hyperbolic distribution will be denoted by $\hat{K}$. This quantity is a measure of the dissimilarity between the two distributions caused by the systematic effect of inaccurate nominal aperture values as well as by random effects. This suggests that the nominal apertures can be adjusted by choosing aperture values that minimize $\hat{K}$. If the extreme sieve values d_1 and d_n are fixed, this minimization problem has a unique solution $(\hat{d}_2, \ldots, \hat{d}_{n-1})$. Let $\hat{H}$ denote the probability distribution function of the fitted hyperbolic distribution. Then $\hat{d}_j$ is given by

$$\hat{d}_j = \exp\left(\hat{H}^{-1}\left(\sum_{i=1}^{j-1} r_i \right) \right), \quad j-2, \ldots, n-1 \qquad (5.4)$$

For this choice of aperture values the observed distribution equals the fitted log-hyperbolic distribution; that is,

$$K(\hat{d}_2, \ldots, \hat{d}_{n-1}) = 0 \qquad (5.5)$$

In the calibration procedure several different samples ought to be used. The method just described can be applied to each sample. A comparison of the corrections obtained from the individual samples can give a good feel as to whether the corrections are caused by inaccurate nominal sieve apertures. If the corrections for the same sieve are close to each other for all samples, this indicates that calibration is in fact needed. If it is decided that calibration is necessary, the mean of the apertures obtained can be used.

We can also calibrate the sieves using all the samples simultaneously by minimizing the sum over the samples of the Kullback–Leibler informations. From (5.3) it follows that the influence of a sample on the adjustment of a particular sieve increases with the mass of its size distribution in the neighbourhood of that sieve. This is a point in favour of the simultaneous procedure. Another point in favour of this procedure is that only the two most extreme sieve apertures for all samples need to be fixed. Since the samples will rarely have identical sieve ranges, the samples are thus used more efficiently than in the individual procedure. If the estimated hyperbolic parameters are taken as known constants, the procedure based on the sum of Kullback–Leibler informations corresponds to maximum likelihood estimation of the d_j's.

It is not, as one might fear, a difficult problem to minimize the sum of Kullback–Leibler informations numerically. The Newton–Raphson method (Kotz & Johnson, 1982, vol. 6, pp. 205–12) works well because the nominal apertures are good initial values for this procedure.

When new apertures are obtained, the hyperbolic parameters can be reestimated using the new aperture values, and then the adjustment procedure can be carried out again. This will often result in a further change of the apertures. The process can be repeated a number of times, but there is no guarantee that the apertures will stabilize as a consequence; therefore it is advisable to make only a few repetitions.

An application

In this section we describe the adjustment of two sets of sieves by the method described above. Parallel calibration of two sieve sets on the basis of the same collection of sand samples makes it possible to check that the adjustments are not due to peculiarities of the size distributions involved.

The sieve sets to be calibrated here are one set of 8-in. certified Endecott sieves conforming to the requirements of ASTM E.11 (1970) and one set of 20-cm certified Endecott sieves conforming to the requirements of B.S. 410 (1976). The certificate means that the sieves are selected and checked manually. Standard sieves must be expected to need larger adjustments than these specially selected precision sieves. The ratio between the nominal apertures of two consecutive sieves in a column is $\sqrt[4]{2}$. The laboratory procedure used is described in the appendix.

In order to ensure that the size distributions used for calibrating the sieves were well fitted by log-hyperbolic distributions, the extreme size fractions were in many cases not used. These fractions were very small. The cut-off points were determined in the following way: Each observed size distribution was plotted together with the fitted hyperbolic distribution in a log-log diagram (see Figure 17.2). If a sequence of size fractions in a particular tail of the distribution deviated in the same direction from the hyperbola, these points were omitted. The cut was made after the first of the deviating points. In a few cases the cutoff points obtained in this way differed between the two sieve sets; in such cases the least extreme cutoff point was chosen. Thus a very satisfactory hyperbolic fit was obtained. The truncated size distributions are shown in Dalsgaard & Sørensen (1985). The distributions differ substantially in their shapes as well as in the size ranges covered.

First the corrections obtained from the individual samples were calculated. These corrections, in percent of the nominal apertures, are given in Figure 5.6 for the 8-in. and the 20-cm sieves. For each sieve the corrections suggested by the nine samples form clusters that are clearly different from one sieve to the other. A substantial part of the considerable scatter in these

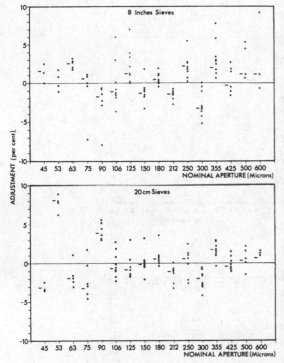

Figure 5.6. The corrections of the 8-in. and 20-cm sieves obtained from the individual samples, in percent of the nominal apertures. The result of the simultaneous calibration procedure (–) is also given.

clusters is caused by points corresponding to the tails of the observed size distributions. The tails are, of course, not so well determined. Because of the variation in the size ranges of the samples used, tail points occur over the entire spectrum of sieve apertures. The clusters are clearly not dispersed around the 0% line, so calibration is needed. Therefore, the simultaneous calibration procedure using all samples was carried out.

The corrections obtained in this way are plotted in Figure 5.6 to the left of the clusters. Note that for sieves with a nominal aperture <125 μm, the corrections are very different for the two sieve sets, whereas they are in the same direction, but still different, for the sieves with nominal apertures <155 μm. This is another indication that the corrections are due to the properties of the sieves rather than those of the sand samples.

The calibration procedure was repeated three times with hyperbolic distributions estimated using the adjusted apertures. Only moderate further changes of the apertures resulted, except for the two largest sieves in both sieve sets. Figure 5.5 gives examples of size distributions before and after calibration.

The two hyperbolic distributions estimated from the size analyses of a particular sand sample on the two sieve sets ought to be more similar after the calibration than before. In order to study this point we consider the hyperbolic parameters υ, τ^{-1}, ξ, and χ. The interpretation of these parameters was given earlier in this chapter (cf. Chapter 17). In Figure 5.7 the hyperbolic parameters from the two sieve sets are plotted against each other before and after the calibration for each sand sample. We find a tendency toward larger similarity after the calibration. In one case the distribution estimated from the 20-cm data was an extreme kind of hyperbolic distribution, the so-called Laplace distribution ($\xi = 1$), before adjustment, whereas the 8-in. data resulted in an ordinary hyperbolic distribution ($0 < \xi < 1$). After the calibration both distributions were ordinary hyperbolic distributions.

A final illustration of the results of our procedure is provided by the size distribution of a sample from a marine deposit analyzed on the 20-cm sieves. Log-log diagrams of this distribution before and after our calibration are given in Figure 5.5. These plots illustrate very well the benefit of sieve calibration before interpreting complex size distributions.

APPENDIX: SIEVING PROCEDURE

The samples are dried at 110 °C over night and cleaned by adding 200 ml of 18% hydrogen peroxide to a 30-sample. After peroxidation the samples are dried at 110 °C and weighed. Next, 200 ml of water is added, and the samples are treated with ultrasound at 300 W for 2 min. Then the samples are transferred to the smallest sieve used, normally 38 μm, and wet-sieved using a spray bottle. The cleaning with hydrogen peroxide and separation on the smallest sieve are done to minimize the sieving loss, and thereby to control the cleaning of the sieves so that sand is not transferred to the next samples. The <38-μm fraction is collected in a bottle via a fun-

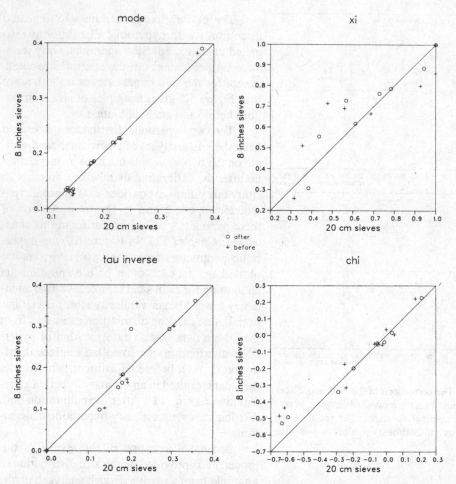

Figure 5.7. The estimates of the hyperbolic parameters from the two sieve sets are plotted against each other before and after calibration.

nel for possible later analysis. The >38-μm fraction is transferred to a glass dish, 110 °C-dried, and weighed. In order to avoid hygroscopic water influencing the sieving loss, the samples are allowed to equilibrate their water content with the air before sieving. The sieving time is 20 min on a Pascall sieving machine type with tapping and rotation motion. The machine works with 300 tappings and 2.6 rotations per minute. The sieves are emptied onto black glossy paper, and each size fraction is transferred to a small glass dish and weighed (to 0.1 mg). The sieves are emptied very carefully, down to 210 μm with a brass brush and from 180 μm to 38 μm with a hair brush. The sieving loss is calculated by taking, for example, the difference between the sum of weights of the sieve fractions and the total weight of the sample, air dried just before the siev-

ing. The sieving loss can be kept <0.1% for 30-g samples and <3% for the smallest samples of 0.05–0.1 g.

REFERENCES

Allen, T. (1981). *Particle Size Measurement*. London: Chapman and Hall.

American Society for Testing Materials (1970). E.11-70 standard specification for wire cloth sieves for testing purposes. In: *1970 Book of ASTM Standards*. Philadelphia: ASTM.

Bagnold, R. A. (1941). *The Physics of Blown Sand and Desert Dunes*. London: Methuen.

Bagnold, R. A., & Barndorff-Nielsen, O. E. (1980). The pattern of natural size distributions. *Sedimentology*, 27: 199–207.

Barndorff-Nielsen, O. E. (1977). Exponentially decreasing distributions for the logarithm of particle size. *Proceedings of the Royal Society of London, Series A*, 353: 401–19.

Barndorff-Nielsen, O. E., Dalsgaard, K., Halgreen,

C., Kuhlman, H., Møller, J. T., & Schou, G. (1982). Variation in particle size distribution over a small dune. *Sedimentology*, 29: 53–65.

Batel, W. (1971). *Einführung in die Korngrüssenmesstechnik*, Berlin: Springer.

British Standards (1976). 410. Specification for test sieves.

Dalsgaard, K., & Jensen, J. L. (1985). A methodological study of sieving small samples. In: *Proceedings of International Workshop on the Physics of Blown Sand*, 609–32. Memoirs no. 8, Department of Theoretical Statistics, Aarhus University.

Dalsgaard, K., & Sørensen, M. (1985). A method of calibrating sieves. In: *Proceedings of International Workshop on the Physics of Blown Sand*, 587–608. Memoirs no. 8, Department of Theoretical Statistics, Aarhus University.

Deigaard, R., & Fredsøe, J. (1978). Longitudinal grain sorting by current in alluvial streams. *Nordic Hydrology*, 9: 7–16.

Emmerling, M., & Tanner, W. F. (1974). Splitting error in replicating size analysis. *Abstracts Program, Geological Society of America*, 6: 352.

Heywood, H. (1946). A study of sizing analysis by sieving. *Transactions of the Institute of Mineralogy and Petrology*, 55: 373–90.

Ingram, R. L. (1971). Sieve analysis. In: *Procedures in Sedimentary Petrology*, ed. R. E. Carver. New York: Wiley.

Janke, N. C. (1973). Sieve load equations and estimated sample size. *Journal of Sedimentary Petrology*, 43: 518–20.

Jensen, J. L., Rasmussen, K. R., Sørensen, M., & Willetts, B. B. (1984). The Hanstholm experiment 1982. Sand grain saltation on a beach. Research Report no. 125, Department of Theoretical Statistics, Aarhus University.

Kaye, B. H. (1981). *Direct Characterization of Fine Particles*. New York: Wiley.

Kotz, S., & Johnson, N. L. (eds.) (1982). *Encyclopedia of Statistical Science*. New York: Wiley.

McManus, D. A. (1965). A study of maximum load for small-diameter sieves. *Journal of Sedimentary Petrology*, 35: 792–6.

Mizutani, S. (1963). A theoretical and experimental consideration on the accuracy of sieving analysis. *The Journal of Earth Sciences, Nagoya University*, 11: 1–27.

Nielsen, H. L. (1985). Shapes of sand grains estimated from grain mass and sieve size. In: *Proceedings of International Workshop on the Physics of Blown Sand*, 677–88. Memoirs No. 8, Department of Theoretical Statistics, Aarhus University.

Rogers, J. J. W. (1965). Reproducibility and significance of measurements of sedimentary size distributions. *Journal of Sedimentary Petrology*, 35: 722–32.

Sahu, B. K. (1965). Theory of sieving. *Journal of Sedimentary Petrology*, 35: 750–8.

Sengupta, S., & Veenstra, H. Y. (1968). On sieving and settling techniques for sand analysis. *Sedimentology*, 11: 83–98.

Sørensen, M. (1985). Estimation of some aeolian saltation transport parameters from transport rate profiles. In: *Proceedings of International Workshop on the Physics of Blown Sand*, 141–90. Memoirs no. 8, Department of Theoretical Statistics, Aarhus University.

Wang, C., & Komar, P. D. (1985). The sieving of the heavy mineral sands. *Journal of Sedimentary Petrology*, 55: 479–82.

Whitby, K. T. (1959). The mechanics of fine sieving. Symposium on particle size measurements. *ASTM Special Technical Publication*, 234: 3–25.

6 Image analysis method of grain size measurement

STEPHEN K. KENNEDY AND
JIM MAZZULLO

Introduction

The extraction of geologic information from the distribution of particle sizes in sediment samples has been attempted for at least one hundred years. Although some workers question the amount of information contained in size and our ability to extract such information (Ehrlich, 1983), the attempt continues. Part of the problem in the analysis of particle size is that it cannot be determined independent of particle shape. As a result, many techniques for particle size determinations have been developed and used, each measuring a different aspect of particle size or of particle behavior that is related to size. The rapid increase in computer and related technologies has spawned new size analyzers. Absorption of x rays, light transmissivity, electrical conductivity, and laser diffraction are examples of new techniques.

Coincidentally, it has become apparent that size distributions are not usually normal distributed, and many workers search for information other than mean and sorting (Klovan, 1966; Visher, 1969; Taira & Scholle, 1979; Kennedy et al., 1981). Such grain size information might be contained in subtle differences in size distributions that would be missed in half- or even quarter-phi intervals. In addition to measuring different aspects of size than do the traditional sieve and pipette analyses, automated size analyzers allow finer intervals or even individual grains to be analyzed, permitting more detailed inspection of the size distributions.

An additional class of sizing techniques is based on the analysis of data obtained in an image, and is referred to herein as image analysis size (IAS). Various instruments use different images and define size in different ways. In fact, a single IAS system may allow for more than one aspect of size to be analyzed: length of one or two axes, length of particle perimeter, or particle projection area.

A procedure using a calibrated ocular to determine the length of a grain long axis (or apparent length in thin section), in fact, contains all the elements of an IAS system: The image is visual, the computer is the brain, and the output is in the form of handwritten numbers. This is essentially the same procedure as the Videoplan, where the image is visual, but the calibrated ocular is replaced by a digitizing tablet and the computation and output are performed by a computer (Schäfer, 1982). At the other extreme is a system in which the analog output of a video camera image, of an entire scene containing many grains, is digitized into an array of points, and grains are automatically identified and sized. An example is the Bausch and Lomb Omnicon.

Each of the above approaches to IAS has disadvantages. The former system requires manual identification of size, which is very slow and prone to error. The latter is faster but has no quality control, and will count all objects of contrasting light intensity as "grains" whether or not they are individual or multiple – and even if they are not mineral grains but a function of poor sample preparation. In addition, precautions must be taken to ensure that all grains, regardless of size, have the same probability of being analyzed.

We describe ARTHUR (a specific system developed by Symbiotic Concepts of Columbia, South Carolina, for the purpose of particle shape analysis) to illustrate the features of an IAS system that is conceptually between these extreme approaches. It is more accurate and faster than visual axis determination, but more selective and therefore slower than completely automatic systems. In ARTHUR the operator points to an acceptable particle, and its size is automatically determined. The hardware for ARTHUR is of the "off-the-shelf" variety, and many technicians in academia and industry can easily construct a similar system. Other commercial systems that are able to return grain outlines can be used as well.

In this chapter we discuss:

1. the number of edge points needed to represent the particle, since calculation of size requires finding the whole particle periphery;

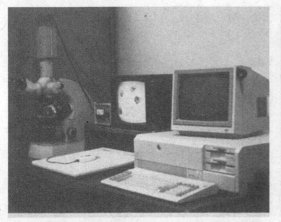

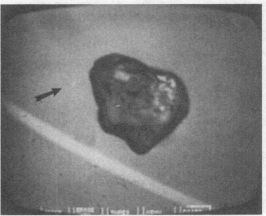

Figure 6.1. The ARTHUR digitizing system, consisting of an RCA TV camera mounted on a petrographic microscope (left), TV monitor (center top), interactive digitizing tablet (center bottom), and microcomputer (right).

Figure 6.2. Illuminated cursor cross (indicated by the arrow), superimposed on the image of a quartz sand grain on the TV monitor. The grain is mounted in glycerine on a glass slide.

2. the number of grains necessary to represent the grain size distribution, because this is a grain-by-grain technique rather than bulk processing; and

3. the relation of our size measure to the long axis.

Edge-point acquisition

The basic components of the ARTHUR image analytic system are a high-resolution, light-sensitive, black-and-white television camera, a visual monitor that receives the camera signal (a "scene"), and a microcomputer with a "frame-grabbing," video-digitizing board that converts the camera analog signal into a digitized array (Fig. 6.1).

The scene scanned by the camera and displayed on the TV monitor is some type of grain mount. In the most commonly used configuration, the camera mounted on a microscope views grains mounted on a glass slide, but the system is versatile enough that any size object can be analyzed: The microscope can be replaced with a camera lens to view large particles (gravel) resting on a background board of contrasting color. The video input can be replaced by the analog signal from an electron microscope to analyze clay-size particles.

The analog signal is relayed to the video digitizer and then to the monitor as a high-

density matrix of "pixels" (picture elements), each of which is represented by x and y coordinates as well as by a light-intensity value in the range 0–15. The density of this matrix depends on the hardware configuration of the system; the ARTHUR system has a density of 512×480 pixels.

The system also includes a digitizing tablet with cursor pen that serves as an interface between the operator and the computer. The surface of this tablet is integrated with the monitor so that each point on the former has a corresponding point on the latter. The tip of the cursor pen appears as an illuminated cross over the digitized scene on the TV monitor (Fig. 6.2), and its movement across the surface of the tablet produces a corresponding movement of the illuminated cross across the monitor. The cursor pen is used to specify grains on the monitor scene for automatic edge finding, or to draw grain edges manually.

An automatic digitizing mode is utilized whenever two conditions are met:

1. The grains are discrete, and do not touch one another.

2. There is a sufficient contrast in light intensity between a grain and its background.

The edge is defined by some threshold value of light intensity separating grain from background. This threshold value is entered by the operator and can be changed at any time to ac-

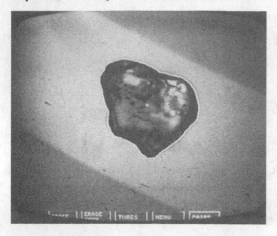

Figure 6.3. Digitized edge of the quartz sand grain from Figure 6.2. This edge consists of a series of discrete, illuminated pixels, and the grain centroid is indicated by an illuminated dot.

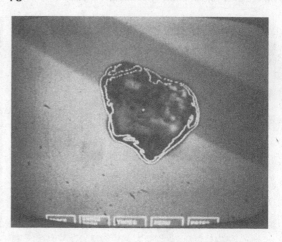

Figure 6.4. Incorrectly digitized grain. This error is caused by a too-low threshold level, which results in a poor light-level contrast between the actual edge of the grain and the background.

commodate changes in light intensity or grain-edge thickness. Sand grains on a glass slide viewed in transmitted light are dark on a light background. Grains, mounted microfossils, and light-colored gravel viewed in reflected light are light on a dark background. The appropriate option (light-on-dark or dark-on-light) is specified to the computer.

In the auto mode, digitization is started by placing the illuminated cursor cross on, or to the left of, a grain and pressing a button on the cursor pen. The edge of the grain is detected using an edge-finding algorithm, and the edge-point pixels are illuminated and projected over the monitor scene (Fig. 6.3) so that the operator can determine if it is correct and acceptable. For some grains, the edge trace will wander into the grain interior or outward into the background when the contrast between the grain and the background is low. In such instances, which are relatively rare, the operator rejects the trace and selects a different threshold value or enhances the grain-to-background contrast with a digital filter available an a menu-driven option on the digitizing tablet. The grain is then redigitized (Figs. 6.4–6.6).

The auto mode is by far the preferred method of edge-point acquisition, but there is also a manual mode that can be used when either of the two conditions of touching grains or insufficient contrast occur. Grains may touch in cases

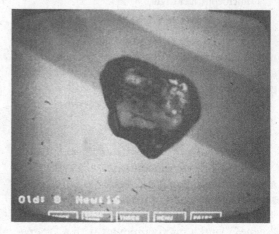

Figure 6.5. Modification of the threshold value for the image shown in Figure 6.4. The old (8) and the new (16) thresholds are shown in the lower left corner.

where the loose grain mount is dense, and is the usual case in the analysis of grains in thin section. In the manual mode, the operator selects a drawing option from the menu and moves the illuminated cursor point around the edge of a grain displayed on the monitor by moving the cursor pen across the digitizing tablet. This illuminates the edge, which can then be digitized in the auto mode.

Using either option, the result is a string of *x–y* coordinates representing the grain boundary. A number of aspects of size can be selected to represent this particle, including the length

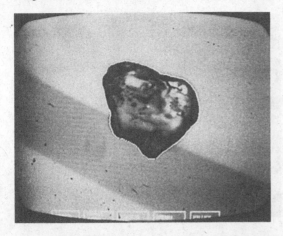

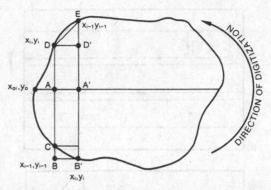

Figure 6.7. Method of area calculation of a digitized grain edge.

Figure 6.6. Correctly digitized edge for the grain shown in Figure 6.4.

of the intermediate or long axis, the perimeter length, or a projection area-related measure. We chose the nominal sectional diameter (NSD) to represent size. The NSD is defined as "the diameter of a circle with the same area as the maximum projection profile of a single particle," and approximates the true nominal diameters of particles that are roughly spherical in form. The concept of the NSD was first introduced by Wadell (1934), and its measurement was described in Mazzullo & Kennedy (1985). Because particles fall with their maximum projection areas perpendicular to the direction of fall, a size measure representing this maximum projection area is most likely to relate to behavioral properties. This argument was used by Sneed & Folk (1958) in their definition of maximum projection sphericity.

Once a single grain's edge has been digitized and identified as x–y coordinates, its area in pixels is approximated using the formula:

$$A = \sum_{i=1}^{N} [(y_0 - y_i)\, \Delta x_i + (\Delta x_i\, \Delta y_i)/2] \qquad (6.1)$$

where N is the number of edge points (Fig. 6.7). A scaling factor, previously determined for this particular configuration (magnification, distance of camera lens to object, etc.), allows the conversion of pixel distances to real distances. At the start of an analysis, the operator inputs the magnification level so that the correct factor is used. At any time during an analysis it is possible to change the magnification level and in-

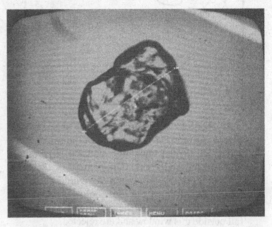

Figure 6.8. Digitized visually determined long axis of a quartz sand grain.

form the computer to that effect. The grain diameter is determined as the diameter of a circle with that maximum projection area.

Other measures can be used to define particle size. ARTHUR has an option that can be used to determine size defined by the length of an axis. The operator enters the visually determined end points of a line such as the long axis (Fig. 6.8). The distance between the two points is the length of that axis.

Sampling procedure

An important aspect of an IAS procedure is the method of selecting grains to be digitized. With respect to the analysis of sand and silt grains, a standard sampling device (a 6.5-oz., 246-cc tuna fish can) contains on the order of eight million medium sand grains, which are to be represented by the size of a relatively small

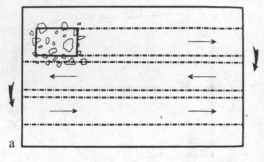

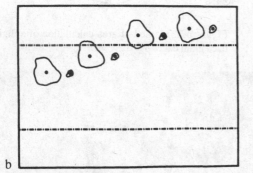

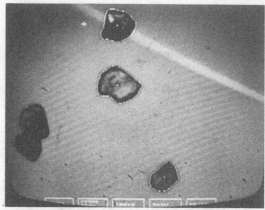

Figure 6.10. Image on the TV monitor showing the digitized edges of three grains. By keeping the digitized edges of all the grains illuminated in this fashion, the operator can ensure that all grains in the image are digitized.

Figure 6.9. Selection of grains for analysis. (a) Search pattern: The slide is scanned in several traverses with equal separation distance. (b) Definition of "in" and "out": The centers of the four grains on the left fall within the window defined by the dotted lines, and thus these grains are accepted. The centers of the remaining four grains fall outside this window and are rejected.

number of grains. The sample volume should be reduced (using a microsplitter or cone-and-quarter methods) so that grains will not touch when strewn on a petrographic glass slide.

To mount grains on the slide, the split should be strewn on the slide slowly using a back-and-forth motion instead of a single sweep. Grains can be mounted in air, glycerine, or a permanent mounting medium. It has been shown that such grains will rest with the short axis perpendicular to the slide (Tilmann, 1973). Because it is impossible to be assured that each and every grain is digitized, and digitized only once, a sample of grains from all portions of the slide can be taken along traverses across the slide at equal separation. The separation between traverses can be determined by the number of grains encountered per transect and the total number of grains desired (Fig. 6.9a).

According to the above procedure, only a portion of the grains mounted on the slide are analyzed: those that fall within a scene as the slide is scanned along traverses. At the individual scene level, a decision must be made as to whether or not a grain is "in" the scene. This decision must not be influenced by grain size: Every grain should have the same probability of being accepted as any other, regardless of size.

The procedure is illustrated in Figure 6.9b, where the dashed lines represent an area within the scene displayed on the TV monitor. Four small and four large grains, and their centroids, are shown. If an entire grain must be in the area in order to be accepted, two small grains but only one large grain will be accepted, underestimating the large grains. If the grain is accepted when any portion of it is in the area, then two small grains and three large grains will be accepted, overestimating the large grains. However, if the definition of acceptability refers to the position of the centroid, the correct proportion of large and small grains will be analyzed.

To facilitate the selection of grains on the basis of centroid, the ARTHUR system displays centroids for each grain outlined. In addition, the edge trace of each grain that has been digitized within a monitor scene remains illuminated so that the operator does not repeat or omit any grains (Fig. 6.10).

Sampling considerations

There are two questions that must be addressed in order to use this method of size analysis confidently:

1. How many edge points represent a grain size?

2. How many grains represent the size distribution?

Number of edge points

An edge trace of a grain can consist of anywhere between one and a very large number of edge points (up to 1,500 using ARTHUR). Magnification and camera distance control the apparent size of particles on the monitor, and these can be adjusted for any sample analysis. Two competing considerations affect the decision with regard to this particle image size: the number of edge points characterizing a grain, and the number of scene changes necessary to obtain the desired number of grains in the size analysis. It should be kept in mind that individual grains can be digitized virtually instantaneously, whereas it takes a relatively long time (several seconds) to change a scene.

A greater number of edge points yields the more accurate estimate of particle size and can be achieved by digitizing at a relatively high magnification. However, a high-magnification scene contains a relatively small number of grains, and the slide must frequently be moved, thereby increasing the time of analysis. Moreover, in a high-magnification image of a poorly sorted sample, there is the increased likelihood that larger grains with centroids within the acceptance area will fall partly outside the scene and thus cannot be digitized without moving the scene. At lower magnification, all grains, large and small, will be entirely within the scene and the digitization procedure will be facilitated. Furthermore, when the magnification is small, there is a greater number of grains in each scene and fewer scene changes are necessary. Consequently, the rate of data collection will increase when the apparent particle size is smaller. However, there is a concomitant loss of accuracy at a lower magnification because there are fewer edge points for each grain.

To test the effect of magnification and number of edge points on the resulting size estimate,

Table 6.1. *Number of edge points and corresponding nominal sectional diameter in microns*

Number of edge-points	NSD
8	10.9
22	10.0
28	10.0
34	10.1
59	10.0
73	10.1
91	10.0
92	10.1
116	10.1
144	10.0
146	10.0

a 10-μm sphere acquired from the National Bureau of Standards was digitized at various magnifications, and the NSD and number of edge points on its trace were recorded at each level (Table 6.1). When the number of edge points is twenty-two or more, the NSD estimate is within 1% of the true diameter. With only eight edge points, the estimated size is larger by almost 10% of the true diameter. To be conservative, our system has been programmed to reject any grain with fewer than twenty-five edge points and to notify the operator to that effect with a loud "beep." In such cases, the operator must use a higher magnification, inform the computer of this change, and redigitize the grain. This requires operator time, but a fully automatic system would either accept incorrect sizes or truncate the size analysis at both fine and coarse extremes.

Number of grains per sample

When several grams or more of sediment are analyzed to characterize the size distribution, it is assumed that a number of grains sufficient to characterize that distribution are analyzed. When the grain sizes are determined on a grain-by-grain basis and the number of grains analyzed is directly proportional to the amount of time required to obtain the size distribution, this question requires more direct attention. For this purpose, we constructed six artificial samples by concatenating varying proportions of sand and silt (see Table 6.2), and then digitized 1,000 grains from each sample. The data were divided

Table 6.2. *Mixing percentages of different size class sediment in six samples*

SIZE CLASS	SAMPLE					
	A	B	C	D	E	F
Coarse sand	20%	40%	10%	35%	10%	10%
Medium sand	20	30	10	10	35	20
Fine sand	20	10	10	10	10	30
Very fine sand	20	10	30	10	35	30
Coarse silt	20	10	40	35	10	10

Table 6.3. *Definition of subsets in eleven runs in mean comparison experiment*

RUN NUMBER	NUMBER OF GRAINS/SUBSET	NUMBER OF SUBSETS IN RUN
1	1	1000
2	2	500
3	5	200
4	10	100
5	25	40
6	50	20
7	100	10
8	200	5
9	300	3
10	400	2
11	500	2

into subsets to determine the difference of grain size among the subsets of each sample.

The question of the number of grains necessary to represent the size distribution of a sample requires one to decide how samples are to be compared, as well as "How close is close?" We first consider the mean as representing the sample, then briefly consider the distribution as represented by quarter-phi intervals.

The question of the representation of the sample by the mean can be addressed by breaking the 1,000-grain sample into subsets of various numbers of grains and comparing the differences of the means of these subsets. The difference of the means of subsets with a large number of grains should be relatively small, but the difference of the means of subsets with a small number of grains might be large. An experiment was set up (as shown in Table 6.3) to determine the mean size in sample subsets:

The first run consisted of 1,000 subsets, each containing a single grain.

The second consisted of 500 subsets, each containing 2 grains.

The third consisted of 200 subsets, each containing 5 grains.

– and so on, until –

The eleventh run consisted of 2 subsets of 500 grains each.

The differences among the subset means is indicated by their standard deviation. The average value of the subset means remains constant because they were all determined by using the same thousand grains (with the exception of runs 9 and 10, which consisted of a total of 900 and 800 grains, respectively). For these tests, it is assumed that the size distributions of the samples are well represented by the diameters of the 1,000 grains.

The results are plotted in Figure 6.11A–F for the six samples. Predictably, these show that the larger subsets better represent the sample, and presumably the population as well. However, if the investigator is satisfied with a mean within 10 μm, these data show that for all samples except no. 5, only about 50 grains are necessary to estimate the sample mean.

Since information may reside in the shape of the frequency distribution, the investigator

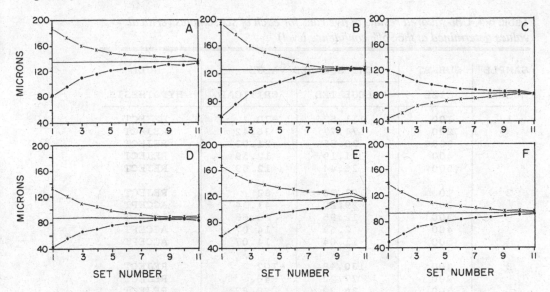

Figure 6.11. Plot showing the effect of subset size and number of subsets (as defined in Table 6.3) on the mean grain size. The (+) is the overall sample mean. The standard deviation of the subset means was calculated and is also plotted.

may not be satisfied with simply representing the mean. The following experiment suggests a method to address this question, but more analyses are needed.

To see how well the entire distribution is approximated by small samples, the data were again divided into subsets. This time,
run 1 consisted of ten sets of 100 grains each,
run 2 consisted of five sets of 200 grains each,
run 3 consisted of three sets of 300 grains each,
run 4 consisted of two sets of 400 grains each, and
run 5 consisted of two sets of 500 grains each.
Number frequency distributions in quarter-phi intervals were constructed for each of the subsets, and these distributions were compared using a chi-square test.

Not surprisingly, a greater number of grains is necessary to represent the entire distribution than to represent the mean, but the results of this experiment do not define that number (Table 6.4). Some samples appear to be well represented using 200–300-grain samples, and others are not well represented using 500-grain samples. It appears that samples on the order of 1,000 grains or more may be required

to represent a complex distribution, although simpler distributions (e.g., of well-sorted sand) may require less. However, since this method is semiautomated, the data collection time is still not prohibitive.

Comparison of techniques

We mentioned above that the various sizing techniques analyze different aspects of size. In image analysis, we have the luxury of being able to analyze more than one aspect simultaneously: long axis, intermediate axis, perimeter length, or the maximum projection area. We use NSD, a function of the maximum projection area, for our own research, but it is informative to compare the NSD to size defined in other ways.

For example, there are other image-analytic techniques (e.g., Videoplan) that measure the long axes of grains. In order to compare NSD data with such long-axis (LA) data, we simultaneously measured the NSD and long axes of approximately 200 quartz grains in six samples of well-sorted sands on the ARTHUR system. Two samples of the St. Peter Sandstone of Minnesota (BC-8 and -10), two of crushed quartz sand (QRF-1 and -2), and two of sandbar deposits of a first-order stream draining granitic rocks in the Llano Uplift of Texas (HB-18 and -20) were used for this study.

In the analysis, we also measured the sphericity of the quartz grains in the six samples

Table 6.4. *Chi-squared results of five runs for each of six samples (critical values determined at the 95% confidence level)*

SAMPLE NUMBER	SUBSET SIZE	CHI-SQUARED VALUE		NULL HYPOTHESIS
		CALCULATED	CRITICAL	
1	100	111.57	~70	REJECT
	200	74.77	36.42	REJECT
	300	30.55	21.03	REJECT
	400	41.19	12.59	REJECT
	500	26.44	12.59	REJECT
2	100	103.96	~82	REJECT
	200	29.20	41.34	ACCEPT
	300	12.95	23.68	ACCEPT
	400	2.64	14.07	ACCEPT
	500	11.04	14.07	ACCEPT
3	100	130.48	~102	REJECT
	200	77.22	~46	REJECT
	300	36.32	28.87	REJECT
	400	28.61	16.92	REJECT
	500	21.56	16.92	REJECT
4	100	116.77	~92	REJECT
	200	49.39	~46	REJECT
	300	23.12	26.30	ACCEPT
	400	12.37	15.51	ACCEPT
	500	18.91	15.51	REJECT
5	100	158.28	~92	REJECT
	200	96.75	~46	REJECT
	300	22.61	26.30	ACCEPT
	400	15.48	15.51	ACCEPT
	500	15.37	15.51	ACCEPT
6	100	129.69	~82	REJECT
	200	77.73	41.34	REJECT
	300	37.27	23.68	REJECT
	400	32.78	14.07	REJECT
	500	20.15	14.07	REJECT

Table 6.5. *Mean size in phi units determined by NSD and LA methods*

	BC-8	BC-10	HB-185	HB-205	QRF-1	QRF-2
NSD	2.28	2.12	2.84	2.80	2.60	2.59
LA	2.04	1.89	2.50	2.48	2.03	2.00
DIFF	0.24	0.23	0.34	0.32	0.57	0.59
Rs	0.85	0.85	0.82	0.82	0.77	0.77

Note: DIFF = difference in the means (NSD − LA); Rs = Rittenhouse sphericity R_s converted from Fourier shape analysis results.

in order to show the effect of variations in shape upon the resultant size data. The sphericity was measured by the amplitudes of the second harmonic of the Fourier series in closed form (Ehr-lich & Weinberg, 1970), which were converted to the equivalent Rittenhouse sphericity value (R_s) by the method described by Haines & Mazzullo (1988). Predictably, the St. Peter sand

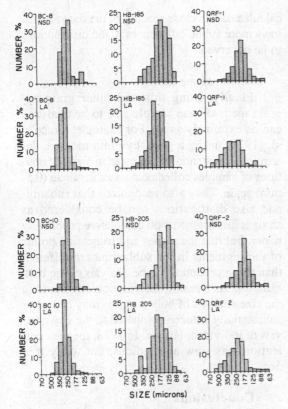

Figure 6.12. Size number frequency distributions of six samples of different shape. Size is determined by NSD and LA image analysis.

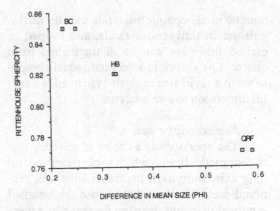

Figure 6.13. The difference between NSD and LA size measures for the BC, HB, and QRF sample pairs plotted against sphericity.

sizes of the relatively nonspherical grains (QRF) differed by almost 0.6ϕ.

Discussion

We have presented information about a semiautomatic IAS technique in which particle size, defined by the nominal sectional diameter, is determined on a grain-by-grain basis. The advantages of this system design over other IAS designs and over other size-analytic systems are discussed below.

Blend of quality and speed

Although the semiautomatic system is slower than the fully automatic systems, it is more accurate than either the fully automatic or manual digitizing systems, and is sufficiently fast to allow analysis of thousand-grain samples. The utilization of operator input has several advantages over the fully automatic systems: Grains can be selected in an unbiased manner, multiple grains are not analyzed, and mineral matter can be distinguished from any "dirt" found on the slide.

Selective sizing

Often, size analysis is performed on samples of varied composition, and it would be useful to determine the size distributions of a specific detrital or biogenic species. By viewing the grains in transmitted or reflected light, targeted species may be visually distinguished with ease. For example, oolites can be identified in a car-

grains had the highest sphericity values, and the crushed quartz sand samples had the lowest values (see Table 6.5).

The resultant size distributions are illustrated in Figure 6.12 and the means shown in Table 6.5. Not surprisingly, the mean NSD for every sample is smaller than the mean LA measure, for the former takes into account the intermediate axis as well as the long axis of the grains, whereas the latter does not. Table 6.5 and Figure 6.13 show that this effect is more apparent when samples are less spherical. In Figure 6.13, R_s is plotted against the difference between the means of the NSD and LA measures. Because the means were determined in phi values, the consistently positive difference indicates that the NSD measure yields a smaller size. However, the difference in size estimated by the two techniques is related to the shape. The mean size of the relatively spherical grains (BC) differed by $\approx 0.25\phi$, whereas the mean bonate sand, opaque minerals can be distinguished in terrigenous sands, and stained or etched feldspars can be distinguished from quartz. The ability to select individual components in a rapid size analysis system adds a useful dimension to size analysis.

Size descriptor option

The operator has a choice of selecting one of four methods with which to characterize size: long axis, short axis, perimeter length, or sectional area. The fact that we use the nominal sectional diameter, because we feel that a measure reflecting the maximum projection area is most likely to relate to particle behavior, does not prevent the investigator from using any or all of the other three measures.

Simultaneous size and shape

Every size analysis technique measures some combination of size and shape. Sedimentologists are periodically reminded that sieve analysis reflects particle shape (Rittenhouse, 1943; Ludwick & Henderson, 1968; Komar & Cui, 1984; Kennedy et al., 1985). Shape is also a complicating factor in settling velocity. For example, Baba & Komar (1981) show that settling velocity is in part a function of grain sphericity. Komar & Cui (1984) state "Baba and Komar (1981) do permit a more refined analysis which includes the actual grain sphericity, but this requires a determination of the Corey Shape Factors of the grains, a laborious process." Using the technique presented herein, size and shape can be determined simultaneously on each and every particle in the analysis, and the interaction of these two parameters can be evaluated.

Narrow size intervals

The fact that individual grains are analyzed by this method allows us to make a continuous frequency distribution, the width of the intervals being the number of decimal places to which the computer can calculate size. As a result, more detailed data summaries can be obtained. For example, it has been shown that grouped moment measures are influenced by the choice of class interval width (Kennedy et al., 1981). With grain-by-grain data, true moments can be calculated. The close spacing of the data also allows more subtle differences in the distributions to be observed.

Small volume samples

Finally, using this and other grain-by-grain methods, the sample size to be analyzed can be extremely small. For example, Grace et al. (1978), using a grain-by-grain method, determined substantial differences in size distributions of samples collected on a sublaminae (0.5-mm) scale. They also recognized that sublaminae size distributions can be considered as straight line segments on cumulative probability paper, but that the slopes and truncation points of the segments in the sublamina are different than those produced by the analysis of the bulk samples. As a result, information contained in the size analysis of bulk samples may be lost or substantially reduced compared to the size analysis of individual lamina. In 1978, the data collection was slow and tedious, but today it is not.

Conclusions

The methods of particle size analysis are numerous, and the choice of the "proper" technique depends on the purpose of the study. The advantages of the semiautomatic image analysis definition of nominal sectional diameter (NSD) have been presented in this chapter. We believe that this method of size analysis is particularly suited to the analysis of small samples, selected analysis of specific components within the sample, and simultaneous measurement of other aspects of size as well as form and roundness. In addition, image analysis systems are relatively immune to problems of wear and miscalibrations, and provide a single means of measuring the sizes of all particles great and small.

Image analyzers were once available in only a few laboratories, but today they can be readily and cheaply adapted to standard personal computers.

REFERENCES

Baba, J., & Komar, P. D. (1981). Measurements and analysis of settling velocities of natural quartz sand grains: *Journal of Sedimentary Petrology*, 51: 631–40.

Ehrlich, R. (1983). Editorial: Size analysis wears no clothes, or have moments come and gone? *Journal of Sedimentary Petrology,* 53(1): 1.

Ehrlich, R., & Weinberg, B. (1970). An exact method for the characterization of grain shape. *Journal of Sedimentary Petrology,* 40: 205–12.

Grace, J. T., Grothaus, B. T., & Ehrlich, R. (1978). Size frequency distributions taken from within sand laminae. *Journal of Sedimentary Petrology,* 48(4): 1193–201.

Haines, J., & Mazzullo, J. (1988). The original shapes of quartz silt grains: A test of the validity of quartz grain shape analysis to determine the sources of terrigenous silt in marine sedimentary deposits. *Marine Geology,* 78: 227–40.

Kennedy, S. K., Ehrlich, R., & Kana, T. W. (1981). The non-normal distribution of intermittent suspension sediments below breaking waves. *Journal of Sedimentary Petrology,* 51(4): 1103–8.

Kennedy, S. K., Meloy, T. P., & Durney, T. E. (1985). Sieve data – size and shape information. *Journal of Sedimentary Petrology,* 55(3): 356–60.

Klovan, J. E. (1966). The use of factor analysis in determining depositional environments from grain-size distributions. *Journal of Sedimentary Petrology,* 36: 115–25.

Komar, P. D., & Cui, B. (1984). The analysis of grain-size measurements by sieving and settling tube techniques. *Journal of Sedimentary Petrology,* 54(2): 603–14.

Ludwick, J. C., & Henderson, P. L. (1968). Particle shape and inference of size from sieving. *Sedimentology,* 11: 197–235.

Mazzullo, J., & Kennedy, S. K. (1985). Automated measurement of the nominal sectional diameters of individual sedimentary particles. *Journal of Sedimentary Petrology,* 55: 593–5.

Rittenhouse, G. (1943). Relation of shape to the passage of grains through sieves. *Industrial Engineering and Chemistry* (annual ed.), 15: 153–5.

Schäfer, A. (1982). The Krontron Videoplan, a new device for determination of grain size distributions from thin sections. *Neues Jahrbuch für Geologie und Paläontologie Monatschefte,* 2: 115–28.

Sneed, E. D., & Folk, R. L. (1958). Pebbles in the lower Colorado River, Texas, a study in particle morphogenesis. *Journal of Geology,* 66: 114–50.

Taira, A., & Scholle, P. A. (1979). Discrimination of depositional environments using settling tube data. *Journal of Sedimentary Petrology,* 49: 787–800.

Tilmann, S. E. (1973). The effect of grain orientation on Fourier Shape Analysis. *Journal of Sedimentary Petrology,* 29: 408–11.

Visher, G. S. (1969). Grain-size distributions and depositional processes. *Journal of Sedimentary Petrology,* 39: 1074–104.

Wadell, H. (1934). Volume, shape and roundness of quartz particles. *Journal of Geology,* 43: 250–80.

7 Quantitative grain form analysis

JULIAN D. ORFORD AND
W. BRIAN WHALLEY

Introduction

Grain size and shape are not truly independent variables for characterizing sedimentary particles. Nearly all measures of grain "size" are associated with some aspect of the form of the envelope enclosing the mass of the grain. This problem can be observed with the role of size–shape interaction on standard sieving for grain size distribution (Pang & Ridgway, 1983). The shape of a grain will affect its passage through a regular square sieve mesh such that spheres are more likely to pass through the mesh than other irregular grains given the same B axis and type of sieving (Ludwick & Henderson, 1968). Alternative methods of sizing (e.g., the use of a settling tube) cannot dispense with the problem of shape effects on a grain's settling rate (Cui & Komar, 1984), and electronic methods of sizing (e.g., Coulter Counter) assume a nominal particle shape.

Sedimentologists usually regard grain size as the most important parameter to be determined in grain characterization. Particle shape has often been dismissed from sediment analysis given that size (and hence mass) is more influential in sediment transport, and because of the practical reality that grain size appears easier to measure than grain shape. Other investigators, notably powder technologists, have been well aware of the need to relate these two aspects of grain structure, and a programme of reference materials including shape aspects for the calibration of sizing techniques is often employed (see, e.g., Scarlett, 1985).

There is presently semantic debate as to the appropriateness of "shape" as a colloquial expression covering all scales of grain morphology. In a search for quantitative measures of shape, basic problems of terminology arise (see, e.g., Whalley, 1972; Barrett, 1980; Winkelmol-

len, 1982). At a general level, Whalley (1972) distinguished between "form" as the "expression of the external morphology of the object" and "shape" (following Krumbein, 1941), as "a measure of the relation between the three axial dimensions of an object." Form is the general term, shape is more precisely defined for three dimensions, although two could be (and often are) substituted when used in the context of quantitative image analysis. We use grain form when considering the total range of grain morphology, while reserving shape for a specific geometrical comparison of grain properties.

Below, we review the range of quantitative grain form indices that can be of help to sedimentology, but concentrate on those methods we believe will be most advantageous in the light of recent innovations in image analysis techniques (Beddow, 1984). This means that any apparent disregard for traditional indices is not a reflection on their validity per se; rather, it means that their implementation and, in effect, discriminatory power have been superseded by new technology based on microcomputer-controlled video imaging. This explains, for example, why rapid form analysis of fine clasts using axial ratios has been overtaken by the use of Fourier analysis of grain outlines.

Quantitative shape analysis can be undertaken at two stages:

1. the interval/ratio measurement of morphological indices (i.e., the reduction of a particle's morphology to numerical form) and

2. the quantitative analysis of a collection of grain form measurements that are usually, but not necessarily, based on objective numerical measurement.

We are concerned here primarily with methods related to the first stage, which can be automated to allow rapid analysis.

This review deals solely with the parameterization of grain morphology. Since 1980, major advances in the discrimination of particle shape populations have been realized by the use of entropic solutions to establish

1. the most informative structure of a distribution of any grain shape parameter (Full, Ehrlich, & Klovan, 1981; Full, Ehrlich, & Kennedy, 1984) and

2. the subsequent decomposition (unmixing) of samples into elements of basic shape populations (provenance) that structure the depositional environment (Full, Ehrlich, & Bezdek, 1982; Full, Ehrlich, & Kennedy, 1984).

Space limitations prevent further discussion of the secondary analysis of grain shape data at the population level, although Syvitski (Chapter 18) provides discussion of some aspects of this problem relative to grain size analysis. Recent studies try to combine these two distinctive strands of population analysis in order to achieve definitive grain shape discrimination (see, e.g., Prusak & Mazzullo, 1987).

Grain form and shape relationships

Size measurement notwithstanding, the need for a rigorous, quantitative measure of form has been a long-standing requirement in most sciences. The concept of shape is often a notion intuitively held:

Shape is probably the most fundamental element property of any particle and is certainly one of the most difficult to specify for any but the simplest shapes; nevertheless, since it implicitly, in terms of habit, forms the basis of identification of many mineral constituents and, explicitly is considered to reflect provenance and process of formation, it is one of the most important properties of sedimentary rocks. (Griffiths, 1967, p. 109)

Following Griffiths (1967), we recognize that form (Orford, 1981) is the total sum of all morphological properties of a grain. These properties can be conceptually split into three scale elements:

1. *macroscale,* the general three-dimensional shape of the particle;

2. *mesoscale,* the general roughness or smoothness of the surface envelope of the particle, traditionally measured by concepts of particle roundness and angularity (which traditionally have been some of the most difficult morphological elements to define and measure); and

3. *microscale,* the surface texture of a grain that is often only identifiable with the aid of a scanning electron microscope.

It is clear that these components of form are interrelated in a complex manner, as well as being functions of the scales of examination. It

is rarely appreciated that, although size itself cannot be defined completely by a single parameter, an assessment of size can be improved by reference to a suitable "form parameter" for a specific purpose. This underlines the strategic importance of scales of morphological variation, which may be dependent on or dictate domains of process influential in grain behaviour.

A wide variety of methods have been derived for characterizing the various aspects of form. This multiplicity makes it both difficult and inadvisable to make specific recommendations. Just as with size measurement, personal choice plays a significant part in the use of any method of grain form characterization.

In this review we are concerned with both form and the gross measures of shape and, in passing, refer to angularity, roundness, and surface texture. As with size, sedimentological measurement of form is used for two main purposes:

1. discrimination of or between sedimentary provenance, given that grain form may be used as a measure of energy involved in transport, and

2. monitoring the change(s) during geological process(es) (e.g., weathering, diagenesis, and lithifaction).

There are other subsidiary uses of grain form:

3. determining grain relationships with soil mechanical parameters (e.g., "friction angle"),

4. determining packing parameters, and

5. determining secondary properties of materials such as specific surface, bulk density, and permeability.

Sampling

As with many sedimentological techniques, sampling problems occur with form analysis. Collecting a representative sample is not easy. This is particularly true when only a relatively few grains can be analyzed. Winkelmollen's (1971) use of particle rollability is the only mass form analysis currently available as the equivalent of the sediment size distribution generated by sieving. Logistically, at best only a few hundred grain images per sediment sample are likely to be analyzed with the types of semi-automatic image analysis equipment most frequently used for form analysis. Hawkins & Davies (1984) have discussed and addressed this problem and suggest that it is difficult to achieve a statistically representative sample. They describe a "spinning riffler" system that allows microscope slides to be loaded with sampled material ready for image analyzer inspection. With care, they suggest, this apparatus was found to be satisfactory. The subject is treated in detail by Allen (1981). Sampling is a particular problem (due to expense constraints) in electron microscopy of grains (see various papers in Whalley [1978b]).

Measures of grain form

Figure 7.1 gives a general view of the way in which many methods have been used to characterize aspects of form. Not all of these are in use, or are indeed useful, in sedimentological studies. Nevertheless, the automation and hence implementation of computer-related measurements can now make some of these methods relevant for specific sedimentological purposes.

This range of methods can be divided into the following principal types by which form parameters are assessed. The order of such types is a reflection of their historical introduction:

1. principal grain-axial ratios;
2. grain comparison with a calibrated visual chart image (a form of pattern recognition);
3. the search for varying scale periodicities in the perimeter angular decomposition (e.g., Fourier analysis);
4. behavioural characterization by which grain transport potential is enhanced or retarded by integrated grain form;
5. the way in which form is related to varying scale estimates of perimeter length (e.g., fractals); and finally
6. polygonal fitting within the shape envelope.

Each of these approaches is examined in an order reflecting historical development as well as approximate relevance to quantitative measurement and automated analysis.

Axial ratios

These are the traditional indices by which grain shape and, in particular, coarse particle shape has been established. The long history of

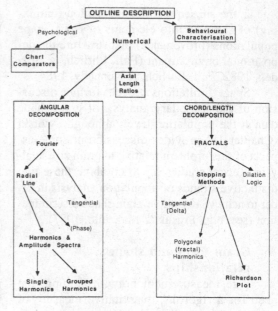

Figure 7.1. Schematic diagram showing the main ways in which particle outines can be described or quantified. Those shown in boldface are discussed in the text.

shape index formulation stems from Wentworth's (1922) initial work on deriving indices of shape from ratios of the three principal particle axes. Since that paper there has been a steady succession of new shape indices that, in general, were dependent on the principal axes. A review of these methods can be found in Barrett (1980). Although some investigators suggested different approaches to shape and roughness (via roundness) measurements using detailed analysis of segments of the particle's convex hull (e.g., Flemming, 1965), they were for the most part impractical to operationalize for other than a few particles at a time. With the advent of microcomputer-controlled video imaging, the means for this type of detailed grain outline analysis has appeared. However, the ability to deal with rapid outline analysis (using methods like Fourier analysis) makes the axial ratios method virtually redundant.

There are two constraints to this last point:

1. The conservatism of the geological community will ensure for the immediate future that axial ratios will persist as a reference frame for sedimentological work.
2. The scale of the grain still defines the

mode of form analysis. The analysis of a sand grain, as opposed to a pebble or boulder, is likely to dictate the use of modern imaging methods, and hence detailed surface outline form analysis, for the former; whereas the latter are still analyzed in terms of axial ratios given the immediacy of axial measurement.

Although shape indices of regular pseudo-ellipsoid particles specify a shape with reasonable accuracy, definitions of roundness and angularity require extra measurements (e.g., radii of curvature) that do not have the same degree of accuracy and precision (e.g., Lisle, 1988).

A surrogate form of axial measurement is to be found with the use of circles to be inscribed (as a maximum radius within the grain outline) and circumscribed (as a minimum radius outside the grain outline). These measures follow the practice of Wentworth (1933) and Wadell (1935), who derived a variety of indices based on them from the two-dimensional particle outline. The problems associated with such measures – notably that of inadequate characterization of gross shape as different shapes can have equivalent areas as well as similar inscribed and circumscribed radii – have been discussed by Medalia (1971), who advocated the use of dynamic shape factors instead of such indices.

Chart comparators

Chart comparisons, for determining "angularity–roundness" (see, e.g., Krumbein, 1941; Powers, 1953) are still widely used, despite their inbuilt psychological complexity, which leads to operator subjectivity. There is nothing wrong with the verbal description of "a sub-rounded clast" for example, or even a histogram of angularity class frequency for initial descriptions and discrimination. However, quantitative methods require precise and replicable methods. The use of multivariate analysis on chart-derived data, for example, does not alleviate the problem of nonobjective methods of form assessment. There have been attempts to calibrate chart comparisons by reference to other objective methods: Grain outlines used by Shepard & Young (1961) have been related to the radial line Fourier method (Whalley, 1978a), and Schäfer & Teyssen (1987) compared the "geometric

prototypes" of Wadell (1935) with the calculation of various shape measures for five angularity–roundness classes from Schneiderhöhn (1954). This was done for two-dimensional projections with an image analyzer. Despite these attempts at reconciliation, there still remains a gap between objective delimitations and the intuitive assessment of angularity–roundness by using charts. Effective quantitative use of charts is not helped by their ordinal scale of measurement. Real number analysis cannot be relied on when analyzing chart comparison results.

Current technology, by which rapid quantitative grain assessment can be undertaken to serve the same purpose as chart methods, has opened up new approaches to formal analysis that now supersede chart comparison methods.

Behavioural methods

Early attempts at form and size characterization considered these elements as separate entities. Integration of both aspects can be observed in the behavioural response of particles in some transport medium. Thus settling velocity, albeit with emphasis on size characterization, is a behavioural measurement. If we can control for size effects when using these methods, then a measure of dynamic form is feasible.

Though of interest to sedimentologists, the behavioural approach is not widely regarded as an important method of grain analysis. This may be due to the complexity of the procedure as well as to the difficulty in relating integrated gross behaviour to specific aspects of form. Perhaps the most widely known behavioural method is that of Winkelmollen (1971), in which a very small sediment sample (in a restricted size range) is allowed to roll the length of a large rotating drum inclined at a low (typically <2.5°) angle. The sample sediment grain form discriminant is a function of the length of time taken for the whole sample to travel the drum, with the median (or other percentile measures) being used to characterize the resulting frequency of sample weight–time distributions. The disadvantage of the method lies in the control on size. A quarter-phi interval is usually taken as the size control, but some behavioural variation may be related to size variation within this interval. Similar methods have been employed, such as the inclined chutes of Glezen & Ludwick (1963), the inclined platform of Krygowski & Krygowski (1965), and the inclined rocking gutter equipment described by Kuenen (1964). In the pharmacological field, Whiteman & Ridgway (1988) describe experiments comparing slotted sieving methods with a vibrating table sorting apparatus. A dynamic sorter is commercially available. Care must be taken with all these methods to avoid overloading the sorting platter or drum (either in quantity or feed rate) such that movement does not become mainly a function of friction angle of the unconsolidated aggregate mass. Some experimentation is required to obtain good, consistent results. It is fair to say that none of these techniques has enjoyed wide popularity. Moreover, their lack of widespread use has constrained standardization of technique, which reinforces lack of use.

A more recent, convenient form of dynamic shape sorting is through the use of the sieve "cascadograph" (Meloy & Makino, 1983; Kennedy et al., 1985). In this, a nest of identical sieves is used (for a given size) to sort the particles. Spherical particles pass through the sieves faster than nonspherical ones. The weight of sediment retained on each sieve after shaking allows a shape histogram to be drawn up. Although this device has been proposed as an indication of poor control on size measurement by sieving, it does seem to have much to commend it as a basic means of particle morphology measurement as a behavioural index. Its low cost is also commendable!

Aggregate tests employ slotted sieves, where sorting is by thickness (C axis), rather than square sieves, where the sorting is by B axis. A combination of square and slotted sieves has been used by some workers to provide B- and C-axis discrimination (Ludwick, 1971).

The difficulty with rollability and cascadograph tests is that they are limited by the sieving problem in obtaining the narrow size range required for each test. A lack of reliable means of controlling size allows variation, and hence error, in behavioural specification.

Dynamic, behavioural methods are potentially valuable, particularly as they are associated with grain populations and characterization

can be obtained in a matter of minutes. This is not the case with other methods, where individual grain outlines must be analyzed and then reanalyzed in population terms. However, it should be remembered that dynamic form measures may not be appropriate for grain characterization in that they are behavioural rather than descriptive of geometry. The pertinent quantitative measurement of a weathered grain does not necessarily need to include its transportational ability – as would be suggested by some measures of dynamic form.

Perimeter angular decomposition by Fourier series and related analysis

Since 1970, geological application of grain morphology has been revolutionized by the use of Fourier analysis, which attempts to decompose the complexities of outlines into a regular trigonometrical series (Beddow, 1980a; M. W. Clark, 1987; Ehrlich et al., 1980a,b). On the evidence of published data, the Fourier approach to form analysis must be regarded as the prime grain shape method for discriminating between grain provenance, as well as for identifying the continuum of variation in mixing of facies that accompanies source and transport variation within and between distinctive environments. Such measures were originally devised for sedimentological applications but were later extended beyond this (see, e.g., Kaesler & Waters, 1972; Jarvis, 1976). The perimeter of the outline is sequentially sampled (at equally spaced intervals) for the varying distance from the perimeter to some central point within the grain's outline. This requires three operational decisions:

1. choice of centre point, usually the centroid or centre of gravity of the outline;

2. an appropriate starting point on the outline; and

3. a function fitting the sampled perimeter to centroid distances, usually Fourier analysis.

Grain outline data have been obtained in a variety of degrees of sophistication, from simple graph paper coordinate measurement to the use of an image analysis approach (e.g., the ARTHUR microcomputer-controlled system [Ehrlich et al., 1987], and the system described by

Granlund & Hermelin [1983]). Although manual methods have been used to obtain a set of perimeter points that can be used in this type of analysis, computer-read and stored points are considered to be essential for efficient use of these methods.

The most frequently used Fourier method is that usually known as the radial line method. One disadvantage of the radial line Fourier technique is that, where the outline is highly indented, a line from the centroid may intersect the perimeter at more than one place (= multiple valued edge points). This clearly defeats the object of the analysis (see, e.g., Orford & Whalley, 1983, 1987; Telford et al., 1987). Fourier techniques that do not use intersecting radials are available but have had little popularity in the field of grain morphology (for specific methods see Zahn & Roskies [1972] and Persoon & Fu [1977]; for discussion see Clark & Clark [1976] and M. W. Clark [1981, 1987]).

Although Schwarcz & Shane (1969) first brought the Fourier outline characterization approach into the sedimentological literature, the main practitioners have been Ehrlich and coworkers from the University of South Carolina, who have developed both the application of Fourier analysis and the subsequent discriminatory methods based on Information Theory.

The Fourier method can be seen by considering a set of Cartesian perimeter points $P_{(x,y)}$, which define the two-dimensional outline of a grain. These can be reexpressed in terms of corresponding polar coordinates $P(R,\beta)$ for $R = 0$ about the centroid. "Unrolling" the shape gives a function $R(\beta)$ with values of the dependent variable, radial length R, associated with the dependent variable, degrees of rotation in the range 0°–360°. A grain profile defined by the polar coordinate pairs can be analyzed to give a Fourier series of the form:

$$R(\beta) = R_0 + \Sigma(A_n \cos_n \beta + B_n \sin_n \beta) \qquad (7.1)$$

where β is the polar angle (usually measured from the centroid), R_0 is the zeroth harmonic, n is the harmonic number, and A_n and B_n are Fourier coefficients (harmonic amplitudes) standardized relative to $R_0 = 1.0$. A mean amplitude (or roughness) coefficient C_n is derived from:

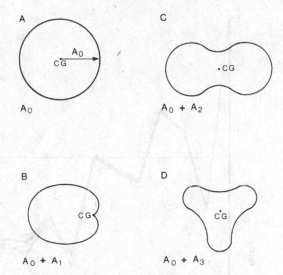

Figure 7.2. Diagram showing the effect of adding harmonic components to a circle, after Meloy (1980). (A) Circle with centre of gravity (CG) defined by a mean radius A_0 (= particle "size"). (B) $A_0 + A_1$, where A_1 is the first coefficient, the "error" in locating the CG because of the asymmetry. (C) $A_0 + A_2$, where $A_2 = 0.2A_0$ (= "elongation"). Note the bilateral symmetry supplied by even coefficients. (D) $A_0 + A_3$, where A_3 is a measure of the "triangularity" of the outline. Note the lack of bilateral symmetry supplied by odd coefficients.

$$C_n = (A_n^2 + B_n^2)^{0.5} \qquad (7.2)$$

which represents the contribution of the nth harmonic to add variation to the outline. The size of C_n decreases rapidly as n increases; that is, the majority of the grain shape of an outline is derived from only a few values of n (Fig. 7.2) As n increases more detail is added to the regenerated outline produced from the polar pairs. A potential outline can thus be characterized by a (mean) amplitude or roughness spectrum, C_n plotted against n (Fig. 7.3). The grain outline can be decomposed into a series of harmonic contributions (= amplitude), some of which can be particularly important in contribution to the structure of the outline. The lower-order harmonics (longer wavelengths) reflect elements of the basic form of the outline, whereas higher-order harmonics (shorter wavelengths) specify the smaller-scale components of morphology (e.g., angularity and surface texture). The relative starting position of each harmonic in its con-

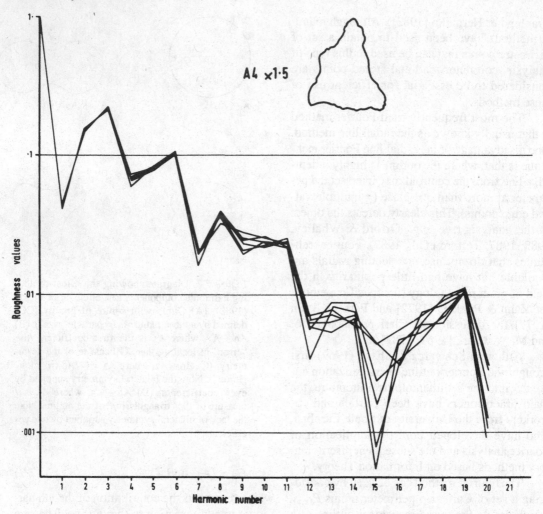

Figure 7.3. Mean harmonic amplitude or roughness plotted against harmonic number to characterize an angular grain (A4, *above*) and a rounded one (F27, *facing*) (outlines from Shepard & Young [1961]). The outlines were digitized six times at a magnification of 1.5× the Shepard & Young original. Notice the scatter as the information is sometimes taken up by slightly different harmonics in different digitizations.

tribution to the total data series is given by the phase angle (arctangent of B_n/A_n). In practice, the phase angle information is rarely used to discriminate between sediments, but has been used occasionally (Beddow, 1980a; Rohlf, 1986). Figure 7.4 shows how adjacent harmonics are sometimes grouped into an average value to provide shape and textural information (Czarnecka & Gillot, 1977; Whalley, 1978a). An alternative method of characterizing the sample is presented as a standard method by Ehrlich (see, e.g., Ehrlich et al., 1980b) based on a shape-frequency histogram associated with a particular

harmonic whose sample variance has been maximized by Information Theory (Full et al. [1984] describe this approach in greater detail). This means that specific single harmonics are used to characterize grain form.

Fourier analysis can be carried out in various ways. Luerkens et al. (1982, 1984) have outlined the three main ways in which information can be obtained from fitting a Fourier series to a grain outline – the (R,β), (l,ϕ), and (R,S) methods:

The (R,β) approach is that of Ehrlich & Weinberg (1970), using the centre of gravity.

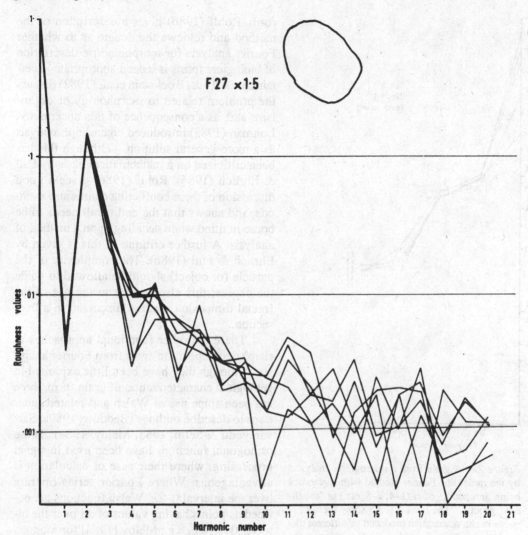

Figure 7.3. (cont.)

This provides area information but cannot be used for convoluted outline shapes, where there may be multiple values of R.

The (l,ϕ) method uses arc lengths l with the change of slope ϕ to characterize the outline shape (Zahn & Roskies, 1972; Fong et al., 1979). This allows complex shape changes to be characterized but does not allow axial information to be derived (although these can be obtained from the perimeter points by other methods).

The (R,S) method (Luerkens et al., 1984) allows analysis of reentrant profiles. Perimeter (x,y) coordinates are transformed into polar form, which are then used to calcu-

late a radius vector R as a function of S, the normalized arc length, by a Fourier expansion.

The (R,S) method would seem to have advantages over the other two methods in that it is more general than the (R,β) method but can also be used to derive areas and moments of inertia, which the (l,ϕ) method cannot. A large amount of profile information is obtained, as upward of 200 profile perimeter positions are required (Luerkens et al., 1984). Full exploitation of the method has yet to be seen in a sedimentological context. Processing time can be speeded up by use of a Fast Fourier Transform (Meloy, 1977). Kanatani (1984) provides a discussion of the

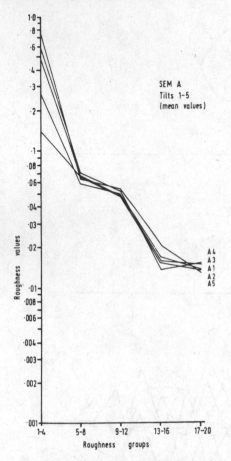

Figure 7.4. A grain from a micrograph analyzed by the radial line Fourier method with harmonics being grouped together (1–4, 5–8, etc.) as "roughness groups." The lines A1–A5 show the differences in characterization produced by different tilts (in 10° steps) of a scanning electron microscope stage. Note that the main variance is in the first group of harmonics, that is, basic shape, due to changing perspective as the grain is tilted. From Orford & Whalley (1983), The use of the fractal dimension..., *Sedimentology*, Blackwell Scientific.

use of the discrete Fourier Transform and appropriate computational methods.

Moellering and Rayner (1981) have employed a more complex approach, termed "dual axis Fourier shape analysis," which employs a reversible transformation. Further discussion of Fourier methods can be found in the literature, some of it somewhat complex and not yet resolved. For instance, Rohlf & Archie (1984) have compared several Fourier methods and found elliptical Fourier analysis to be a useful

form. Rohlf (1986) gives a description of the method and reviews the debate as to whether Fourier analysis for morphometric description of biological forms is indeed appropriate in certain cases. Thus, Bookstein et al. (1982) discuss the problem related to morphology of organisms and, as a consequence of this uncertainty, Lohman (1983) introduced eigenshape analysis as a more general solution – although this has been criticized on a mathematical basis by Full & Ehrlich (1986). Rohlf (1986) gives a good discussion of these conflicting claims and methods, and shows that the end result needs to be borne in mind when deciding upon a method of analysis. A further critique of this is given by Ehrlich & Full (1986). The complexity of the particle (or object) should be allowed to guide the choice; this also applies to the use of the fractal dimension method discussed in a later section.

There are other functional approaches to shape decomposition apart from Fourier analysis. Although they have been little exploited in geological characterization of grain form, there has been some use of Walsh and related functions to describe outlines (Beddow, 1980a; Sarvarayudu & Sethi, 1983; Meloy, 1984). These orthogonal functions have been used in signal processing, where their ease of calculation is advantageous. Where Fourier series operate over the interval $0–2\pi$, Walsh functions are defined as points having values of ±1 over the interval 0–1 (see, e.g., Meloy [1984] for more).

Practical use of Fourier radial line method

Sedimentological use of the radial line Fourier method has become reasonably widespread in the past fifteen years, although this has been somewhat constrained generally by the need to use specialized image analysis equipment. Ehrlich et al. (1987) give a recent summary of the technique and some applications where, in particular, the basic radial line method is used. In general for this type of investigation, Ehrlich and coworkers base discrimination on an average radius (R_0) and twenty-four harmonic amplitudes – although, given the high degree of redundancy between adjacent harmonics and subharmonics, the analyses usually depend on only a few of the harmonics selected by means of Information Theory techniques. The population is then discriminated according to proportional contributions to these harmonic amplitudes.

As there is no consolidated list, several applications of radial line Fourier analysis can be mentioned briefly. Wagoner & Younker (1982) achieved a separation of sources by the "conventional" shape frequency histogram method. Riester et al. (1982) distinguished three beach zones on the basis of the technique by examining the proportion of abraded and irregular components. Mazzullo et al. (1986) linked scanning electron microscopy (SEM) work with Fourier analysis to examine sorting and abrasion contributions to aeolian grain roundness. Mazzullo & Magenheimer (1987) specifically tested ideas on the use of SEM as a source indicator by means of Fourier analysis. They collected weathered bedrock samples from a transect across Texas covering three source terrains and found that mechanical and chemical mechanisms did leave a characteristic imprint, although there were some difficulties. Not surprisingly, the relict set of textures found on grains released from water-transported sedimentary rocks (with no secondary quartz) were not discriminatory. Byerly et al. (1975) used a simple Fourier technique in determining the shape of zircon grains. Dowdeswell (1982) used a combination of Fourier and SEM to examine sediment sources from a glaciomarine core, an approach similar to that taken by Mazzullo & Anderson (1987). Dowdeswell et al. (1985) extend their analysis of Arctic marine cores with the use of a "multiparameter" approach linking SEM, size, and shape discrimination. In terms of linking other types of determinative technique to Fourier methods, Schultz (1980) has investigated the effects of hydrofluoric acid etching on quartz and feldspar grains, and Kennedy et al. (1985) have looked at the "cascadograph" (see section on "Behavioural methods") in comparison with Fourier analysis.

Czarnecka and Gillott (1980) have used a "modified Fourier method" to investigate roughness, on limestone and quartzite pebbles, and surface textures examined by SEM. Their meth-od is to use the summated value of C_n (see eq. (7.2)) to obtain first a parameter P_n, where

$$P_n = [\ 0.5 \sum (A_n{}^2 + B_n{}^2)]^{0.5} \qquad (7.3)$$

which can be restricted to a value for selected harmonic ranges ($i < n \leq j$). Then a Total Roughness Coefficient T is introduced, defined by

$$T = K(L/P_n)$$

where L is the difference between the normalized grain periphery and 2π, and K is an empirical constant designed to reduce the contribution of elongation masking the effects of texture (see Fig. 7.2). A fuller explanation is given by Czarnecka & Gillott (1977). Histograms showing profile frequency for given harmonic values are used by Czarnecka & Gillott (1980) to show differences in gross shape, as well as the contribution of finer-scale surface texture information. This approach has not been pursued to any great extent, however, in linking SEM with Fourier-derived data. The problems associated with highly complex outlines and high harmonic variance have not yet been fully worked out.

Problems with Fourier methods

M. W. Clark (1987) has made some useful general comments about imaging systems and Fourier methods in particular, but some further comments are pertinent. The Fourier method relies upon a periodic algorithm (although phase information is rarely used), and this may not be appropriate to the type of data to be analyzed. This is related to the discussions by Ehrlich & Full (1986), Rohlf (1986), and others mentioned above.

Although it would seem that a single fitting of a Fourier series provides an unambiguous result, this is in fact not so, as Figure 7.3 illustrates (see also M. W. Clark, 1987). This variability needs to be taken into account for high harmonics. Just where this variance sets in seems to depend upon the overall shape of the particle as well as possible operator and machine variance. The calculation of amplitudes from twenty-four harmonics requires a minimum of forty-eight edge points. However, for very complex particles this may mean that one or both information sets may be missed by not utilizing all the outline, or that information is swamped by variance noise. Thus the radial line Fourier method appears problematic for complex angular particles, as outline explanation at low frequencies will also be related to overtones at higher frequencies resulting in leakage. In other words, Fourier "roughness coefficients" are not independent. This problem is not avoided by bandwidth averaging or grouping of Fourier harmonic terms (Whalley 1978a; Boon et al., 1982; see also Fig. 7.4). Thus, while the Fourier method works well for a large variety of sedimentological grains (which are, on the whole, rather rounded), where there are sharp corners and indentations it may encounter difficulties. This effect is in addition to the inability of the radial line method to cope with highly crenulated outlines. The difficulties surrounding the radial line Fourier method (Whalley & Orford, 1982; Orford & Whalley, 1983; M. W. Clark, 1987) suggest this method is best avoided for shapes that are very angular, distorted, or indented. However, Fourier methods have led to effective source and facies discrimination in most cases where it has been used in analysis of clastic deposition.

Despite the proven discriminatory power of the Fourier method (see, e.g., Ehrlich et al., 1980a; Mazzullo & Withers, 1984; Kennedy & Ehrlich, 1985; Prusak & Mazzullo, 1987), the technique has not been employed on a widespread basis. This appears to be due to perceived problems relating to:

1. the seemingly complex mathematics of Fourier analysis;

2. the cost of the data capture equipment (although this appears to be decreasing, given the latest automated imaging methods becoming more readily available); and

3. the complexity of the modes of subsequent analysis of Fourier data – in particular, the use of Information Theory (Full et al., 1984) to facilitate effective facies discrimination.

Scale estimates of perimeter length

The use of varying estimates of perimeter length for morphological analysis of particles has been available for over a decade. The principal technique in this form of analysis is known as *fractal analysis*. This provides a complementary method to Fourier analysis by enabling investigation of complex forms otherwise intractable to the radial line Fourier method (Orford & Whalley, 1983, 1987). Fractal analysis is an approach to the general problem of providing a function to describe nonregular polygons or closed loops, which are not capable of approximation by Euclidean geometry. Mathematically, this leads to the concept of continuous but nondifferentiable functions, which require new methods of characterization.

Early work on grain morphology using fractals stems from powder technology, particularly from the work of Kaye (1978, 1984, 1986, 1987), while Schwarz & Exner (1980) present a method and computer code for deriving fractals using a semiautomatic digitizer. Kaye's perimeter-stepping technique is the basis for most work on fractals. Further information on the use of computer programs and hardware for the derivation of fractals can be found in Kennedy & Lin (1986) and Telford et al. (1987).

Fractal analysis is basically limited at this stage to analysis of grain form in two dimensions. Calculations are simple but highly repetitious, and hence exceedingly tedious unless undertaken with the aid of a computer. The advent of microcomputer-controlled video analysis as a means of obtaining image analysis has unshackled this technique such that there is no technical reason why fractal analysis cannot be regarded as one more standard means of grain form analysis. Further enhancement of the procedure is likely with the widespread use of "transputer processing," given that fractal derivation is based on simple but repetitive measurement steps.

Given a grain outline, as progressively shorter unit lengths are used to measure the perimeter length (known as perimeter stepping), this length estimate as a function of the stepped-off length unit correspondingly increases. This effect is of significance for the form characterization of complicated, crenulated, or indented outlines. The methods for estimating perimeter lengths are given by Schwarz & Exner (1980), N. N. Clark (1986a), and Kennedy & Lin (1986); fractal derivation in general is covered by Kaye (1978, 1986), Orford & Whalley (1983), and Whalley & Orford (1989).

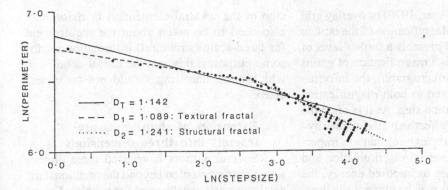

The figure shows a Richardson plot with axes LN(PERIMETER) versus LN(STEPSIZE), with legend entries:

$D_T = 1 \cdot 142$

$D_1 = 1 \cdot 089$: Textural fractal

$D_2 = 1 \cdot 241$: Structural fractal

Figure 7.5. A Richardson plot of log (perimeter) versus log (step size = stride) showing the total fractal D_T, the textural fractal D_1, and the structural fractal D_2. Notice the scatter of points as the stride increases.

Figure 7.5 shows the relationship between step length and perimeter length, known as the Richardson plot to commemorate the pioneer of this analysis (Mandelbrot, 1982).

Denote the variable step length as the stride s with the corresponding perimeter estimate as p. A plot of $\log(p)$ against $\log(s)$ usually gives a scatter of points best characterized by an inverse linear function with a gradient of b ($b \leq 0$). The fractal dimension D is defined by $D = 1 - b$ ($D \geq 1$). A linear regression model is generally used to fit the function to these data points (Fig. 7.5). The value of D can fall between 1 and 2. $D = 1$ reflects a simple Euclidean figure that can be functionally characterized by axial ratios. As D approaches 2, the form becomes increasingly complicated at all scales, such that axial ratios are inapplicable to characterize form and the crenulations of the grain outline create problems for Fourier analysis (Orford & Whalley, 1983). There is an intuitive appeal for this formal method, perhaps associated with the inherent simplicity of derivation. The theoretical straight line function of the Richardson plot relates to a central ideal of fractals, that of self-similarity (introduced by Mandelbrot [1967]), which means that the morphological nature of the fractal form is repeated at all scales.

Kaye (1978) indicated that certain closed loops did not behave so well as the Mandelbrot ideal. The Richardson plot for a number of fine

particles showed the need for two nonoverlapping regression elements to generalize the function $\log(p)$ versus $\log(s)$. These two elements were denoted as the "structural" (large-scale stride $\approx$ shape) and "textural" (small-scale stride $\approx$ edge detail) fractal components (Fig. 7.5). Schwarz & Exner (1980) used visual inspection to suggest where the breakpoint (intersection) between the two regression elements should occur. They also restricted the stride length of analysis to between 0.085 and 0.5 of the maximum diameter of the outlines as they were concerned that multifractal values could be invalidated. In particular, they suggested that the wide scatter of step length–perimeter estimates at large strides ($s > 0.5A$ axis) was an artifact of the digitizing and analytical systems they used. Subsequent analysis by N. N. Clark (1986b, 1987) suggests that this scatter is related to the basic shape of the grain rather than the measuring equipment per se.

The Richardson plot can be structured to express absolute or relative analyses. An appropriate relative measure could be stride length as a ratio to the longest axis of the grain. Thus, information can be gained about the size of the feature relative to a grain, or perhaps more important, the absolute size of a feature (Kaye, 1986, 1987). The measuring system must be calibrated for absolute determinations.

Although it may be considered that Fourier and fractal techniques are independent of size, this is not so because there is a minimum level of outline digitization possible. This is governed by the minimum step, the size or distance between pixels if using an automatic digitizing system (Telford et al., 1987), or is a function of the digitizing table (Piper, 1970) or overlay grid (Håkanson, 1978). Magnification of the outline itself when digitized presents a further level of consideration. For high magnification of grains (e.g., an SEM photomicrograph), the information obtained is related to both magnification used and the digitization step. As it is often the small-scale features (effectively, the angularity–roundness and surface texture) that are important both in environmental discrimination and in assessing responses to imposed energy, the large-scale size effects may be numerically over-dominant. It is usual to restrict attention to a scale of irregularities substantially less than the grain diameter (see, e.g., Orford & Whalley, 1983; Whalley & Orford, 1986).

Peter Schweitzer and Pat Lohman (pers. comm., 1984) have suggested some ways in which the problems produced by scale effects can be countered. One particular problem is the way in which small changes on the perimeter (texture) may become swamped by the general circularity of many grains. Schweitzer and Lohman unroll the outline in a manner analogous to the use of the radius (zeroth harmonic normalization) in the radial line Fourier technique. Effectively, one can imagine a circle unrolled to give a straight line (such unrolling has been suggested for Fourier analysis by Lohman, 1983). This is done geometrically by Schweitzer and Lohman; the calculation for the outline points is trivial on a computer (where the points will be stored anyway). Unrolling is done by calculating the tangent-angle function of the outline and then detrending it to "subtract" a circle. Deviations from the circle are then given by reconstruction from the detrended tangent-angle function. The perimeter and stride are then calculated in the normal way.

If unrolling procedures are used in fractal analysis (see, e.g., Frisch et al., 1987), then it should be noted that the shape variance is being thrown away in order to maximize the finer texture or detail. The typical values and variance of fractals obtained in this way will be lower than recognized in examples from Richardson plots published so far. In fact, we need not worry too much about the scatter of points in the structural fractal domain, but should concern ourselves more with the value of the pseudofractal dimen-

sion of the textural element. A decision may also need to be taken about the requirement for large-scale (structural) information as, for some purposes, this may be useful detail – in which case, unrolling should not be undertaken.

Extension of shape measure fractals into three dimensions

Several authors have used three-dimensional characterization beyond the traditional triaxial approach usually used for pebbles. For instance, Hulbe (1955) embedded sand grains in a resin rod to allow measurement of three dimensions under a microscope. However, this is not easy, especially where the particles are very small or consist of aggregates. Aggregation can be countered by dispersal techniques in a liquid for size analysis, but this is more difficult to do for form analysis. Medalia (1980) discusses possible methods for small particles; he points out that, although automated equipment performs measurements relatively easily, subjective judgments are still necessary where there are aggregates. Radii of gyration can be calculated from two dimensions, but there are clearly difficulties where the particles are flat or flakey, especially if particles have settled with their maximum projection planes parallel to the microscope slide. A useful discussion of size and three-dimensional characteristics of powders is given by Heywood (1963), who suggests approaches little used in sedimentological work. Gotoh et al. (1984) discuss the way in which shape can be characterized by packing density determinations and treated by statistical methods used in thermodynamics.

Generally, a two-dimensional viewpoint is used to exploit fractal relationships even though three dimensions might be desirable (Medalia, 1980). Logistically, it is easier to manage two-dimensional than three-dimensional projections. In many cases, the former may suffice for characterization. Griffiths & Smith (1964) suggest that >95% of the three-dimensional explanation of grain form could be obtained by using the two-dimensional maximum projection plane. If required, the three-dimensional characterization may be approximated in certain instances. At its simplest, this could be of two outlines, one in the maximum projection plane and the other at 90° to it. As mentioned above, Medalia (1971) has discussed methods of obtaining dynamic shape factors by utilizing radii of gyration, although this is concerned with basic shape of particles. Conversely, tilting grains on a scanning electron microscope stage can produce markedly different results according to the view taken (Fig. 7.4). This has also been noted by Tilmann (1973) on a universal stage experiment, also using Fourier analysis, and by Whalley & Orford (1982) using fractals.

A more complex way might involve characterizing a series of outlines $n°$ apart on a particle rotated either systematically or randomly. Thus the magnitude of the projected area varies according to the aspect presented. The results from such rotations are plotted to give a signature waveform (Kaye et al., 1983). These authors have suggested two main types of waveform that are usefully diagnostic of the solid morphology of particles. These waveforms are the "Cauchy convex hull" and the "projected area waveform." The former is found by systematic rotation for 90° around one axis and then for 90° around an axis at right angles to the first. Kaye relates this waveform to settling of a grain under laminar fluid flow, thereby linking morphodescriptive with dynamic-behavioural methods in a manner similar to the triaxial ratio "Sneed and Folk sphericity" (Sneed & Folk, 1958), widely used in a geological context.

As well as simple area-related waveforms, Kaye et al. (1983) also show how fractal methods could be used to give three-dimensional information. This initial work derives data from single particles, but Kaye et al. suggest practical methodology for large numbers of particles using a modified optical stream counter/sizer.

Although much still needs to be done further to refine methods of grain characterization at scales ranging from basic shape to fine texture, it would appear that there are advantages in extension into three dimensions, especially if a high degree of automation can be incorporated. For many puposes, however, work in two dimensions would seem to be adequate as long as its limitations are recognized.

Fractal–Euclidean transitions

We have treated fractals so far in the pure sense, following Mandelbrot (1982), in that the fractal dimension is associated with "self-similarity" (Mandelbrot, 1967, 1982). The notion of self-similarity is central to theoretical fractals and, in particular, to the generation of fractal forms. In Figure 7.5, where two regression lines are used to minimize the variance, it is, of course, possible to supply a single line. As yet, it is unclear whether we can really apply a rigorous fractal approach and use a single line or a split regression into two or even three fractal elements as in Figure 7.5: If the latter, where should the break(s) be (Kennedy & Lin, 1986)? Instead of worrying about the strictness of the breakpoint methodology per se we should further explore the meaning, in physical terms, of such breaks on morphological development.

Just as the self-similarity principle suggests a unified structure to an outline, so the fact that there are two linear portions indicates that there may be two mechanisms providing such self-similar structures and that each gives a characteristic signature in the Richardson plot.

The break point in a Richardson plot itself is significant as it suggests a rapid transition from one fractal domain to another, where domain may relate to scale of morphological structure. We might also believe that a gradual change (easily computer-calculated by seeking the second derivative of perimeter with respect to stride – or, perhaps just as easily, by eye) might indicate a gradual change in the dominance of one mechanism over another. As yet, we do not know enough about such changes, either sharp or gradual, to be able to say just what is the importance of break points.

Statistically, it is not easy to decide how to weight-fit linear portions or where the transition zone starts and ends. For this reason, at least in the present state of knowledge, it may be best to fit break points by eye (the method of Schwarz & Exner [1980] and also used by Orford & Whalley [1983]). Accepting the need for a pragmatic division into textural and structural fractal elements means that we should call the reduced domain-fractals "pseudofractals."

It should be noted that the "unrolling" technique discussed above tends to use a single fractal value for discrimination, whereas high-lighting structural aspects of an outline is accomplished by the use of two pseudofractals. Nevertheless, once the perimeter and stride data have been obtained there is no reason why both methods should not be employed to explore the properties of outlines.

Polygonal harmonics

This approach is the latest of the methods to be considered but has yet to be tested for effective shape specification on grains. N. N. Clark (1986b, 1987) has suggested a method of obtaining some basic grain shape information in a very simple way by adapting the perimeter dividers method but using large strides; however, instead of changing the stride to obtain a Richardson plot, he continued stepping around the perimeter with the same stride length. Clark used an outline of the State of West Virginia as his closed loop. His procedure involves an arbitrary protocol: Choose any point on the perimeter, place the point of the dividers on this and, with an arbitrary stride, swing the other point inward in a clockwise direction until it intersects the perimeter. Repeat progressively around the perimeter but, instead of ending with some closure error, continue the process. Somewhat surprisingly, it is usual to find that the dividers start to walk, often quite rapidly, in their own footsteps and that a regular polygon is formed. If the stride on the dividers is changed markedly, then another polygon solution is obtained.

There are further interesting aspects of this work. The structure of the Richardson plot, such as in Figure 7.5, causes analytical problems with fractal evaluation. When stride values approach half the length of the principal axis of the outline they generate significant "jumps" in estimates of the perimeter. Clark (1986b), at the end of his note, suggests:

The nature and stability of fractal harmonics [later called polygonal harmonics] will play an important role in future characterization of rugged particle shapes and curves. Since a change in solution is generally accompanied by a significant jump in polygon perimeter, fractal harmonics can also be used to explain error and scatter of data in performing conventional and digital hand-and-dividers frac-tal analysis, where fractal dimension is deduced from a plot of perimeter versus step length [Richardson plot].

Subsequently, Clark et al. (1987) have explored some of the idiosyncrasies of polygonal harmonics and suggest that:

1. polygons are sensitive to the starting point;

2. polygons are sensitive to rotation direction (for West Virginia, at a certain stride, anti-clockwise rotation gave a square but clockwise gave a triangle); and

3. similar polygons will persist with small changes in position as step length varies within limits (starting point retained) – at the limit, a jump occurs to a new polygonal form.

A plot of the ratio of polygon perimeter : number of traverses to achieve stability (as ordinate) versus stride shows these "jumps" quite clearly. Lines of similarity correspond to the stability of particular polygons. We are now in a position to see that the scatter of points and the envelope encompassing them in a Richardson plot is not (necessarily) the result of measurement errors but is a real consequence of the decomposition of that particular outline. As yet, the potential of these findings has fully to be explored. In comparison to Fourier techniques, it would seem that the errors, that is, stability jumps, involved are likely to be more easily reconciled with the actual behaviour of the outline than the variability of amplitude values for given harmonics (Whalley, 1978a).

Grain surface textures

Surface texture analysis has provided a fruitful means of analysis with a considerable volume of literature since the advent of the scanning electron microscope (SEM). Useful reference material can be found in Krinsley & Doornkamp (1973), Whalley (1978b), Smart & Tovey (1981), and a bibliography in Bull et al. (1986). The SEM provides a mass of detail, much of it in addition to that of form parameters. It can give information on all two-dimensional components together with that added by the view of the surface texture over half the solid surface. A height parameter (C axis) can also be obtained with extra effort. The technique has been used to identify erosional-transport environments in several ways. Marshall (ed., 1987) provides a variety of techniques and applications for form analysis of clastic particles, many of which use the SEM; Culver et al. (1983) discuss statistical evaluation of the general technique of surface texture feature logging as a crude measure of texture. A general difficulty is that, despite a wealth of published accounts of types of surface textures, there are few orderly and controlled accounts of how such textures are formed, and hence which textures can be grouped together as environmental indicators.

In spite of the wealth of information provided by the SEM, there are problems in quantifying it. This mainly stems from the relative ease with which the brain can process visual information compared with the difficulty in providing algorithms that can be implemented automatically. There are, however, ways in which information of textures (as well as angularity–roundness on a visual scale) can be aided by computers, especially in the processing of the data. "Fuzzy" methods (Beddow, 1980b), using linguistic truth values and approximate rules of inference, are one such way.

Outline information can be processed by digitizing the outline of an SEM-obtained micrograph. In the main, this has been done with Fourier techniques as discussed above; Czarnecka & Gillott (1980), Ehrlich et al. (1987), Mazzullo & Anderson (1987), and Haines & Mazzullo (1988) provide good examples. Fractal analysis (see, e.g., Whalley & Orford, 1982) has also been applied in a similar way. Thus, the two techniques of Fourier/fractal quantification and SEM can go hand in hand for many investigations. For some time, image analysis microscopes have been used to extract quantitative information (Kennedy & Mazzullo, Chapter 6 in this volume), but it is now possible to couple an SEM (or TEM) with image analysis frame grabbers. The data can be analyzed directly on line or – perhaps more conveniently – recorded by video for subsequent analysis.

New technology in grain form analysis

The plethora of devices for measuring grain size over the past few years has not been matched by a similar number for characteriz-

ing form components. However, a new device (Galai CIS-1; Brinkman PSA model 2010 in the USA) promises to measure both size and shape, the latter as biaxial as well as densitometric information. The instrument uses a laser as a flying spot scanner, with the interaction between the spot and particles in a small volume (Aharonson et al., 1986). This time domain information is processed by computer into 300 segments in four ranges from 0.5 to 1,200 μm. Axis measurement is taken by a frame grabber, via a charge-coupled device (CCD) camera, from stroboscopic illumination of the grain volume. The outline information can then be used for a variety of measurements in a microcomputer. How well this instrument performs, for both size and shape measurement, is not yet known.

Access to equipment similar to the ARTHUR system (Ehrlich et al., 1987) is likely to be easier in the near future as interface cards become freely available for microcomputers. In the past, rather slow computers have been used (see, e.g., Apple II and BBC machines [Telford et al., 1987; Hayward et al., 1989]) but fast IBM-compatible and Macintosh systems can now be fitted with such boards. A trend likely to increase is the use of "transputers" to speed calculation by parallel processing techniques. This is especially useful for fractals, which require many repetitive calculations. Transputers can be fitted to IBM-type machines but require a suitable language; these may be specific for parallel processing (e.g., OCCAM for INMOS transputers) or parallel FORTRAN of more general application from existing program suites. Computational speeds can also rise with the use of RISC chips or machines. What remains as the slowest part in any shape analysis investigation is the necessity to hand-feed grain sets to the microscope–camera setup for data capture.

Conclusions

We suggest, as others have, that grain form is an important sedimentological parameter and that grain size analysis should include some consideration of form. The sedimentological literature shows a surprising lack of their combination, which we surmise is due to the lack of suitable rationale by which form can be considered. Modern analytical methods should now make it possible to assess both size and form and hence allow combination of such measures. Some image analysis methods indeed make it possible to make both measurements together, and this is now undertaken on a conventional basis in the field of powder technology. Form–size interactions in sedimentology play a crucial role in certain areas such that their combination might best be measured by a dynamic response through behavioural measures. Unfortunately, even these methods are not without pitfalls, not least in their lack of standardization. Sedimentologists should be in the process of debating some standard methods of grain form characterization.

The many years of methodological development on form characterization show that simple and generally applicable methods specifying grain form are unlikely to be available. However, the relative cheapness of modern computing power and plethora of image analysis techniques can now provide us with the means of fast grain form evaluation. We further suggest that, given the technology, investigators should concentrate on the use of Fourier and fractal analysis for geometric description of grains, whereas the behavioural methods clearly need enlargement if consideration of dynamic grain form is required. At this stage the relative values of Fourier and fractal methods have still to be debated, but we do not believe that form analysis requires the exclusive use of either technique. It seems likely that as grain form becomes a more standardized need in analysis, sedimentologists will wish to apply such methods to specify increasingly complex formal structures. We suspect the future differing use of Fourier and fractal analysis will reflect the need to investigate all morphological aspects of a full continuum of simple to increasingly complex crenulated grain morphologies.

ACKNOWLEDGMENTS

We thank Pat Lohman, Peter Schweitzer, Fred Brookstein, David Evans, Adam Frisch, Nigel Clark, and Brian Kaye for verbal and/or written discussions, and Maura Pringle for the figure artwork. The critical comments of Don Forbes and Jim Mazzullo were also greatly appreciated.

REFERENCES

Aharonson, E. F., Karasikov, N., Roitberg, M., &. Shamir, J. (1986). GALAI-CIS-1 – A novel approach to aerosol particle size analysis. *Journal of Aerosol Science*, 17: 530–6.

Allen, T. (1981). *Particle Size Measurement*. London: Chapman & Hall, 3rd ed., 678 pp.

Barrett, P. J. (1980). The shape of rock particles, a critical review. *Sedimentology*, 27: 291–303.

Beddow, J. K. (1980a). Particle morphological analysis. In: *Advanced Particulate Morphology*, eds. J. K. Beddow & T. P. Meloy. Boca Raton, Florida: CRC Press, pp. 31–50.

(1980b). Fuzzy sets in property representation. In: *Testing and Characterization of Powders and Particles*, eds. J. K. Beddow & T. P. Meloy. London: Heyden, pp. 176–89.

(1984). Recent applications of morphological analysis. In: *Particle Characterization in Technology*, ed. J. K. Beddow. Boca Raton, Florida: CRC Press, pp. 149–72.

Bookstein, F. L., Strauss, R. E., Humphries, J. M., Chernoff, B., Elder, R. L., & Smith, G. R. (1982). A comment upon the uses of Fourier methods in systematics. *Systematic Zoology*, 31: 85–92.

Boon, J. D., III, Evans, D. A., & Henningar, H. F. (1982). Spectral information from Fourier analysis of digitized grain profiles. *Mathematical Geology*, 14: 589–605.

Bull, P. A., Whalley, W. B., & Magee, A. W. (1986). *An annotated bibliography of environmental reconstruction by SEM, 1962–1985*. British Geomorphological Research Group, Technical Bulletin no. 35. Norwich, U.K.: Geo Books, 45 pp.

Byerly, G. R., Mrakovich, J. V., & Malcuit, R. J. (1975). Use of Fourier shape analysis in zircon petrographic studies. *Geological Society of America Bulletin*, 86: 956–8.

Clark, M. W. (1981). Quantitative shape analysis: A review. *Mathematical Geology*, 13: 303–20.

(1987). Image analysis of clastic particles. In: *Clastic Particles*, ed. J. R. Marshall. New York: Van Nostrand Reinhold, pp. 256–66.

Clark, M. W., & Clark I. (1976). A sedimentological pattern recognition problem. In: *Quantitative Techniques for the Analysis of Sediments*, ed. D. F. Merriam. Oxford: Pergamon, pp. 121–41.

Clark, N. N. (1986a). Three techniques for implementing digital fractal analysis of particle shapes. *Powder Technology*, 46: 45–52.

(1986b). Fractal harmonics and rugged materials. *Nature*, 319: 625.

(1987). A new scheme for particle shape characterization based on fractal harmonics and fractal dimensions. *Powder Technology*, 51: 243–9.

Clark, N. N., Diamond, H., Gelles, G., Bocoum, B., & Meloy, T. (1987). Polygonal harmonics of silhouettes: Shape analysis. *Particle Characterization*, 4: 38–43.

Cui, B., & Komar, P. D. (1984). Size measures of the ellipsoidal form of clastic sediment particles. *Journal of Sedimentary Petrology*, 54: 783–97.

Culver, S. J., Bull, P. A., Campbell, S., Shakesby, R. A., & Whalley, W. B. (1983). Environmental discrimination based on quartz grain surface textures: A statistical investigation. *Sedimentology*, 30: 129–36.

Czarnecka, E. T., & Gillott, J. E. (1977). A modified Fourier method of shape and surface texture analysis of planar sections of particles. *Journal of Testing and Evaluation*, 5: 292–8.

(1980). Roughness of limestone and quartzite pebbles by the modified Fourier method. *Journal of Sedimentary Petrology*, 50: 857–68.

Dowdeswell, J. A. (1982). Scanning electron micrographs of quartz sand grains from cold environments examined using Fourier Shape Analysis. *Journal of Sedimentary Petrology*, 52: 1315–1432.

Dowdeswell, J. A., Osterman, L. E., & Andrews, J. T. (1985). Scanning electron microscopy and other criteria for distinguishing glacial and non-glacial episodes in a marine core from Frobisher Bay, Baffin Island, N.W.T., Canada. *Sedimentology*, 32: 119–32.

Ehrlich, R., Brown, P. J., & Yarus, J. M. (1980a). The origin of shape frequency distributions and their relationship between size and shape. *Journal of Sedimentary Petrology*, 50: 475–84.

Ehrlich, R., Brown, P. J., Yarus, J. M., & Eppler, D. T. (1980b). Analysis of particle morphology data. In: *Advanced Particulate Morphology*, eds. J. K. Beddow & T. P. Meloy. Boca Raton, Florida: CRC Press, pp. 101–19.

Ehrlich, R., & Full, W. E. (1986). Comments on "Relationships among eigenshape analysis, Fourier analysis, and analysis of coordinates." *Mathematical Geology*, 18: 855–7.

Ehrlich, R., Kennedy, S. K., & Brotherhood, C. D. (1987). Respective roles of Fourier and SEM techniques in analyzing sedimentary quartz. In: *Clastic Particles*, ed. J. R. Marshall. New York: Van Nostrand Reinhold, pp. 292–301.

Ehrlich, R., & Weinberg, B. (1970). An exact method for the characterization of grain shape. *Journal of Sedimentary Petrology*, 40: 205–12.

Flemming, N. C. (1965). Form and function of sedimentary particles. *Journal of Sedimentary Petrology*, 35: 381–90.

Fong, S.-T., Beddow J. K., & Vetter, A. F. (1979). Refined method of particle shape representation. *Powder Technology*, 22: 1–14

Frisch, A. A., Evans, D., Hudson, J. P., & Boon, J. (1987). Shape discrimination of sand samples using the fractal dimension. In: *Coastal Sediments '87*. ASCE, vol. 1, pp. 138–53.

Full, W. E., & Ehrlich, R. (1986). Fundamental problems associated with "eigenshape analysis" and similar "factor" analysis procedures. *Mathematical Geology*, 18: 451–63.

Full, W. E., Ehrlich, R., & Bezdek, J. C. (1982). FUZZY QMODEL: A new approach for linear unmixing. *Mathematical Geology*, 14: 259–70.

Full, W. E., Ehrlich, R., & Kennedy, S. K. (1984). Optimal configuration and information content of sets of frequency distributions. *Journal of Sedimentary Petrology*, 54: 117–26.

Full, W. E., Ehrlich, R., & Klovan, J. E. (1981). EXTENDED QMODEL – Objective definition of external end members in the analysis of mixtures. *Mathematical Geology*, 13: 331–44.

Glezen, W. H., & Ludwick, J. C. (1963). An automated grain-shape classifier. *Journal of Sedimentary Petrology*, 33: 23–40.

Gotoh, K., Kumamoto, M., & Meloy, T. P. (1984). Particle shape characterization from packing density. In: *Particle Characterization in Technology*, ed. J. K. Beddow. Boca Raton, Florida: CRC Press, pp. 15–22.

Granlund, A. H., & Hermelin, J. O. R. (1983). MIAS – a microcomputer-based image analysis system for micropaleontology. *Stockholm Contributions to Geology*, 39: 127–37.

Griffiths, J. C. (1967). *Scientific Method in Analysis of Sediments*. New York: McGraw-Hill, 508 pp.

Griffiths, J. C., & Smith, C. M. (1964). Relationship between volume and axes of some quartzite pebbles from the Olea Conglomerate, Rock City, New York. *American Journal of Science*, 262: 497–512.

Haines, J., & Mazzullo, J. (1988). The original shapes of quartz silt grains: A test of the validity of the use of quartz grain shape analysis to determine the sources of terrigenous silt in marine sedimentary deposits. *Marine Geology*, 78: 227–40.

Håkanson, L. (1978). The length of closed geomorphic lines. *Mathematical Geology*, 10: 141–67.

Hawkins, A. E., & Davies, K. W. (1984). Sampling powders for shape description. *Particle Characterization*, 1: 22–7.

Hayward, J., Orford, J. D., & Whalley, W. B. (1989). Three implementations of fractal analysis of particle outlines. *Computers in Geosciences*, 15: 199–207.

Heywood, H. (1963). The evaluation of powders. *Journal of Pharmacy and Pharmacology*, 15: 56T–73T.

Hulbe, C. W. H. (1955). Mounting technique for grain size and shape measurement. *Journal of Sedimentary Petrology*, 25: 302–3.

Jarvis, R. S. (1976). Classification of nested tributary basins in analysis of drainage basin shape. *Water Resources Research*, 12: 1151–64.

Kaesler, R. L., & Waters, J. A. (1972). Fourier analysis of the ostracode margin. *Geological Society of America Bulletin*, 83: 1169–78.

Kanatani, K. (1984). Fast Fourier Transform. In: *Particle Characterization in Technology*, ed. J. K. Beddow. Boca Raton, Florida: CRC Press, vol. 2, pp. 31–50.

Kaye, B. H. (1978). Specification of the ruggedness and/or texture of a fine particle profile by its fractal dimension. *Powder Technology*, 21: 1–16.

(1984). Multifractal description of a rugged fine particle profile. *Particle Characterization*, 1: 14–21.

(1986). The description of two-dimensional rugged boundaries in fine particle science by means of fractal dimensions. *Powder Technology*, 46: 245–54.

(1987). Fine particle characterization aspects of predictions affecting the efficiency of microbiological mining techniques. *Powder Technology*, 50: 177–91.

Kaye, B. H., Leblanc, J. E., & Clark, G. (1983). A study of the physical significance of three-dimensional signature of waveforms. *Fine Particle Characterization Conference*, Hawaii, August, 1983.

Kennedy, S. J., & Ehrlich, R. (1985). Origin of shape changes of sand and silt in a high gradi-

ent stream system. *Journal of Sedimentary Petrology*, 5: 57–64.

Kennedy, S. J., & Lin. W.-H. (1986). FRACT – a FORTRAN subroutine to calculate the variables necessary to determine the fractal dimension of closed forms. *Computers in Geosciences*, 12: 705–12.

Kennedy, S. R., Meloy, T. P, & Durney, T. E. (1985). Sieve data – size and shape information. *Journal of Sedimentary Petrology*, 55: 356–60.

Krinsley, D. H., & Doornkamp, J. C. (1973). *Atlas of Quartz Sand Surface Textures*. Cambridge, U.K.: Cambridge University Press, 91 pp.

Krumbein, W. C. (1941). Measurement and geological significance of shape and roundness of sedimentary particles. *Journal of Sedimentary Petrology*, 11: 64–72.

Krygowski, B., & Krygowski, T. M. (1965). Mechanical method of estimating the abrasion grade of sand grains (mechanical graniformametry). *Journal of Sedimentary Petrology* 35: 496–502.

Kuenen, P. H. (1964). Pivotability studies of sand by a shape sorter. In: *Developments in Sedimentology*, ed. L. M. J. U. Van Straaten. Amsterdam: Elsevier, vol. 1, pp. 208–15.

Lisle, R. J. (1988). The superellipsoidal form of coarse clastic sediment particles. *Mathematical Geology*, 20: 879–90.

Lohman, G. P. (1983). Eigenshape analysis of microfossils: a general morphometric procedure for describing changes in shape. *Mathematical Geology*, 15: 659–72.

Ludwick, J. C. (1971). Particle shape classification of thickness using slotted screens. *Journal of Sedimentary Petrology*, 41: 19–29.

Ludwick, J. C., & Henderson, P. L. (1968). Particle shape and inference of size from sieving. *Sedimentology*, 11: 197–235.

Luerkens, D. W., Beddow, J. K., & Vetter, A. F. (1982). A generalized method of morphological analysis (the (R,S) method). *Powder Technology*, 31: 217–20.

(1984). Theory of morphological analysis. In: *Particle Characterization in Technology*, ed. J. K. Beddow. Boca Raton, Florida: CRC Press, vol. 2, pp. 3–14.

Mandelbrot, B. (1967). How long is the coastline of Britain? Statistical self-similarity and fractional dimension. *Science*, 156: 636–8.

(1982). *The Fractal Geometry of Nature*. San Francisco: Freeman, 461 pp.

Marshall, J. R. (ed). (1987). *Clastic Particles*. New York: Van Nostrand Reinhold, 346 pp.

Mazzullo, J., & Anderson, J. B. (1987). Grain shape and surface texture analysis of till and glacial-marine sand grains from the Weddell and Ross Seas, Antarctica. In: *Clastic Particles*, ed. J. R. Marshall. New York: Van Nostrand Reinhold, pp. 314–27.

Mazzullo, J., & Magenheimer, S. (1987). The original shapes of quartz sand grains. *Journal of Sedimentary Petrology*, 57: 479–87.

Mazzullo, J., Sims, D., & Cunningham, D. (1986). The effects of eolian sorting and abrasion upon the shapes of fine quartz sand grains. *Journal of Sedimentary Petrology*, 56: 45–56.

Mazzullo, J., & Withers, K. D. (1984). Sources, distribution and mixing of Late Pleistocene and Holocene sands on the South Texas continental shelf. *Journal of Sedimentary Petrology*, 54: 1319–34.

Medalia, A. J. (1971). Dynamic shape factors of particles. *Powder Technology*, 4: 117–38.

(1980). Three-dimensional shape parameters. In: *Testing and Characterization of Particles and Fine Powders*, eds. J. K. Beddow, & T. P. Meloy. London: Heyden, pp. 66–76.

Meloy, T. P. (1977). Fast Fourier transforms applied to shape analysis of particle silhouettes to obtain morphological data. *Powder Technology*, 17: 27–35.

(1984). Particulate characterization: Future approaches. In: *Handbook of Powder Science and Technology*, eds. M. E. Fayd & L. Otten. New York: Van Nostrand Reinhold, pp. 69–98.

Meloy, T. P., & Makino, K. (1983). Characterizing residence times of powder samples on sieves. *Powder Technology*, 36: 253–8.

Moellering, H., & Rayner, J. N. (1981). The harmonic analysis of spatial shapes using dual axis Fourier shape analysis. *Geographical Analysis*, 13: 64–77.

Orford, J. D. (1981). Particle form. In: *Geomorphological Techniques*, ed. A. S. Goudie. London: George Allen & Unwin, pp. 86–90.

Orford, J. D., & Whalley, W. B. (1983). The use of the fractal dimension to quantify the morphology of irregular–shaped particles. *Sedimentology*, 30(5): 655–68.

(1987). The quantitative description of highly irregular sedimentary particles: The use of the fractal dimension. In: *Clastic Particles*, ed.

J. R. Marshall. New York: Van Nostrand Reinhold, pp. 267–80.

Pang, H. M., & Ridgway, K. (1983). Mechanism of sieving: Effect of particle size and shape. Institute of Chemical Engineering Symposium Series, no. 69 (Powtech, 83), pp. 163–9.

Persoon, E., & Fu, K.-S. (1977). Shape discrimination using Fourier descriptors. *IEEE Transactions on Systems, Man, and Cybernetics*, 7: 170–9.

Piper, D. J. W. (1970). The use of the D-Mac pencil follower in routine determinations of sedimentary particles. In: *Data Processing in Biology and Geology*, ed. J. L. Cutbill. Systematics Association, special vol. 3, pp. 97–103.

Powers, M. C. (1953). A new roundness scale for sedimentary particles. *Journal of Sedimentary Petrology*, 23: 117–19.

Prusak, D., & Mazzullo, J. (1987). Sources and provinces of late Pleistocene and Holocene sand and silt on the mid-Atlantic continental shelf. *Journal of Sedimentary Petrology*, 57: 278–87.

Riester, D. D., Shipp, R. C., & Ehrlich, R. (1982). Patterns of quartz sand shape variation, Long Island littoral and shelf. *Journal of Sedimentary Petrology*, 54: 1307–14.

Rohlf, F. J. (1986). Relationships among eigenshape analysis, Fourier analysis, and analysis of coordinates. *Mathematical Geology*, 18: 845–55.

Rohlf, F. J., & Archie, J. W. (1984). A comparison of Fourier methods for the description of wing shape in Mosquitoes (Diptera: Culicidae). *Systematic Zoology*, 33: 302–17.

Sarvarayudu, G. P. R., & Sethi, I. K. (1983). Walsh descriptors for polygonal curves. *Pattern Recognition*, 16: 327–36.

Scarlett, B. (1985). Measurement of particle size and shape, some reflections on the BCR reference material programme. *Particle Characterization*, 2: 1–6.

Schäfer, A., & Teyssen, T. (1987). Size, shape and orientation of grains in sands and sandstones – image analysis applied to rock thin sections. *Sedimentary Geology*, 52: 251–71.

Schneiderhöhn, P. (1954). Eine vergleichende Studie über Methoden zur quantitativen Bestimmung von Abrundung und Form an Sandkörnern (im Hinblick auf die Verwendbarkeit an Dünnschliffen). *Heidelberger Beitrage der Mineralogie und Petrologie*, 4: 172–91.

Schultz, D. J. (1980). The effect of hydrofluoric

acid on quartz shape – Fourier grain shape analysis. *Journal of Sedimentary Petrology*, 50: 644–5.

Schwarcz, H. P., & Shane, K. C. (1969). Measurement of particle shape by Fourier analysis. *Sedimentology*, 13: 213–31.

Schwarz, H. B., & Exner, H. E. (1980). The implementation of the concept of fractal dimension on a semi-automatic image analyser. *Powder Technology*, 27: 207–13.

Shepard, F. P., & Young, R. (1961). Distinguishing between beach and dune sand. *Journal of Sedimentary Petrology*, 31: 196–214.

Smart, P., & Tovey, N. K. (1981). *Electron Microscopy of Soils and Sediments: Examples.* Oxford, U.K.: Oxford University Press, 178 pp.

Sneed, E. D., & Folk, R. L. (1958). Pebbles in the lower Colorado River, Texas, a study in particle morphogenesis. *Journal of Geology*, 66: 114–50.

Telford, R. W., Lyons, M., Orford, J. D., Whalley, W. B., & Fay, D. Q. M. (1987). A low-cost, microcomputer-based image analyzing system for characterization of particle outline morphology. In: *Clastic Particles*, ed. J. R. Marshall. New York: Van Nostrand Reinhold, pp. 281–9.

Tilmann, S. E. (1973). The effect of grain orientation on Fourier shape analysis. *Journal of Sedimentary Petrology*, 43: 867–9.

Wadell, H. (1935). Volume, shape and roundness of quartz particles. *Journal of Geology*, 43: 250–80.

Wagoner, J. L., & Younker, J. L. (1982). Characterization of alluvial sources in the Owens Valley of eastern California using Fourier shape analysis. *Journal of Sedimentary Petrology*, 52: 209–19.

Wentworth, C. K. (1922). The shape of pebbles. *Bulletin of the U.S. Geological Survey*, 730C: 91–114.

(1933). Shapes of rock particles, a discussion. *Journal of Geology*, 41: 306–9.

Whalley, W. B. (1972). The description of sedimentary particles and the concept of form. *Journal of Sedimentary Petrology*, 42: 961–5.

(1978a). An SEM examination of quartz grains from subglacial and associated environments and some methods for their characterization. *Scanning Electron Microscopy*, 1978/I: 335–60.

(ed.) (1978b). Scanning Electron Microscopy in the Study of Sediment. *Geo-Abstracts*, Norwich: U.K., 414 pp.

Whalley, W. B., & Orford, J. D. (1982). Analysis of SEM images of sedimentary particle form by fractal dimension and Fourier analysis methods. *Scanning Electron Microscopy*, 1982/II: 639–47.

(1986). Practical methods for analysing and quantifying two-dimensional images. In: *The Scientific Study of Flint and Chert*, eds. G. de G. Sieveking & M. B. Hart. Cambridge, U.K.: Cambridge University Press, pp. 235–42.

(1989). The use of fractals and pseudo-fractals in the analysis of two-dimensional outlines: Review and further exploration. *Computers in Geosciences*, 15: 185–97.

Whiteman, M., & Ridgway, K. (1988). A comparison between two methods of shape-sorting particles. *Powder Technology*, 56: 83–94.

Winkelmollen, P. J. F. (1971). Rollability, a functional shape property of sand grains. *Journal of Sedimentary Petrology*, 41: 703–14.

(1982). Critical remarks on grain parameters, with special emphasis on shape. *Sedimentology*, 29: 255–66.

Zahn, C. T., & Roskies, R. Z. (1972). Fourier descriptors for plane closed curves. *IEEE Transactions Computers*, C21: 269–81.

8 Electroresistance particle size analyzers

T. G. MILLIGAN AND KATE KRANCK

Introduction

Electroresistance particle size analyzers (e.g., the Coulter Counter or the Elzone [Particle Data Inc.]) have long proven suitable for grain size analysis of sediment in the size range 0.5–1,000 μm. Very small amounts of material are required to perform a highly accurate size analysis based on the volume of the individual particles in the sample. The same analytical method can be applied both to unconsolidated bottom sediments and to material sampled from a suspension, thus providing a single method of analysis that can be used to study sediment from source materials to deposition.

This chapter describes the principles of electroresistance counters and discusses the techniques developed by the Sediment Dynamics Lab, Physical and Chemical Sciences Branch, Department of Fisheries and Oceans, Bedford Institute of Oceanography for the analysis of naturally occurring sediments of divers origin. It does not discuss detailed operation and calibration of the instrument or its accuracy and precision; such information is available in manufacturers' handbooks and in earlier works on electroresistance techniques (see, e.g., Coulter, 1957; Berg, 1958; Wales & Wilson, 1961, 1962; Sheldon & Parsons, 1967; Allen, 1968; Walker & Hutka, 1971; McCave & Jarvis, 1973; Shideler, 1976; Kranck & Milligan, 1979).

Our methods were designed for a model TAII Coulter Counter, with population control accessory, interfaced to an HP85 microcomputer using software developed in our laboratory (Fig. 8.1). Interfacing of the multichannel particle analyzers with computers has now become standard, and manipulation of the raw data has been facilitated (see, e.g., Muerdter et al., 1981). The techniques described can be used with other counters when used in conjunction with the correct operating procedures for the particular instrument.

Operational principle

Electroresistance particle size analyzers determine the number and volume of particles held in an electrolytic suspension. A dilute suspension of sediment is stirred and drawn by means of a vacuum through a small aperture of known diameter. A constant current is maintained between two electrodes located on either side of the aperture. As each particle is drawn through the aperture it displaces its equivalent volume of electrolyte, and the particles are detected by the fluctuations in impedance. The changes in impedance are detected as voltage pulses proportional to the volume of the particle. The pulses are amplified, sized, and counted according to a preset number of equivalent spherical diameter size classes or channels, which are calibrated with particles of known diameter (Coulter Electronics Inc., 1979).

For the analysis of sediment, multichannel instruments use logarithmic size classes in which the diameter in each channel is based on an expansion of $2^{i/n}$, where i is the channel number and n is the phi interval. The sixteen-channel TAII Coulter Counter used in our studies has a one-third-phi interval. Most instruments have two possible methods to express the results of an analysis:

1. *percent volume mode:* the normalized volume fraction in each channel calculated as a percent of the total volume accumulated within the range analyzed by a given aperture; and

2. *population mode:* the number of discrete particles counted in each channel.
In the TAII Coulter Counter an aperture tube is theoretically capable of analyzing particles 2%–50% of its nominal diameter. This limit has been extended in the new Coulter Multisizer. However, the usable size range of an aperture tube is in fact reduced due to electrical noise often occurring in the small-diameter channels and low concentrations of large particles in the upper channels.

Electroresistance size analysis has also been combined with a fluorescence analyzer in the FACS from Becton–Dickinson. This instru-

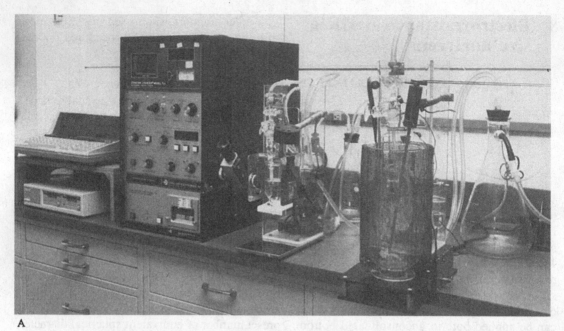

A

Figure 8.1. (A, *above*) Photograph of Model TAII Coulter Counter showing large and small glassware units and HP85 microcomputer. (B, *facing*) Close-up of large glassware unit with shielding removed showing vacuum gauges, remotely powered stirrer, 2,000-ml round-bottom beaker, and the drop-away platform.

ment's prime role is in the analysis of biological material, and the small sample size and dynamic range make it of limited use in sediment studies at this time. Other applications have been in the in situ size analysis of suspended material in the oceans, such as described by Dauphinee (1977) and Herman & Dauphinee (1980).

Coincidence

The concentration of particles suspended in the electrolyte must be such as to preclude two particles occurring in the sensing zone at the same time. This condition, known as *coincidence,* can be avoided by observing the allowable limits in counts per second for each aperture. These limits are listed in the operating manuals or can be calculated using the formula described by McCave & Jarvis (1973). Exceeding coincidence limits will cause data loss in the small size classes, thereby narrowing the observed distribution (Fig. 8.2).

The population distribution of a sample decreases logarithmically as size increases; hence

the number of particles in the large size classes will be extremely low when correct coincidence levels are maintained. To avoid poor analytical resolution in these size ranges, multiple-tube and screening techniques, along with long counting times, must be used to size statistically significant numbers of these particles. The difficulty in sizing a sufficient number of particles in the large-diameter size classes is multiplied as the number of channels increases. This is a critical problem with the 128- and 256-channel instruments.

Electrolyte

The electrolyte used for resuspension and dilution of samples can be varied to suit the material being analyzed, provided the resistance through the aperture is within the limits set for it. Seawater, 6% $(NaPO_3)_6$, 3% NaCl, and commercially available solutions, such as Isoton, can be used in the analysis of sediments of up to 100-μm equivalent diameter. Larger grains must be suspended in more viscous solutions. For the techniques described, a solution of 3% NaCl by weight in water is used for apertures of up to 400 μm, and 30% glycerol by volume in a 1% NaCl solution is used for sediments of greater diameter. A recirculating filtration system may be used to provide an adequate supply of particle-free 3% NaCl. The glycerol

B

Figure 8.1 (cont.)

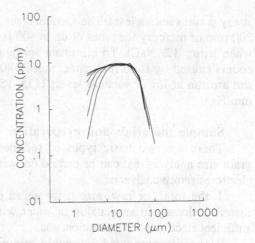

Figure 8.2. Inorganic suspended-sediment size spectra showing the effect of coincidence on the size distribution of a marine glacial clay. Concentration doubles in each curve from top to bottom.

solution can be reused by vacuum filtering it through an 8-μm nominal pore size Millipore filter between samples.

Modifications to glassware

To use electroresistance analyzers for the full range of sizes 0.5–1,000 μm, alterations to the standard glassware units are recommended. The instruments were designed to accurately size particles with narrow distributions, suspended in a small volume of electrolyte. Due to the broad range of particle sizes that occur in sediments, counting times of up to 5 min and over a litre of electrolyte are needed to count sufficient numbers of particles when using the largest-aperture tubes.

Analysis of sediment in the 100–1,000-μm size range is carried out using the standard large apertures available mounted on a glassware unit modified to accommodate the large volumes of suspension required. Large glassware units are now available from some manufacturers. Kranck & Milligan (1979) also describe a large glassware unit for use with 560–2,000-μm apertures. This unit accommodates 2,000-ml round-bottom beakers and consists of a specially designed drop-away platform, remotely powered large stirrer, and appropriate shielding (Fig. 8.1A).

The mercury manometer present in most instruments was designed to sample small preset volumes of suspension and to monitor the vacuum level when the time-sampling mode was selected. For sediment analysis in our laboratory, the manometer has been removed from the Coulter glassware unit and replaced by a vacuum gauge. The connection to the manometer has been sealed by reforming the glass, but this can also be accomplished by clamping a short length of flexible tubing connected in place of the manometer. The vacuum gauge allows precise calibration of the flow rate for each aperture at any preset vacuum level. The vacuum is set at 200 mm of mercury for tubes of up to 400 μm when using 3% NaCl. To eliminate spurious counts caused by fluid resonance, tubes >400 μm are run at lower vacuum levels (125–150 mm Hg).

Sample materials and preparation

There are two basic types of sediment grain size analysis that can be carried out with electroresistance analyzers:

1. the total, or raw, size distribution of material already in suspension in water with sufficient electrolyte concentration; and

2. the disaggregated inorganic mineral grain size distribution of either unconsolidated bed sediment or suspended sediment concentrated by filtration.

Total suspended sediment

Several recent studies have pointed out the problems involved with the analysis of suspended sediment by methods other than in situ techniques (Kranck & Milligan, 1980; Gibbs, 1981; Eisma, 1986; Bale & Morris, 1987; Kranck & Milligan, 1988; see also Chapters 14 and 15 in this volume). The fragile, ephemeral nature of flocs present in water does not allow a true particle size spectrum to be obtained by using electroresistance counters. A pseudosize distribution for the material in suspension can be obtained provided the electrolyte concentration is within acceptable limits for the instrument. The shape of the resultant distribution depends on the way the sample was handled and the aperture with which it was counted. Kranck & Milligan (1979) show that increased stirrer speed during counting results in a decrease in the modal size of the distribution (Fig. 8.3). The size distributions of material in suspension, obtained by image analysis of photos taken using a Benthos 373 Plankton Camera, differ drastically from those obtained using a Coulter Counter (see also Chapter 15). For example, the in situ modal size of samples from St. John Harbour, N.B., is an order of magnitude greater than that obtained using the Coulter Counter (Fig. 8.4). Bale & Morris (1987) found similar results using a submersible laser diffraction instrument in the Tamar Estuary.

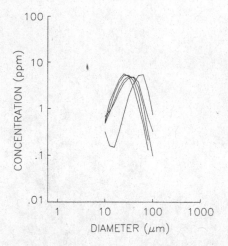

Figure 8.3. Total suspended-sediment size spectra showing how the size distribution of artificially created flocs changes due to stirring. Stirring time increases from right to left.

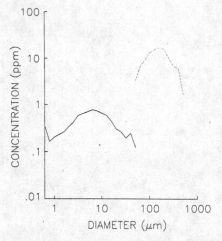

Figure 8.4. Total suspended-sediment size spectra of a sample from Saint John Harbour, N.B., using the Coulter Counter (solid) and image analysis of the in situ photograph from the same station and depth (dashed).

Counting of raw water samples has been used extensively in the study of plankton and the description of bodies of water using particle size distributions. The ease of obtaining particle size spectra using this method and its ability to count individual plankton cells make it a useful tool in the study of material in suspension, provided that the resultant spectra are not considered to be the true in situ size distributions.

Inorganic suspended sediment

For inorganic suspended sediment analysis, a known volume of water is filtered through a preweighed filter, and the inorganic sediment is recovered by oxidizing the filter and associated organic material. Sediments are collected on preweighed 8.0-μm nominal pore size Millipore SCWP filters, or equivalent cellulose acetate filters, using standard gravimetric methods (Winneberger et al., 1963). It has been shown that this type of filter traps particles far below its nominal pore size (Sheldon, 1978; Kranck & Milligan, 1979), and the increased loading capacity allows a greater amount of sediment to be collected. Eight-micron filters are appropriate in coastal and estuarine water where sediment concentrations are high. In waters with low suspended particulate matter (SPM) concentrations, 0.8-μm HAWP Millipore filters or equivalent are used.

After reweighing the loaded filters to obtain the SPM concentration, the filters are then ashed using a low-temperature asher (LTA-504, LFE Corp. Waltham, Mass., USA), which oxidizes the filter and most of the organic material. In this type of instrument, samples are ashed at <60 °C, which is sufficiently low to prevent fusing together of the mineral grains. The samples are then treated with an excess of 30% hydrogen peroxide to remove any remaining organic residue. An ash weight can be obtained if preweighed beakers are used in this procedure. The sample is then resuspended in a known volume of electrolyte in preparation for analysis.

By combining the results from both the total and inorganic suspended sediment analyses, other properties such as wet density can be obtained. Gartner & Carder (1979) describe a method to determine the specific gravity of particles in suspension by comparing the total particulate volume with the dry weight of sample.

The 0.4-μm Nucleopore membrane filters frequently used in SPM analysis have also been used in inorganic grain size analysis. In this case the material is removed from the filter before digestion with hydrogen peroxide (Tramontano & Church, 1984). However, it has been found that some material remains on the filter after dispersal, and in areas with low SPM concentrations the proportion of material that remains on the filter could become significant. With estuarine samples, Tramontano & Church report recoveries of >97%.

Bed sediments

Bed sediments – with equivalent diameters of up to 1 mm, and in which individual grains may be separated and then suspended in electrolyte – can be examined using electroresistance analyzers. Sources can range from individual laminae in fine-grained turbidites, sampled by scalpel, to the <1-mm fraction of weathered material from scree slopes. For an equivalent weight-percent size distribution, small subsamples of material are digested in an excess of 30% hydrogen peroxide, dried, split, and weighed. Final weights can vary from 0.05 g in fine sediments to 0.5 g in coarse sediments (McCave & Jarvis, 1973). In very finely divided sediment with low organic content, a small amount of dry sediment is weighed and then digested. When material larger than the maximum size for the analysis is present, the percent greater than that maximum is determined by sieving and then used to correct the total concentration, so that the data are still expressed in terms of the total sample. To obtain a percent total volume distribution for muddy bottom sediments, a known volume of sediment is sampled and suspended in particle-free water. Volumetric subsamples are then removed for digestion in hydrogen peroxide. Prepared samples are suspended in particle-free water and then subsampled and suspended in either 3% NaCl or glycerol.

Analytic procedures

Dilution

Dilution of raw samples for total suspended sediment analysis is usually not required in most waters. However, in areas of high particle concentration where coincidence levels are exceeded, known volumes of filtered 3% NaCl solution should be used to dilute the samples to correct levels.

To resuspend and adjust the concentration in inorganic and bed sediment analyses, aliquots extracted and volumes of electrolyte added are measured by weight, using tared beakers on a top-loading balance. This allows dilutions to be performed quickly and accurately. Samples are first suspended in either electrolyte or particle-free water and subsampled by pipette while being stirred. Inorganic grains >100 μm in diameter are difficult to keep in suspension, and aliquots by pipette cannot be taken accurately for counting. In this case pipette subsamples are taken for the small-diameter tubes (<200 μm), and the remaining sample is screened as described below. The dilution values for each sample are recorded as final volume/sample volume and used to calculate the equivalent final volume of the suspension analyzed.

Disaggregation

Disaggregation of inorganic suspended and bed sediments is required to separate the material into individual grains prior to analysis on the counter. After resuspending a sample in electrolyte, a sapphire-tipped ultrasonic probe (Heat Systems Ultrasonics Inc., Farmingdale, N.Y.) is used to deflocculate the suspension. Although disintegration of the probe tip introduces contaminating particles into the sample, it has proved to be more efficient in the disaggregation of sediments than ultrasonic baths (Rendigs & Commeau, 1987). Rendigs & Commeau also showed that the contamination of samples using a sapphire-tipped horn has no significant effect on the sediment distribution in fine-grained marine muds. A blank sample of sonified clean electrolyte is counted, and the number of particles in each channel introduced by the probe is subtracted from the total counts accumulated in each channel of the sample prior to calculating the concentration. The number of particles subtracted is adjusted for dilution with clean, unsonified electrolyte and the counting time for the sample. The proportion represented by the contamination will vary with the sample concentration. With the sapphire-tipped horn, the variation in contamination among samples treated for 3 min and analyzed using a 200-μm aperture tube is 0.89% of the population in each channel when maximum concentration levels are used. Each new sample is initially sonified for up to 4 min prior to counting with the first aperture of the series, and examined using an inverted microscope to verify that disaggregation is complete. The length of time required to disaggregate a sample varies but, in general, will be constant for samples from the same environment. Approximately 1 min is required to eliminate bubbles from normal-viscosity suspensions before counting. After initial disaggregation and analysis using the first aperture, the original sample suspensions are resonified for 30 s prior to counting on each subsequent aperture.

Screening

Screening of samples is often required to prevent the blockage of an aperture during counting by particles greater than its diameter. If screening of the sample is required, the selected screen size should be half the difference between the size of the aperture and the maximum size of particles it can analyze (Sheldon & Parsons, 1967).

The exponential nature of the population distribution of sediment samples means that the greatest influence on the coincidence level comes from the smallest particles. Screening samples to remove small particles from the distribution analyzed by the largest-diameter aperture makes it possible to reduce substantially the dilution required to reach correct coincidence levels. This procedure effectively increases the number of large particles counted. Screen size will depend on the region of overlap with the smaller-diameter aperture tube and should allow at least five unaffected channels to overlap (Fig. 8.5). When using this enhancement procedure, aliquots for the small-diameter tubes are removed from the suspension first. In the analysis of <100-μm-material, pipette draw-offs from a stirred suspension can be used for screening. For material >100 μm all the remaining suspension material is screened and resuspended in glycerol. This eliminates the problems normally encountered with subsampling coarse particles from an aqueous suspension. When the population method is used to calculate the concentration, the results are unaffected by this technique. It cannot be used with the percent total volume mode.

Counting

When performing sediment analysis using electroresistance particle sizers, it is most efficient to count all samples on one aperture at a time, reducing the amount of time spent chang-

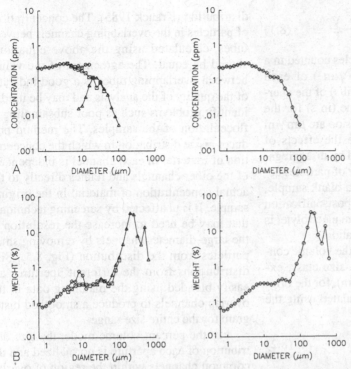

Figure 8.5. Raw and edited particle size spectra from Minas Basin, N.S. (A) Inorganic suspended sediment showing three-tube analysis: 30-μm (circles), 200-μm (squares), and 400-μm (triangles). (B) Bottom sediment from the same station using 1,000-μm tube (triangles). Note the low values for the first channels of the largest aperture, which result from screening.

ing tubes. If a blank sample is used, this beaker is counted first so that it can be subtracted from each subsequent sample. In the newer microprocessor-based instruments, a blank-subtraction function has been built in. However, in the case of the contamination from the ultrasonic probe, these instruments are unable to adjust the blank values to account for any dilution of the samples with particle-free electrolyte after sonification. Modification to the program is required to correct for different dilutions and counting times.

For inorganic suspended and bed sediments each sample is sonified, checked for disaggregation, then counted for a short period (e.g., 5 s) to ensure that coincidence levels are correct. If the coincidence values are exceeded, the dilution required to meet the correct levels can be easily calculated from the total number of particles accumulated. When analyzing coarse sediment with large apertures, we have found it advantageous to switch the reservoir and vacuum lines on the glassware unit so that particles are removed up the narrow-diameter filling tube as soon as they pass through the sensing zone. Samples are otherwise counted in accordance with the methods in the operator's manuals.

Expression of results

The standard output of most instruments, including the model TAII Coulter Counter, is the percent volume mode displayed as either differential or cumulative frequency versus log diameter. This method of expressing the results is appropriate for industrial applications where the entire size distribution of the material being analyzed lies within the range of a single aperture tube. For the analysis of naturally occurring sediments, whose size distributions normally extend over the range of several different apertures, we have found it more suitable to use the population mode to calculate the concentration of material in each size-class channel in relation to the volume of the original sample. In each channel, the concentration of particles measured C_m is expressed by:

$$C_m = \frac{\text{volume in channel}}{\text{volume sampled}} = \frac{N * V}{F * T} \qquad (8.1)$$

where N is the number of particles counted in a channel, V is the volume (in μm^3) of each particle, F is the flow rate (in ml/s) of the aperture, and T is the counting time (in s) for the sample. The units of the expression are $\mu m^3/ml$ or 1 part per 10^{12}. To remove the effects of noise and contamination caused by disaggregation of the samples, the number of particles accumulated in each channel of a blank sample, adjusted for counting time and postsonification dilution of the sample with clean electrolyte, is subtracted from N before calculation.

For suspended sediments, the volume concentration of particles in each size class, expressed as parts per million (ppm), for the original water volume (V_o) is calculated using the equation:

$$C_{ppm} = C_m * \frac{V_f}{V_o} * 10^6 \qquad (8.2)$$

V_f is the equivalent final volume of electrolyte calculated from the dilutions required to reach coincidence levels before counting.

For unconsolidated bed sediments the total sample volume V_o is calculated from a known weight of sample using an appropriate specific gravity. The volume of particles in each channel is then related to this volume and an equivalent weight-percent in each size class, similar to conventional sediment analysis, is calculated. This method gives an indication of the amount of material that occurs beyond the range of the analysis. Alternatively, a measured volume of sediment can be used for V_o to calculate particle concentrations as a percent of the total volume of sediment. This method gives a rough approximation of the amount of pore space in the sample. For these calculations the equation becomes:

$$C_\% = C_m * \frac{V_f}{V_o} * 10^{10} \qquad (8.3)$$

Because of the wide range of particle sizes that occurs in natural sediment distributions, three or four aperture tubes are required to obtain a complete and accurate size analysis of both the well-sorted and fine-tail portions of the distribution (Kranck 1985). The concentrations of particles in the overlapping channels between tubes, calculated using the above equations, should be equal. The agreement of the values between overlapping tubes is a good indicator of the quality of the analysis and may be used to identify problems such as poor subsampling or flocculation of the samples. The method produces a size distribution in which the concentration of material in each channel is independent of the other channels and relates directly to the actual concentration of material in the original sample. It is unaffected by screening techniques that may be used to increase the resolution in the large-diameter channels by removing small particles from the distribution (Fig. 8.5). Size distributions from the different apertures are easily blended using the matching data in the overlap channels to produce a smoothed histogram for the entire size range.

In the percent volume mode, the size distribution of each aperture is normalized and the common channels within the region of overlap will not be equal. The percent volume mode of the TAII, and of many other counters, can also lose data from channels with low concentrations when a single dominant mode in a distribution accounts for most of the volume. This further complicates merging of the tube data when the overlapping channels do not occur on the peak.

Results from our analyses are displayed as log-log volume frequency distribution spectra. This form of presentation is used rather than the cumulative size frequency or semilog differential histogram, to allow direct comparison of different samples with similar relative distributions over parts of their size range (Kranck, 1980; Kranck & Milligan, 1980; Hunt, 1982). These plots also emphasize the basic exponential nature of sediments as shown by Bagnold (1937, 1941), Bader (1970), Bagnold & Barndorff-Nielsen (1980), and others (see also Chapter 17).

Discussion

Numerous studies have been carried out using electroresistance counters in both field and laboratory settings. The instruments provide rapid (up to twenty samples per day) analysis of material from almost any source. They have been available for the grain size analysis of sediments for many years but have often been set aside for new and more complex instruments that, in fact, do not perform as well when analyzing sediments with anything but well-sorted distributions (see McCave et al., 1986; Singer et al., 1988). Perhaps the most difficult material to analyze is a till in which the slope of the distribution is constant in all size classes from submicron to several millimetres [*sic*]. In this case, methods such as laser diffraction, which assume the material being analyzed conforms to a known distribution, will distort the distribution to obtain a best fit. This results in a false mode being generated in a flat distribution.

The procedures used for obtaining accurate grain size spectra for sediments using electroresistance particle sizers depend on precise sampling and careful preparation of the material for counting. Errors can usually be attributed to sample handling rather than to the instruments. The instruments do, however, require considerable knowledge and care on the part of the operator to ensure that the results are reliable.

REFERENCES

Allen, T. (1968). The Coulter Counter. In: *Particle Size Analysis*. London: Society of Analytical Chemistry, 143–53.

Bader, H. (1970). The hyperbolic distribution of particle sizes. *Journal of Geophysical Research*, 75(15): 2831–55.

Bagnold, R. A. (1937). The size grading of sand by wind. *Proceeds of the Royal Society of Australia*, 163: 250–64.

(1941). *The Physics of Blown Sands and Desert Dunes*. London: Methuen.

Bagnold, R. A., & Barndorff-Nielsen, O. (1980). The pattern of natural size distributions. *Sedimentology*, 27: 199–207.

Bale, A. J., & Morris, A. W. (1987). In situ measurement of particle size in estuarine waters. *Estuarine and Coastal Shelf Science*, 24: 253–63.

Berg, R. H. (1958). Electronic size analysis of subsieve particles by flowing through a small liquid resistor. *ASTM Special Technical Publication*, 234: 245–58.

Coulter, W. H. (1957). High speed automatic blood cell counter and cell size analyzer. *Proceeds of the National Electronics Conference*, 1034–42.

Coulter Electronics Inc. (1979). *Coulter Counter model TAII operator's manual*. PN 4201044, Hialeah, Florida: Coulter Electronics Inc.

Dauphinee, T. M. (1977). Zooplankton measurements using a conductance cell. MTS–IEEE, Oceans '77 Conference Record, Session 39B, pp. 1–5.

Eisma, D. (1986). Flocculation and deflocculation of suspended matter in esuaries. *Netherlands Journal of Sea Research*, 20: 183–9.

Gartner, J. W., & Carder, K. L. (1979). A method to determine specific gravity of suspended particles using an electronic particle counter. *Journal of Sedimentary Petrology*, 49: 631–3.

Gibbs, R. J. (1981). Floc breakage by pumps. *Journal of Sedimentary Petrology*, 51: 670–2.

Herman, A. W., & Dauphinee, T. M. (1980). Continuous and rapid profiling of zooplankton with an electronic counter mounted on a "Batfish" vehicle. *Deep–Sea Research*, 27: 79–96.

Hunt, J. R. (1982). Self-similar particle size distributions during coagulation: Theory and experimental verification. *Journal of Fluid Mechanics*, 122: 169–85.

Kranck, K. (1980). Experiments on the significance of flocculation in the settling of fine grained sediment in still water. *Canadian Journal of Earth Science*, 17: 1517–26.

(1985). Origin of grain size spectra of suspension deposited sediment. *Geo–Marine Letters*, 5: 61–6.

Kranck, K., & Milligan, T. (1979). The use of the Coulter Counter in studies of particle-size distributions in aquatic environments. *Bedford Institute Report Series BI–R*, 79: 7.

(1980). Macroflocs: Production of marine snow in the laboratory. *Marine Ecology Progress Series*, 3: 19–24.

(1988). Macroflocs from diatoms: In situ photography of particles in Bedford Basin, Nova Scotia. *Marine Ecology Progress Series*, 44: 183–9.

McCave, I. N., Bryant, R. J., Cook, H. F., & Goughhanowr, C. A. (1986). Evaluation of a laser-diffraction size analyzer for use with natural sediments. *Journal of Sedimentary Petrology*, 56: 561–4.

McCave, I. N., & Jarvis, J. (1973). Use of the model T Coulter Counter in size analysis of fine to coarse sand. *Sedimentology*, 20: 305–15.

Muerdter, D. R., Dauphin, J. P., & Steele, G. (1981). An interactive computerized system for grain size analysis of silt using electroresistance. *Journal of Sedimentary Petrology,* 51: 647–50.

Rendigs, R. R., & Commeau, J. A. (1987). Effects of disaggregation on a fine-grained marine mud by two ultrasonic devices. *Journal of Sedimentary Petrology,* 57: 786–7.

Sheldon, R. W. (1978). Sensing zone counters in the laboratory. In: *Phytoplankton Manual,* ed. A. Sournia. UNESCO Monographs in Oceanographic Methodology, Paris.

Sheldon, R. W., & Parsons, T. R. (1967). *A Practical Manual on the Use of the Coulter Counter in Marine Science.* Toronto: Coulter Electronics, 66 pp.

Shideler, G. L. (1976). A comparison of electronic particle counting and pipette techniques in routine mud analysis. *Journal of Sedimentary Petrology,* 46: 1017–25.

Singer, J. K., Anderson, J. B., Ledbetter, M. T., McCave, I. N., Jones, K. P. N., & Wright, R. (1988). An assessment of analytical techniques for the analysis of fine-grained sediments. *Journal of Sedimentary Petrology,* 58: 534–43.

Tramontano, J. M., & Church, T. M. (1984). A technique for the removal of estuarine seston from Nucleopore filters. *Limnology and Oceanography,* 29(6): 1339–41.

Wales, M., & Wilson, J. N. (1961). Theory of coincidence in Coulter Particle Counters. *Review of Scientific Instrumentation,* 32: 1132–6.

(1962). Coincidence in Coulter Counters. *Review of Scientific Instrumentation,* 33: 575–6.

Walker, P. H., & Hutka, J. (1971). Use of the Coulter Counter (model B) for particle size analysis of soils. Division of Soils Technical Paper no. 1, Commonwealth Science and Industrial Research Organization, Australia.

Winneberger, J. H., Austin, J. H., & Klett, C. A. (1963). Membrane filter weight determination. *Journal of the Water Pollution Control Federation,* 35: 807–13.

9 Laser diffraction size analysis

Y. C. AGRAWAL, I. N. McCAVE,
AND J. B. RILEY

Introduction

Several authors in the 1970s suggested ways of inferring the size distributions of fine particles from the angular distribution of the intensity of forward-scattered coherent light (Cornillault, 1972; Weiss & Frock, 1976; Swithenbank et al., 1977). This work led to a first generation of commercial laser diffraction sizers from CILAS (France), Leeds and Northrup (USA), and Malvern Instruments (UK). Subsequently several more manufacturers have produced instruments – Horiba, Coulter, and Fritsch – while the original makers have improved their products. The original machines used the Fraunhofer diffraction approximation which de Boer et al. (1987) point out is only applicable to particles that are large relative to the wavelength of light. McCave et al. (1986) reported rather poor performance of one instrument at the lower end of the size range with samples containing a significant clay fraction. Nowadays instruments are mostly (not all) using a combination of Fraunhofer and the full Mie scattering theory, which deals properly with fine particles. However, many instruments still have detectors with relatively few elements (e.g., 18 in one machine) whose results after processing may be presented as multielement histograms (e.g., up to 56 bars) conferring a somewhat spurious air of precision to multimodal size distributions. However, the method is fast, reproducible, nondestructive for weak particles (e.g., flocs), and is constantly being improved.

One potential advantage of the method that has not yet been much exploited is that suspended particles can be sized in situ in the air, sea, or rivers. Bale & Morris (1987) have packaged a Malvern instrument for use in the sea and Agrawal & Pottsmith (1989) have constructed a prototype instrument for unattended long-term observation of particle size spectra near the sea floor. This system, being specifically designed for the sea is more flexible and can be programmed to record for a long time. A future development will be a device that encloses a volume of water with its suspended particles and allows them to settle, from which the change in concentration and size distribution as a function of time would allow the settling velocity distribution *and* the particle density distribution to be derived. For natural suspensions of aggregated fine particles this would be a real advance. In the laboratory the main analytical problem lies in sizing multimodal, broad-spectrum size distributions with significant clay (i.e., muds with sand).

In this chapter we outline the theory, instruments, problems, and future of laser diffraction size analysis with particular reference to geological and natural environmental materials.

Principles of laser diffraction size analysis

Sizing

The scattering by spheres at small angles is nearly equivalent to diffraction by apertures of equal diameter (see Riley, 1987). Under this approximation the scattering due to a single particle of radius a (or, for nonspheres, of equivalent radius a) at an angle θ takes the following form (see Fig. 9.1):

$$I(\theta) = Ca^4[J_1^2(ka\theta)/(ka\theta)^2] \qquad (9.1)$$

where $k = 2\pi/\lambda$, λ is the wavelength of light, C is a constant of proportionality (which we shall set to 1 for simplicity), and J_1 is the first-order Bessel function of the first kind.

In the presence of a continuum of particulate sizes, equation (9.1) is modified to

$$I(\theta)= \int n(a)\, a^4[J_1^2(ka\theta)/(ka\theta)^2]\, da \qquad (9.2)$$

Referring to Figure 9.1, $I(\theta)$ is the intensity observed in the focal plane of the lens, of focal length f, and θ is the angle at which the intensity is measured, given by y/f, where y is the distance from the lens axis to a point in the focal plane. The size distribution function $n(a)$ is defined from the number of particles in size range a and $a + da$ being $n(a)da$. To obtain $n(a)$ from

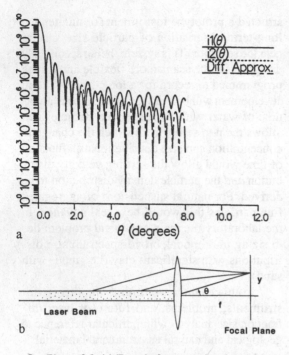

a

b   **Laser Beam**

Focal Plane

Figure 9.1. (a) Fraunhofer approximation (broken line) compared with exact Mie formulation. Scattering intensity plotted against angle. (b) Geometry defining scattering angle.

observations of $I(\theta)$, one needs to invert equation (9.2).

While it is not our intent to make a detailed survey of the methods of inverting equation (9.2) to obtain $n(a)$, we will present an essential discussion to develop a physical understanding of the inversion process. The purpose of this discussion is to include an understanding of the information content of the data, especially to illuminate why only a limited amount of size data can be extracted from the observations. We feel this discussion is warranted by a number of apparently irreconcilable results in information extracted from similar data by various investigators.

Inverse theory

Equation (9.2) is a Fredholm's integral equation of the first kind. An analytical solution to it exists and was reported by Chin et al. (1955). Whereas the analytical solution has much to recommend it, its use is very limited due to the lack of a constraint ensuring positivity of $n(a)$. Much physical insight was, howev-

er, obtained into the inverses by Fymat & Mease (1978) and Agrawal & Riley (1984) by looking at the asymptotic properties of the analytic inverse. For example, Agrawal & Riley defined a "Nyquist size" – an upper limit to the size of observable particles and the theoretical limit to the resolution of particle sizes from noise-free data over a given range of angles. These results followed from noting the similarity of the analytical inverse to the Fourier sine transform and the well-known properties of Fourier transforms. The introduction of noise modifies these conclusions in an as yet unquantified manner. However, it appears that there is renewed faith in this approach. For example, recently Bayvel et al. (1987) analyzed data from a Malvern instrument and reported that the results from this analytic method were superior to those from the "black box" algorithm of Malvern, which is presumably based on matrix inversions (see below). Their conclusion was not discussed in the light of the effects of measurement noise in the inversion, particularly bearing in mind that the noise is amplified by the third power of the scattering angle.

The practical situation is usually one where the intensity data may be available over some range of angles θ_{min} and θ_{max}. The data are contaminated by noise from sources such as ambient light, electronics, and statistical fluctuations in the number of particles in the sample volume, as well as from laser speckle. One then asks the question: Given a data set, how much information can I extract? Specifically, what is the number of sizes at which $n(a)$ may be estimated (i.e., resolution), and how sensitive is the inversion to noise in the data (the mathematical terminology: "uniqueness" and "stability")? The subject is a continued topic for mathematical research. Here, we address the questions in a mathematical-intuitive way for the scientist who may not be an expert inverse-theorist.

One begins with rewriting equation (9.2) in its finite-sum form:

$$I(\theta_i) = \sum a_j^4 n(a_j) J_1^2 (ka_j\theta_i)^2 - w(a_j) \qquad (9.3)$$

where w is a weight factor depending upon the quadrature rule applied for integration. Rewriting equation (9.3) in matrix form, we have

$$I = K \cdot N \qquad (9.4)$$

where I is the observation vector, K is the kernel represented by the right-hand side of equation (9.1), and N is the size distribution vector, discretized from $n(a)$, to be recovered. Weight factors w_j are set to 1 throughout on the premise that the errors due to numerical quadrature are not serious in this case. Evidently, if the matrix equation can be inverted stably, N can be recovered. In general, however, nature appears determined not to present physical problems with matrices of such desirable properties. The matrix K in its simplest form is also nearly singular, or equivalently, ill conditioned.

In order to invert equation (9.4), the properties of the kernel matrix K may be made "desirable" to some extent by choosing the kernel sample points judiciously. One may ask how the matrix K may be constructed to obtain the most noise-insensitive inverses. A brief foray into the world of linear algebra is inevitable.

We note that the finite-sum form, equation (9.4), represents a series of linear algebraic equations. The intensity $I(\theta)$ at each angle θ is obtained as the sum of contributions from all sizes. Algebraically, this is obtained from the product of a row of the matrix K with the column vector N; that is,

$$I_i = \sum_j K_{ij} N_j \qquad (9.5)$$

where the index i corresponds to the angle θ and j to size a. To help visualize, we have shown the kernel plotted in Figure 9.2 for two values of i. Also shown is a fictitious size distribution curve $n(a)$. It is possible now to see that according to equation (9.4) the kernel K_{ij} samples a part of the size function, weighting some parts highly, some low. The different rows of the kernel matrix weight different parts of the function $n(a)$. The illustrative figure is a case where the kernel is "well-conditioned" (i.e., the different rows of the kernel "sample" different parts of the size distribution function). If K_{ij} were Dirac functions, there would be no overlap in the rows K_{ij}, producing completely "orthogonal" rows, giving excellent estimates of the function $n(a)$. (In matrix algebra, this is the case of K being a diagonal matrix.) Thus,

the kernel rows sample different parts of the size distribution function, and the ability to invert can be expected to depend upon the "orthogonality" of the rows of K. Obviously, the curves (arrays) K_i and K_j are othogonal when the sum $\langle K_i \cdot K_j \rangle = 0$, indicating no overlap. In general the matrix K will not be of this type; nevertheless, this physical insight is helpful in choosing "weight factors" for constructing the kernel matrix in a more stable form. A matrix comprised of the quantities $\langle K_i \cdot K_j \rangle$ is the covariance matrix, and its eigenvalues are of interest in establishing the number of independent measurements in (or the information content of) the data.

The properties of the kernel matrix can be examined through its eigenvalues and eigenvectors. This is based on the observation that the eigenvectors form an orthogonal set (for details see Twomey [1977]; Syvitski, Chapter 18, this volume). It is helpful to draw an analogy between the eigenvectors and the unit vectors of a multidimensional space and to visualize that the matrix K transforms the size distribution vector N into the intensity distribution vector I. Then, any intensity vector is a weighted sum of the unit vectors – eigenvectors – in this space. Consequently, any data such as $I(\theta)$ can be expressed as the sum of weighted eigenvectors of the matrix K; that is,

$$I = \sum_j \xi_j \, u_j \qquad (9.6)$$

where the ξ_j are constants and u_j are eigenvectors of K.

Then it is known that the solution to equation (9.4) can be written as

$$N = \sum (1/\lambda_i) \, \xi_i \, u_i \qquad (9.7)$$

(Twomey, 1977, p. 75), where the λ_i terms are the eigenvalues of K. It can be seen from this that the smallest eigenvalues will magnify any errors in the data. Evidently, the inversions will be more difficult when the ratio of the largest to the smallest eigenvalues is large. This ratio is the "condition number" of the matrix, which determines the inherent properties of the matrix with respect to stability of the inversions to noise; a small condition number is desirable. To some extent, the stability of the inversion can be

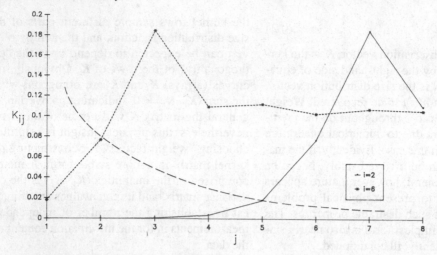

Figure 9.2. Two rows of the kernel for the light scattering formulation of equation (9.4). A fictitious $n(a)$ is also sketched in. The product of a kernel row curve with the $n(a)$ curve gives one value of $I(\theta)$.

improved by simply dropping those eigenvectors in (9.7) that have small eigenvalues λ_i. The effect in the solution is loss of structure in $n(a)$. This is the concept of techniques called *singular-value decomposition*.

Whereas the conditioning of the matrix is examined in this manner, the "dimension of independence" of the data is related to the eigenvalues of the covariance matrix discussed above. The examination of the dimension of independence begins with estimating the number of equations contained in the matrix formulation that are independent within the measurement accuracy; that is, is it possible to synthesize an equation in the set from a linear combination of any of the others to an accuracy contained in the data? The dimension of independence must be reduced by 1 for each such possibility. Mathematically, this leads to the result that the dimension of independence equals the number of eigenvalues of the covariance matrix, which exceeds a known constant times the mean-square noise or error in the data.

A number of investigators have used some version of a minimization scheme to achieve a higher-resolution inversion of the matrix equation. Most frequently, these investigators seem to overlook that the least-squares methods do not provide a trick to defeat the fundamental

conclusions just presented. The general category of constrained solutions produces smoothing, an example being the method of Lagrange multipliers.

A survey of matrix methods applied to this problem is presented by Riley (1987) in his thesis. This focused on the properties of the kernel matrix when it was constructed by weighting the intensity vector by the third power of the scattering angle. Shortly thereafter, Hirleman (1987) published a generalized formulation by weighting the kernel by a power α of the scattering angle as follows:

$$d(\theta) = \theta^\alpha I(\theta) = \int a^2 n(a)\theta^{\alpha-2} J_1{}^2(ka\theta)\, da$$

(9.8)

The matrix K was constructed to be symmetric for two cases, integrating the above equation over linear- and log-spaced size and angle ranges. As a result, an inverse was sought not for the size distribution itself but for some moment of the size distribution function, that is, the area or volume distribution. The conditioning of the matrix depends on two nondimensional constants: the dynamic range of angles over which the data are available and the degrees of freedom (or size classes). Hirleman found that for a dynamic range in angles ($\theta_{max}/\theta_{min}$) of 100, between ten and twelve size classes were the maximum that could be recovered, only with log spacing of size bins. An increase can be achieved only by increasing the dynamic range of angles, stressing the value of making observations at small angles θ_{min}. This

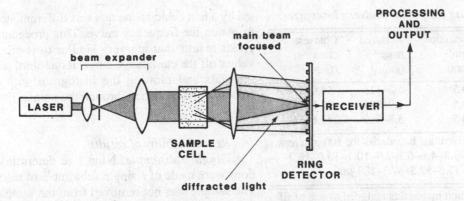

Figure 9.3. Schematic diagram showing the arrangement of the Malvern 3600E laser particle sizer from McCave et al. (1986).

result may be less dismal than it first appears: Ten size classes for a 100:1 size range gives individual bins with a 1.6:1 size range (a little over one-half phi).

The conclusions of Hirleman's work are not to be regarded lightly. Whereas his results guide matrix methods, no such simple, equivalently robust result exists for the analytical inversions (Chin et al., 1955). For example, the Fourier-transform-based results of Agrawal & Riley (1984) link the range of angles ($\theta_{max} - \theta_{min}$) with the size resolution, and θ_{min} and θ_{max} with the high and low end of sizes, respectively. There is still the need for a theory unifying the information content of data based on matrix and analytical methods.

Having stressed the importance of log spacing and matrix conditioning, it is important to note that there is one factor that casts significant doubt on the inverses obtained using log binning of the data: The typical data have a strong curvature in scattering angle. As a result, depending on how many degrees of freedom are built into the analysis, the averaging of $I(\theta)$ data may produce different size information simply as a consequence of averaging.

Problems with geological samples

We have not tested the range of available instruments, but the problems we have encountered with two suggest that it might be worth examining reported data closely and with a skeptic's eye. Comparative data have been reported

by Cooper et al. (1984), McCave et al. (1986), and Singer et al. (1988) that suggest problems of detection at the fine (submicron) end and resolution throughout the particle size range – problems whose basis in the procedures of inversion and use of appropriate theory have been examined in the preceding section of this chapter, by de Boer et al. (1987), and by Riley & Agrawal (1989). A survey of modern instruments and their manufacturers is given by Bunville (1984), but in this field almost any list is obsolete by the time it is published.

McCave et al. (1986) examined samples of clayey silt soil using a Malvern 3600E laser particle sizer. The setup of the Malvern 3600E involves a laser source, beam expander, sampler chamber, focusing lens and ring detector with associated electronics, and a microcomputer (Fig. 9.3). In the present study the samples were dispersed ultrasonically in Calgon solution, and then introduced to a small volume cell containing a magnetic stirrer. Analysis time in the cell is <1 min, and the full printout can be obtained in about 5 min. The rate-determining step is the speed with which the cell can be cleaned and a new dispersion prepared. Up to forty samples a day can be run if disaggregation is not difficult.

The size range detected by the machine depends upon the focal length of the focusing lens. A total range of 0.5–560 μm is claimed by the manufacturers. Three lenses are available, each one resolving sizes into fifteen size classes with an overall hundredfold size range (Table 9.1). For example, the 100-mm lens sizes material of 1.9–188 μm and records the data in the fifteen classes whose size boundaries are given

Table 9.1. *Size ranges for Malvern laser sizer*

Focal length (mm)	Detection limit (μm)	Analyzed range[a] (μm)	Coarsest class (μm)
63	0.5	1.2–118	54.9–118.4
100	0.5	1.9–188	87.2–188
300	0.5	5.8–560	261.6–564

Sequence of size class boundaries for 100-mm lens: 1.9–2.4–3.0–3.8–4.8–6.2–7.9–10.1–13.0–16.7–21.5–28.1–37.6–53.5–87.2–188 μm

[a]Range in which material is assigned to one of fifteen classes. An amount below the analyzed range but above the detection limit is also given by the instrument (see also Fig. 9.3).

in Table 9.1. However, the size range within which the resolution is between one-half and one-third phi is only about fiftyfold in the first fourteen classes, the coarsest class being very wide. The sequence is nearly but not exactly logarithmic (Table 9.1); thus, curve fitting and interpolation are necessary for sedimentological use. The lower limit of 0.5 μm is approximate and is the region where the particles do not diffract light in the manner required for valid application of Fraunhofer diffraction theory because their diameter d approaches the wavelength of light λ. In fact, this approximation becomes increasingly poor below 7 μm or $d \approx 10\lambda$ (Bayvel & Jones, 1981; de Boer et al., 1987).

We also report analyses of polymodal standards 1B, 2A, and 2B of Singer et al. (1988) by the Fritsch Analysette 22 laser particle sizer. From 10.17ϕ to 5.62ϕ the data are reported in classes of about 0.20ϕ. For present purposes the frequency per half phi was calculated for comparison with the Malvern 3600E. The amount of clay (finer than 8ϕ, ~4 μm as reported by Singer et al.) was also obtained.

Methods and sample analyses
General procedure
Samples were examined with interest focused on the silt and upper-clay size range. For this reason, most were dispersed by ultrasonic energy in Calgon, and sand was removed on a 63-μm sieve. Five minutes of ultrasonic treatment was the norm. Because the size classes are of unequal width, the cumulative curve was fit-

ted by a taut cubic spline that was differentiated to obtain the frequency curve. This procedure yields results that are very similar to reading values off the cumulative curve at one-third phi intervals and plotting the histogram with a smooth curve sketched through the points (Fig. 9.3).

Reproducibility of results
Single subsample. Nine size determinations were made of a single subsample of mud. The sample was not removed from the sample chamber during the analysis, and the median size clearly showed coarsening from 5.60 μm to 6.65 μm over a 3-h period, presumably due to flocculation.

Multiple subsamples. Twenty subsamples were taken from a single sample of the same mud. The average median diameter was 5.82 μm with a standard deviation of 0.17 μm. This very high precision gives confidence that the coarsening with time ascribed above to flocculation is real.

Comparison with Coulter Counter: Modal structure
Samples previously analyzed using a Coulter Counter model TAII with 70- and 200-μm apertures were analyzed by the Malvern sizer. Several consistently recurring modes on the Malvern frequency curve make comparison with the Coulter Counter difficult (Fig. 9.4A–C). The modes are most pronounced in data from the 63-mm lens but are also present in data from the 100-mm lens. In some cases there appears to be little correspondence between the Malvern and Coulter Counter data (Fig. 9.4A); in others, a good correspondence with one lens (Fig. 9.4B, 100 mm) or another (Fig. 9.4C, 63 mm). What was particularly disturbing was persistence of the modal structure and the close similarity of results before and after removal of the 44% carbonate for the sample in Figure 9.4D. In this context it is difficult to evaluate the results for lack of confidence intervals provided by the manufacturer.

Amount of clay
The Malvern sizer determines some quantity of material in suspension below the lowest

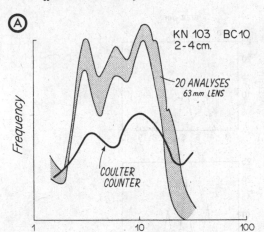

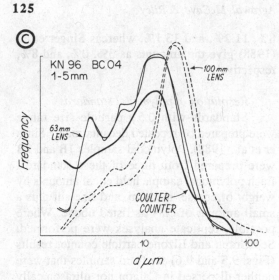

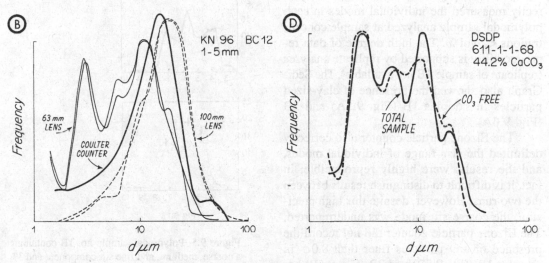

Figure 9.4. (a) Twenty subsamples of a single sample analyzed by Malvern sizer compared with analysis by Coulter Counter. (b) Replicate analyses of a sample using the 63-mm lens and the 100-mm lens compared with analysis by Coulter Counter. Note lack of correspondence of main Malvern peaks with themselves and with those from the Coulter Counter. (c) Replicate analyses similar to (b) but with (fortuitously?) close correspondence of the major mode from the 63-mm lens with that from Coulter Counter analysis. (d) Sample analyzed before and after removal of carbonate (from McCave et al. [1986]).

size class limit. For the 100-mm lens that limit is 1.9 μm, but for the 63-mm lens it is 1.2 μm. In both cases it is straightforward to determine the amount the machine determines as being below 1.9 μm. From the pipette analyses of the soils we know the percentage less than 2 μm of the silt plus clay population. The results show that, with both the 100- and 63-mm lenses, the Malvern sizer sees only a small fraction of the

clay present (16%–21%). For the same samples, the two lenses indicate amounts of clay that are not at all closely related (correlation coefficient only 0.62). The Malvern is not alone in showing results with a poor relation to those from the pipette method: Cooper et al. (1984) give poor correlations using a Leeds & Northrup Microtrac. The Fritsch is no better, as the recorded clay in standards 1B, 2A, and 2B was

8%, 11.2%, and 13.1%, whereas Singer et al. (1988) give the amounts as 3%, 0%, and 8%, respectively.

Resolution of polymodal standards

Standards with 0.5ϕ particle size range were prepared by repeated decantation by Singer et al. (1988). Polymodal samples 1B and 2B were prepared from three of these standards. Each polymodal sample held equal amounts by weight of coarse, medium, and fine silt plus a small amount of clay as listed above. Where possible, replicate analyses were performed. SediGraph and Elzone particle counter results (Figs. 9.5 and 9.6) are from samples that were neither dispersed in Calgon nor ultrasonically treated prior to analysis.

The SediGraph correctly delimited and correctly measured the individual modes in each polymodal sample analyzed at sample concentrations <2 vol %. The high degree of data reproducibility is supported by replicate analyses (replicate of sample 1B not available). The SediGraph also showed the presence of clay-sized particles in samples 1B (Fig. 9.5A) and 2B (Fig. 9.6A).

The Elzone particle counter also correctly delimited the size range of individual modes, and the results were highly reproducible; in fact, it is difficult to distinguish results between the two runs. However, despite this high precision, the coarse-silt mode was underreported. The Elzone particle counter did not record the presence of any particles finer than 8.0ϕ in samples 1B (Fig. 9.5B) and 2B (Fig. 9.6B) because those particles were below the detection limit of the 120-μm aperture (2.4 μm).

Results from the Malvern laser sizer show broadening of the individual modes and, in sample 2B with 8% clay (Fig. 9.6C), a lack of reproducibility in replicates. The clear modal structure seen by the other analyzers, and present in the standard, is discernible but blurred. In the Fritsch machine this tendency becomes quite extreme (Figs. 9.5D, 9.6D) and the standards are scarcely recognizable as polymodal. Advertisements for laser sizers often display results from bimodal samples, but these generally have well-separated modes that do not pose the problem encountered here.

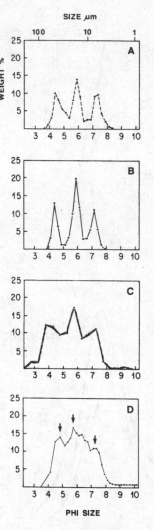

Figure 9.5. Polymodal sample no. 1B containing a coarse, medium, and fine silt component and 3% clay (by weight) analyzed on the (A) SediGraph, (B) Elzone particle counter, (C) Malvern laser sizer, and (D) Fritsch laser sizer. The Malvern laser sizer results are for the 63-mm lens, with one curve (solid line) corresponding to results obtained following dispersion and ultrasonic treatment. SediGraph, Elzone, and Fritsch results do not include a replicate analysis.

Discussion and conclusion

This method of size analysis is very powerful and has many features to commend it, particularly its speed, precision, and capacity to analyze undisturbed samples from the environment. For geologists who wish to know a *total* fine sediment distribution – including the per-

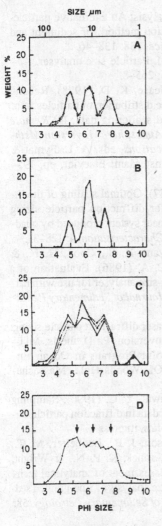

Figure 9.6. Polymodal sample no. 2B containing a coarse, medium, and fine silt component, and 8% clay (by weight) analyzed on the (A) Sedi-Graph, (B) Elzone particle counter, (C) Malvern laser sizer, and (D) Fritsch laser sizer. Replicate analyses are also shown except for (D). The Malvern laser sizer results are for the 63-mm lens, the solid line showing results following dispersion and ultrasonic treatment.

centage of clay, as might be obtained from a pipette or Sedigraph – there are difficulties. However, this is true of other devices in common use (e.g., electrical sensing zone particle counters). The resolution of the latest generation of machines still leaves much to be desired, and one would be wise to test them with a polymodal standard that has been analyzed by other

means. If one's goal is dissection of polymodal size distributions and the tracing of populations through modal structure, it may be worth waiting for the next generation.

The discussion of inverse theory can be summarized also. In the absence of the knowledge of algorithms and methods used by any commercial manufacturer, scientists should continue to view the results with skepticism, remembering that:

1. inconsistent results observed with a single instrument but using different lenses imply algorithmic problems;

2. the absence of error bounds and confidence limits on particle volume data is poor experimental procedure; and

3. results at small sizes are almost always suspect due to the worsening nature of the Fraunhofer approximation.

It is worth reiterating that the Fraunhofer approximation predicts intensity to behave as θ^{-3} at large scattering angles, whereas Mie theory *never* does. At the range of angles covered by most laser sizers, the Fraunhofer approximation is definitely not valid even for large particles. New work by the authors, obtaining detailed scattering patterns with a CCD array, combined with the use of full Mie theory should improve results in the future. Progress in obtaining improved information can be made by:

1. better formulation of the scattering model;

2. measurement at smaller range of angles; and

3. measurements of $I(\theta)$ at smaller minimum angles θ_{min}.

A remark about averaging is also in order. This technique requires averaging of $I(\theta)$ to eliminate speckle and particle statistics. Of these two factors the latter is a little more serious. In order to achieve proper averaging, the samples of $I(\theta)$ must be *statistically independent*. The implication, very often overlooked, is that the time between samples should exceed the integral scales in the flow: the ratio of typical cuvette dimensions to typical flow velocities in a laboratory experiment, or, in the field, the time it takes for a dust or turbidity cloud to be advected past the instrument. In the field, say, on the seafloor, the integral time scales will typically be z/u, where z is elevation of the instrument above the seabed and u is mean advection. This interval clearly can be of order 1 m/(10 cm/s) $\approx$ 10 s in the field, and of order 1 s in the laboratory.

Finally we observe that the potential of this principle for determining size and settling velocity distributions in situ (Agrawal & Pottsmith, 1989) is most exciting and will lead to a new generation of instruments for sediment transport research.

REFERENCES

Agrawal, Y. C., & Pottsmith, H. C. (1989). Autonomous long-term in-situ particle sizing using a new laser diffraction instrument. *Oceans '89 Proceedings Conference IEEE,* 5: 1575–80.

Agrawal, Y. C., & Riley, J. B. (1984). Optical particle sizing for hydrodynamics based on near-forward scattering. *Society Photo-optical Instrumentation Engineers,* 489: 68–76.

Bale, A., & Morris, A. W. (1987). In situ measurement of particle size in estuarine waters. *Estuarine Coastal and Shelf Science,* 24: 253–63.

Bayvel, L. P., & Jones, A. R. (1981). *Electromagnetic Scattering and Its Applications*: London: Applied Science, 289 pp.

Bayvel, L. P., Knight, J. C., & Robertson, G. N. (1987). Application of the Shifrin inversion to the Malvern particle sizer. *Proceedings International Sizing Conference, Rouen, France,* June 1987.

de Boer, G. B. J., de Weerd, C., Thoenes, D., & Goossens, H. W. J. (1987). Laser diffraction spectrometry: Fraunhofer versus Mie scattering. *Particle Characterisation,* 4: 14–19.

Bunville, L. G. (1984). Commercial instrumentation for particle size analysis. In: *Modern Methods of Particle Size Analysis,* ed. H. G. Barth. *Chemical Analysis,* vol. 73. New York: J. Wiley, pp. 1–42.

Chin, J. H., Sliepcevich, C. M., & Tribus, M. (1955). Particle size distributions from angular variation of forward scattering at very small angles. *Journal of Physical Chemistry,* 59: 841–4.

Cooper, L. R., Haverland, R. L., Hendricks, D. M., & Knisel, W. G. (1984). Microtrac

particle-size analysis: An alternative particle-size determination method for sediment and soils. *Soil Science,* 138: 138–46.

Cornillault, J. (1972). Particle size analyser. *Applied Optics,* 11: 265–8.

Fymat, A. L., & Mease, K. D. (1978). Reconstructing the size distribution of particles from angular forward scattering data. In: *Remote Sensing of the Atmosphere: Inversion Methods and Applications,* eds. A. L. Fymat & V. E. Zuev. Amsterdam: Elsevier, pp. 195–231.

Hirleman, E. D. (1987). Optimal scaling of the inverse Fraunhofer diffraction particle sizing problem: The linear system produced by quadrature, *Particle Characterisation,* 4: 128–33.

McCave, I. N., Bryant, R. S., Cook, H. F., & Coughanowr, C. A. (1986). Evaluation of a laser-diffraction-size analyser for use with natural sediments. *Journal of Sedimentary Petrology,* 56: 561–4.

Riley, J. B. (1987). Laser diffraction particle sizing: Sampling and inversion. Ph.D. thesis, MIT–Woods Hole Joint Program in Ocean Engineering, WHOI, Woods Hole, Massachusetts.

Riley, J. B., & Agrawal, Y. C. (1990). Sampling and inversion of data in diffraction particle sizing. *Applied Optics,* in press.

Singer, J. K, Anderson, J. B., Ledbetter, M. T., McCave, I. N., Jones, K. P. N., & Wright, R. (1988). An assessment of analytical techniques for the size analysis of fine-grained sediments. *Journal of Sedimentary Petrology,* 58: 534–43.

Swithenbank, J., Beer, J. M., Taylor, D. S., Abbot, D., & McGreath, G. C. (1977). A laser diagnostic technique for the measurement of droplet and particle size distributions. In: *Experimental Diagnostics in Gas Phase Combustion Systems: Progress in Astronautics and Aeronautics,* ed. B. J. Zinn. 53: 421–47.

Twomey, S. (1977). *Introduction to the Mathematics of Inversion in Remote Sensing and Indirect Measurements.* New York: Elsevier, 243 pp.

Weiss, E. L., & Frock, H. N. (1976). Rapid analysis of particle-size distributions by laser light scattering. *Powder Technology,* 14: 287–93.

10 SediGraph technique

JOHN P. COAKLEY AND
JAMES P. M. SYVITSKI

Introduction

Size analysis of silt and clay-sized sediment (i.e., particles <63 μm in diameter) can be accomplished by a variety of automated methods based on their electrical properties, transport properties (e.g., gravitational sedimentation, centrifugal sedimentation, hydrodynamic chromatography, and aerodynamic transport) and nonimaging optical methods (optical blockage technique, time-averaged light scattering, photon correlation spectroscopy) (for details, see Bunville [1984]). The most widely used method in the earth sciences is based on the sedimentation principle and Stokes's Law. The physics of sediment settling in a fluid medium have already been discussed elsewhere (Chapters 1 and 2), and the concept is readily applicable to dilute suspensions of silt and clay. Before the development of the SediGraph, the most commonly used sedimentation techniques were those of the pipette and the hydrometer (Chapter 1), both of which used manual sampling and recording. Though effective and reasonably accurate, these techniques had the disadvantage of being highly labour intensive and time consuming, taking hours to days for a routine analysis.

The entry onto the market of the SediGraph in the early 1970s was therefore welcome in that, while maintaining comparable accuracy, it offered an increased level of automation of the analysis, as well as a reduction in the time of analysis to minutes (Berezin & Voronin, 1981; Duncan & Rukavina, 1982). The first model sold was the 5000, which was followed by the 5000D, E, and ET model variations. In 1988 the latest model, the 5100, incorporating a built-in computer interface, was brought onto the market. In this chapter we review the operational and theoretical principles of the SediGraph, present suggestions for practical analysis, and discuss calibration and intercomparison of results with other methods.

Operational principle

The SediGraph method assumes that the particles are dispersed in a fluid and settle in accordance with Stokes's Law. Thus by monitoring the rate at which particles settle and are removed from the monitored volume (i.e., the rate at which the particles fall below a certain depth in the sedimentation column), an accurate measure of the cumulative size distribution of the sediments in suspension can be obtained.

The precursor techniques to the SediGraph, the pipette and the hydrometer, respectively used a series of timed, measured draw-offs to monitor the sediment concentration at a given depth in the column, and changes in the flotation height (displacement volume) of a calibrated object immersed in the suspension. The basic innovation that the SediGraph brought to size analysis was the use of a collimated beam of x rays to sense the changing concentration of fine sediments settling in a suspension with time, and thus, the size distribution of the settling particles. In contrast to the more invasive sensing procedures used in the earlier techniques mentioned above, the x-ray beam used in the SediGraph does not disturb or change the suspension. This technique made automation possible, thus eliminating much of the potential for operator error that plagued earlier methods. The closed sensor-housing chamber, required by the x-ray beam for safety reasons, also has the advantage of isolating the sample from temperature fluctuations, contamination, and physical disturbances, thus increasing the analytical accuracy.

The instrument is able to minimize the analytical time required through a controlled downward movement of the sedimentation cell with time. This continuously reduces the depth of the x-ray sensor below the cell surface, so the effective settling depth becomes inversely proportional to time. This innovative procedure allowed the analysis time to be reduced to <10 min for analysis down to 1 μm. Particle diameters are calculated in terms of the Stokesian diameter, otherwise known as the equivalent spherical sedimentation diameter (ESSD). Due to the small dimensions of its sedimentation cell, the SediGraph requires only a relatively small sample (≤3 g) to achieve a suitable concentration level. Also, because the suspension is not physically altered, multiple replications can be made on a sample for even higher accuracy and precision. A final advantage of the SediGraph is the relative ease of computation of accurate size-analysis statistics because of the digital data output. This also facilitates the further manipulation of the size frequency data by computer.

In summary, the main advantages of the SediGraph are as follows:

1. speed of analysis with no loss of accuracy;

2. ease of automated or unattended operation;

3. small size of sample required compared to pipette and hydrometer techniques;

4. isolation from environmental fluctuations; and

5. adaptability to computerized data processing.

Theoretical basis of x-ray sensing

The theory relating the difference in x-ray beam transmission to the size distribution of solid particles in a fluid is discussed in detail in Oliver, Hicken, & Orr (1970–1), Vitturi & Rabitti (1980), Micromeritics (1982), and Stein (1985). In brief, from Stokes's Law, it can be shown that a particle of spherical equivalent diameter D will settle a distance h in time t according to the expression:

$$D = K(h/t)^{0.5} \qquad (10.1)$$

where $K = [18\eta/(\rho_s - \rho_l)g]^{0.5}$, η is the liquid viscosity (in poise [P] for cgs units), ρ_s and ρ_l are the densities of the solid and liquid phases, respectively, and g is the acceleration due to gravity. Consequently, after a given time t_i all particles larger than the corresponding diameter D_i will have fallen below a given distance h from the surface of the suspension. If the initial uniform concentration of the material was C_0 (g/ml), and the measured concentration after time t_i at distance h is C_i (g/ml), then the weight percent P_i of material finer than D_i is given by:

$$P_i = 100 \, (C_i/ \, C_0) \qquad (10.2)$$

By obtaining values of C_i after various times, the corresponding values of P_i and D_i may be calculated and plotted to present a cumulative mass distribution of particle sizes in terms of Stokes's ESSD.

To relate the record of changes in x-ray transmittance measured by the SediGraph to sediment concentrations, use must be made of theoretical relationships between transmittance and turbidity (Micromeritics, 1982). If a container or cell of rectangular cross section is irradiated from a direction perpendicular to one of its sides by a collimated x-ray beam, then the fraction of incident radiation transmitted by the cell when filled with the suspension under study is given by:

$$I/I_0 = \exp[-(a_1\phi_1 + a_s \phi_s)L_1 - a_cL_2] \qquad (10.3)$$

where I and I_0 are the transmitted and incident intensities; a_1, a_s, and a_c are the x-ray absorption coefficients of the liquid, solid, and the cell walls, respectively; ϕ_1 and ϕ_s are the weight fractions of the liquid and solid present in the suspension; L_1 is the internal cell thickness in the direction of irradiation; and L_2 is the total thickness of the cell walls. By using the relation ($\phi_s = 1 - \phi_1$), and defining a transmittance T as the ratio of the transmission of the cell when filled with sample to that when filled with pure suspending liquid (most often, water), we obtain:

$$T = \exp[-\phi_s(a_s - a_1)L_1] \qquad (10.4)$$

or

$$\ln T = -A\phi_s \qquad (10.5)$$

where A is a constant for the particular apparatus and suspension components.

By collimating the x-ray beam through horizontal slits with a vertical dimension that is small compared to the sedimentation depth h, the instantaneous values of T (i.e., T_i) can be used in calculating P, the cumulative-percent-finer distribution, as follows:

$$P = 100 \, (\ln T_i/ \ln T_0) \qquad (10.6)$$

where T_0 refers to the transmittance of the initial suspension.

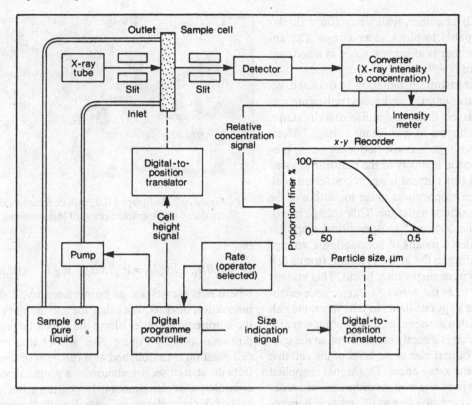

Figure 10.1. Simplified schematic diagram of the SediGraph instrument. (Reproduced with permission from Kalita et al. [1985].)

Physical description of the SediGraph

SediGraph 5000

The SediGraph is manufactured by Micromeritics Instrument Corporation, 1 Micromeritics Drive, Norcross, Georgia 30093-1877, USA. The original SediGraph 5000 remains the model found in most laboratories as of mid-1988. The 5000 model has undergone some changes over the years, the revised versions being sold as 5000D, 5000E, and 5000ET. Many research labs have modified their 5000 series model (through computer interfacing, adding larger circulating pumps, etc.). The following general description is abbreviated from Micromeritics (1982). The basic construction is illustrated schematically in Figure 10.1.

The radiation from an x-ray tube with its associated power supply is collimated by horizontal slits to a beam 0.0051 cm high and 0.9525 cm wide. The small, air-cooled x-ray tube has a tungsten target inclined at about 30° to a thin beryllium window. The x-ray tube is usually operated at a power level of ~14 W. The x-ray beam passes through a sample cell located midway between the collimating slits. The cell is of rectangular cross section, and typically has internal dimensions of 1.27 cm wide, 3.5 cm high, and 0.53 cm thick, for a volume of 2.36 cm³. The windows are normally 0.1587-cm-thick homalite®, but other materials may be employed. The sample cell is constructed such that its inlet and outlet do not interfere with the x-ray transmission. To eliminate meniscus effects, the cell is closed at the top and, in use, is completely filled with the dispersed sample. The cell is connected to a reservoir containing the sample in suspension by flexible plastic tubing. A circulating pump is used to keep the sample suspension flowing through the cell until the analysis begins. The reservoir usually contains ~30 ml of suspension, and is equipped with a magnetic stirrer to maintain the material in suspension while it is being circulated. Both the reservoir and the sample cell are kept isolated from the laboratory environment by the access door to the analysis chamber, which contains a shielding glass panel to block stray x rays. The enclosed chamber is also maintained at a constant temperature of ~30 °C.

The transmitted radiation is detected via scintillation counter. The detected pulses are amplified, passed through a noise discriminator, and then clipped to a constant voltage. Afterward, the voltage is converted to a current proportional to the intensity of the transmitted x-ray beam, and this current is again converted back to a voltage proportional to the logarithm of the change in current with time. This voltage is produced at the output of an operational amplifier that provides a means of subtracting a voltage corresponding to the analysis output from a cell filled with pure suspending liquid. This voltage is referred to as the *zeroing voltage;* once established, it is kept constant, so that when the cell is filled with a suspension, a net output signal is produced that is directly proportional to the concentration of particles at the level of the cell that intercepts the x-ray beam. This signal is applied to the x-axis of the $x–y$ recorder, the recorder zero point corresponding to 0% particle concentration (i.e., with the cell filled with pure water and dispersion). During analysis, the instrument automatically converts the sedimentation rates to cumulative mass frequency distribution (finer than a given diameter), and plots the results onto a strip-chart graph. The initial signal produced when the cell is filled with homogeneously dispersed sample can be scaled to a reading of 100% by adjusting the recorder sensitivity control dial. Thus, it is not essential that suspensions for size analysis be prepared at an accurately known concentration.

It can be shown, using Stokes's Law, that to resolve particles at the coarse end of the size distribution (e.g., particles 50 μm in diameter) accurately, the settling depth in the cell must be ≥ 2.54 cm (1 in.). However, to measure the fine end (e.g., 0.2 μm) at such a depth would take more than 200 h. In the SediGraph, the sample cell is driven downward with respect to the x-ray beam as a function of time. The nature of the time-related change in depth is selected to conform with equation (10.1); that is,

$$h_i = B^2/t_i \qquad (10.7)$$

Figure 10.2. Sedigraph 5100, with dedicated microcomputer. (Photograph courtesy of Micromeritics.)

or

$$D_i = KB/t_i, \quad \log D_i = \log(KB) - \log t_i \qquad (10.8)$$

where B is a constant of proportionality. This innovation also requires that the movement of the automatic recorder along the x-axis be proportional to $\log t$. Figure 10.1 shows that the cell position is established as a function of time from the start of sedimentation by a programmed controller that simultaneously positions the x-axis of the recorder to indicate directly the particle size D_i corresponding (through Stokes's Law) to the instantaneous value of h_i and t_i. The programmed controller is a solid-state digital controller with one output controlling the cell-drive mechanism so that equation (10.7) is satisfied, and another output satisfying equation (10.8), producing the required displacement of the x-axis of the recorder as a function of time. Both output rates may be simultaneously adjusted to take into account various particle and liquid densities and liquid viscosities as related by the K factor in equation (10.1). The time for an analysis is inversely proportional to the square root of the density differences concerned (i.e., $(\rho_s - \rho_l)^{0.5}$).

SediGraph 5100

The SediGraph 5100 (Fig. 10.2) is Micromeritics' latest generation of fine-particle size analyzers. Though based on principles similar to those of the 5000 series, this model incorporates the latest in automated operation and advanced data processing. This is in direct response to the successful user attempts to digitize existing 5000 model outputs (e.g., Jones, McCave, & Patel

[1988]). An important operational difference between this SediGraph and its precursors is the expanded range of particle sizes measured. The 5100 has a range of 300–0.1 μm, compared to 100–0.2 μm for the 5000 series; this means that there often is no need to combine SediGraph analysis with sieve or settling tube to resolve the fine sand fraction.

Analytical procedure

The procedure for SediGraph analysis adopted by individual laboratories depends on the type of samples generally analyzed, the accuracy desired, the SediGraph model used, and the mix of other techniques used for analyzing the coarse fractions, among other factors. As well, general procedures recommended by the manufacturer are included with any instrument purchased. The idealized procedure (Fig. 10.3) can be used with either the SediGraph 5000 or 5100, and for sediments having both fine and coarse components. However, it is intended only as an example, and individual users might well develop techniques and approaches that better fit their laboratories. The steps below give a summary of how an analysis is carried out.

Sample preparation

The sample is first split to obtain a subsample of approximately 3 g for analysis. An effort is made, for reasons discussed in the next section, to obtain an eventual suspension concentration of 0.02–0.1 g/ml (1% or 2% in aqueous suspensions). The subsample can first be treated to remove shells and organic matter, prior to dispersion. For instance, it has been noted (J. S. Mothersill, Royal Roads Military College, pers. commun., 1988) that samples with high organic carbon (>10%) or carbonate (>2%) can hinder proper dispersion of the sample. Chemical dispersion of the sample is followed by mechanical or ultrasonic stirring. The dispersed suspension is then placed in the sample reservoir, from which it is circulated by pumping through the SediGraph sedimentation cell. Inside the reservoir, the suspension is maintained using a magnetic stirrer, taking care to avoid introducing air bubbles to the suspension. To begin analysis the pump is shut off and the automatic size analysis proceeds.

Prior to analysis on the SediGraph 5000, it is necessary to select and switch into the instrument the value to be used for the rate at which the sedimentation cell is to be lowered. This rate setting is based on the relative densities of the solid and liquid phases (ρ_s and ρ_l), the fluid viscosity η, and the maximum particle diameter D_m to be resolved as follows:

$$\text{Rate} = [211.8(\rho_s - \rho_l)]/(50/D_m)^2\eta \qquad (10.9)$$

The maximum diameter is set on the recorder chart prior to beginning an analysis. If the power-line frequency is other than 60 Hz, then the right-hand term in the above equation must be multiplied by the ratio 60/line frequency. The particle Reynolds number ($D\omega\rho_l/\eta$) should be <0.1 for organic liquids and <0.3 for aqueous dispersions (where D is the particle diameter and ω is the equilibrium settling velocity). With the use of the SediGraph 5100, it is especially important that the rate be set properly to reflect the higher maximum diameters possible with this new model.

Analysis

As stated previously, after the pump is turned off the analysis proceeds automatically. However, the user can choose to have the results from a SediGraph analysis output in the form of raw voltages or as digitized discrete frequency values from the strip record. In the case of the former, it is necessary to use an interfacing device to convert the analog voltage output to digital values, prior to forwarding them to a computer for computation of the changing concentration and cumulative-percent-finer distribution (e.g., Jones et al., 1988). This procedure is now done directly by the 5100 model, and the output can either be a printout of size statistics or in a graphic form specified by the analyst.

Assessment of the accuracy and precision of the results

In discussing the accuracy of the SediGraph in producing grain size data from geological samples, one must keep in mind that sediment particles are not perfect spheres but rather are irregularly shaped. Thus, without particle shapes accounted for, Stokesian diameters

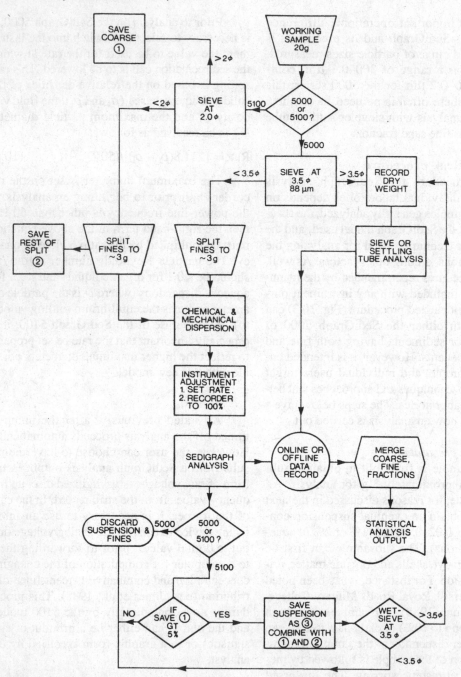

Figure 10.3. Generalized flowchart for size analysis using the 5000 and 5100 SediGraph models. The expanded upper limit of the 5100 model (300 μm) gives an opportunity for better overlapping and merging of coarse fraction data (sieve or settling tube) with fine-fraction results from the Sedi-Graph.

are only equivalent spherical sedimentation diameters (ESSD). Oliver et al. (1970–1) estimate the combined error from mechanical and electrical components at <±1%. Coates & Hulse (1985) concluded that while the precision of the instrument on the same sample was very good (well within the ±1% standard deviation of the cumulative percent value for each size subclass),

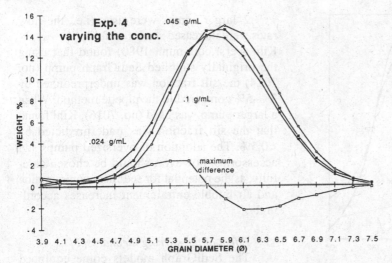

Figure 10.4. Size frequency distributions of a marine sediment sample run at ever decreasing and more accurate suspension concentrations. The maximum difference is shown to be ≤2% of a class interval. The grain diameters are given in the geological units of phi ($\phi = -\log_2 D_{mm}$).

that for separate subsamples of the same sample was worse than for other analytical techniques such as the pipette and hydrometer. Coates & Hulse believed the scatter to be caused by splitting a sample down to 2 g while retaining homogeneity. However, there are other important factors that affect the accuracy and precision of SediGraph analysis. These are treated in detail in Oliver et al. (1970–1), Welch, Allen, & Galindo (1979), and Syvitski (1986, 1987) and are summarized below.

Concentration effect

The settling rate of particles settling in a viscous medium depends greatly on intergrain effects. The effective settling rate is affected by increased viscous forces between closely spaced particles, by streaming of particles, and by turbulent eddies created by the larger grains. The topic of mass settling has received a rigorous mathematical treatment in Batchelor (1972). The ideal, although impractical, situation for accurate particle size results is for the settling to take place as individual grains. The best compromise is to have the suspension as dilute as possible. On the other hand, the SediGraph requires a certain level of difference between a clear and turbid suspension (i.e., the y-axis should be ad-

justed to read over 50%). The SediGraph manufacturer recommends a sediment concentration of <3% by volume to bring mass settling errors to a negligible level. However, results of comparative studies indicate that a concentration of <2% by volume is even better (Stein, 1985; Singer et al., 1988). Figure 10.4 demonstrates the effect on varying the concentration of the liquid suspension: As the concentration is decreased from 0.1 to 0.024 g/ml, the size frequency distribution shifts toward a coarser and more accurate mode.

Effect of diffusion

Oliver et al. (1970–1) have demonstrated that diffusion effects are caused by the tendency of very fine particles to move randomly within the fluid (Brownian motion) in addition to their settling due to gravity. This combined movement has the net result of "smearing" the theoretical settling velocities and calculated ESSD sizes of the particles (Fig. 10.5). From the figure, it is clear that this effect becomes significant only where particles <1 μm, or of reduced density, dominate the sample.

Effect of the container wall

This is the drag effect exerted on a particle settling close to a boundary. For a rigorous, detailed treatment of this phenomenon, the reader is referred to Brenner (1964) and Caswell (1970). Although the cell walls of the SediGraph exert an appreciable effect (the SediGraph has a cell thickness of only 0.53 cm), especially

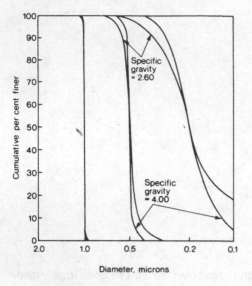

Figure 10.5. Cumulative frequency plot of three grain size distributions showing the effect of diffusion. This effect begins to be detectable only for particles <1 μm and is also more pronounced for particles with low specific gravity. (Reproduced with permission from Oliver et al. [1970–1].)

for the larger particles, Oliver et al. (1970–1) estimate the error introduced by the presence of the two cell walls causes an underestimation of the true ESSD by <0.1%.

Effect of pump/hose size

One important effect that is related more to operational considerations than to the physics of the process is the effect of undersized pumps and hoses in the system circulating the suspension between the cell and the reservoir. If these mechanical components are too small, it is possible for the heavier or coarser fraction of a size distribution to be left behind to a greater or lesser degree. Rolf Kihl (Institute of Arctic and Alpine Research) was one of the first to recognize that the early models of the SediGraph were incapable of maintaining sufficient velocity and turbulence to keep the suspension homogeneous until the analysis begins. The result is a systematic bias toward the finer end. Therefore, two solutions were subsequently employed:

1. many labs simply began the analysis at a finer maximum particle diameter (e.g., 40 μm), and

2. larger pumps were used (i.e., the flow rates were increased some fourfold).

Kihl (pers. commun., 1980) found that using the originally supplied SediGraph pump (no. 7014) the silt fraction was underpredicted by 3%–5% compared to the pipette method. When a larger pump was used (no. 7016), Kihl found that the silt fraction was underpredicted by <0.5%. The adoption of oversized pumps and accessories, however, should be chosen carefully, as the potential for sediment comminution and air-bubble entrainment increases accordingly.

Effect of the stirring mechanism

The SediGraph models come equipped with a magnetic stirring rod device. If the geological sample contains magnetic minerals, they either must be removed prior to analysis, or the stirring mechanism supplied must be substituted with a mechanical rotor and blade mechanism.

X-ray absorption effects

Some of the x rays produced by the primary beam are absorbed due to interactions with the atoms of various elements in the sample. Thus the intensity of the x-ray radiation finally reaching the detector is reduced in magnitude. The SediGraph method assumes that there is uniform x-ray absorption by all particles in suspension (see equation (10.3)). (To obtain the mass absorption coefficients the reader is referred to equation 7.16 and Table 14.5b in Goldstein et al. [1981]. Minerals containing significant amounts of Mg, Fe, and Ti have very high absorption coefficients.) Jones et al. (1988) recognized that particles with high absorption coefficients could "blind" the machine to the presence of other particles in the suspension less effective at blocking x rays. They conducted a single-sample experiment using a mixture of pure chamosite and quartz, and reported little effect between the observed SediGraph distribution and theoretical (expected) distribution based on mathematically combining the pure mineral size frequency distributions. As their test was based on a one-sample experiment, we conducted more detailed tests, described below.

The first experiment consisted of mixing pure quartz (SiO_2) and pure ilmenite ($FeTiO_3$)

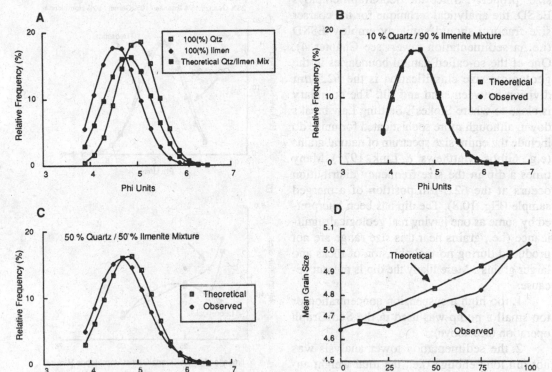

Figure 10.6. Experiments on the effect of mineralogy on x-ray absorption to the SediGraph technique. (A) Pure and mathematically mixed quartz and ilmenite standards of the 25–63-μm size range. (B, C) Two examples of size frequency distributions comparing theoretical with observed quartz–feldspar silt mixtures. (D) A shift of 0.1ϕ in the mean grain size is seen in many of the mixtures due to the increased x-ray absorption of Fe and Ti (ilmenite) compared to Si (quartz).

standards, both in the 25–63-μm particle range (Fig. 10.6A). The standards were combined in the following quartz : ilmenite weight mixtures – 10 : 90, 25 : 75, 50 : 50, 75 : 25, and 90 : 10. Little change occurred when the mixed sample was dominated by one standard (i.e., 10 : 90 or 90 : 10 [Fig. 10.6B,D]). However, systematic changes in the modal position and the mean grain size occurred in the more equitable mixture ratios (Fig. 10.6C,D). Effectively, the ilmenite particles absorbed x rays, shifting the mean grain size by ~0.1 ϕ. The second experiment involved four sedimentary minerals (quartz, ilmenite, garnet, and hornblende), again

analyzed in their pure form and then in five different weight mixtures. Although one mixture (25% qtz : 15% ilm : 10% gar : 50% hor) had a similarly large shift between the expected and the observed distribution (Fig. 10.7A), in general, the more complex the mineral mixture is, the less evident is the absorption effect (Fig. 10.7B).

The third and fourth experiments were identical to those above, only the mineral standards were <25 μm in the particle size range. Again a shift in the size frequency distributions of similar magnitude was observed (Fig. 10.7C).

Merging of SediGraph data

One of the most difficult aspects of analyzing geological samples is that no one method is capable of analyzing the entire spectrum of sedimentary particles (i.e., from gravel to clay-sized particles). Thus, unless the natural sample is of a sufficiently narrow size frequency distribution, merging/overlap techniques must be employed, and should measure the same "grain size" property. Since the SediGraph employs ESSD, the analytical technique for the coarser size fraction should similarly employ ESSD (i.e., a sedimentation tower: see Chapter 4). One of the so-called natural boundaries in the geological size classification is the 62.5-μm division between sand and silt. The boundary is close to where Stokes's Settling Law breaks down, although more sophisticated formulae do include the entire size spectrum of natural grains (e.g., Gibbs, Matthews, & Link, 1971). Many times a dip in the size frequency distribution occurs at the 62.5-μm position of a merged sample (Fig. 10.8). The dip has been interpreted by some as one having real geological significance (i.e., grains near this size range are not produced during normal abrasion of rocks and larger grains). More likely the dip is present because:

    1. too high a suspension concentration or too small a pump was used in the SediGraph operation (see above);

    2. the sedimentation tower analysis was not run long enough (i.e., the total weight entered into the tower was not analyzed); and/or

    3. the distributions of the two techniques are not overlapped and renormalized but rather are abutted.

    To demonstrate that the dip is not real, an experiment was run by changing the placement of the overlap. In the first instance a marine sample was wet-sieved at 53 μm (the tower had a lower limit of 44 μm, the SediGraph an upper limit of 63 μm). Next the same sample was reanalyzed using a wet sieve at 25 μm (the tower had a lower limit of 20 μm, the SediGraph an upper limit of 53 μm). The large dip observed in the first and more typical procedure was shifted from the 53-μm wet-sieve break toward the 25-μm wet-sieve break (Fig. 10.8B).

Calibration

    There are three forms of practical calibration:

    1. internal tests with standards,

    2. calibration between like instruments, and

    3. calibration between different instruments that provide similar results.

The SediGraph comes equipped with a narrow-

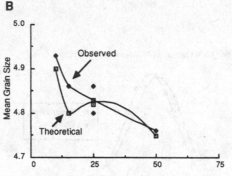

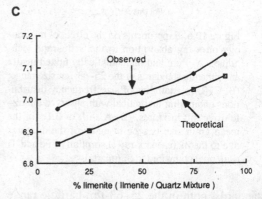

Figure 10.7. Experiments on the effect of mineralogy on x-ray absorption to the SediGraph technique. (A) Size frequency distributions of more complex silt-sized mineral standards. (B) Shifts in mean grain size as observed using complex standard mixtures. (C) Shift in the mean grain size in the <25-μm fractions of quartz and feldspar mixtures. (Cf. Fig. 10.6D for the same effect in the coarse silt range.)

ranged garnet standard (μ = 12 μm or 6.34ϕ, σ = 0.9 ϕ) that is useful as an internal test to see if one's instrument is working well – it is not a good geological standard. The garnet standard

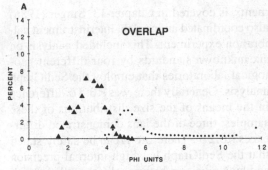

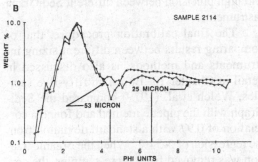

Figure 10.8. (A) A marine sediment sample analyzed with the AGC settling tube (triangles) for the sand fraction and the computerized AGC SediGraph (circles) for the mud fraction. (B) The resulting "break" between the merged size frequency distribution changed, depending where the wet-sieving between the two methods was undertaken (solid, 53 μm; open, 25 μm).

is such that the weight frequency mode occurs at the median or fiftieth percentile of the cumulative distribution. We suggest that laboratories develop their own internal standards based on natural mud samples.

Standards of various size range and complexity can be prepared using repeated washings and timed clearing rates in a standard 1,000-ml graduated cylinder. The following are the basic steps:

1. Introduce a mud sample (<2% by weight) into a cylinder containing a dispersing agent and ensure that the sample solution is homogeneous.

2. Decide what you want the upper size limit of your standard to be.

3. That size will be associated with a settling velocity (you will need to assume a singular particle density even if your sample contains a mineral mixture) that can be obtained from

equation (10.1), from which a settling time may be obtained.

4. Once this time has elapsed, decant the liquid into another cylinder.

5. Decide what the lower size limit of your standard is to be, and calculate the settling time for that particle size.

6. Starting with the decanted liquid rehomogenized, wait until the settling time has elapsed, and then decant the remaining solution (which may be discarded).

7. Resuspend the sediment that had settled during step 6 with a new dispersing solution within the cylinder, and repeat step 6.

Step 6 is known as a *washing* and should be repeated until the fines have been removed and a pure standard with known natural size boundaries has been obtained.

Once the size frequency distribution of your unique standard has been established, it may be mixed by weight with other standards of different size characteristics to obtain new and complex mixtures that may be polymodal (Fig. 10.9A). These artificial mixtures may now be run on the SediGraph. The same sample can be reanalyzed numerous times without operator disturbance. Such analysis will yield a machine precision value for the SediGraph, which, as discussed above, is very good. Replicate runs of similar weight mixtures will allow users to establish some estimation of the precision of their sample preparation (e.g., Fig. 10.9B). If the lab preparation precision is sufficiently small, then the actual or observed distribution can be compared to the theoretical or calculated distribution by adding the various standards biased by their weight contribution (Fig. 10.9C). Such a procedure can be used to test SediGraph methodological performance. We have already demonstrated their use in defining the effect of mineralogy and thus x-ray absorption on size statistics (cf. Figs. 10.6 and 10.7). Many experiments, such as that provided in Figure 10.9C, were used to confirm the poor performance of the SediGraph in the coarse silt range. Changes in the pump size and reduction in the initial sample concentration eliminated this methodological problem.

Details of the second calibration procedure, that of comparing results between like instru-

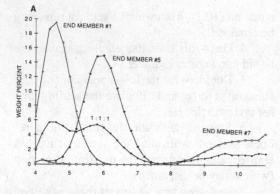

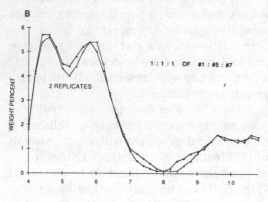

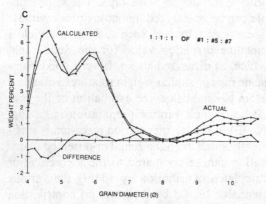

Figure 10.9. (A) Laboratory standards of known upper and lower size limits produced using the washing–settling techniques outlined in the text. Note the calculated distribution from mixing each of the three end members in a 1 : 1 : 1 ratio. (B) Repeat analysis of the mechanically mixed three end-member standards shown in (A). (C) The difference between the calculated and the observed mixed distributions given in (A) and (B). Note the observed distribution underpredicts the coarse mode and thus overpredicts the fine mode. This relates directly to an excessively high concentration used in the mechanically mixed distribution (cf. Fig. 10.4).

ments, is covered in Chapter 13. Singer (1986) also coordinated an interlab, interinstrument calibration experiment. This included analysis on six unknown standards by four different geological laboratories that employ the SediGraph analysis. Generally there was $<0.5\phi$ difference in the means of the size distributions of these samples; three of the labs demonstrated differences $<0.2\phi$. Therefore it can be safely stated that the SediGraph has high internal precision and high precision between different SediGraph instruments.

The final calibration procedure, that of comparing results between different sizing instruments and methods, is also discussed in detail in Chapter 13. Based on fifty-five samples, Welch et al. (1979) compared the SediGraph with the pipette method and found a correlation of 0.97 with a standard deviation from the regression of 5.54%. When the organic fraction was removed from these samples, the correlation coefficient was even higher ($r = 0.99$). Stein (1985) also found very good agreement between the SediGraph and the Atterberg method (Fig. 10.10A) as well as between the SediGraph and the Coulter Counter technique (Fig. 10.10C). Stein noted that a noisy relation in the latter comparison was due to the presence of biogenic opal (i.e., where the particle density between the opal and mineral grains was significant – the Coulter Counter is not based on settling velocity of the particle). Stein also noted that montmorillonite-rich samples were difficult to analyze on the SediGraph (Fig. 10.10B). The thixotropic properties of the montmorillonite were enough to change the viscosity and hinder grain settling. Singer et al. (1988) compared the SediGraph results to those obtained from Malvern Laser Sizer (E3600), Electozone Particle Counter (112), and a hydrophotometer, and found small differences mostly related to the different particle properties being measured by each instrument.

Conclusions

Since its arrival onto the market in the early 1970s, the SediGraph has become one of the most widely used instruments in the earth sciences for the grain size analysis of fine-grained sediments. This trend is due largely to:

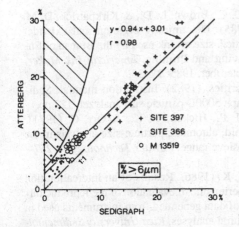

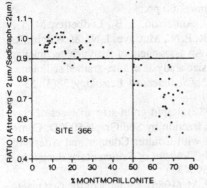

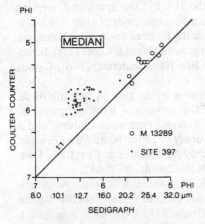

Figure 10.10. (A) Comparison of grain size data derived from SediGraph and Atterberg analyses. (B) Correlation of the deviation between the SediGraph and Atterberg results (<2-μm values) and the content of montmorillonite. (C) Comparison of grain size data (median values) obtained from SediGraph and Coulter Counter analyses (reproduced with permission from Stein [1985]).

1. its ease of automation and unattended operation versus other more labour-intensive techniques (pipette and hydrometer), and

2. the reduction in analysis time required for a routine analysis to a matter of minutes instead of hours.

This automation also allowed very fine digitization of the grain size cumulative curve for input into sophisticated sedimentological applications and statistics. In addition, the size of sample required was reduced to a few grams. However, its relatively high capital cost (>$30,000 U.S.) is a considerable financial outlay for all but the larger research laboratories.

Apart from the above operational advantages, the SediGraph also provided improvements in the precision (reproducibility) of fine sediment size analysis, so long as some important conditions were met. The most noteworthy of these is maintenance of the suspended sediment concentration below 3%. Comparison testing (Singer et al., 1988) has concluded that, for sorted silts, the SediGraph matches or outperforms the other techniques. However, if the clay content is high, then the SediGraph results differ significantly from the others. Also, the presence of low-density minerals in the silt fraction (e.g., biogenic opal) causes deviations between the SediGraph and the Coulter Counter, for instance.

In summary, since the SediGraph is based on sound theoretical principles, the results accurately describe the grain size attributes of a sediment sample in terms of the diameter of equivalent spheres. With minimum calibration, the statistical results are comparable to those obtained from other techniques using the same settling principle. Differences might occur with respect to other techniques – such as those based on grain imaging – but in general these are negligible.

ACKNOWLEDGMENTS

The authors are grateful to Micromeritics for supplying descriptive material on the SediGraph, and to the authors of the reports from which figures were reproduced for their kind permission to publish them here. The technical assistance of K. W. G. LeBlanc, D. A. Clattenburg, and K. W. Asprey (Atlantic Geoscience Centre) in preparing the manuscript and in laboratory experiments is gratefully acknowledged. G. A. Duncan, N. A. Rukavina, J. S. Mothersill, and R. Kihl provided valuable comments during review of the manuscript. This report comprises Geological Survey of Canada contribution no. 34489.

REFERENCES

Batchelor, G. K. (1972). Sedimentation in a dilute dispersion of spheres. *Journal of Fluid Mechanics*, 32(2): 245–68.

Berezin, P. N., & Voronin, A. D. (1981). Use of a SediGraph for the particle-size analysis of soils. *Soviet Soil Science*, 13(2): 101–9.

Brenner, H. (1964). Effect of finite boundaries on the Stokes resistance of an arbitrary particle. Part 2. Asymmetrical orientations. *Journal of Fluid Mechanics*, 18(1): 144–58.

Bunville, L. G. (1984). Commercial instrumentation for particle size analysis. In: *Modern Methods of Particle Size Analysis*, ed. H. G. Barth. New York: John Wiley & Sons, pp. 1–42.

Caswell, B. (1970). The effect of finite boundaries on the motion of particles in non-Newtonian fluids. *Chemical Engineering Science*, 25: 1167–76.

Coates, G. F., & Hulse, C. A. (1985). A comparison of four methods of size analysis of fine-grained sediments. *New Zealand Journal of Geology and Geophysics*, 28: 369–80.

Duncan, G. A., & Rukavina, N. A. (1982). Grain-size analysis with the SediGraph analyser. Abstr. *11th. International Congress on Sedimentology (IAS)*, Hamilton, Canada, 11: 165.

Gibbs, R. J., Matthews, M. D., & Link, D. A. (1971). The relationship between sphere size and settling velocity. *Journal of Sedimentary Petrology*, 41: 7–18.

Goldstein, J. I., Newbury, D. E., Echlin, P., Joy, D. C., Fiori, C., & Lifshin, E. (1981). *Scanning Electron Microscopy and X-ray Microanalysis*. New York: Plenum Press, 673 pp.

Jones, K. P. N., McCave, I. N., & Patel, P. D. (1988). A computer-interfaced sedigraph for modal size analysis of fine-grained sediment. *Sedimentology*, 35(1): 163–72.

Kalita, C. C., Brown, L. D., & Kirkpatrick, D. M. (1985). Determining pigment and extender particle size for ink used by the Bureau of Engraving and Printing. *American Ink Maker*, September, 1985: 27–38.

Micromeritics. (1982). Instruction manual: SediGraph 5000D particle size analyzer, 122 pp.

Oliver, J. P., Hicken, G. K., & Orr, C. (1971). Rapid, automatic particle-size analysis in the subsieve range. *Powder Technology*, 4: 257–63.

Singer, J. K. (1986). Results of an intercalibration experiment: An evaluation of the reproducibility of data generated from instruments used in textural analyses. *Rice University Sedimentology Report.*, 60 pp.

Singer, J. K., Anderson, J. B., Ledbetter, M. T., Jones, K. P. N., McCave, I. N., & Wright, R. (1988). An assessment of analytical techniques for the size analysis of fine-grained sediments. *Journal of Sedimentary Petrology*, 58(3): 534–43.

Stein, R. (1985). Rapid grain-size analyses of clay and silt fraction by SediGraph 5000D: Comparison with Coulter Counter and Atterberg methods. *Journal of Sedimentary Petrology*, 55(4): 590–3.

Syvitski, J. P. M. (1986). Minutes of the first meeting of the IUGS–COS sponsored working group on modern methods of grain size analysis, held at the Bedford Institute of Oceanography, Dartmouth, Canada. Copies may be obtained at Box 1006, Dartmouth, N.S., Canada, B2Y 4A2.

(1987). Minutes of the second meeting of the IUGS–COS sponsored working group on modern methods of grain size analysis, held at the University of Heidelberg, West Germany. Copies may be obtained at Box 1006, Dartmouth, N.S., Canada, B2Y 4A2.

Vitturi, L. M., & Rabitti, S. (1980). Automatic particle-size analysis of sediment fine fraction by SediGraph 5000D. *Geologia Applicata e Idrogeologia*. 15: 101–8.

Welch, N. H., Allen, P. B., & Galindo, D. J. (1979). Particle-size analysis by pipette and SediGraph. *Journal of Environmental Quality*, 8(4): 543–6.

11 Size, shape, composition, and structure of microparticles from light scattering

Miroslaw Jonasz

Introduction

When light impinges on a small particle, a fraction of the incoming light power is redirected in a process called the scattering of light. The angular distribution of intensity of the scattered light depends on the size, shape, orientation, composition, and structure of the particle. The scattered light thus carries information about these characteristics of the particle.

The use of light as a probe permits nondestructive characterization of a particle. The light intensity (photon flux) and wavelength (photon energy) are usually too low to damage the particles, even delicate biological cells. The scattered light can be measured in situ; thus the particles need neither be isolated from their natural environment nor specially prepared. In contrast, the preparation required by many other methods often modifies the particles and may render them unusable for further studies. Since the scattered light can be measured continuously, the temporal changes in particle characteristics can also be determined.

As with any other characterization method, the use of light scattering has its limitations. The information about the particle characteristics carried by the scattered light is encoded in a complex manner and often cannot be fully retrieved using currently available techniques. In addition, the scattered light can be extremely weak and may need sophisticated detection systems and time to be measured accurately.

The purpose of this paper is to introduce the reader to light scattering by small particles, its dependence on the particle characteristics, and its possible uses in oceanography and marine geology.

How light scattering occurs

From the classical "point of view" of atoms, light is an oscillating electric field (as opposed to quantum mechanical). Electrons in the atoms vibrate in response to this field. The response of an electron depends on its natural frequency in a particular atomic configuration and on the frequency of the electric field oscillations (see, e.g., Weiskopff, 1970). Vibrating electrons radiate light in all directions; hence a fraction of the directed power of the incoming light is scattered.

Refractive index

The refractive index of a medium characterizes the willingness of electrons in a particular atomic configuration to follow the oscillations of the electric field. The refractive index n is represented by the complex number:

$$n = n_r - in_i \tag{11.1}$$

where $i = (-1)^{1/2}$. The real part n_r of the refractive index determines the phase delay (or advancement) of the light wave in the medium. The phase change translates into the reduction (or increase) of the velocity of light v in the medium as compared with the velocity c of light in vacuum:

$$v = c/n_r \tag{11.2}$$

The imaginary part n_i characterizes the absorption (or amplification if n_i is negative, as in laser media) of light in the medium. The imaginary part of the refractive index relates to the absorption coefficient a of the medium as follows (see, e.g., Bohren & Huffman, 1983):

$$a = n_i \, 4\pi/\lambda \tag{11.3}$$

where λ is the wavelength of light. The light power incoming onto the slab of matter with absorption coefficient a and thickness z will be reduced from $F(o)$ to $F(z)$ according to Lambert law:

$$F(z) = F(o) \exp(-az) \tag{11.4}$$

[For elementary treatment of the refractive index see Feynman et al. (1963), Crawford (1968), and Weiskopff (1970). More elevated discussion is provided by Bohren & Huffman (1983).]

The refractive index can vary greatly with

the wavelength in wavelength bands for which the medium absorbs light. The imaginary part of the refractive index has a significant magnitude only in these wavelength bands. The refractive index for other wavelengths varies slowly with the wavelength and can be thought of as a real constant. The refractive index of the particle medium will be expressed in this text relative to the refractive index of water (i.e., as a ratio of the refractive index of the medium to that of water).

The refractive index depends on the chemical composition of matter and its density. The refractive indices of minerals common in seawater in particulate form range from ~1.08 to ~1.24, increasing roughly linearly with the density of the mineral (Carder et al., 1974). The imaginary part n_i of the refractive index of minerals is negligible in the visible region of the spectrum, which has no absorption bands.

Phytoplankton cells have indices n_r between 1.01 and 1.10 (Carder et al., 1972; Aas, 1981; Jonasz, 1986; Spinrad & Brown, 1986; Ackleson & Spinrad, 1988; Stramski et al., 1988). The refractive index of phytoplankton may change in response to environmental conditions (Ackleson & Spinrad, 1988). Bacteria have refractive indices similar to those of the phytoplankton (Ross & Billing, 1957). Generally, the more water is contained in a particle, the closer to 1 is the refractive index of the particle. Biological cells contain a number of pigments, most notably chlorophylls, that have strong absorption bands in the visible. For example, chlorophyll *a* absorbs around 475 nm and 680 nm (Rabinovitch & Govindjee, 1969). Fairly high values (0.001–0.01) of the imaginary part n_i of the refractive index are reported for biological particles suspended in seawater. A 10-μm-thick phytoplankton cell with $n_i = 0.01$ transmits at 0.475 μm about 7% of the incident light power (eqs. (11.3) and (11.4)). Practically no light passes through millimeter-thick slabs of matter with such high n_i values. However, microscopic observations do show low transmission through some phytoplankton cells (Ackleson & Spinrad, 1988).

Particle size and shape

If the atoms or molecules of the medium are uniformly distributed in space – that is, if the medium is optically homogeneous – then the light radiated in all directions interferes destructively except in the original direction of light propagation. Thus such a medium does not change the direction of light propagation (Weiskopff, 1970).

In some media, such as air, only the time-averaged spatial distribution of the molecules is homogeneous. As a function of time, the number of molecules of air per unit volume fluctuates around the average value. Light waves radiated by the molecules of such medium in directions different from the propagation direction of the incoming light are not canceled and scattering occurs. Water molecules are packed more orderly than are molecules of air. Thus the light scattered per unit length by water (green light, Morel [1974]) is only 100 times greater than that by air (McClatchey et al., 1978), although water is 1,000 times more dense.

If the infinite and homogeneous medium contracts to a finite particle, the waves scattered by the atoms at the particle walls do not cancel and the particle scatters light. The angular scattering pattern of light by the particle depends on the spatial arrangement of the atoms (i.e., on the particle shape).

The particle size and density of the particle material also influences the scattered light power. Waves scattered by identical atoms arrive and add at the observation point with phases that depend on the locations of the atoms. A light detector responds to the square of the sum of the wave amplitudes (Yariv, 1985, p. 346). The square of this sum is another sum containing squares of the amplitudes of individual scattered waves and the interference terms. Each such term is a product of the amplitudes of waves from two different atoms and of the cosine of the related phase difference. If the atoms are randomly distributed in space and are separated by distances that are large as compared with the wavelength of light, then large and random phase differences result. Thus the cosines assume effectively random values from the range −1 to +1. Consequently the interference terms cancel and the intensity of light scattered by N atoms is only N times greater than that scattered by one atom. However, if the atoms form a particle, the average distance between the atoms is much less than the wavelength of light. Each cosine is now nearly 1 because the phase differences are very small. There are N^2-N interference terms and N amplitude squares. Since they are all nearly equal, the total intensity of scattered light is approximately N^2 times greater than the light intensity scattered by one atom.

The definition of particle size presents a formidable problem (Allen, 1975) because particles are not always spherical. The definition of size thus depends on the sizing technique used. The *electrosizing* or *resistive* technique is represented by Coulter (Sheldon & Parsons, 1967) and Elzone (Snyder & Carson, 1986) particle counters. This technique, often used in marine sciences, defines the particle "diameter" as the diameter of a sphere with volume equal to that of the particle. Also in use is the *hydrodynamic* size defined by sedimentation techniques (Allen, 1975). The sedimentation technique is often combined with the optical detection (Zaneveld et al., 1982).

The suspended marine particles range in size from colloids ($<1\ \mu$m) to whales (>10 m). Based on the data obtained mostly with resistive particle counters, the concentration of particles (number of particles per unit volume) is approximately inversely proportional to the fourth power of particle size (Sheldon et al., 1972; Jonasz, 1983) in the size range of about 1–100 μm. There is some experimental evidence that this trend reverses for particles ≤ 0.5 μm (Lambert et al., 1981) (i.e., just below the lower resolution limit of the resistive counters).

Particles with sizes ranging from a few hundredths of a micron to several tens of microns can be characterized using visible light because it propagates through water with little attenuation (Jerlov, 1976). Also there are relatively simple methods of generating, guiding, and detecting visible light. Large particles can be more easily characterized using other methods: If the reader wanted to determine the size, shape, and composition of a fish, the light scattering would probably be of little use.

Marine particles come in an infinite variety of shapes and compositions. Spherical but inhomogeneous particles do exist in seawater – mostly phytoplankton cells. However, a typical marine particle is nonspherical.

The angular pattern of light scattering depends on the relative particle size x, defined as:

$$x = \pi D / \lambda \qquad (11.5)$$

where D is the geometrical particle size (e.g., the diameter of a sphere). The light scattering pattern of a particle is the same as that of an f-fold larger replica of the particle at an f-fold larger wavelength, if the refractive index and orientation of the particle in relation to the incident wave are the same in both cases. The effect of particle shape and orientation on light scattering can be studied more easily and in more detail when the particle is large. This inspired experiments in which the scattering of microwaves from large replicas simulates visible light scattering by micron-sized particles (Greenberg et al., 1961; Zerull, 1976).

Models of light scattering

Small particles can be characterized more effectively using light scattering, if mathematical models link the size, composition, shape, and structure of the particle with the scattered light. A brief presentation of the existing mathematical models of light scattering should give the reader a feeling for the approximations simplifying the problem of light scattering.

Rayleigh: Very small particle

Light scattering by very small (relative to the wavelength of light) particles was first explained by Lord Rayleigh in 1881. In the Rayleigh approach (van de Hulst, 1957; Bohren & Huffman, 1983) the electric field of the incident wave is assumed homogeneous within the particle. This requires that the particle be much smaller than the wavelength:

$$x \ll 1 \qquad (11.6)$$

It is also required that the entire particle respond "instantaneously" to the temporal variations in the electric field – that is, light does not slow down (eq. 11.2) appreciably inside the particle:

$$|n|x \ll 1 \qquad (11.7)$$

The particle then behaves as a single electric dipole oscillating in phase with the electric field of the incident light wave. As a rule of thumb, a particle that is smaller than one-tenth of the light wavelength is the Rayleigh particle. The reader should be on guard when researching Rayleigh scattering because there are other meanings in use (Young, 1981).

Rayleigh–Debye–Gans: Small particle with the refractive index close to that of the surrounding medium

The electric field is not homogeneous in a particle that is large as compared with the wavelength of light. Such a particle, however, can be thought of as composed of fragments that are so small that each "feels" a homogeneous field. Individual waves scattered by the fragments interfere at the observation point according to their phases.

The light scattered by a particle fragment depends on the field of the incident wave and of the waves radiated by other fragments. These waves depend, in turn, on the light scattered by the fragment in question. Thus we must know in advance the field we want to find. This difficulty can be overcome if the refractive index of the particle is only slightly different than that of the surrounding medium:

$$|n-1| \ll 1 \qquad (11.8)$$

The unity inside the absolute value symbol in eq. (11.8) is the refractive index of the medium relative to itself, because n is the refractive index of the particle relative to that of the medium. If the condition expressed by eq. (11.8) is fulfilled, then the particle interacts very weakly with the incident light. Thus the wave scattered by a fragment is so weak that the neighboring fragments "feel" mostly the incident wave. This assumption leads to the Rayleigh–Debye–Gans model of light scattering (van de Hulst, 1957; Bohren & Huffman, 1983). It is also required that, relative to a common origin, the phase differences between waves scattered by the fragments depend only on the relative position of the fragments and not on the refractive index of the particle; that is, the phase change caused by the particle material is small:

$$|n-1| \, x \ll 1 \qquad (11.9)$$

As one suspects, the shape, structure, and orientation of a Rayleigh–Debye–Gans particle influences the angular light scattering pattern.

Van de Hulst (anomalous diffraction): Large particle with the refractive index close to that of the surrounding medium

This model was proposed by van de Hulst (1957). When the particle is very large – that is, when

$$x \gg 1 \qquad (11.10)$$

we can trace a light ray through the particle. If the condition expressed by eq. (11.8) is fulfilled, then the particle merely shifts the phase of the ray. The phase shift is the product of the ray path length inside the particle and of the difference $n-1$ between the refractive indices of the particle and the surrounding medium. Thus the electric field just behind the particle can be easily calculated.

Mie: Homogeneous sphere of arbitrary size

Gustav Mie gave a unifying mathematical description of light scattering by spheres in 1908. The Mie model (van de Hulst, 1957; Kerker, 1969; Born & Wolf, 1980; Bohren & Huffman, 1983) is a solution of the Maxwell equations (Born & Wolf, 1980) describing propagation of the electromagnetic wave of light in space. Mie obtained the solution for a plane wave (i.e., the wavefronts of which are planes) incident on a homogeneous sphere. The Mie model is valid for any diameter and refractive index. The Mie solution of Maxwell equations is a slowly converging series. Before the advent of computers, Mie computations were rarely done, and then only for small particles. Today a PC can be used to calculate the Mie series, especially when improved Mie algorithms are used (Lentz, 1976; Wiscombe, 1980). Computer code is listed, for example, by Bohren & Huffman (1983).

Particles that are large in relation to the wavelength of light are sometimes called *Mie particles*. However, the Mie model applies to spheres with diameters between 0 and infinity.

Others: Arbitrary particles

Aden & Kerker (1951) provided a rigorous solution for a layered sphere. Bhandari (1985) extended that solution to include a multilayered sphere, permitting calculations of light scattering by a sphere with radially varying refractive index.

Asano & Yamamoto (1975) presented the rigorous model of light scattering for spheroids of arbitrary size and refractive index. Barber & Yeh (1975) developed an Extended Boundary Condition Model for calculation of light scattering by a particle of arbitrary shape. Such models lead to calculations that are neither easy nor quick (Mugnai & Wiscombe, 1980).

Purcell & Pennypacker (1973) developed a conceptually simple model of light scattering by an arbitrary particle. They approximated the particle by a lattice of small fragments, much as in the Rayleigh–Debye–Gans model, and added waves radiated by the fragments. However, in the Purcell–Pennypacker approach the interaction between the fragments is taken into account. Their model runs into numerical difficulties for large particles. Singham & Bohren (1987, 1989) proposed an improvement that eases these difficulties and extends the applicable size range.

Chylek (1976) and Chylek et al. (1976) pointed out that one of the major differences between the scattering by spheres and nonspheres is in the support of the surface waves (van de Hulst, 1957) by the spheres. The interference of these waves with the incident light causes sharp peaks in the scattered light intensity plotted against the relative particle size. Thus Chylek et al. (1976) suggested the modeling of light scattering of nonspherical particles by suppressing the surface waves in the Mie model. Acquista (1978) criticized this approach as leading to a fictitious absorption of light in the particle.

Pollack & Cuzzi (1980) presented a semiempirical, three-parameter composite model. In their model the scattering pattern of a particle smaller than some critical size x_0 is calculated using the Mie model. Light scattering by larger particles is modeled using a combination of light diffraction, refraction, and reflection. The components of this combination are influenced by two parameters depending on the average shape irregularity of the particles. One of the parameters adjusts the intensity of the total scattered light to account for the excess of the projected area (averaged over all orientations) of nonspherical particles as compared with that of the equal-volume spheres. The remaining parameter accounts for the influence of the particle shape on the angular asymmetry of the refracted light.

At present we know in principle how to calculate the light scattering properties of a relatively small arbitrary particle. However, the realistic models of light scattering by an arbitrary particle are complicated and require substantial computational power. Numerical problems prevent us from performing the light scattering calculations for large arbitrary particles. Calculations of light scattering by ensembles of many different, randomly oriented particles that realistically simulate natural particle ensembles are impractical because of the computational power requirements. These difficulties still drive many to disregard any shape and structure of real particles and to follow the physicist (referred to by Kerker et al. [1979]) who says about the horse: "Let's assume that it is a sphere!" Such a simplification may not always work (Latimer, 1984a), especially when one attempts to characterize particles using light scattering. Irregular particles do scatter light differently than do spheres.

Characteristics of scattered light
Single particle

The angular pattern of light scattered by a particle is characterized by the differential scattering cross section σ, defined in eq. (11.11). This equation relates the power of light scattered into a small solid angle $d\Omega$ about direction (Θ, ϕ) (Fig. 11.1) to the irradiance E_i (power/area) of the incident light:

$$dF(\Theta,\phi,n,\lambda) = \sigma(\Theta,\phi,n,\lambda)\, E_i\, d\Omega \qquad (11.11)$$

The differential cross section has the dimension of [area · (solid angle)$^{-1}$] and is interpreted as an effective projected area of the particle for light scattering at direction (Θ, ϕ). The scattered light power $dF/d\Omega$ per unit solid angle about direction (Θ, ϕ) equals the light power σE_i that falls upon the effective area σ of the particle.

The scattering angle Θ is measured from the direction of the incident light to the direction

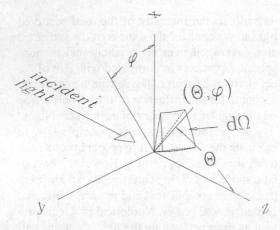

Figure 11.1. Geometry of light scattering by a single particle. The particle is located at the beginning of the reference frame *xyz*. The incident light travels along the *z*-axis. The observation direction (Θ, ϕ), the axis of the solid angle $d\Omega$, is specified by the scattering angle Θ and the azimuth angle ϕ.

of observation of the scattered light (Fig. 11.1). The azimuth angle ϕ is measured between the scattering plane and a reference plane. The scattering plane contains the directions of the incident light and of the observation. The reference plane (*xz* in Fig. 11.1) contains the direction of the incident light. Because of symmetry, the differential cross section of a homogeneous sphere is independent of the azimuth angle.

The differential cross section of a single particle is measured with a nephelometer capable of "freezing" the particle motion. In one approach, the particle is immobilized with a levitation device (Ashkin & Dziedzic, 1981; Weiss-Wrana, 1983). A flow-cytometric approach involves either a fast angle-scanning detector (Marshall et al., 1976) or a multiangle detector (Bartholdi et al., 1980), both of which are capable of measuring "instantaneously" the differential cross section of moving particles. Recent advances in flow cytometry (Steinkamp, 1984; Shapiro, 1988) make possible routine measurements of light scattering by single particles.

The incident light may be linearly polarized – that is, the plane of vibrations of the electrical vector of the light wave may be fixed (Born & Wolf, 1980; Hecht, 1987). Even if the incoming light is not polarized, the scattered light may be at least partly polarized as a result of interaction with the particle. The full description of

scattering of polarized light requires a 4×4 Mueller matrix (Hecht, 1987) instead of one scalar σ. The scalar E_i (eq. (11.11)) is then replaced by the Stokes vector with four components called the *Stokes parameters* (Born & Wolf, 1980; Hecht, 1987). The Stokes vector uniquely characterizes the power and polarization of the incoming light. The Mueller matrix transforms the Stokes vector of the incoming light into the Stokes vector of the scattered light (corresponding to dF in eq. (11.11)). Each of the sixteen components of the Mueller matrix is a function of two angles, Θ and ϕ, and of the wavelength. In the usual formulation the element m_{11} of the Mueller matrix corresponds to the differential scattering cross section for the unpolarized light. If the scattering medium behaves symmetrically, then some of the elements of the Mueller matrix vanish. Full treatment of polarized light scattering is given by Bohren & Huffman (1983) and van de Hulst (1957).

Integration of the differential cross section σ over all directions yields the integral cross section s. The integral cross section has the dimension of [area]. The light power sE_i falling onto s is scattered. The effective area can differ substantially from the projected area of the particle (see the section on "Light scattering by spheres").

Many particles: The scattering medium

A light scattering experiment usually involves a large number of particles. We thus need to relate light scattering properties of individual particles to the light scattering properties of a medium composed of many such particles. In the following text we assume that the particle concentration is low enough that the light scattered by a particle is not rescattered by another particle. The scattered light power is then proportional to the particle concentration. This is referred to as *single scattering,* as opposed to the multiple scattering that occurs in media with high particle concentrations. *Multiple scattering* can significantly modify the interaction of light with the medium as compared to the interaction applicable for the limiting case of the single scattering (Bohren, 1987).

The light power scattered by a small volume dV of a medium into solid angle $d\Omega$ about

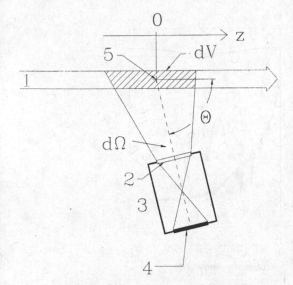

Figure 11.2. Geometry of light scattering by a medium with axial symmetry about the z-axis: (1) incident light beam; (2) entrance aperture of the nephelometer detector; (3) detector housing; (4) detector; (5) center of the detector rotation about the axis perpendicular to the paper plane. The field of view of the detector is determined by apertures 2 and 4 as well as by the distance between them, the distance between aperture 2 and rotation center 5, and by the refractive indices of the media inside and outside the detector housing.

a direction-making angle Θ with the direction of the incident light (Fig. 11.2) is given by the following equation:

$$dF(\Theta) = \beta(\Theta) \, E_i \, d\Omega \, dV \qquad (11.12)$$

where β is the volume scattering function of the medium. Equation (11.12) is the operational definition of the volume scattering function (Jerlov, 1976). The particles are assumed to be randomly oriented; hence the azimuth angle Θ (Fig. 11.1) is omitted. The volume dV is determined by the intersection of the incident light beam and the field of view of the detector (Fig. 11.2). The volume scattering function has the dimension of [length^{-1} · (solid angle)$^{-1}$].

Particles suspended in seawater are randomly distributed in space and are usually far enough from each other so that single scattering dominates. The volume scattering function of such a medium is a linear combination of the differential cross sections of all particles within the volume dV:

$$\beta(\Theta) = \left\langle \sum_i \sigma_i(\Theta,\phi,n_i,\lambda) \, N_i \right\rangle \text{ average over } \phi$$

$$(11.13)$$

where i numbers the collections of N_i identical particles per unit volume that are located within the scattering volume. The concentration of particles is usually large enough to justify replacement of the sum in eq. (11.13) by the integral. If the particles are homogeneous spheres, then N_i can be replaced by $dN = FD(D) \, dD$, where D is the particle size and $FD(D)$ is the size distribution (i.e., a relationship between the concentration of the particles and particle size).

The volume scattering function is determined using a nephelometer (Kullenberg, 1968; Jerlov, 1976; Jonasz & Prandke, 1986), which measures power of the scattered light relative to that of the incident light (eq. (11.12)). Thus the nephelometer requires no light-sensitivity calibration. However, the product of volume dV and the acceptance angle $d\Omega$ of the detector (Fig. 11.2) is usually determined experimentally (Pritchard & Elliot, 1960; Fry, 1974; Holland, 1980) because it depends in a complex way on the nephelometer geometry. Systematic errors of the volume scattering function measurements have been analyzed recently (Jonasz, 1990). The measurements of the Mueller matrix are similar to those of the scattering function, except that the polarized beam of the incident light is used and the scattered light is filtered with polarizers (Holland & Gagne, 1970; Hunt & Huffman, 1973; Bickel & Stafford, 1981).

The scattering coefficient of the medium is the integral of the volume scattering function over the 4π solid angle. The scattering and absorption coefficients add to form the attenuation coefficient that one may use in the Lambert law (eq. (11.4)).

Light scattering by spheres: Effects of size and refractive index

The differential cross section σ of a Rayleigh sphere for unpolarized light is proportional to $1 + \cos^2 \Theta$. The differential cross section is symmetrical about $\Theta = 90°$, where it assumes the minimum (Fig. 11.3). The differential cross section of a homogeneous sphere much smaller than the wavelength of light does not change

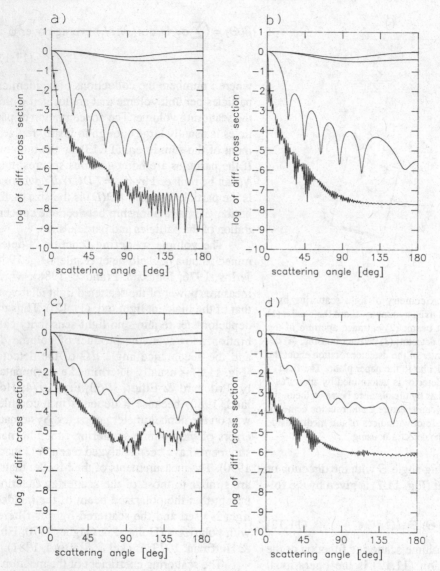

Figure 11.3. The differential scattering cross sections of spheres versus the scattering angle. The refractive indices of spheres are (a) 1.03, (b) 1.03–0.01*i*, (c) 1.2, and (d) 1.2–0.01*i*. The refractive indices were selected to show the difference in light scattering by a phytoplankton cell (*n* = 1.03) and a mineral particle (*n* = 1.2). The curves in each box represent the differential cross sections of spheres with the size *x* of 0.1, 1, 10, and 100 (top to bottom in each box). All cross sections are normalized to 1 at the scattering angle of 0°. The differential cross section is calculated using the Mie model. The scattering angle ranges from 0° to 180° with the increment of 1°.

with the sphere diameter. However, the particle shape or anisotropy influences σ. For randomly oriented nonspherical or anisotropic particles, σ is proportional to $A + B\cos^2\Theta$, where $1 \leq A \leq 1.1$ and $0.7 \leq B \leq 1$, depending on the particle shape and anisotropy (Bohren & Huffman, 1983).

Light scattered by a small homogeneous

sphere is completely polarized at $\Theta = 90°$. However, light scattered at $\Theta = 90°$ by a small anisotropic sphere or a nonsphere is only partly polarized (Bohren & Huffman, 1983).

As the particle size increases, the forward scattering becomes more pronounced (Fig. 11.3). The half-width of the differential cross section (the scattering angle at which σ is reduced to $0.5\sigma(0)$) decreases for large spheres as the inverse of the relative particle size (Kattawar & Plass, 1967). The differential cross section of large spheres oscillates as a function of the scattering angle (Fig. 11.3). These oscillations are caused by angle-dependent interference between the waves scattered by various fragments of the particle.

A sphere with a high refractive index scatters more light at the large scattering angles than a particle with a low refractive index: The angular light scattering pattern flattens (Fig. 11.3a,c). The flattening becomes more obvious as particle size increases. Absorption dampens the oscillations in the differential cross section and lowers the back-scattering (compare Fig. 11.3a,b, as well as Fig. 11.3c,d).

The integral cross section of small particles increases with relative size as x^6 (Fig. 11.4) for a constant wavelength of light. If the geometrical particle size is constant, the cross section varies with the wavelength as $D^6\lambda^{-4}$. The scattering of blue light is thus much stronger than the scattering of red light. Due to the wavelength selectivity of light scattering, the sky is blue.

The integral cross section begins to oscillate as function of the relative size for an intermediate size range (Fig. 11.4b), sometimes called the *resonance size region*. At the maxima, the integral cross section can be several times greater than the projected area of the sphere. The oscillations, most distinct for spheres, are suppressed for irregular particles (Hodkinson, 1963; Proctor & Barker, 1974). The oscillations decay for large spheres, and the cross section continues to increase with the relative size more slowly as x^2 (Fig. 11.4b), becoming approximately proportional to the projected area p. The integral cross section tends to a limit of $2p$ for the sphere diameter approaching infinity (van de Hulst, 1957, p. 107). Thus a very large particle scatters two times more light power than falls on its projected area. This is referred to by van de Hulst (1957) as the *extinction paradox*. The factor of 2 is a sum of two constants, both of which equal 1. The first constant represents the scattering or absorption of the light falling onto the particle; the second represents the diffraction of light by the particle.

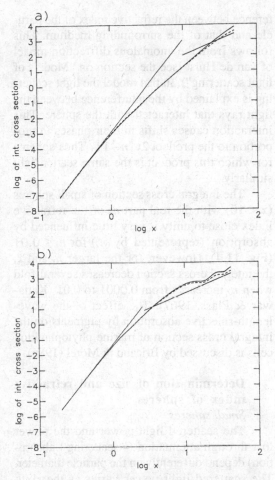

Figure 11.4. The integral scattering cross section of spheres versus the relative size x. The refractive indices of the spheres are (a) 1.03 (solid), 1.03–0.01i (dashed), and (b) 1.2 (solid), 1.2–0.01i (dashed). The sections of the straight lines represent the asymptotes of the curves for the small values of x (the asymptote proportional to x^6) and the large values of x (the asymptote equal to x^2). The integral cross section is calculated using the Mie model. The relative particle size x ranges from 0.1 to 100. The increment of x equals 0.1 in the range of $0.1 \leq x \leq 1$. The increment of x equals 1 in the range of $1 \leq x \leq 100$.

The resonance region moves toward the larger particle sizes when the refractive index of the particle decreases (Fig. 11.4a,b). However, the curves in Figure 11.4a,b become essentially identical if the integral cross section is divided by the projected area of the sphere and the result is plotted against the product $2x|n-1|$, where x is the relative particle size and $n-1$ is the difference between the refractive index of the particle and that of the surrounding medium. This follows from the anomalous diffraction model of van de Hulst (see the section on "Models of light scattering"). In that model the light scattering is explained by the interference between the light rays that interacted with the sphere. This interaction causes shifts in their phases in proportion to the product $2x|n-1|$. Thus spheres for which this product is the same scatter light similarly.

The integral cross section of small spheres ($x < 10$) with the real part n_r of the refractive index close to unity is very little influenced by absorption (represented by n_i) for $n_i \leq 0.01$ (Fig. 11.4). However, for the larger n_r and x, the integral cross section decreases several-fold when n_i increases from 0.0001 to 0.01 (Kattawar & Plass, 1967). The effect of the wavelength-selective absorption by pigments on the integral cross section of marine phytoplankton cells is discussed by Bricaud & Morel (1986).

Determination of size and refractive index of spheres
Small spheres
The scattered light power and the power lost through attenuation (= scattering + absorption) depend differently on the particle diameter. The scattered light power varies as the sixth power of the particle diameter D. The attenuation of light by absorbing particles, which is almost entirely due to the absorption, varies as the third power of the diameter. Thus the size and number of identical, small, absorbing particles can be determined simultaneously by measuring the scattering and attenuation (= scattering + absorption) of light by the particles at a single wavelength (Flower, 1984). The wavelength of light and the refractive index of the particles must be known.

If the particles do not absorb light and their concentration is unknown, their size or the size distribution can be determined by measuring the spectrum of light scattered by the particles (Elicabe & Garcia-Rubio, 1989).

Small spheres ($<1 \mu$m) can also be sized using dynamic light scattering or photon correlation spectroscopy (Berne & Pecora, 1976). This technique utilizes the influence of random

motion of the particles on the wavelength of light they scatter. The particles in suspension move randomly following collisions with the molecules of the medium. The wavelength of the light wave scattered by a moving particle differs from the wavelength of the incident wave. This phenomenon is referred to as the *Doppler shift* (Crawford, 1968, p. 429). If the particles are illuminated with a monochromatic light, the light scattered by randomly moving particles becomes polychromatic. This "spectral line broadening" depends on the mobility of the particles, which decreases with increasing particle size. The technique of dynamic light scattering can also be used to size small nonspherical particles. However, the estimate of particle size is then influenced by the rotations and flexing of the particles (Coulter Electronics, 1984). Paul & Jeffrey (1984) used the technique of dynamic light scattering to determine the size distributions of small marine particles. They found that the apparent size of live motile bacteria was several times smaller than the size of the killed bacteria. This difference was caused by the mobility of the live bacteria.

Large spheres
The angular light scattering pattern of large spheres is a function of the particle size and refractive index. This functional relationship can be used to size spheres using the Mie model.

In a typical optical particle sizer the angular scattering pattern is measured in a limited range of the scattering angles. The forward scattering is often chosen to diminish the effect of the refractive index of the particles. If the scattered light is measured at a single angle and a monochromatic light source, such as the helium–neon laser, is used, then the scattered light power is a nonmonotonic function of the particle size. By measuring the scattered light in an optimized angular range (Barnard & Harrison, 1988) or using a white light source (Cooke & Kerker, 1975), the nonmonotonicity can be minimized.

Commercially available laser diffractometers (Cornillault, 1972; Frock, 1987; Bale, Chapter 14, this volume) also use the forward scattering range. The particle size is derived from the angular width of the scattering pattern. This pattern is measured using a multielement circular array photodetector, and then equation (11.13) is solved numerically for the particle size distribution assuming a diffraction approximation for the differential scattering cross sections of individual spheres. Burkholz & Polke (1984) and McCave et al. (1986) critically assess the precision and accuracy of several laser diffractometers applied to sizing of nonspherical particles.

A comment on the difficulties in solving equation (11.13) for the particle size distribution of spheres is appropriate. The solution is very sensitive to the errors in the volume scattering function (Twomey, 1977). Chow & Tien (1976) and Twomey (1977) review techniques that reduce this sensitivity. Equation (11.13) can be solved more easily if the refractive index of spheres is known, because the differential cross section of spheres can then be calculated. If the refractive index is unknown, the trial-and-error method can be used (Kullenberg & Olsen, 1972; Brown & Gordon, 1974; Jonasz & Prandke, 1986). In this approach a refractive index and a size distribution are assumed, and the volume scattering function is calculated. The result is compared with the experimental data. The refractive index and size distribution are varied until the difference between the measured and calculated volume scattering functions reaches a minimum. This search does not guarantee that the global minimum has been reached.

The decrease in the half-width of the scattering pattern with the particle size is also utilized in a simple transmission photometer for particle sizing (Bryant & Latimer, 1985). The apparent transmission of light by a suspension of particles increases with the detector acceptance angle, because the detector collects a fraction of the scattered light power besides the light transmitted through the suspension. The rate of change of the apparent transmission with the detector acceptance angle increases with the particle size because the half-width of the scattering pattern of the particles decreases. The rate of the increase in the apparent transmission can be calculated for spheres and other simple particle shapes.

Tycko et al. (1985) developed a flow-cytometric method to determine simultaneously the size and refractive index of red blood cells

that were chemically sphered. In this method the scattering of light by individual cells is measured at two optimized forward scattering angles. Each pair of the scattered light intensities corresponds to a pair of the cell diameter and its refractive index. The correspondence is established using the Mie model. Ackleson & Spinrad (1988) applied this method to marine phytoplankton.

Light scattering by a sphere at a particular angle exhibits very sharp peaks (called *optical resonances*) when plotted against the wavelength. The sequence and shapes of these peaks is a unique signature of the particle and varies very strongly with the particle diameter and refractive index (Ashkin & Dziedzic, 1981). Chylek et al. (1983) used this phenomenon to determine simultaneously the size and refractive index of oil droplets to 1 part in 10^5. The particle size and refractive index, however, has to be known a priori with fairly high accuracy so that the measured particle signature can be matched with that computed using the Mie model. Arnold et al. (1985) used the technique of optical resonances to examine the absorption of infrared light by single liquid droplets. This could be done because absorbed power of the infrared light heats the droplet and some of the liquid evaporates, causing a decrease in the droplet size. The sensitivity of the technique is so high that evaporation of the monomolecular layer of liquid can be detected.

Light scattering by nonspheres: Effects of shape, structure, and orientation

Many particles that take part in important natural and industrial processes are nonspherical. Significant research effort has been undertaken to examine the features of light scattering by nonspherical and nonhomogeneous particles and to determine the differences between light scattering by spheres and nonspheres. Major directions of research in this field are as follows.

The scattering cross sections have been calculated for various particle shapes and orientations in relation to the direction of the incident light. The cross sections are often averaged over the particle orientation since the particles are randomly oriented in a typical light scattering experiment. Among the particle shapes considered are the following:

absorbing sphere with rough surface (Bahar & Chakrabarti, 1985),

spheroids (Asano, 1979, 1983; Asano & Sato, 1980; Latimer et al., 1978),

aggregates of colloidal particles (Latimer & Wamble, 1982),

dielectric helix (Chiappetta & Torresani, 1988),

hexagonal cylinders (Liou et al., 1983),

spheres with projections (Latimer, 1984a) and holes (Latimer, 1984b),

ice crystals (Muinonen, 1989) and randomly oriented ice crystals (Muinonen et al., 1989),

snow crystals (O'Brien & Goedecke, 1988), and

particles with shapes synthesized using Chebyshev polynomials (Mugnai & Wiscombe, 1980, 1986, 1989; Wiscombe & Mugnai, 1988).

Schiffer & Thielheim (1982) developed a physical-optics model of the reflection of light by randomly oriented convex particles with rough surfaces.

The scattering of microwaves by single models of small particles has been measured to determine variations of the scattering cross sections with particle shape and orientation. The scattering cross sections have been measured for:

spheroids and cylinders (Greenberg et al., 1961; Schuerman et al., 1981),

stacks of cylinders (Greenberg & Wang, 1971), and

cubes and particle aggregates (Zerull, 1976).

The scattering of visible light has been measured for:

cubes (Perry et al., 1978),

single ice crystals (Pluchino, 1986), and

single irregular particles (Weiss-Wrana, 1983).

The volume scattering functions and Mueller matrices of irregular particles with sizes in a wide range have been measured and compared with the results of calculations performed for equivalent spheres. The particle shapes examined include disklike talc powders (Holland & Gagne, 1970), nearly spherical aggregates of sodium chloride and potassium sulfate particles (Pinnick et al., 1976). Patitsas (1988) reported results of measurements of light scattering by

needlelike asbestos particles and attempted to explain the experimental data using the model of Rayleigh–Debye–Gans. Marine particles also received deserved attention (Kullenberg & Olsen, 1972; Brown & Gordon, 1974; Kullenberg, 1974; Kadyshevitch et al., 1976; Reuter, 1980; Fry & Voss, 1985; Jonasz & Prandke, 1986).

The influence of the particle structure on the light scattering by the particle has also been investigated. The scattering of light by layered spheres was studied theoretically (Brunsting & Mullaney, 1972; Pyle et al., 1979; Latimer & Wamble, 1982; Aas, 1984). Kerker et al. (1979) found that dense inclusions in small spheres influence mainly the back-scattered light. Light scattering by composite particles, modeled by water inclusions in acrylic, was investigated by Chylek et al. (1988), who found that the volume averaged refractive index is among the worst approximations to use. Bohren & Huffman (1983) discuss the problem of averaging the refractive index of a nonhomogeneous particle and summarize additional experimental results.

Although the references given here are far from complete, a quick glance in each will uncover more publications of note. A collection of papers on light scattering by nonspheres (Schuerman, 1980) and a recent compilation of outstanding contributions on light scattering (Kerker, 1987) are recommended. The findings of these investigations are summarized in the following text (partly based on conclusions of Wiscombe & Mugnai [1988]).

There is a fair agreement between the angular scattering patterns of spheres and nonspheres up to the relative size $3 < x < 5$. This agreement worsens rapidly for larger sizes because light with a wavelength much smaller than the geometrical size of the particle can resolve particle shape. However, the forward scattering by large particles continues to be relatively insensitive to the particle shape. This can be qualitatively explained using the Rayleigh–Gans–Debye model as follows: The angular scattering pattern of a large particle is determined by the interference of light scattered by small fragments into which the particle can be subdivided. This interference depends on the phase differences between light scattered by the fragments. The phase differences are determined by the differences in the paths of light waves from the source to the observer *through* various particle fragments. The path differences are less dependent on the location of the fragments (i.e., on the particle shape) for the forward scattering than are the path differences for scattering in other directions. In the Rayleigh–Gans–Debye approximation the differences vanish for the scattering angle of 0. The relative insensitivity to the particle shape of the angular scattering pattern in the forward direction is consistent with the domination of diffraction in this angular range (Hodkinson & Greenleaves, 1963) and the relative insensitivity of diffraction to the shape of the diffracting obstacle (Latimer, 1978).

The nonspheres scatter more light than spheres in the middle range of the scattering angle (60°–120°). This side-scattering by nonspheres does not exhibit strong oscillations, such as those for spheres, and can be greater than the side-scattering by spheres up to an order of magnitude. In the back-scattering range (120°–180°) the nonspheres with relative sizes $8 < x < 12$ scatter less than spheres. The randomly oriented concave particles scatter more light than randomly oriented convex particles and spheres at most angles. The side- and back-scattering is more sensitive to the particle structure than is the forward scattering.

These conclusions should be applied with care. The transition from spheres to nonspheres opens infinite possibilities, only a fraction of which have been experimentally and theoretically explored.

Characterization of nonspherical and nonhomogeneous particles
Experimentally determined relationships between light scattering and particle properties
Even without the use of a model of light scattering, Bickel & Stafford (1981) were able to relate changes in the scattering of polarized light by spores and bacteria to subtle variations in particle morphology that may go undetected with other methods. Also, without the need for a model of interaction of light with particles, light scattering is used to determine the con-

centration of various proteins in human serum in precipitation immunoassays (Killingsworth, 1978). When antibody proteins are added to a sample with antigen proteins, the protein molecules form aggregates with the size increasing as the reaction progresses. The increase of light scattering by the sample caused by the increasing particle size outweighs the decrease of light scattering caused by the decreasing particle concentration. The net effect is the increase with time of the light scattered by the sample up to a plateau reached when all molecules have reacted. The reaction rate for a given pair of proteins depends in the first approximation only on the initial concentration of the antigen molecules. This concentration can thus be determined from the reaction rate using a calibration curve obtained for a series of known antigen protein concentrations.

Light scattering has been used to map the concentration of suspended particles in the ocean (Biscaye & Eittereim, 1977; McCave, 1983; Richardson, 1987) because the concentration of suspended particles correlates reasonably well with light scattering. It follows from equation (11.13) that light scattering is proportional to particle concentration (factored out from N_i's) if the form of the size distribution and the particle composition (refractive index) do not change. If the particles are irregular, their orientation must be fixed or random. Studies using resistive particle counters (Sheldon et al., 1972; Jonasz & Zalewski, 1978; Jonasz, 1983) confirm that the form of the size distribution of marine particles and of particles suspended in lake waters (Osmann-Sigg & Stumm, 1982) does not change in the first approximation. The variability of the form of the size distribution and of the particle composition is often more limited in smaller water masses or layers (Prandke & Jonasz, 1984; Richardson, 1987), and a better correlation between light scattering and particle concentration can be expected than that characteristic for the whole water bodies.

Application of models of light scattering by nonspheres
In some cases all particles may have essentially the same shape and size. An appropriate model can then be applied to deduct the particle

characteristics from light scattering (spheroidal bacterial cells *E. coli* [Cross & Latimer, 1972]). Latimer et al. (1977) determined the average ax- ial ratio of disk-shaped blood platelets from the change of light transmission caused by the change in platelet orientation.

If the particles are nonspherical, some ap- proximations usually have to be made before at- tempting the solution of equation (11.13); these will obviously decrease the accuracy of the so- lution. The summary of differences – spheres versus nonspheres – suggests that the replace- ment of the nonspherical particles by "equiv- alent" spheres may cause errors in the charac- teristics of the particles obtained using light scattering. One questioned finding (C. F. Boh- ren, pers. commun.) is the high absorption (n_i of the order of 0.01) of light by marine parti- cles determined with the use of light scattering (Brown & Gordon, 1974; Morel & Bricaud, 1981; Jonasz & Prandke, 1986; Stramski et al., 1988). Computations show that absorption in- fluences the angular pattern of light scattering especially at medium and large angles (Kerker et al., 1979), where the differences between the spheres and nonspheres are greatest. Thus the absorption estimate may be wrong if the non- spheres are replaced with spheres in solving equation (11.13).

Forward scattering by nonspheres: selection of the equivalent spheres

As already mentioned, forward scattering is less dependent on the particle shape than is side- and back-scattering; it is thus usually mod- eled using equivalent spheres. Bohren & Koh (1985) discuss the limitations of this approach. What equivalent spheres should be chosen? Typically a resistive particle counter is used to determine the particle size distribution in experi- ments aimed at comparing measured and calcu- lated light scattering. The size of a particle deter- mined using a resistive counter is approximately the size of a sphere with volume equal to that of the particle. C. M. Boyd & G. W. Johnson (pers. commun.) assess the validity of this ap- proximation. In this section I discuss experi- mental evidence that spheres with equal project- ed area are a better approximation than those

with equal volume when characterizing marine particles with light scattering.

The forward scattering of light by marine particles is well represented by the scattering coefficient, which is related to the integral cross sections of individual particles through an equa- tion equivalent to (11.13). The large particles are the major contributors to the forward scat- tering, and hence to the scattering coefficient (Brown & Gordon, 1974; Jonasz & Prandke, 1986). If a particle is large, its integral cross section is proportional to the projected area of the particle. For nonspheres this approximation is valid starting from a size smaller than that for spheres (Hodkinson, 1963; Proctor & Barker, 1974). The projected area of a nonspherical particle averaged over the particle orientation is larger than that of a sphere with equal volume. Strictly speaking this is a hypothesis, since the rigorous proof exists only for the convex parti- cles (Vouk, 1948; van de Hulst, 1957, pp. 110– 11). The orientation-averaged projected area of a small nonspherical particle (equivalent spher- ical diameter of ~0.1 μm) suspended in clear ocean water has been experimentally determined to be about the same as that of an equal volume sphere (Jonasz, 1987a). However, the project- ed area may be about twice that of the equal- volume sphere for a 10-μm particle (Jonasz, 1987a). Thus the large, randomly oriented non- spheres should scatter more light than do equal- volume spheres (Pollack & Cuzzi, 1980).

In an experiment (Jonasz & Prandke, 1986), the volume scattering function of marine particles was measured. The particle size distri- bution was also determined using a Coulter Counter. The volume scattering function of the particles was then computed using this size distribution and assuming the particles to be equal-volume spheres. Good agreement between measured and calculated light scattering for all scattering angles was obtained for the summer samples, which contained biological (low- refractive-index) particles. For the winter sam- ples, however, with mostly mineral (high- refractive-index) particles, no realistic refractive index could be used to increase the computed forward scattering to the level of the measured forward scattering. Comparison of Figure 11.4a to Figure 11.4b suggests that the scattering cross section for a given particle size increases with the refractive index of the particles up to a limit roughly depicted by the line representing x^2 for the large x. This may explain the failure of Mie modeling using the equivalent volume spheres. After the projected area in the model was increased to account for the nonsphericity of marine particles, a much better fit was achieved between the calculated and measured light scattering (Jonasz, 1987a). Volume-equivalent spheres frequently replace actual particles in algorithms for inferring particle properties from light scattering (see, e.g., Spinrad [1986]). The discrepancy between the projected area of nonspherical particles and that of equal-volume spheres may be one of the sources of errors in the inferred particle properties (Baker & Lavelle, 1984; Jonasz, 1987c).

Refractive index and nonsphericity of particles: A determination method

The simultaneous determination of the particle size distribution with a resistive and an optical particle counter can be used to determine the refractive index of the particles relative to that of the particles used to calibrate the optical counter (Jonasz, 1986). The resistive counter is sensitive to the particle volume, whereas the optical counter is sensitive to the projected area weighted by the scattering (or attenuation) cross section. The latter depends on the refractive index of the particles and on their shape. During calibration of the optical counter, its "optical size" scale is transformed into a geometric size scale by using the calibration spheres. As a result the size distributions of the calibration spheres obtained with each of the counter types are equal. When particles with refractive index and shape that differ from those of the calibration particles are analyzed with the optical counter, the "optical" size distribution differs from that obtained with the resistive counter (Jonasz, 1987b). This is because the particle cross sections are now different from those sensed in the optical counter calibration. If the analyzed particles, such as some phytoplankton species, can be approximated with spheres, the difference in the cross sections is caused only by the difference between the refractive indices of the analyzed particles and calibration spheres. Con-

versely, if the refractive index of the particles is known, their nonsphericity (measured as the ratio of the average projected area of a particle to that of an equal-volume sphere) can be determined.

Conclusions

Light scattering can be successfully used to characterize size, shape, structure, and composition of small particles nondestructively. The accuracy of such characterization depends on the quality of the physical model relating light scattering to the properties of the particles. Such a model is not, however, necessary for monitoring deviations in the particle characteristics from a norm in a repeatable particle-generating process.

The greatest successes in characterizing the particles with light scattering are still being obtained for spheres because the accurate Mie model of light scattering by spheres exists. The accuracy of characterization of nonspherical and nonhomogeneous particles is poorer, and substantial approximations must be made to make the problem mathematically tractable.

ACKNOWLEDGMENTS

I thank Dr. P. Mulqueen, F. Gregory, M. Celentano, and the referees for critical remarks.

REFERENCES

Aas, E. (1981). *The refractive index of phytoplankton.* University of Oslo, Institute of Geophysics, ref. no. 46: Oslo.

(1984). *Influence of shape and structure on light scattering by marine particles.* University of Oslo, Institute of Geophysics, ref. no. 53: Oslo.

Ackleson, S. G., & Spinrad, R. W. (1988). Size and refractive index of individual marine particulates: A flow-cytometric approach. *Applied Optics,* 27: 1270–7.

Acquista, Ch. (1978). Validity of modifying Mie theory to describe scattering by nonspherical particles. *Applied Optics,* 17: 3851–2.

Aden, A. L., & Kerker, M. (1951). Scattering of electromagnetic waves from two concentric spheres. *Journal of Applied Physics,* 22: 1242–6.

Allen, T. (1975). *Particle size measurements.* London: Chapman and Hall.

Arnold, S., Murphy, E. K., & Sageev, G. (1985). Aerosol particle molecular spectroscopy. *Applied Optics*, 24: 1048–53.

Asano, S. (1979). Light scattering properties of spheroidal particles. *Applied Optics*, 18: 712–22.

  (1983). Light scattering by horizontally oriented spheroidal particles. *Applied Optics*, 22: 1390–6.

Asano, S., & Sato, M. (1980). Light scattering by randomly oriented spheroidal particles. *Applied Optics*, 19: 962–74.

Asano, S., & Yamamoto, G. (1975). Light scattering by a spheroidal particle. *Applied Optics*, 14: 29–49.

Ashkin, A., & Dziedzic, J. M. (1981). Observation of optical resonances of dielectric spheres by light scattering. *Applied Optics*, 20: 1803–14.

Bahar, E., & Chakrabarti, S. (1985). Scattering and depolarization by large conducting spheres with rough surfaces. *Applied Optics*, 24: 1820–5.

Baker, E. T., & Lavelle, J. W. (1984). The effect of particle size on the light attenuation coefficient of natural suspensions. *Journal of Geophysical Research*, 89: 8197–203.

Barber, P., & Yeh, C. (1975). Scattering of electromagnetic waves by arbitrarily shaped dielectric bodies. *Applied Optics*, 14: 2864–72.

Barnard, J. C., & Harrison, L. C. (1988). Monotonic responses from monochromatic optical particle counters. *Applied Optics*, 27: 584–92.

Bartholdi, M. G., Salzman, G. C., Hiebert, R. D., & Kerker, M. (1980). Differential light scattering photometer for rapid analysis of single particles in flow. *Applied Optics*, 19: 1573–81.

Berne, B. J., & Pecora, R. (1976). *Dynamic light scattering*. New York: J. Wiley.

Bhandari, R. (1985). Scattering coefficients for a multilayered sphere: Analytic expressions and algorithms. *Applied Optics*, 24: 1960–7.

Bickel, W. S., & Stafford, M. E. (1981). Polarized light scattering from biological systems. *Journal of Biological Physics*, 9: 53–66.

Biscaye, P. E., & Eittereim, S. L. (1977). Suspended particulate loads and transports in the nepheloid layer of the abyssal Atlantic Ocean. *Marine Geology*, 23: 155–72.

Bohren, C. F. (1987). Multiple scattering of light and some of its observable consequences. *American Journal of Physics*, 55: 524–33.

Bohren, C. F., & Huffman, D. R. (1983). *Absorp-tion and scattering of light by small particles*. New York: Wiley–Interscience.

Bohren, C. F., & Koh, G. (1985). Forward-scattering corrected extinction by nonspherical particles. *Applied Optics*, 24: 1023–9.

Born, M., & Wolf, E. (1980). *Principles of optics*. Oxford: Pergamon Press.

Bricaud, A., & Morel. A. (1986). Light attenuation and scattering by phytoplanktonic cells: A theoretical modelling. *Applied Optics*, 25: 571–80.

Brown, O. B., & Gordon, H. R. (1974). Size-refractive index distribution of clear coastal water particulates from light scattering. *Applied Optics*, 13: 2874–881.

Brunsting, A., & Mullaney, P. F. (1972). Light scattering from coated spheres. *Applied Optics*, 11: 675–80.

Bryant, C. F., & Latimer, P. (1985). Real-time particle sizing by a computer controlled transmittance photometer. *Applied Optics*, 24: 4280–2.

Burkholz, A., & Polke, R. (1984). Diffraction spectrometers: Experience in particle size analysis. *Particle Characterization*, 1: 153–61.

Carder, K. L., Betzer, P. R., & Eggiman, D. W. (1974). Physical, chemical and optical measures of suspended particle concentrations: Their intercomparison and application to the West African shelf. In: *Suspended solids in water*, ed. J. Gibbs. New York: Plenum Press, pp. 174–93.

Carder, K. L., Tomlinson, R. D., & Beardsley, G. F., Jr. (1972). A technique for the estimation of indices of refraction of marine phytoplankton. *Limnology and Oceanography*, 17: 833–9.

Chiappetta, P., & Torresani, B. (1988). Electromagnetic scattering from a dielectric helix. *Applied Optics*, 27: 4856–60.

Chow, L. C., & Tien, C. L. (1976). Inversion techniques for determining the droplet size distribution in clouds: Numerical examination. *Applied Optics*, 15: 378–83.

Chylek, P. (1976). Partial-wave resonances and ripple structure in the Mie normalized extinction cross section. *Journal of the Optical Society of America*, 3: 285–7.

Chylek, P., Grams, G. W., & Pinnick, R. G. (1976). Light scattering by irregular, randomly oriented particles. *Science*, 193: 480–2.

Chylek, P., Ramaswamy, V., Ashkin, A., & Dziedzic, J. M. (1983). Simultaneous determination of refractive index and size of spherical dielectric particles from light scattering data. *Applied Optics*, 22: 2302–7.

Chylek, P., Srivastava, V., Pinnick, R. G., & Wang, R. T. (1988). Scattering of electromagnetic waves by composite spherical particles: Experiment and effective medium approximation. *Applied Optics*, 27: 2396–404.

Cooke, D. D., & Kerker, M. (1975). Response calculation for light scattering aerosol particle counters. *Applied Optics*, 14: 734–9.

Cornillault, J. (1972). Particle size analyzer. *Applied Optics*, 11: 265–8.

Coulter Electronics, Inc. (1984). *Coulter N4 Technical Information Booklet*. Hialeah, Florida.

Crawford, F. S., Jr. (1968). *Waves*. New York: McGraw–Hill.

Cross, D. A., & Latimer, P. (1972). Angular dependence of scattering from *E. coli* cells. *Applied Optics*, 11: 1225–8.

Elicabe, G. E., & Garcia-Rubio, L. H. (1989). Latex particle size distribution from turbidimetry using inversion techniques. *Journal of Colloid and Interface Science*, 129: 192–200.

Feynman, R. P., Leighton, R. B., & Sands, M. (1963). *The Feynman lectures on physics*. New York : Addison–Wesley.

Flower, W. L. (1984). Optical diagnostics for particles less than 0.1 micrometer. In: *Advanced diagnostics for particle-laden combustion flow*. Sandia National Laboratories: Albuquerque.

Frock, H. N. (1987). Particle size determination using angular light scattering. *American Chemical Society Symposium Series*, 332: 147–59.

Fry, E. (1974). Absolute calibration of a scatterance meter. In: *Suspended solids in water*, ed. R. J. Gibbs. New York: Plenum Press, pp. 101–9.

Fry, E. S., & Voss, K. J. (1985). Measurement of the Mueller matrix for phytoplankton. *Limnology and Oceanography*, 30: 1322–6.

Greenberg, J. M., Pedersen, N. E., & Pedersen, J. C. (1961). Microwave analog to the scattering of light by nonspherical particles. *Journal of Applied Physics*, 32: 233–42.

Greenberg, J. M., & Wang, R. T. (1971). Extinction by rough particles and the use of Mie theory. *Nature*, 230: 110–12.

Hecht, E. (1987). *Optics*. Reading, Mass.: Addison–Wesley.

Hodkinson, J. R. (1963). Light scattering and extinction by irregular particles larger than the wavelength. In: *Proceedings of the International Conference on Electromagnetic Scattering*, ed. M. Kerker. Oxford: Pergamon Press.

Hodkinson, J. R., & Greenleaves, I. (1963). Computations of light scattering and extinction by sphere according to diffraction and geometrical optics and some comparison with the Mie theory. *Journal of the Optical Society of America*, 53: 577–88.

Holland, A. C. (1980). Problems in calibrating a polar nephelometer. In: *Light scattering by irregularly shaped particles*, ed. D. W. Schuerman. New York: Plenum Press.

Holland, A. C., & Gagne, G. (1970). The scattering of light by polydisperse systems of irregular particles. *Applied Optics*, 9: 1113–21.

Hunt, A. J., & Huffman, D. J. (1973). A new polarization-modulated light scattering instrument. *Review of Scientific Instruments*, 44: 1753–62.

Jerlov, N. G. (1976). *Marine Optics*. Amsterdam: Elsevier.

Jonasz, M. (1983). Particle size distribution in the Baltic. *Tellus*, 35B: 346–58.

(1986). Role of nonsphericity of marine particles in light scattering and a comparison of results of its determination using SEM and two types of particle counters. *Proceedings of the Society of Photo-Optical Instrumentation Engineers*, 637: 148–54.

(1987a). Nonsphericity of suspended marine particles and its influence on light scattering. *Limnology and Oceanography*, 32: 1059–65.

(1987b). Nonspherical sediment particles: Comparison of size and volume distributions obtained with an optical and a resistive particle counter. *Marine Geology*, 78: 137–42.

(1987c). The effect of nonsphericity of marine particle on light attenuation. *Journal of Geophysical Research*, 92C: 14637–40.

(1990). Volume scattering function measurement error: Effect of angular resolution of the nephelometer. *Applied Optics*, 29: 64–70.

Jonasz, M., & Prandke, H. (1986). Comparison of measured and computed light scattering in the Baltic. *Tellus*, 38B: 144–57.

Jonasz, M., & Zalewski, M. S. (1978). Stability of the shape of particle size distribution in the Baltic. *Tellus*, 30: 569–72.

Kadyshevitch, Ye. A., Lyubovtseva, Yu. S., & Rozenberg, G. V. (1976). Light-scattering matrices of Pacific and Atlantic waters. *Journal of the Academy of Sciences of the USSR (Izv. Akad. Nauk SSSR): Atmospheric and Oceanic Physics*, 12: 106–11.

Kattawar, G. W., & Plass, N. G. (1967). Electromagnetic scattering from absorbing spheres. *Applied Optics*, 6: 1377–82.

Kerker, M. (1969). *The scattering of light.* New York: Academic Press.

(ed.) (1987). Selected Papers on Light Scattering. *Proceedings of the Society of Photo-Optical Instrumentation Engineers*, vol. 951.

Kerker, M., Chew, H., McNulty, P. J., Kratohvil, J. P., Cooke, D. D., Sculley, M., & Lee, M.-P. (1979). Light scattering and fluorescence by small particles having internal structure. *Journal of Histochemistry and Cytochemistry*, 27: 250–63.

Killingsworth, L. M. (1978). Analytical variables for specific protein analysis. In: *Automated immunoanalysis (I)*, ed. R. E. Ritchie. New York: Marcel Dekker.

Kullenberg, G. (1968). Scattering of light by Sargasso Sea water. *Deep-Sea Research*, 15: 423–32.

(1974). Observed and computed scattering functions. In: *Optical aspects of oceanography*, eds. N. G. Jerlov & E. Steeman-Nielsen. London: Academic Press.

Kullenberg, G., & Olsen, N. B. (1972). *A comparison between observed and computed light scattering functions (II)*. Report no. 19, Institute of Physical Oceanography, University of Copenhagen: Copenhagen.

Lambert, C. E., Chesselet, R., Jehanno, C., Silverberg, N., & Brun-Cottan, J. C. (1981). Lognormal distribution of suspended particles in the open ocean. *Journal of Marine Research*, 39: 77–98.

Latimer, P. (1978). Determination of diffractor size and shape from diffracted light. *Applied Optics*, 17: 2162–70.

(1984a). Light scattering by a homogeneous sphere with radial projections. *Applied Optics*, 23: 442–7.

(1984b). Light scattering by structured particle: The homogeneous sphere with holes. *Applied Optics*, 23: 1844–7.

Latimer, P., Born, G. V., & Michal, F. (1977). Application of light scattering theory to the optical effects associated with the morphology of blood platelets. *Archives of Biochemistry and Biophysics*, 180: 151–9.

Latimer, P., Brunsting, A., Pyle, B. E., & Moore, C. (1978). Effect of asphericity on single particle scattering. *Applied Optics*, 17: 3152–8.

Latimer, P., & Wamble, F. (1982). Light scattering by aggregates of large colloidal particles. *Applied Optics*, 21: 2447–55.

Lentz, W. J. (1976). Generating Bessel functions in Mie scattering calculations using continuous fractions. *Applied Optics*, 16: 668–75.

Liou, K. N., Cai, Q., Barber, P. W., & Hill, S. C. (1983). Scattering phase matrix comparison for randomly oriented hexagonal cylinders and spheroids. *Applied Optics*, 22: 1684–7.

McCave, I. N. (1983). Particulate size spectra behavior and origin of nepheloid layers over the Nova Scotia continental rise. *Journal of Geophysical Research*, 88C: 7647–66.

McCave, I. N., Bryant, R. J., Cook, H. F., & Caughanowr, C. A. (1986). Evaluation of a laser-diffraction-size analyzer for use with natural sediments. *Journal of Sedimentary Petrology*, 56: 561–4.

McClatchey, R. A., Fenn, R. W., Selby, J. E. A., Volz, F. E., & Garing, J. S. (1978). Optical properties of the atmosphere. In: *Handbook of optics*, eds. W. G. Driscoll & W. Vaughan. New York: McGraw–Hill.

Marshall, T. R., Parmenter, C. S., & Seaver, M. (1976). Characterization of polymer latex aerosols by rapid measurement of 360° light scattering patterns from individual particles. *Journal of Colloid and Interface Science*, 55: 624–36.

Morel, A. (1974). Optical properties of pure water and pure sea water. In: *Optical aspects of oceanography*, eds. N. G. Jerlov & E. Steeman-Nielsen. London: Academic Press.

Morel, A., & Bricaud, A. (1981). Theoretical results concerning absorption in a discrete medium and application to absorption of phytoplankton. *Deep-Sea Research*, 28: 1375–93.

Mugnai, A., & Wiscombe, W. J. (1980). Scattering of radiation by moderately nonspherical particles. *Journal of the Atmospheric Sciences*, 37: 1291–1307.

(1986). Scattering from nonspherical Chebyshev particles. I: Cross sections, single-scattering albedo, asymmetry factor and backscattering fraction. *Applied Optics*, 25: 1235–44.

(1989). Scattering from nonspherical Chebyshev particles. 3: Variability in angular scattering patterns. *Applied Optics*, 28: 3061–73.

Muinonen, K. (1989). Scattering of light by crystals: A modified Kirchoff approximation. *Applied Optics*, 28: 3044–50.

Muinonen, K., Lumme, K., Peltoniemi, J., & Irvine, W. M. (1989). Light scattering by randomly oriented crystals. *Applied Optics*, 28: 3051–60.

O'Brien, S. G., & Goedecke, G. H. (1988). Scattering of millimetre waves by snow crystals and equivalent homogeneous symmetric particles. *Applied Optics*, 27: 2439–44.

Osmann-Sigg, G. K., & Stumm, W. (1982). Particle size distribution and natural coagulation in Lake Zurich. *Schweizerische Zeitschrift für Hydrologie*, 44(2): 405–22 (in German).

Patitsas, A. J. (1988). Size characterization of asbestos fibres using the Rayleigh–Debye–Gans theory. *Journal of Colloid and Interface Science*, 122: 15–23.

Paul, J. H., & Jeffrey, W. H. (1984). Measurement of diameters of estuarine bacteria and particulates in natural water samples by use of a submicron particle analyzer. *Current Microbiology*, 10: 7–12.

Perry, R. J., Hunt, A. J., & Huffman, D. R. (1978). Experimental determination of Mueller scattering matrices for nonspherical particles. *Applied Optics*, 17: 2700–10.

Pinnick, R. G., Carroll, D. E., & Hoffman, D. J. (1976). Polarized light scattered from monodisperse randomly oriented nonspherical aerosol particles: Measurements. *Applied Optics*, 15: 384–93.

Pluchino, A. (1986). Observations of halo scattering from single ice crystals. *Optics Letters*, 11: 276–8.

Pollack, J. B., & Cuzzi, J. N. (1980). Scattering by nonspherical particles of size comparable to a wavelength: A new semi-empirical theory and its application to tropospheric aerosols. *Journal of the Atmospheric Sciences*, 37: 868–81.

Prandke, H., & Jonasz, M. (1984). The correlation between particle concentration and light scattering intensity in the Baltic. *Proceedings of the XII Conference of Baltic Oceanographers, Leningrad, April 14–17, 1980*, Leningrad: Gidrometeoizdat.

Pritchard, B. S., & Elliot, W. G. (1960). Two instruments for atmospheric optics measurements. *Journal of the Optical Society of America*, 50: 191–202.

Proctor, T. D., & Barker, D. (1974). The turbidity of suspensions of irregularly shaped diamond particles. *Aerosol Science*, 5: 91–9.

Purcell, E. M., & Pennypacker, C. R. (1973). Scattering and absorption of light by nonspherical dielectric grains. *The Astrophysical Journal*, 186: 705–14.

Pyle, B. E., Brunsting, A., & Latimer, P. (1979). Detection of the vacuole of yeast cells in suspension by transmittance radiometry. *Applied Optics*, 18: 3615–19.

Rabinovitch, E., & Govindjee (1969). *Photosynthesis*. New York: Wiley.

Reuter, R. (1980). Characterization of marine particle suspensions by light scattering (II). Experimental results. *Oceanologica Acta*, 3: 325–32.

Richardson, M. J. (1987). Particle size, light scattering and composition of suspended particulate matter in the North Atlantic. *Deep-Sea Research*, 34: 1302–29.

Ross, K. F. A., & Billing, E. (1957). The water and solid content of living bacterial spores and vegetative cells as indicated by refractive index measurements. *Journal of General Microbiology*, 16: 418–25.

Schiffer, R., & Thielheim, K. O. (1982). Light reflection from randomly oriented convex particles with rough surfaces. *Journal of Applied Physics*, 53: 2825–30.

Schuerman, D. W. (ed.) (1980). *Light scattering by irregularly shaped particles*. New York: Plenum Press.

Schuerman, D. W., Wang, R., Gustafson, B., & Schaefer, R. (1981). Systematic studies of light scattering. I: Particle shape. *Applied Optics*, 20: 4039–50.

Shapiro, H. M. (1988). *Practical flow cytometry*. New York: Alan R. Liss, Inc.

Sheldon, R. W., & Parsons, T. R. (1967). *Practical manual on the use of the Coulter Counter in marine sciences*. Toronto: Coulter Electronics, Inc.

Sheldon, R. W., Prakash, A., & Sutcliffe, W. H., Jr. (1972). The size distribution in the ocean. *Limnology and Oceanography*, 17: 327–40.

Singham, B. S., & Bohren, C. F. (1987). Light scattering by an arbitrary particle: A physical reformulation of the coupled dipole model. *Optics Letters*, 12: 10–12.

(1989). Hybrid method in light scattering by an arbitrary particle. *Applied Optics*, 28: 517–22.

Snyder, G. W., & Carson, B. (1986). Bottom and suspended particle size: Implications for modern sediment transport in Quinault submarine canyon. *Marine Geology*, 71: 85–105.

Spinrad, R. W. (1986). A calibration diagram of specific beam attenuation. *Journal of Geophysical Research*, 91: 7761–4.

Spinrad, R. W., & Brown, J. F. (1986). Relative real refractive index of marine microorganisms: a technique for flow-cytometric estimation. *Applied Optics*, 25: 1930–4.

Steinkamp, J. A. (1984). Flow-cytometry. *Review of Scientific Instruments*, 55: 1375–1400.

Stramski, D., Morel, A., & Bricaud, A. (1988). Modelling the light attenuation by spherical phytoplanktonic cells: A retrieval of the bulk refractive index. *Applied Optics*, 27: 3954–6.

Twomey, S. (1977). *Introduction to the mathematics of inversion in remote sensing and indirect measurements*. Amsterdam: Elsevier.

Tycko, D. H., Metz, M. H., Epstein, E. A., & Grinbaum, A. (1985). Flow-cytometric light scattering measurement of red blood cell volume and hemoglobin concentration. *Applied Optics*, 24: 1355–65.

van de Hulst, H. C. (1957). *Light scattering by small particles*. New York: Dover.

Vouk, V. (1948). Projected area of convex bodies. *Nature*, 162: 330–1.

Weiskopff, V. F. (1970). How light interacts with matter. In: *Physics in the twentieth century*. Cambridge, Mass.: MIT Press.

Weiss-Wrana, K. (1983). Optical properties of interplanetary dust: Comparison with light scattering by larger meteoritic and terrestrial grains. *Astronomy and Astrophysics*, 126: 240–50.

Wiscombe, W. J. (1980). Improved Mie scattering algorithms. *Applied Optics*, 19: 1505–9.

Wiscombe, W. J., & Mugnai, A. (1988). Scattering from nonspherical Chebyshev particles. 2: Means of angular scattering patterns. *Applied Optics*, 27: 2405–21.

Yariv, A. (1985). *Optical electronics*. New York: Holt, Rinehart and Winston.

Young, A. T. (1981). Rayleigh scattering. *Applied Optics*, 20: 533–5.

Zaneveld, J. R. V., Spinrad, R. W., & Bartz, R. (1982). An optical settling tube for the determination of particle-size distributions. *Marine Geology*, 49: 357–76.

Zerull, R. H. (1976). Scattering measurements of dielectric and absorbing nonspherical particles. *Beitrage zur Physik der Atmosphare*, 49: 166–88.

12 Textural maturity of arenaceous rocks derived by microscopic grain size analysis in thin section

SONG TIANRUI

Introduction

Graton & Fraser (1935) studied the relationship between permeability, porosity, and grain packing in sedimentary rocks according to the distribution of voids of spheres on a random plane section. Later Krumbein (1935) experimented with embedding lead shot of the same diameter in sealing wax and grounding to produce a polished surface. An average observed radius was computed to equal 0.763 of the actual radius. Friedman (1958) was the first to determine a sieve size distribution from thin-section data. A plot of thin-section quartile parameters against sieve quartile parameters indicates a linear relationship between the two methods. The mean of a converted thin-section size distribution equals $0.381 + 0.90$ times the mean of an observed thin-section size distribution (Textoris, 1971).

For a representative grain size distribution, the thin section should be cut from a rock specimen so as to represent the macroscopic characteristics of the sedimentary rock in outcrop. Generally an appropriate cutting direction is perpendicular to the depositional laminations. According to Friedman (1958), 300 grains should be counted per thin-section grain size analysis. It is also necessary that measurments should be taken along the long-axis direction. Care must be taken so as to discriminate the original particle grain, avoiding interference from grain dissolution or overgrowths due to diagenesis. Generally, the original grain shape can be determined based on residual traces of clay and/or iron oxide particles along boundaries between original grains and secondary growths.*

*Editor's note: Thermoluminescence provides another simple technique for determining overgrowth effects.

The concept of maturity in sediments was first discussed by Plumley (1948) and Pettijohn (1949). They considered that maturity to be related to grain roundness and sphericity, and to the abundance of chemically or mechanically produced matrix. For example, a very mature clastic sediment consists primarily of highly rounded, spherical grains and little matrix. Folk (1951) designated four textural maturity stages to indicate processes that work to remove clay and to increase grain sorting and grain rounding. These include an initial stage of poorly sorted immature sediment high in clay content with angular grains, progressively changing to a supermature sediment with little or no clay and extremely well-sorted and rounded grains at the final stage (Fig. 12.1).

The objective of my study is to advance Folk's scheme by proposing a comprehensive textural formula for arenaceous rocks, using as a basis the sorting and roundness of grains and matrix content in rock-forming components. The formula is particularly appropriate for automated thin-section analysis.

Sorting index

A sorting index was proposed by Trask (1932) as a numerical expression of the geometric spread of the central half of the particle size distribution of a sediment (also see Krumbein & Pettijohn, 1938). It was expressed as the square root of the ratio of Q_1, the diameter having 25% of the cumulative size curve, to Q_3, the diameter having 75% of the cumulative size frequency distribution. Folk & Ward (1957) suggested a more appropriate sorting index that included 90% of a size frequency spectra, for example, using the diameters having 5%, 16%, 84%, and 95% of the cumulative size frequency distribution.

Roundness class

The degree of abrasion of a clastic particle, as shown by the sharpness of its edges and corners, was expressed by Wadell (1932). It is the ratio of the average radius of curvature of the edges or corners of the particle versus the radius of curvature of the maximum sphere inscribed inside the particle. Many sedimentologists use visual comparisons with standards as a measure-

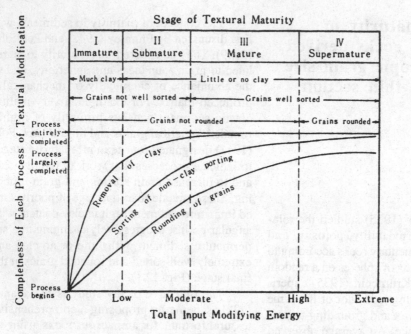

Figure 12.1. Four stages of textural maturity in sediments (after Folk, 1951).

ment of grain roundness. Pettijohn (1957) suggested five roundness classes: angular, subangular, subrounded, rounded, and well-rounded. Powers (1953) added a sixth class: very angular.

Matrix

Matrix is the fine-grained background material of <30 μm (Pettijohn et al., 1972) filling the interstices among larger grains or particles that provide the structural stability of the sediment. Dott (1964) and Pettijohn et al. (1972) stated that the abundance of matrix is related to the maturity of arenaceous rocks, and can be used as a criterion in rock classification. For example, arenite rocks have a matrix of 15%, wakes contain a matrix of 15%–75%, and mudrocks contain a matrix of >75%. Thus wakes are an immature type among arenaceous rocks.

Comprehensive textural formula

The comprehensive textural formula of arenaceous rocks was suggested by Song (1979), and later applied by Cheng et al. (1982), Liang et al. (1983), and Song & Gao (1987). The comprehensive textural coefficient T can be determined by the following formula:

$$T = (S_o\phi P_o)/(P\phi C) \tag{12.1}$$

where $S_o\phi$ is the sorting index, P_o is the average roundness, $P\phi$ is the grain distribution index, and C is the content of matrix. As $S_o\phi$ increases toward 1, sorting improves. Sorting is generally proportionate to maturity. $S_o\phi$ is obtained by the following formula:

$$S_o\phi = P_{25}/P_{75} \tag{12.2}$$

P_{25} is the diameter of the first quartile in the cumulative size frequency curve in units of ϕ ($=-\log_2 D$, where D is the grain diameter in millimeters), and P_{75} is the third quartile diameter. In order to support the sorting index, the grain size distribution index ($P\phi$) is used in the comprehensive textural formula, where:

$$P\phi = (P_{84} - P_{16})/(P_f - P_c) \tag{12.3}$$

P_{84} is the diameter at 84% of the cumulative size frequency distribution in units of ϕ, P_{16} is the diameter at 16% of the cumulative size frequency distribution, P_f is the finest diameter at 100% of the cumulative size frequency distribution, and P_c is the coarsest fraction at 0% of the cumulative size frequency distribution. In practice, the above data can be obtained from Visher's (1969) lognormal population in the probability frequency figure, because the finest grain and the coarsest grain might not be measured in

Table 12.1. *Rounding classes*

Power's Verbal Classes	Wadell's Classes Interval	Folk's rho (ρ) Classes	Author's Average Rounding Classes P_o	Verbal
Very angular	0.12 - 0.17	0.00 - 1.00	(0) 0.00	Very angular
Angular	0.17 - 0.25	1.00 - 2.00	(1) 0.00 - 1.00	Angular
Subangular	0.25 - 0.35	2.00 - 3.00	(2) 1.00 - 2.00	Subangular
Subrounded	0.35 - 0.49	3.00 - 4.00	(3) 2.00 - 3.00	Subrounded
Rounded	0.49 - 0.70	4.00 - 5.00	(4) 3.00 - 4.00	Rounded
Well Rounded	0.70 - 1.00	5.00 - 6.00		

thin section. The ratio $S_o\phi/P\phi$ reflects the relationship between textural maturity and degree of sorting in arenaceous rocks.

In the comprehensive textural formula, roundness class is very similar to Folk's rho (ρ) scale (Boggs, 1987), but with different intervals (Table 12.1). Hence, there are five classes of roundness: (0) very angular, (1) angular, (2) subangular, (3) subrounded, and (4) rounded (Fig. 12.2). The average roundness value can be obtained by:

$$P_o = (0n_0 + 1n_1 + 2n_2 + 3n_3 + 4n_4)/\Sigma n \quad (12.4)$$

where P_o is the average roundness, 0–4 refer to the five rounding classes of grains, and $n_0, \ldots, n_4$ refer to the number of grains counted for each roundness grade.

Power's verbal rounding classes, Wadell's roundness, Folk's rho (ρ), and the present author's roundness classes and their relationship are shown in Table 12.1. In many studies, the discrimination of rounding classes of sand grains is estimated arbitrarily by visually comparing the sedimentary particles with standard figures. Therefore, the more classes used, the more difficult it is to distinguish them.

Estimating the roundness of grains in thin section is different than observation of grains under a binocular microscope for the same sample. According to experiments, the estimated values of roundness of grains observed in thin section are always smaller than the values determined from binocular observation due to the cutting plane of grains showing more angular corners for angular arenaceous rocks. A corrected roundness value is given as

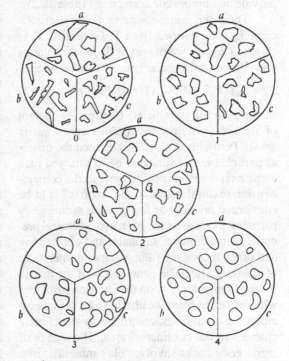

Figure 12.2. Rounding classes of grains: (0) very angular; (1) angular; (2) subangular; (3) subrounded; (4) rounded (a = quartz grains, b = feldspar grains, c = grains formed of rock fragments).

$$(P_o^2 + (R - A)^2)^{0.5} \quad (12.5)$$

where

$$R = (3n_3 + 4n_4)/\Sigma n$$
$$A = (0n_0 + 1n_1 + 2n_2)/\Sigma n \quad (12.6)$$

Based on the statistical data of the roundness of beach sands from Hainan Island, Southern China (Song & Liang, 1978), the stereoscopic

Table 12.2. *Comparison of observed roundness data from binocular and microscopic analysis*

ROUNDNESS CLASSES	0	1	2	3	4	TOTAL	P_o	R	A	$(P_o^2 + (R-A)^2)^{0.5}$
BY BINOCULAR	8	18	29	68	16	139	2.5	1.9	0.6	
BY MICROSCOPE IN THIN SECTION	10	27	33	60	9	139	2.2	1.6	0.6	2.4

observation of grains under binocular measured roundness is always higher than that of thin section with the same grains. But the observed roundness corrected by function (12.5), may provide an appropriate correction (Table 12.2).

Thus, the comprehensive textural coefficient T_d is developed from T. Generally, T_d is close to T except when very well-rounded grains prevail over the angular grains in rocks, when:

$$T_d = S_o \phi (P_o^2 + (R - A)^2)^{0.5} / (P \phi C) \qquad (12.7)$$

In the formula parameter, C refers to content of matrix (= matrix area/(grains area + matrix area)). Pettijohn et al. (1972) defined the matrix as particles with a diameter of 30 μm, and such a criterion is recommended here. In the comprehensive textural formula, the "matrix" is to be considered not only as fine-grained sedimentary particles, but also as the content of chemical precipitates (carbonate, sulphate, and chalcedony and other secondary fillings) among interstices of clastic grains. The constitution of the matrix is dependent mainly upon the abrasion resources with influence from weathering agents and precipitation conditions, controlled by the climate changes in the continental area. The matrix of arenaceous rocks involves clay minerals, iron oxides, and organic particles under the influence of humid climatic conditions. Under arid climatic conditions calcite, gypsum, and sulphates are more common matrix material. The matrix of marine sediment excluding chemical cements is composed mainly of phyllosilicates and biogenic fragments, and is less responsive to climatic changes.

In most cases T_d is similar to T, with the rare exception of well-rounded grains prevailing over angular grains. T value can be computed with unconsolidated sediment, but T_d is used principally for consolidated rocks, especially supermature sandstones.

Analyses of more than fifty rocks from a range of sedimentary environments (recognized by megascopic facies indicators) show a close relationship between graphic standard deviation σ_1 (using the formula of Folk & Ward [1957]) and the second moment m_2 (Fig. 12.3). In contrast, there is no relationship between the second moment measure m_2 and the comprehensive textural coefficient T_d (Fig. 12.4) and, although there are no mature sandstones ($T > 20$) in the area of poor sorting ($m_2 > 1$), better sorted sandstones ($m_2 > 1$) include all four maturity stages.

Usage of the comprehensive textural coefficient

Since there are many factors that influence the sedimentary rocks (mechanical, geophysical, and biological), it is impossible to determine a definite sedimentary environment by a single parameter measured from sedimentary rocks, including the comprehensive textural coefficient. Undoubtedly, the value of the comprehensive textural coefficient increases with the progressive maturity of arenaceous rocks. Therefore, the comprehensive textural coefficient can be used in study of sedimentary environments in a limited area or a vertical section of strata.

Four stages of maturity

Using the comprehensive textural coefficient, four stages of maturity are distinguishable:

1. The *immature* stage is characterized by a very low comprehensive textural coefficient ($T < 10$), high matrix content, poor sorting, and very angular to angular grains in rocks.

2. The *submature* stage is defined as having $10 < T < 40$, a relatively low matrix content, and increasing roundness and sorting.

3. The *mature* stage is reflected by higher roundness, good sorting, and little matrix in the rocks, with $40 < T < 100$.

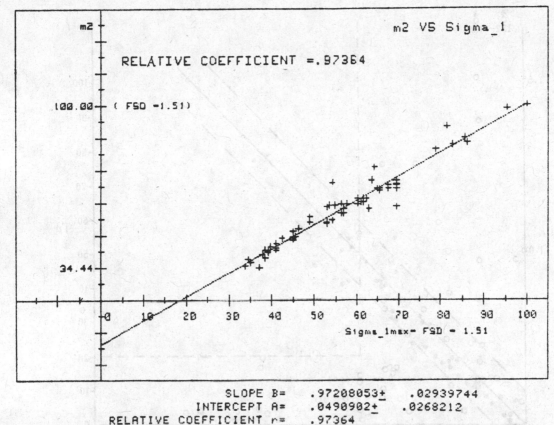

SLOPE B=   .97208053±   .02939744
INTERCEPT A=   .0490902±   .0268212
RELATIVE COEFFICIENT r=   .97364

Figure 12.3. Relationship between graphic standard deviation σ_1 (using Folk and Ward's formula) and second moment m_2: slope $B = 0.972$, intercept $A = 0.049$, relative coefficient $r = 0.974$.

4. The *supermature* stage is the highest stage of maturity, characterized by very low matrix content, very good roundness and sorting of grains, and a high comprehensive textural coefficient (T or $T_d > 100$).

Maturity systems and sedimentary environments

Maturity systems are divided by different fluid flow influences into three types:

In the first maturity system, unidirectional flow of river water is the dominant factor in reworking clastic grains.

The second maturity system consists of bidirectional flows from tidal currents that rework clastic grains.

Complex flows appear in the third maturity system, such as shelf transport supported by both storm waves and Ekman flow, and turbidite currents.

Twelve important sedimentary environments are listed in Figure 12.5, including subtypes of some of these environments. The clastic materials first appear in the piedmont area, then they are washed into the braided river, followed by the meandering river, the distributary channels, and finally onto the delta. This process is a result of a single direction of water flow characterized by the direct influence of flow competence, and can be included in the first maturity system. The second maturity system includes lakes or coastal basins, the coastal plain, intertidal, subtidal, and barrier environments. These environments involve characteristic agitational hydrodynamic currents. The sand bar yields the highest comprehensive textural coefficient among all of these sedimentary environments. The third maturity system takes place over continental shelves and the deep sea. Arenaceous rocks in this system and the maturity may be extremely variant.

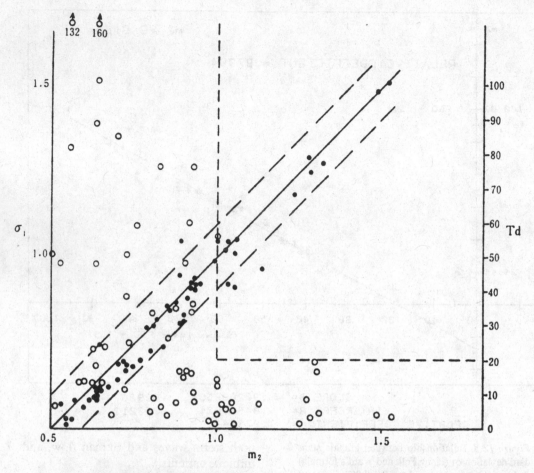

Figure 12.4. Relationship among standard deviation σ_1 and second moment m_2 (points): comprehensive textural coefficient T_d and second moment m_2 (circles).

There is a trending curve through the first to the third maturity system using the comprehensive textural coefficient, but with variants in branches within each sedimentary environment (Fig. 12.5). The connection of twelve sedimentary environments marks the progressive change of geographic and tectonic controls through different areas and geological ages.

Application of the comprehensive textural coefficient

Example 1: Comparison of T, T_d, and other grain size parameters with six arenaceous rocks from different facies columns (Fig. 12.6):

A. glacial sediments, tillite, from Tibet, an immature rock, $T = 9.50$, $T_d = 11.91$;

B. alluvial sediments, muddy sandstone, from Gangshu Province, an immature rock, $T = 5.54$, $T_d = 6.67$;

C. braided river sediments, arkose, from Xiangjiang Autonomous Region, a submature rock, $T = 15.18$, $T_d = 16.48$;

D. aeolian sediments, carbonate lithic sandstone, from Shaanxi Province, a submature rock, $T = 32.45$, $T_d = 34.23$;

E. distributary channel sediments, lithic chert-quartz sandstone, from Guangdong Province, a submature rock, $T = 32.30$, $T_d = 35.42$.

F. sand bar sediments, pure quartz sandstone, from Ningxia Province, a supermature rock, $T = 120.7$, $T_d = 160.7$.

Clearly, the comprehensive textural coefficient of the above six rocks can strongly reflect their rock properties as well as the differences of sedimentary environments by measured data, but not from other grain size parameters (see Fig. 12.6 and Table 12.3).

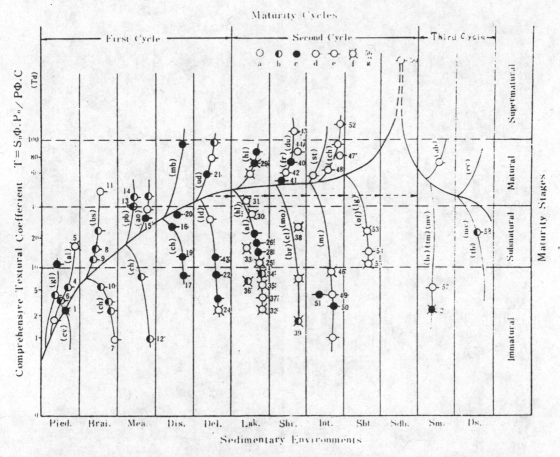

Figure 12.5. Relationship among comprehensive textural coefficient, maturity stages, maturity systems, and sedimentary environments: (a) quartz grains >80%; (b) feldspar grains >20%; (c) grains formed of rock fragments >20%; (d) grain arrangement with tendency to single direction; (e) grain arrangement with tendency to double direction; (f) grain arrangement random; (g) number of thin section from standard collections. *Abbreviations:* Pied. = Piedmont (av, alluvial; gl, glacial; cv, continental volcanic); Brai. = braided (bs, branch spit; ch, channel); Mea. = meander (pb, point bar; eo, aeolian; ch, channel); Dis. = distributary (mb, mouth bar; ch, channel); Del. = delta (ud, upper delta plain; ld, lower delta plain); Lak. = lakes (hl, humid lake; al, arid lake); Shr. = shore (du, beach dunes; fr, foreshore; br, backshore; cl, collapse; mo, moraine); Int. = intertidal (st, sandy tidal flat; tch, tidal channel; mt, muddy tidal flat); Sbt. = subtidal (of, offshore; lg, lagoon); Sdb. = sandbar; Cs. = Continental shelf (ab, along beach current; bi, bioclastic; tm, tempestite; mv, marine volcanic); Ds. = deep sea (tb, turbidite; mv, marine volcanic; ct, counter current).

Example 2: The comprehensive textural coefficient T or T_d can be used in the study of vertical stratigraphic section for facies interpretation. Take the example of a lower part of Proterozoic strata in the Ming Tombs District, Beijing, where there are gradual facies changes from the bottom to the top part of the Changzhougou Formation. For example, from distributary channel sediment directly overlying Archean metamorphic rocks to tidal flat sediment, and to sand bar–coastal plain sandstone beds, corre- sponding to the facies, change the T and T_d coefficients, increasing from immature through to supermature from the bottom to the top of the Changzhougou Formation (Fig. 12.7).

Summary
A formula to define the textural maturity of arenaceous rocks based on microscopic grain size analysis of thin sections establishes a numerical relationship among maturity, sorting, roundness of grains, and matrix content in rock.

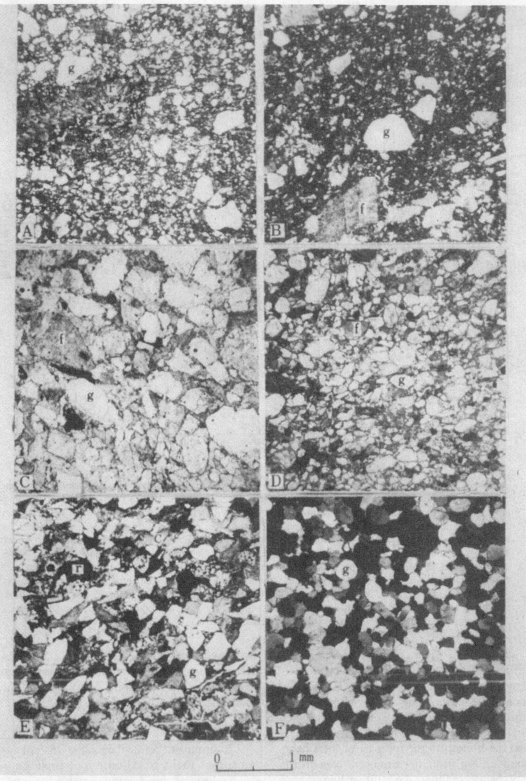

0   1 mm

Figure 12.6. Thin-section microphotographs: (A) glacial (Carboniferous, Tibet) using parallel polarized light; (B) alluvial (Jurassic, Gangshu) using parallel polarized light; (C) braided river (Tertiary, Xingjiang) using parallel polarized light; (D) aeolian (Cretaceous, Shaanxi) using parallel polarized light; (E) distributary channel

Table 12.3. *Pertinent statistical data from thin-section microphotographs shown in Figure 12.6*

Sediment Type	T	Td	P_o	R-A	$S_o\emptyset/P\emptyset$	md	d_1	SK_1^*	KG^*	m_1'	m_2'	m_3'	m_4'
glacial: (Carboniferous)	9.30	11.90	2.00	1.60	1.61	4.22	1.05	-0.06	0.75	4.43	1.00	-0.3	2.27
alluvial (Jurassic)	5.54	6.67	1.4	-0.2	1.66	3.55	0.97	0.04	0.95	3.84	1.13	-0.07	2.47
braided river (Tertiary)	15.18	16.48	2.25	-0.92	1.62	2.3	0.94	0.09	0.96	2.56	0.94	-0.5	2.92
aeolian (Cretaceous)	32.45	34.23	3.10	1.54	1.57	3.37	1.05	0.35	1.29	3.32	0.89	0.26	3.42
distrib. channel (Permian)	32.30	35.42	2.26	0.5	1.56	2.37	0.93	0.37	0.90	2.58	0.92	0.69	3.09
sand bar (Carboniferous)	120.7	160.7	3.3	2.9	3.13	2.48	0.58	-0.02	0.97	2.77	0.63	-0.2	4.31

* Graphic measures, by Folk and Ward's formula.

' moment measures.

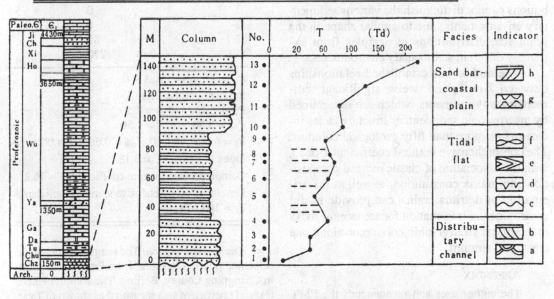

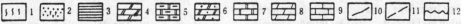

Figure 12.7. Correlation of sedimentary facies change with variance of comprehensive textural coefficient of arenaceous rocks in the lower part of Proterozoic Strata (Changzhougou Formation) in the Ming Tombs District, Beijing. *Left:* (1) gneiss; (2) sandstone; (3) shale; (4) muddy dolostone; (5) silicified dolostone; (6) sandy dolostone; (7) dolomite; (8) marl; (9) limestone; (10) conformity; (11) disconformity; (12) unconformity. *Right:* (a) festoon bedding; (b) trough cross bedding; (c) waves ripples; (d) mud cracks; (e) herring-bone bedding; (f) flaser-lenticular bedding; (g) lingoidal ripples; (h) coastal cross-stratification.

Caption to Figure 12.6 (cont.)
(Permian, Guangdong) using parallel polarized light; (F) sand bar (Carboniferous, Mingxia) using cross-polarized light. *Abbreviations:* g, quartz grain; f, feldspar; r, rock fragment; c, chert. See Table 12.3 for pertinent statistics.

The comprehensive textural coefficient can be obtained by the formula $T = (S_o \phi P_o)/(P \phi C)$. Since the roundness of grains as determined in the petrographic microscope is always smaller in thin sections, due to the grain projection reflecting its angular corners, grain roundness is corrected by a developed textural formula, $T_d = S_o \phi (P_o^2 + (R - A)^2)^{0.5}/(P \phi C)$. In most cases T_d is similar to T with the rare exception of very well-rounded grains, where T_d is more useful in checking the supermature sandstones. Four maturity stages are recognized: immature ($T < 10$), submature ($10 < T < 40$), mature ($40 < T < 100$), and supermature (T or $T_d > 100$).

Thin-section analysis is the only choice in determining the grain size distribution of lithified arenaceous rocks, which cannot be dispersed without breakage of the original grain shapes. Classical analyses of grain size distributions cannot distinguish the various sedimentary environments, due to similar shape in the grain size distributions of samples from extremely different sedimentary environments.

The author has established relationships between T or T_d and twelve significant sedimentary environments, which are determined by macroscopic sedimentary structure facies indicators or more than fifty geological columns. The comprehensive textural coefficient T or T_d, with the association of clastic mineral composition and matrix constitution, as well as the orientation of detritus grains, can provide useful environmental information for arenaceous rocks in a vertical stratigraphic column or along one correlated stratum.

APPENDIX

The author uses and recommends the PMT Thin Section Grain Size Analyzer, which is designed by Liu Xiufen and manufactured by Pule Electronic Company (Fig. 12.8). This is a third generation of China-made microimage analyzer. It uses thin sections for the determination of the grain frequency distribution, the fabric of grains, and information on fractures in rock. The analyzer uses a CAA computer system, and MOT optical system and digital processing table, which can be moved along the grain image outline in the observation field under the microscope. Measured grains can be divided every 0.25ϕ from 2ϕ to 6ϕ with a sensitivity of $1 \mu m$. Any common petrologic microscope can be used.

Figure 12.8. Photograph of a PMT Thin Section Grain Size Analyzer using a CAA computer system and MOT optical system and digital processing unit.

Table 12.4. *Example calculation*

Roundness class no.	No. of grains
0	0
1	0
2	0
3	30
4	270

$P_o = 3.9$, $R = 3.9$, $A = 0$
Index of grain distribution:
$P_{25} = 1.65$, $P_{75} = 2.50$,
$P_{84} = 2.80$, $P_{16} = 1.50$
$P_f = 6.0$, $P_c = 1.00$, $S_o \phi = 0.66$, $P \phi = 0.26$
Index of matrix: $C = 0.13$
Comprehensive textural coefficient: $T = 76.2$
Developed comprehensive textural coefficient: $T_d = 107.698$

Example calculation: The sample is a dolomitic cemented quartz sandstone from Proterozoic strata in Changping County, Beijing. The comprehensive textural coefficient was measured as shown in Table 12.4.

Interpretation: The dolomitic quartz sandstone belongs to supermature sandstone formed in a high-energy coastal environment with high roundness and sorting indices cemented by saline sediments (dolomite) within sabkha-lagoon area to be a matrix-supporting texture.

ACKNOWLEDGMENTS

The author most gratefully acknowledges Dr. J. P. M. Syvitski for his editorial help in improving the manuscript before and after reviews, and for his lectures on the theories of grain size analysis in my institute in November, 1988. The author is much obliged to Drs. D. J. W. Piper and L. F. Jansa for their careful review of the manuscript and for their opinions on some of the concepts, terminology, and language. The author also thanks Mr. Bill LeBlanc and Ms. Wendy Gregory for their comments on style.

REFERENCES

Boggs, S., Jr. (1987). *Principles of sedimentology and stratigraphy.* Columbus, Ohio: Merrill, pp. 105–34.

Chen B., Ai C., & Zhaxi W. (1982). Latest geological observations, Southeastern Xizang (Tibet). In: *Contribution to the Geology of Qinghai-Xizang (Tibet) Plateau,* no. 10, CGQXP Editional Committee, pp. 159–210.

Dott, R. H., Jr. (1964). Wakes, greywake and maturity with approach to immature sandstone classification? *Journal of Sedimentary Petrology,* 34: 625–32.

Folk, R. L. (1951). Stages of textural maturity in sedimentary rocks. *Journal of Sedimentary Petrology,* 21: 127–30.

Folk, R. L., & Ward, W. (1957). Brazos River bar: A study in significance of grain-size parameters. *Journal of Sedimentary Petrology,* 27: 394–416.

Friedman, G. M. (1958). Determination of sieve-size distribution from thin section data for sedimentary studies. *Journal of Geology,* 66: 394–416.

Graton, L. C., & Fraser, H. J. (1935). Systematic packing of spheres with particular relation to porosity and permeability. *Journal of Geology,* 43: 785–909.

Krumbein, W. C. (1935). Thin-section mechanical analysis of indurated sediments. *Journal of Geology,* 43: 489–96.

Krumbein, W. C., & Pettijohn, F. J. (1938). *Manual of sedimentary petrography.* New York: Appleton–Century–Crofts, pp. 228–67.

Liang D., Nie Z., Guo T., Xu B., Zhang Y., & Wang W. (1983). Permo-Carboniferous Gongwana–Tethys Facies in Southern Karakoran Ali, Xizang (Tibet). *Journal of the Wuhan College of Geology,* 1(19): 9–27.

Pettijohn, F. J. (1949). *Sedimentary rocks.* New York: Harper & Row, pp. 24–89, 209–12.

Pettijohn, F. J. (1957). *Sedimentary rocks,* 2nd ed. New York: Harper & Row, 718 pp.

Pettijohn, F. J., Potter, P. E., & Siever, R. (1972). *Sand and sandstone.* Berlin/Heidelberg/New York: Springer–Verlag, pp. 85–6.

Plumley, W. J. (1948). Black Hills terrace gravels: A study in sediment transport. *Journal of Geology,* 56: 526–77.

Powers, M. C. (1953). A new roundness scale for sedimentary particles. *Journal of Sedimentary Petrology,* 23: 117–19.

Song T. (1979). A new formula for counting "Predominant textural coefficient" of sandstone in thin section. *Geological Review,* 25(1): 43–7.

Song T., & Gao J. (1987). *Precambrian sedimentary rocks in the Ming Tombs District, Beijing.* Beijing: Geological Publishing House, pp. 19–25.

Song T., & Liang B. (1978). Two types of recent littoral sediments in South China. *Acta Geologica Sinica,* 52(2): 123–33.

Textoris, D. A. (1971). Grain-size measurement in thin section. In: *Procedures in sedimentary petrology,* ed. Robert E. Carver. New York: John Wiley and Sons, pp. 95–107.

Trask, P. D. (1932). *Origin and environment of source sediments of petroleum.* Houston: Gulf Publication Co., 323 pp.

Visher, G. S. (1969). Grain-size distribution and depositional processes. *Journal of Sedimentary Petrology,* 39: 1074–106.

Wadell, H. A. (1932). Volume, shape and roundness of rock particles. *Journal of Geology,* 40: 443–60.

13 Interlaboratory, interinstrument calibration experiment

JAMES P. M. SYVITSKI,
K. WILLIAM G. LEBLANC, AND
KENNETH W. ASPREY

Introduction

There has been a historical need for sedimentologists to refine the characterization of both particle size and shape. Automated size-characterization instruments have improved on "classical" techniques, such as the sieve and pipette method, in terms of speed and precision. However, many of these instruments provide a size frequency distribution of particle diameters from a population of sediment grains by proxy (i.e., by converting a particle's cross-sectional area, surface area, volume, settling velocity, or some other form of particle behaviour, to particle diameter). Some techniques, such as the settling tube, may involve user-built instruments not yet calibrated in the traditional analytical sense. To this end, the International Union of Geological Sciences – Committee on Sedimentology (IUGS–COS) sponsored this study to compare the results from automated instruments that measure the frequency distribution of grain diameters in geological samples. This chapter reviews some of the previous attempts at interlaboratory or interinstrument calibration, discusses the philosophy and preparation of geological standards (silts and sands), and presents new results with recommendations for future experiments.

Grain size and calibration standards

The size of a particle is not uniquely defined, except for the most simple of geometric objects – the sphere. For natural and irregularly shaped particles, size depends on the method of measurement. Allen (1968) provides a number of differing definitions of particle size, including surface diameter, volume diameter, drag diameter, projected area diameter, free-falling diameter, Stokes's diameter, sieve diameter, and specific surface diameter. As a particle becomes less equant, the discrepancy between these different measures becomes progressively greater.

Pettijohn, Potter, & Siever (1973) note that the method of choice depends on the object of study. For instance, a sedimentologist studying the behaviour of grains in a fluid might find the drag diameter more meaningful than volume diameter. A geochemist studying the adsorption kinetics of cations onto clay particles might consider the specific surface diameter a more useful measure of size. Although the objective of a study should influence what instruments are to be used, an instrument's availability is usually the determining factor.

Modern techniques sense grain size in ever-increasingly different ways. Aware of this, many geologists tend to express instrumental results as "equivalent" diameters. Equivalency allows the measurement of differing grain size parameters to be expressed as a common value for comparative purposes. The practice is considered enlightened by some and perplexing by others.

For instance, a sand sample may have its size frequency distribution in the form of equivalent spherical sedimentation diameters (ESSD), such as determined from a settling tube (see Chapter 4). In this manner, the diameter of a mineral grain is expressed as the equivalent diameter of a sphere of standard density (typically that of quartz, the most common siliclastic mineral), settling within a liquid of common viscosity and temperature (typically 20 °C). However, the sand may have been emplaced by seawater and not fresh water, and at a temperature much different from the standard. The natural sand grains may also have consisted of minerals having particle densities different from quartz; or they may be very angular (e.g., a micaceous sand). In other words, a "fine-sand" sample identified with the concept of equivalency may behave in the original sedimentary environment as a "coarse-silt" sample.

Calibration standards are of three types:

1. those that test for the accuracy and precision of a particular size measure, say, particles of a narrow density in the case of settling tubes,

or spherical volume in the case of electroresistance counters;

2. those that test for size equivalency using spherical constant-density standards; and

3. those that test for the behaviour and natural characteristics of geological samples.

Many of the type 1 standards are provided by the manufacturers of size-analytical equipment. For instance, Coulter Counter provides a variety of narrow-sized latex beads for the calibration of its resistance pulse counters; Micrometrics likewise supplies narrow-sized garnet standards for calibrating its SediGraph. Similar standards may be purchased from the U.S. National Bureau of Standards (NBS) or the European Communities (BCR) reference materials.

Type 2 standards include spherical constant-density glass beads. The U.S. NBS provides excellent and certified standards. Glass beads sold by other companies may include an unknown proportion of nonspherical particles fraught with inclusions of steel and glass.

Type 3 standards are developed to test an instrument's ability to discern modes in a multimodal sample, or the size properties of poorly sorted samples that have a size range over some nine orders of magnitude (e.g., some debris flows and tills). BCR standards of quartz powders and sands offer a large variety to choose from. Many geological laboratories are capable of developing their own standards, and a brief description of one such analytical technique is included (see p. 176).

Past studies
Individual instruments
Most previous studies have concentrated on the capabilities of individual instruments (Jordan et al., 1971; McCave & Jarvis, 1973; Komar & Cui, 1984; Stein, 1985; McCave et al., 1986; Jones et al., 1988; see Chapters 4–12 in this volume). In most cases, the new instrument/technique is compared with the classical methods of sieving and/or pipetting. Such comparisons indicate whether the new technique is relatively accurate, assuming that these classical methods provide accurate results. A measure of true accuracy, however, is seldom obtained as the classical techniques are themselves rather imprecise (see "Results" section) and very de-

pendent on laboratory technique and operator error.

ASTM equivalency
A different approach, and one adopted by the American Standards for Testing Materials (ASTM, 1983) is to test for equivalency of particle size. In this procedure, individual automated counters of particle size are first calibrated with standard, spherical particles according to manufacturer procedures. An unknown powder may then be used as the testing standard. A known mass of particles is used – either as a known mass within a field of view (in the case of an image analyzer) or as a known mass concentration within a liquid suspension. The instrumental results are then presented as a cumulative curve of number of particles per unit mass of sample versus particle diameter. A plot from one instrument is used to provide the frame-of-reference measure of the diameter. In other words, the size spectrum generated from one instrument is placed over a similar plot from another kind of instrument, and the second plot moved along the particle-diameter axis until the two curves coincide. The magnitude of the shift from one diameter scale to the other provides the scale-conversion factor. If the shape of the curves are dissimilar, this technique is not considered valid.

Based on a test using a quartz standard determined on four separate instruments, each using three separate investigators in different laboratories, ASTM (1983) found the following:

$$d_e = 1.30 d_0 = 1.72 d_v = 1.60 d_s \qquad (13.1)$$

where d_e is longest end-to-end diameter measured using image analysis, d_0 is the diameter of a circle of equal area as measured on an optical counter, d_v is the diameter of a sphere of equal volume as measured on an electrical resistance counter, and d_s is the sedimentation diameter (Stokes's in this case) as measured on sedimentation instruments. ASTM (1983) is careful to point out that the three significant digits shown in (13.1) are not warranted by the precision obtained among the various investigators. They are also careful to note that these conversion factors may change depending on the size range of the test standard used.

Interinstrument mud experiment

Singer et al. (1988) compared the results of four instruments capable of characterizing the size distribution of mud-sized standards (i.e., <63 μm). The principal instruments used in the experiment included the SediGraph (Stein, 1985; cf. Chapter 10), the Hydrophotometer (Jordan et al., 1971), the Electrozone particle counter (Muerdter et al., 1981; cf. Chapter 8), and the Malvern laser sizer (McCave et al., 1986; cf. Chapter 9). The equivalent diameters obtained from these instruments respectively include spherical sedimentation diameters (Sedi-Graph and Hydrophotometer), spherical volume diameters, and spherical area diameters.

The standards consisted of a series of narrow-size-ranged glacial silt and milled Ottawa sand. Each standard had a size range of 0.5 ϕ (i.e., 4.5–5 ϕ, 5–5.5 ϕ , etc.) prepared by repeated decanting. A version of this method of preparing calibration standards can be found in Chapter 10, and is repeated in part below:

Standards of various [mud] size range and complexity can be prepared using repeated washings and timed clearing rates in a standard 1,000-ml graduated cylinder. The following are the basic steps:

1. Introduce a mud sample (<2% by weight) into a cylinder containing a dispersing agent and ensure that the sample solution is homogeneous.

2. Decide what you want the upper size limit of your standard to be.

3. That size will be associated with a settling velocity (you will need to assume a singular particle density even if your sample contains a mineral mixture) that can be obtained from [Stokes's Settling Law], from which a settling time may be obtained.

4. Once this time has elapsed, decant the liquid into another cylinder.

5. Decide what the lower size limit of your standard is to be, and calculate the settling time for that particle size.

6. Starting with the decanted liquid rehomogenized, wait until the settling time has elapsed, and then decant the remaining solution (which may be discarded).

7. Resuspend the sediment that had settled during step 6 with a new dispersing solution within the cylinder, and repeat step 6.

Step 6 is known as a *washing* and should be repeated until the fines have been removed and a pure standard with known natural size boundaries has been obtained.

Once the size frequency distribution of your unique standard has been established, it may be mixed by weight with other standards of different size characteristics to obtain new and complex mixtures that may be polymodal.

Singer et al. (1988) make the implicit assumption that their standards are known and correct. In other words, a standard of 4–4.5 ϕ has no particles beyond these size limits, and the boundaries of the size limits are correct. However, this assumes that the equivalent spherical sedimentation diameter is the true descriptor for particle size, and that particle size boundaries determined by mass settling/segregation are without bias or error. This puts size analytical instruments that do not sense particle size by settling at a distinct disadvantage.

For comparative purposes, we have assumed that the midpoint of each of these 0.5 ϕ standards is the expected mean, and have plotted these values with the observed mean for each of the above four instruments (Fig. 13.1) for one of their experiments. Compared to the standards, the most accurate instrument is the electroresistance particle counter (Fig. 13.1B), followed closely by the Hydrophotometer (Fig. 13.1D), which shows slightly coarser (0.2 ϕ) values, and the SediGraph (Fig. 13.1A), which had observed mean particle diameters 0.3 ϕ finer. The Malvern laser sizer had very good agreement with the coarse standards (<6 ϕ); for the finer standards the agreement is held to be unacceptable. Considering that the expected range of particle sizes for any one standard was 0.5 ϕ, the Hydrophotometer generated very narrow size spectra so that the one-standard-deviation values are possibly too narrow (0.25 ϕ). The Electrozone particle counter showed larger standard deviations (0.4 ϕ) that increased with decreasing particle size (from 0.22 ϕ for a 4–4.5 ϕ standard, to 0.66 ϕ for a 7.5–8 ϕ standard: Fig. 13.1B). The SediGraph had standard deviations in the expected range (0.57 ϕ); the Malvern sizer showed larger values (0.78 ϕ), possibly too large.

Singer et al. (1988) also noted that the Malvern laser sizer and the Hydrophotometer had a poor ability to resolve polymodal samples.

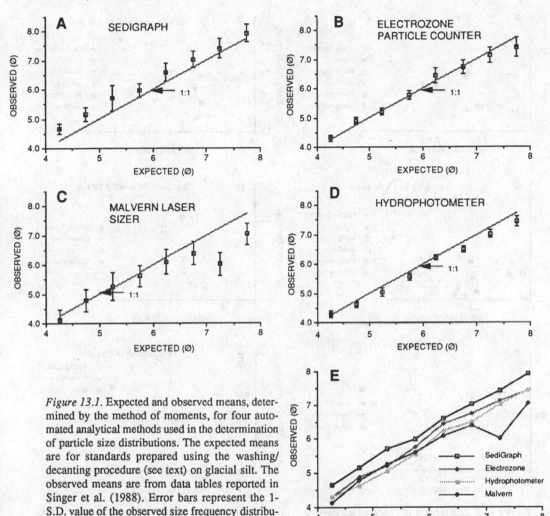

Figure 13.1. Expected and observed means, determined by the method of moments, for four automated analytical methods used in the determination of particle size distributions. The expected means are for standards prepared using the washing/decanting procedure (see text) on glacial silt. The observed means are from data tables reported in Singer et al. (1988). Error bars represent the 1-S.D. value of the observed size frequency distributions (method of moments).

Interinstrument, interlaboratory experiment

Singer (1986) reported on an experiment that involved twenty size-analytical laboratories. The experiment was meant to assess the standardization of instruments used in textural analyses. The results include those of four Coulter Counters, one videodigitizer, two Malvern laser sizers, four SediGraphs, two Hydrophotometers, one Sonic Sifter/Laser Granulometer, two Spectrex laser particle counters, seven settling tubes, and four labs that used the classic pipette and sieve method.

The absolute size range of the standards used in the calibration experiment was unknown. The standards were mostly unimodal samples of natural sediments or fractions thereof. The standards had a size range that varied from 0.5ϕ to 5.0ϕ. Singer (1986) provided overlapping plots of the size frequency distributions from the various instruments and labs, but as the distributions were provided at dissimilar class intervals, the curves are difficult to compare visually.

Figure 13.2 provides scatter plots of the observed means and standard deviations (based on the method of moments) from this experiment. We report the results of only six of the nine standards used (Singer, 1986), and only seven of the nine methods/instruments (Fig.

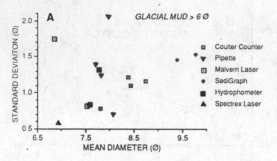

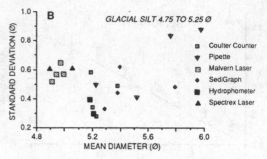

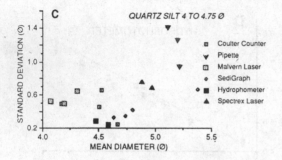

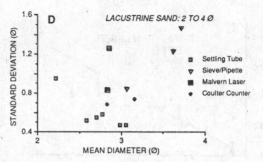

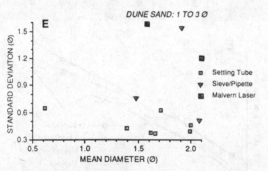

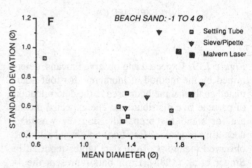

Figure 13.2. Scatter plot of observed standard deviations and mean particle size of six of the nine sediment standards reported in the Singer (1986) calibration experiment involving twenty laboratories and nine size-analytical techniques. Values are in terms of phi grain size (ϕ); statistics were determined from the method of moments.

13.2). The scatter between instruments and labs is surprisingly large, particularly with the silt and mud standards. For instance, the range in observed means for one glacial mud standard was 3ϕ (6.84–9.78ϕ), the range in standard deviations was 1.5ϕ (0.47–2.06ϕ: Fig. 13.2A). Thus, depending on the instrument used, this standard could be described as "well-sorted fine silt" or "poorly sorted clay."

The method showing the most scatter in the results (i.e., lowest precision) and the results most divergent from those of all the other instru-

ments investigated, is the pipette (Fig. 13.2A–C). This is an important finding because many of the previous calibration tests had used the pipette method as the standard on which to base comparisons. In fact, all other instruments tested outperformed the pipette method in terms of reproducibility.

With respect to the standards within the mud range, the Malvern laser sizer provided invariably coarser grained results. The Spectrex laser counter also produced results mostly coarser in terms of particle size. The Coulter Counter and Hydrophotometer provided consistent results with a high degree of precision. Excepting one of the four SediGraph labs that consistently provided finer-grained results, SediGraph results were similar to those generated from Coulter Counters (Fig. 13.2A–C).

Laboratory results from the sand standards

were mostly similar (Fig. 13.2D–F). One of the seven settling tubes did provide more poorly sorted and 0.5–1.0 ϕ coarser distributions. Two of the four sieve and pipette labs provided widely variable results. The Malvern laser sizer also showed a bias toward producing more poorly sorted distributions, although the mean particle size was similar to the settling tube results. Depending on the lab and instrument used, a dune sand (Fig. 13.2E) could be labelled as anything from a "very well-sorted medium sand" to a "poorly sorted fine sand."

IUGS–COS "precision" experiment
Objective
Before an instrument can be evaluated for its accuracy of measurement, its precision must be determined. A very precise instrument (i.e., one that provides consistent values of a given standard) need only provide a single measurement of a "known" standard, from which its accuracy may be deduced. An imprecise instrument may need to average several measurements of a given standard in order to provide an accurate result. Commercial size-analytical instruments tend to be very precise; laboratory-built instruments (one of a kind) generally have unknown precision. The objective of this portion of the IUGS–COS experiment was to determine the precision of size-analytical instruments, particularly "one of a kind" settling tubes, that were to be used in the following accuracy experiment.

Preparation of standards
Standards of sand- and silt-sized glass beads were prepared, at cost to IUGS–COS participants, by Granulometry (West Germany). Three sand standards were prepared using a mixture (by weight) of three of seven nonoverlapping sieve fractions: The final size distributions were therefore polymodal in nature. Representative samples from each of the narrow sieve fractions were obtained using a high-quality rotational sample splitter. Two silt standards were prepared by wet-sieving commercial glass beads into unimodal distributions with known upper and lower size boundaries. Participants were asked:

1. to use all of the standard provided in one analysis, or, where this amount was too large,
2. to perform separate analysis on subfractions of the standard until the entire standard was analyzed, and appropriately mesh these individual results into one final result.

Participants were only told if the standards were in the sand or silt size range.

The glass beads used in the preparation of these standards consisted of a significant proportion of nonspherical beads, even though the bead manufacturer claimed to yield "almost perfect spheres with homogeneous density." [Scientists beware of uncertified standards!] The percentage of spherical particles was modestly improved using an inclined vibrating table. The seven sieve fractions were also found to have slightly different density, due to inclusions of gas and steel. Particle densities were obtained using heavy liquids.

Methods
Each instrument is described briefly in the appendix. As the shape and density of the beads composing these standards were inconsistent, the results could not be used to test for intermethod accuracy. The results do provide a good measure of precision, as four replicates of each standard were analyzed. In the case of the sand standards, we present the results of five settling tubes, with particle size data equally reported in the form of equivalent spherical sedimentation diameters (ESSDs).

Four instruments were used in the silt experiment: two SediGraphs, one Galai computerized inspection system (CIS-1), and a Coulter Counter (a resistance pulse counter). These methods do not sense particle size in a consistent manner: The SediGraph senses particle size based on a particle's settling characteristics; in the latter two instruments, the size of a particle is based on its surface area or volume, respectively.

Results
Standards were sent out to fifteen laboratories, nine of which returned results. Tables 13.1 and 13.2 summarize the results of our precision experiment.

Table 13.1. *Precision (in ϕ) of instruments used in the determination of particle size spectra of silt-sized samples*[a]

Std.	Instrument code[b]					
	SG2		SG3		GC1	CC1
	x	σ	x	σ	x	x
4A	5.21	0.56	5.19	0.53	5.10	5.06
4B	5.35	0.56	5.21	0.54	5.07	5.06
4C	5.34	0.64	5.16	0.54	4.97	5.05
4D	5.33	0.58	5.19	0.54	5.01	5.11
5A	6.11	2.25	5.24	0.50	5.06	5.09
5B	5.99	2.02	5.24	0.49	4.90	5.13
5C	6.27	2.38	5.23	0.51	4.94	5.09
5D	6.27	2.58	5.22	0.51	5.03	5.06

[a]Based on four replicate analyses of specially prepared (see text) glass bead standards: x = mean diameter, σ = standard deviation, $\phi = -\log_2$ mm; statistics determined using the method of moments.
[b]SG, SediGraph; GC, Galai CIS-1; CC, Coulter Counter.

Table 13.2. *Precision (in ϕ) of settling tube instruments used in the determination of particle size spectra of sand-sized samples*[a]

Std.	Instrument code								
	ST1		SB1		ST2		ST3		ST4[b]
	x	σ	x	σ	x	σ	x	σ	x
3A	3.05	0.76	2.94	0.52	2.84	0.42	2.89	0.50	2.77
3B	3.01	0.75	2.84	0.59	2.82	0.41	2.93	0.50	2.80
3C	2.96	0.74	2.89	0.58	2.80	0.38	2.83	0.49	2.78
3D	2.82	0.56	2.82	0.54	2.85	0.38	2.89	0.52	
2A	2.02	0.38	2.08	0.37	2.10	0.34	2.20	0.32	2.09
2B	2.02	0.40	2.08	0.36	2.10	0.35	2.20	0.34	2.08
2C	2.03	0.38	2.06	0.37	2.10	0.31	2.20	0.34	
2D	2.01	0.39	2.10	0.36	2.10	0.34	2.20	0.34	
1A	1.10	0.46	1.17	0.41	1.26	0.41	1.15	0.38	1.18
1B	1.09	0.49	1.17	0.41	1.24	0.40	1.17	0.44	1.14
1C	1.10	0.47	1.14	0.42			1.19	0.38	1.16
1D	1.07	0.460	1.13	0.42			1.14	0.40	

[a]As per Table 13.1.
[b]Although not presented, σ is also very precise.

For silt standard no. 4 (Table 13.1), the four instruments provide very similar results (mean ±0.2 ϕ; standard deviation ±0.1 ϕ). The precision of any given instrument was extreme-ly good. Excluding analysis SG2-4A, the precision of each instrument is within ±0.05 ϕ around the mean and ±0.03 ϕ around the standard deviation.

For silt standard no. 5 (Table 13.1), the interinstrument results were within ±0.2 ϕ around the mean, except for SediGraph SG2. The results of SG2 were so very different (1 ϕ finer and 1.75 ϕ more poorly sorted) that we wonder if the standards analyzed were the same. In all cases excepting SG2, instrument precision was extremely good for this standard. SG3 had a precision of ±0.01 ϕ around the mean and standard deviation; GC1 and CC1 had a more typical precision of ±0.05 ϕ around the mean and ±0.03 ϕ around the standard deviation; SG2 showed a more modest precision of ±0.2 ϕ around the mean and ±0.3 ϕ around the standard deviation.

The five settling tubes provided very consistent and precise results from the analysis of sand standards nos. 1–3 (Table 13.2). For the finest sand standard (no. 3), the instruments provide similar results (mean ±0.24 ϕ; S.D. ±0.2 ϕ). Excluding analysis ST1-3D, the precision of these instruments on standard no. 3 is within ±0.03 ϕ around the mean and ±0.03 ϕ around the standard deviation. The interinstrument precision for coarser sand standards (nos. 2 and 1) was much improved (mean ±0.1 ϕ; S.D. ±0.05 ϕ). Single-instrument precision improved to ±0.02 ϕ around the mean and standard deviation.

Figure 13.3 provides examples of replicate size spectra for two of the settling tubes: one that uses an analytical balance with a "weighing below the pan" option (ST2), and another that uses an underwater balance (ST4). The two curves are for two different standards (nos. 1 and 3, respectively). Although standard no. 1 comprised three distinct grain populations or modes, ST2 was only able to recognize two modes. The resulting size spectra are very similar. Therefore we may characterize ST2 as providing very precise results but with a limited ability to discern modes. Although ST4 is capable of recognizing the three grain populations in each of the standards, the proportioning of modal weights was of moderate (±10%) reliability.

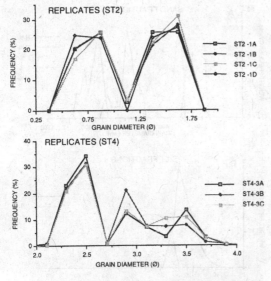

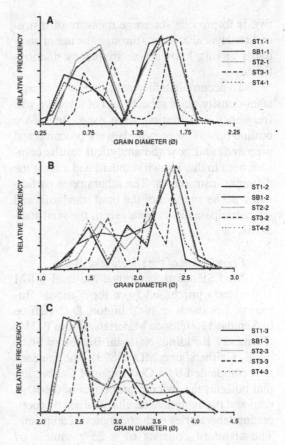

Figure 13.3. Four replicate size spectra of IUGS–COS sand standard no. 1 as analyzed by settling tube participant ST2 (upper graph) showing a high degree of precision. The original standard was composed of three modes (cf. Fig. 13.4A), although only two modes can be identified. The lower graph shows three replicate size spectra of IUGS–COS sand standard no. 3 as analyzed by settling tube participant ST4. Although the instrument is precise in terms of reported means and standard deviations of the size spectra (cf. Table 13.2) and all three modes were identified, there is considerable variation in the contribution of the two finer modes.

Figure 13.4. A comparison of size spectra of the three IUGS–COS sand standards (A) no. 1, (B) no. 2, and (C) no. 3, as determined from the five settling tubes tested in this experiment. In all three cases, the standard was composed of a mixture of three narrow-sized populations of glass beads, but only with standard no. 3 did all instruments identify all three modes. Although the means and standard deviations for these standards, as determined by each instrument, were very similar (cf. Table 13.2), there remains a fair amount of disagreement between the plotted size spectra.

Figure 13.4 provides a comparison of one of the four replicate size spectra (chosen at random) from each of the five settling tubes and for each of the three sand standards. The settling tubes that employed underwater balances (ST3 and ST4) were able to discern the three modes present in each of the standards. ST1, using a "weighing below the pan option," was capable of discerning three modes in the two finer sand standards (Fig. 13.4B,C). All five settling tubes were able to discern the three modes in the finest of the sand standards (Fig. 13.4C). This suggests that the increased sampling rate and response time of the settling tubes fitted with underwater balances may be necessary for subtle and detailed size characterization of coarse ($<1\,\phi$) sand particles. The general population characteristics (mean and variance) are, however, well described by all of the settling tubes.

IUGS–COS "accuracy" experiment
Objective
Earlier attempts at calibrating size-analytical methods have provided useful information on instrument performance. However, these calibration exercises did not employ *known* standards, and thus only provide a measure of instrument precision with some sense as to how a particular instrument performs vis-à-vis other instruments that sense particle size in different ways. Therefore, the second IUGS–COS objec-

tive is to provide for some measure of instrument analytical *accuracy* through the use of standards having known size frequency distributions.

To accomplish this task, standards of constant-density glass spheres and of a known size frequency distribution were used. Below we outline how these standards were procured and prepared, and how the analytical results compare both to this known standard and among the different instruments. Ten laboratories participated in the analysis of the mud standards, and six participated in the analysis of the sand standards.

Preparation of the silt standards

Mud Standard Reference Material (SRM 1003a) was purchased from the National Bureau of Standards in Washington, D.C. (Office of Standard Reference Materials, Room B-311, Chemistry Building, National Bureau of Standards, Gaithersburg, MD 20899). The standard is accompanied by a Certificate of Calibration that outlines the specific properties of the standard and the analytical results obtained by cooperating laboratories with commercial equipment. The standard consists of a 25-g sample of spheres in the 8–58-μm size range. The certified size distribution is provided with uncertainties for class intervals based on microscopic measurements of nearly 8,000 individual beads, of which <1% of the beads are nonspherical. The density of the glass spheres is 2.414 g cm^{-3} as determined by the classical volume displacement method using 25-ml pycnometers and xylene as the displacement fluid (value is provided uncertified).

SRM 1003a was subdivided by our laboratory into three size ranges (Fig. 13.5B):
  (A) 38–58 μm,
  (B) 25–38 μm, and
  (C) 8–25 μm.
These fractions were prepared by wet sieving, using calibrated "precision" sieves. Fractions A and B were dried in a standard drying oven at 60 °C. Fraction C was freeze-dried to ensure that the spheres would dry individually. These fractions were then mixed by weight as follows:
  1. a 35:65 mixture of fractions A and C;
  2. a 55:45 mixture of fractions A and C.

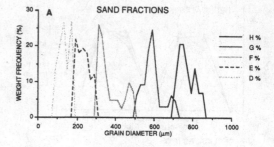

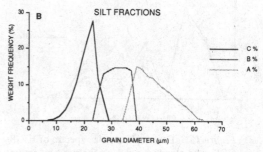

Figure 13.5. The predetermined (certified) size spectra of the seven fractions used in the preparation of the IUGS–COS standards used in the "accuracy" experiment. These fractions were later mixed by weight in the preparation of four sand and three silt standards.

A third standard,
  3. 100% of fraction B,
was sent to participants unmixed. Each standard sent to participating laboratories had a total weight of ≈1.506 g (range 1.500–1.564 g). Laboratories were not told the density of the glass beads but assumed 2.65 g cm^{-3}, a 9% increase over the actual density. Thus the size distribution determined from methods involving sedimentation should be slightly finer (but <0.1ϕ) than those of other measures of size.

The expected size distributions were determined by subdividing the certified size frequency distribution into the same three fractions as outlined above (A–C). These new distributions were then normalized and recombined with each other in the same mixture ratios that were sent to the participants (i.e., 0.35A + 0.65C, 0.55A + 0.45C, and 1.0B). Thus each laboratory was given two bimodal and one unimodal standards having known size frequency distributions.

Preparation of the sand standards

Sand Standard Reference Material (SRM 1017a and 1018a) was purchased from the U.S.

National Bureau of Standards. Each standard is accompanied by a Certification of Calibration. The standards were developed to help evaluate the effective opening of wire-cloth sieves in the range of 100–310 μm and 225–780 μm, respectively. SRM 1017a calibration is based on microscopic measurement of over 60,000 glass beads; SRM 1018a calibration is based on similar measurements on over 18,000 beads. While most of the beads are spherical, about 6% by number for SRM 1018a and 8% by number for SRM 1017a range from nearly spherical beads to ellipsoidal or fused beads. The density of the glass spheres averaged 2.4 g cm^{-3}.

One bottle of each standard (SRM 1017a and 1018a) was combined by our laboratory, and this mixture subdivided into five size fractions (Fig. 13.5A):
- (D) 80–180 μm,
- (E) 180–300 μm,
- (F) 300–500 μm,
- (G) 500–710 μm, and
- (H) 710–850 μm.

These fractions were prepared by dry sieving, using pristine and calibrated "precision" sieves. These fractions were then mixed by weight as follows:

SAND 1 – a mixture of fractions D, F, and H in the ratio of 60:15:25;

SAND 2 – a mixture of fractions E and G in the ratio of 60:40;

SAND 3 – a mixture of fractions D and H in the ratio of 65:35; and

SAND 4 – 100% of fraction F.

Each sample of these sand standards had a total weight of ≈1.5 g. Laboratories were not told the density of the glass beads, and all assumed 2.65 g cm^{-3}, a 9.3% increase over the actual density.

The expected distributions were determined first by combining the certified size frequency distributions biased by the actual weight in each of the bottles. Standard 1017a had an original weight of 83.9 g; that for Standard 1018a was 74.1 g. The meshed size frequency spectra were next subdivided into the same five fractions outlined above (D–H). The individual size spectra of these five fractions were each found to be bimodal (Fig. 13.5A). These new size spectra were normalized and then recombined with each other in the same mixture ratios that made up our four sand standards. Thus each laboratory was given four polymodal size standards that could be described primarily as having trimodal, bimodal, or unimodal populations, although even the essentially unimodal distribution had two very closely spaced modes.

Methods

A brief description of each instrument used in this IUGS–COS experiment is provided in the appendix. The ten instruments used in our silt accuracy experiment include three Coulter Counters, two SediGraphs, two Malvern laser sizers, one Lumosed photosedimentometer, one Galai computerized inspection system, and one image analyzer. The sand experiment instruments included three settling tubes, a sieve analysis, one image analyzer, and one Coulter Counter.

Except for the image analyzer, all instruments provided size spectra results in the form of weight or volume percent. The image analyzer provided size spectra in the form of number frequency spectra. We converted these number data to weight data by calculating the weight of a single grain located at the midpoint of logarithmic size intervals and multiplying the number of grains falling within that class interval by this weight. We used the volume of a sphere ($\frac{4}{3}\pi r^3$) and assumed a constant particle density of 2.65 g cm^{-3}.

As the different analyses provided size spectra in a variety of class intervals, we normalized each spectrum to the largest peak height obtained by any of the analytical methods. This way, the spectra are more directly comparable.

Errors affecting accuracy may be generated by the operator, instrument, or sample preparation. Even when the results of like instruments are compared, it is not easy to proportion total error to these separate components. For example, good sample preparation and a careful operator on a less-than-optimal machine may yield results as accurate as an optimal machine operating on a standard that did not have the same operator or sample preparation care. Our sample preparation error can be considered a constant for each of the analyses if and only if all of the standard provided was analyzed in total. This

Table 13.3. *Mixture standard A (38–60 μm) and standard C (8–25 μm), 35:65 by weight[a]*

Lab ID[b]	Mean (μm)	S.D. (μm)	Mode Std. A (μm)	Std. A (%)	Std. C (μm)	Std. C (%)
CC1-10	26.8*	11.4*	20.2*	71.6	40.4*	28.4
CC2-37	31.1*	11.1*	25.4*	59.5	40.3*	40.5
CC3-11	27.8*	10.6	20.2*	69.6*	40.3*	30.4*
SG1-4	24.5	9.2	21.4*	68.2*	37.2*	31.8*
SG2-9	25.8	10.1	20.3*	68.6*	40.5*	31.4*
L1-7	32.0*	13.4*	27.0*	66.0*	46.6	33.0*
L2-12	28.0*	12.2*	25.0*	100.0	—	—
LS1-1	30.9*	13.8*	23.3*	66.5*	46.0	33.5*
GC1-5	28.2*	10.0	23.7*	85.8	51.5	15.2
IA1-3	<u>31.9*</u>	<u>12.2*</u>	<u>24.1*</u>	<u>18.3</u>	<u>48.2</u>	<u>81.5</u>
Expected	29.5	13.1	23.0	65.0	39.0	35.0

[a]See text for details; asterisk = acceptable (compared to expected) at a tolerance of ±3 μm mean or mode, ±2 μm standard deviation (S.D.), and ±5% of deconvoluted modal weight.
[b]CC, Coulter Counter; SG, SediGraph; L= Malvern laser diffractometer; LS, Lumosed photosedimentometer; GC, Galai CIS-1; IA, image analysis.

Table 13.4. *Mixture standard A (38–60 μm) and standard C (8–25 μm), 55:45 by weight[a]*

Lab ID	Mean (μm)	S.D. (μm)	Mode Std. A (μm)	Std. A (%)	Std. C (μm)	Std. C (%)
CC1-22	29.4	12.1*	20.2*	58.6	40.4*	41.4
CC2-38	33.6*	11.6*	25.4*	47.4*	40.3*	53.6*
CC3-23	32.5*	11.6*	20.2*	46.2*	40.3*	53.8*
SG1-16	27.1	9.2	21.4*	56.2	34.7	43.6
SG2-21	29.0	10.3	24.1*	52.3	40.5	47.7
L1-19	37.2*	14.8*	27.0*	47.3*	46.6	52.3*
L2-24	37.1*	18.7	—	—	45.0	100.0
LS1-13	38.0*	12.0*	24.0*	49.0*	46.4	51.0*
GC1-17	31.9*	11.1	23.7*	50.8	42.4*	49.2
IA1-15	<u>32.6*</u>	<u>10.4</u>	<u>24.1*</u>	<u>23.9</u>	<u>48.2</u>	<u>76.1</u>
Expected	34.7	13.8	23.0	45.0	39.0	55.0

[a]See text for details and Table 13.3 for notation.

was not always the case, particularly for image analysis, where only a small number of grains from the supplied standard was analyzed (see discussion on p. 188).

Table 13.5. *Standard B (25–38 μm), 100% by weight[a]*

Lab ID	Mean (μm)	Mode (μm)	S.D. (μm)
CC1-34	30.4*	31.9*	7.8
CC2-39	34.3	32.0*	5.1*
CC3-33	31.7	32.0*	5.8
SG1-26	27.7*	26.3	5.1*
SG2-28	30.6*	34.1*	7.2
L1-31	37.4	34.6*	6.7
L2-36	37.0	34.0*	6.2
LS1-25	31.1*	36.0	4.4*
GC1-29	30.8*	38.5	6.9
IA1-27	<u>35.9</u>	<u>34.1*</u>	<u>4.0*</u>
Expected	29.2	33.0	4.2

[a]See Table 13.3 for abbreviations; asterisk = acceptable (compared to expected) at a tolerance of ±2 μm mean or mode, ±1 μm standard deviation (S.D.).

Results

Tables 13.3–13.9 summarize the results of our accuracy experiment. We present the results of the silt standards in microns (Tables 13.3–13.5), as this is a common unit of expression for many researchers. (The associated size spectra [Figs. 13.6–13.8] are presented in phi units in deference to those more familiar with this mode of presentation.) The results of the sand standards are presented in traditional phi units in both associated figures and tables.

Knowing the errors specified with the NBS standards, we were able to set tolerance levels for the comparison of the analytical results with our "expected" values. Tolerance categories included the mean and standard deviation of the size spectra, modal peak diameters, and modal proportions (by weight or volume). For the two mixtures of fractions A and C (Tables 13.3 and 13.4), the tolerance levels were set at ±3 μm of the mean or mode, ±2 μm of the standard deviation, and ±5% of deconvoluted modal weight. For the unimodal standard B (Table 13.5), the tolerance levels were set at ±2 μm of the mean or mode and ±1 μm of the standard deviation. The four sand standards had tolerance levels set at ±0.2 φ of the mean or mode, ±0.1 φ of the standard deviation, and ±2% of deconvoluted modal weight.

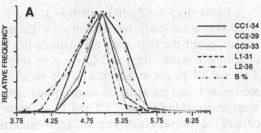

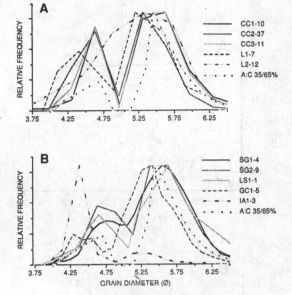

Figure 13.6. Interinstrument comparison of size spectra as determined by three Coulter Counter resistance particle analyzers (CC1–3), two Malvern laser sizers (L1,2), two SediGraph x-ray analyzers (SG1,2), a Lumosed photosedimentometer (LS1), a Galai computerized inspection system (GC1), and an image analyzer (IA1), and as compared to the 35 : 65 mixture by weight of certified fractions A and C (cf. Fig. 13.5).

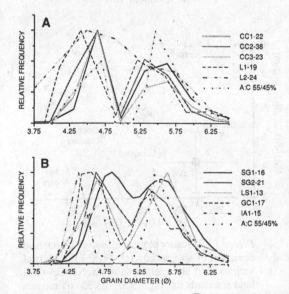

Figure 13.7. Interinstrument comparison of size spectra (see Fig. 13.6 for abbreviations), and as compared to the 55 : 45 mixture by weight of certified fractions A and C (cf. Fig. 13.5).

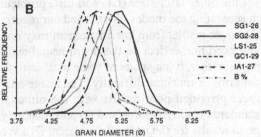

Figure 13.8. Interinstrument comparison of size spectra (see Fig. 13.6 for abbreviations), and as compared to the certified size fraction B (cf. Fig. 13.5).

For the 35 : 65 mixture of fractions A and C, no instrumental results were able to meet all characterization requirements (Table 13.3). For the 55 : 45 mixture of these fractions, two of the Coulter Counters (CC2 and CC3) provided accurate results, and the third (CC1) failed only one of the six tolerance categories. Laser sizer L1 and photosedimentometer LS1 also met all but one accuracy requirement for both of these silt standards. No instrument was able to meet all tolerance categories for unimodal silt standard B (Table 13.5).

We find it interesting that all the techniques provided data in the tails of these silt distributions where none were supposed to be. This can be best seen by comparing the size spectra generated analytically with those of the theoretical spectra of standard B (Fig. 13.8). For this particular standard, both tails were physically truncated, yet only the Lumosed photosedimentometer and the image analyzer came close to ascertaining this characteristic of the standard. As there are sedimentological methods that employ details of the tails of the distribution (e.g., the Passega [1964] method), it is unfortunate that most of the analytical methods spread out the particle size distribution.

Relatedly, the SediGraph was found to spread out a sample's distribution such that the separation of the two grain populations in the mixtures of standards A and C was poor (Figs. 13.6 and 13.7); but at least the modes could be recognized, and the assigned modal proportions were reasonable (Tables 13.3 and 13.4). In the case of the Malvern laser sizer (L2), the bimodal nature of these standards was not discernible, yet the other laser sizer (L1) had little problem in separating the modes. We remind our readers that poor results from one instrument may not, and need not, reflect on the instrumental technique; rather, it may point to some local analytical error or machine difficulty. The image analyzer provided poor results with the bimodal standards (mixtures of A and C) and relatively good results for the unimodal standard (B). We attribute this to a subsampling problem and discuss this problem separately.

Figure 13.9 summarizes the silt experiment in terms of an instrument's ability to discern the mean and standard deviation of a particular standard. Although there might be a tendency to view how well an instrument performs compared to other instruments, the average of many instruments should not be taken as an indication of what the true or expected results ought to be. In the case of standard B (Fig. 13.9C), only two instruments came close to describing the standard exactly (i.e., SG1 and LS1).

In dealing with bimodal distributions, the SediGraph provided results that were too fine-grained. This was true even though both laboratories proceeded with low-concentration analysis (thus eliminating the problem identified in many previous calibration experiments) and, in the case of SG1, a more powerful pump was used (see Chapter 10). Yet in the analysis of the unimodal standard B, both SediGraphs provided accurate prediction of sample mean.

The results on the sand standards (Tables 13.6–13.9, Fig. 13.10) show a higher degree of conformity and accuracy, although the number of instruments involved in this portion of the experiment was more limited. The image analyzer results were again poor with the multimodal standards (SAND 1–3) but highly accurate with the unimodal standard F.

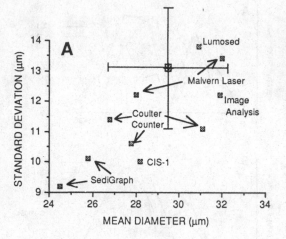

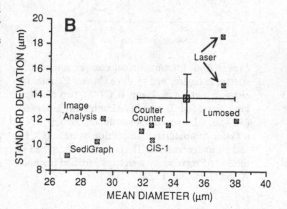

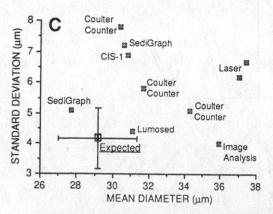

Figure 13.9. Scatter plots of mean versus standard deviation of size spectra (cf. Figs. 13.6–13.8) of various mud analyzers for the three IUGS–COS mud standards (cf. Fig. 13.5): (A) 35 : 65 mixture by weight of certified fractions A and C; (B) 55 : 45 mixture by weight of certified fractions A and C; and (C) certified fraction B. Highlighted on all three plots are the expected mean and standard deviation.

Table 13.6. *SAND-1, a mixture of standards D (80–180 µm), F (300–500 µm), and H (710–850 µm), 60 : 15 : 25 by weight*[a]

| Lab ID[b] | Mean (φ) | S.D. (φ) | Mode | | | | | |
| | | | Std. D | | Std. F | | Std. H | |
			(φ)	(%)	(φ)	(%)	(φ)	(%)
ST1-1	2.00*	1.16	2.90*	59.5*	1.10	14.3*	0.30*	26.2*
ST2-2	2.14*	1.15	2.88*	61.7*	1.50*	13.8*	0.38*	24.1*
S1-9/10	2.03*	1.12	2.88*	59.6*	1.62*	15.1*	0.38*	25.3*
IA1-4	0.96	1.02	3.10*	16.3	1.30	25.4	0.30*	58.2
Expected	2.07	0.75	2.94	60.0	1.69	15.0	0.42	25.0

[a]See text for details; asterisk = acceptable (compared to expected) at a tolerance of ±0.2 φ mean or mode, ±0.1φ standard deviation (S.D.), and ± 2% of deconvoluted modal weight.
[b]ST, settling tube; S, sieve analysis; IA, image analysis; CC, Coulter Counter.

Table 13.7. *SAND-2, a mixture of standards E (180–300 µm) and G (500–710 µm), 60 : 40 by weight*[a]

| Lab ID | Mean (φ) | S.D. (φ) | Mode | | | |
| | | | Std. E | | Std. G | |
			(φ)	(%)	(φ)	(%)
ST1-11	1.57*	0.68*	2.10*	59.5*	0.90*	40.5*
ST2-12	1.60*	0.62*	2.12*	61.7*	0.88*	38.3*
ST5-13	1.48*	0.63*	2.10*	59.3*	0.90	40.7*
S1-19/20	1.55*	0.66*	2.12*	59.4*	0.88*	40.6*
IA1-14	1.01	0.54	1.90	19.4	0.90*	80.6
CC2-41	1.13	0.71*	2.30*	53.7	0.97	46.3
Expected	1.55	0.69	2.12	60.0	0.76	40.0

[a]See text for details and Table 13.6 for notation.

Table 13.8. *SAND-3, a mixture of standards D (8 –180 µm) and H (71 –850 µm), 65 : 35 by weight*[a]

| Lab ID[b] | Mean (φ) | S.D. (φ) | Mode | | | |
| | | | Std. D | | Std. H | |
			(φ)	(%)	(φ)	(%)
ST1-21	1.95*	1.23	2.70	63.6*	0.30*	36.4*
ST2-22	2.00*	1.19	2.62	64.9*	0.38*	35.1*
ST5-23	2.01*	1.26	3.00*	62.5	0.50*	37.5
S1-29/30	2.00*	1.23	2.90*	64.8*	0.38*	35.2*
IA1-24	0.41	0.58	2.90*	5.5	0.30*	94.5
Expected	2.04	0.75	2.94	65.0	0.42	35.0

[a]See text for details and Table 13.6 for notation.

Table 13.9. *Standard F (300–500 µm), 100% by weight*[a]

Lab ID	Mean (φ)	Mode (φ)	S.D. (φ)
ST1-31	1.44*	1.50*	0.25*
ST2-32	1.46*	1.62*	0.19*
ST5-33	1.44*	1.60*	0.18*
S1-39/40	1.70	1.88*	0.21*
IA1-34	1.42*	1.50*	0.21*
CC2-42	1.64*	1.69*	0.26*
Expected	1.46	1.69	0.23

[a]See Table 13.6 for notation.

SAND 1 and SAND 3 are coarse-grained standards having widely separated modal populations of grain size (Tables 13.6 and 13.8). The settling tubes and the sieve analysis produced accurate predictions of the population mean, modal median, and frequency proportions, but produced a population standard deviation that was much too high (1.2φ vs. 0.75φ for the standard).

The Coulter Counter is only capable of analyzing the two finer sand standards (Tables 13.7 and 13.9). Its results for the bimodal standard (SAND 2) were too coarse (by 0.4φ) but accurate for the essentially unimodal standard F. While producing data at a 0.33φ class interval, the Coulter Counter did have some problems accurately defining the tails of the distribution (Fig. 13.10D).

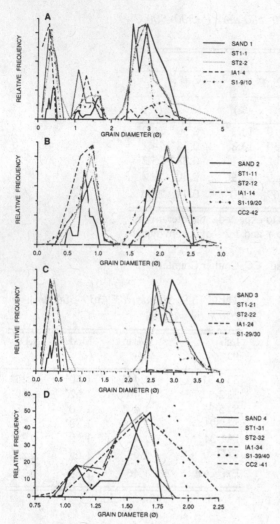

Figure 13.10. Interinstrument comparison of size spectra as determined by two settling tubes (ST1, 2), a Coulter Counter resistance particle analyzer (CC2), a precision sieve analysis (S1), and an image analyzer (IA1), and as compared to the certified sand fractions (cf. Fig. 13.5A): (A) SAND 1, a mixture of fractions D, F, and H, 60:15:25 by weight; (B) SAND 2, a mixture of fractions E and H, 60:40 by weight; (C) SAND 3, a mixture of fractions D and H, 65:35 by weight; and (D) SAND 4, size fraction F, 60:40 by weight. See Tables 13.6–13.9 for more background information.

Number frequency distributions

Imagine a seafloor environment dominated by the deposition of hemipelagic mud, but that also receives occasional dropstones of ice-rafted gravel. Imagine further that a typical 1-kg grab

sample of this seafloor contains one gravel-sized stone that may weigh more than the muddy matrix that accompanies the stone. Is the sample a muddy gravel in terms of its texture? Or do you ignore the stone and concentrate on the mud fraction? Glacial geologists may suggest that very large samples (>1 tonne) should be collected so that a statistically reliable number of gravel particles are analyzed. Oceanographers interested in the effect of local currents on the dispersion of the mud fraction might very well study a few grams of the mud and completely ignore the gravel fraction. The lab technician, however, knows that the one gravel-sized stone can strongly alter the nature of the final weight frequency distribution.

Those who conduct particle size measurements on a Coulter Counter are familiar with the differences between a volume frequency and a number frequency distribution, for the instrument produces both distributions simultaneously. These distributions are related but not comparable: The number frequency is based directly on particle radius; the weight or volume frequency is based on the cube of the particle radius. One large grain may have the same volume (or weight, if we assume constant particle density) as thousands of smaller particles. Thus, very few large grains provide little significance to a number frequency distribution but dominate a weight frequency distribution. Many studies have demonstrated the poor comparison of number size spectra to weight or volume size spectra (Schäfer & Teyssen, 1987).

Image-analyzer (IA) analyses conducted on the IUGS–COS-produced standards (Tables 13.3–13.9) were based on 300–500 individual grain measurements, although tens of thousands of grains comprised the supplied standard. These IA measurements involved a calibrated image processing unit such that each grain size can be considered exact for the purposes of particle size measurement. Image analysis is already considered the standard for particle shape analysis; yet, except for the very well-sorted standards (Tables 13.5 and 13.9), the results are very disappointing. How can such an exact method of measurement produce such poor volume frequency distributions?

Figure 13.11 provides a comparison be-

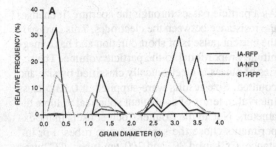

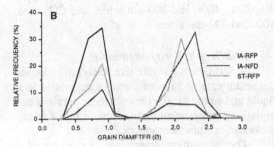

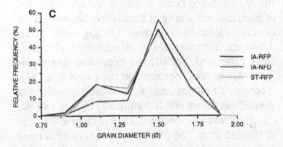

Figure 13.11. A comparison of the volume size spectra (IA–RFP) and the number size spectra (IA–NFD) as determined by an image analyzer (IA) for three IUGS–COS sand standards: (A) SAND 1 (cf. Fig. 13.10A); (B) SAND 2 (cf. Fig. 13.10B); and (C) SAND 4 (cf. Fig. 13.10D). Also shown for comparison are the size spectra determined by one of the settling tubes (ST).

tween the size spectra of an IA-generated number frequency distribution and its converted weight frequency distribution. For comparison, the weight frequency distribution generated from one of the settling tubes is provided as an "acceptable" size spectrum, considering the tolerance levels set by our error analysis of the standards. The size spectra for SAND 1 (Fig. 13.11A) demonstrate the extreme in results. The coarse mode representing standard H (710–850 μm) should comprise 25% of the total distribution, yet the associated IA mode shows <3% by grain number and >58% by sample

weight. The reverse is true for the fine mode, representing standard D (80–180 μm). Neither the number frequency nor the weight frequency distribution is close to the expected size distribution. The same patterns can be seen with the IA size spectra for SAND 2 (Fig. 13.11B); yet for the size spectra of the narrow standard F (300–500 μm), the number frequency and the weight frequency distribution are very similar to each other and to the expected distribution (Fig. 13.11C).

Statistically, counting and sizing 300–500 grains is sufficient to define the size population of the standards if and only if the grains were chosen in an unbiased manner. We suggest that not enough of the smaller size grains were counted in relation to the very few coarse particles sized. We do not know whether the problem lay in the technique of microsplitting the supplied standard to obtain the 500 grains, or whether bias entered into the choice of which grains to count. Past intercalibration experiments whereby *all* of a standard was analyzed by an image analyzer did not show poor results (ASTM, 1983).

We conclude that image analysis is not a viable technique for producing size spectra, as either number or weight frequency, unless the problem in microsplitting or bias in counting is solved. The problem is not with the data produced or number of particles counted, but with the choice of particles counted.

Summary

We have presented a detailed analysis of the accuracy and precision of modern particle size instruments commonly employed in sedimentology laboratories. Each particle size instrument is associated with both positive and negative attributes that must be considered before purchase. We have *not* in this chapter considered equipment purchase price, repair and service record, cost of analysis (in terms of expendables), speed of analysis, sample size, ease of analysis, or other factors germane to instrument choice.

Most analytical instruments were found to be very precise. However, no one instrument was able to provide results on all of our standards that fell within tolerance levels set to test and approve an instrument for its ability to provide truly accurate results. Thus all instruments are associated with some bias. We have specifically avoided a ranking of instruments in terms of their reported results. Readers are encouraged to reflect on the enclosed results in terms of their research requirements. We are pleased to report that the six settling tubes tested, although essentially laboratory built, do provide consistent and accurate results.

On behalf of the International Union of Geological Sciences – Committee on Sedimentology, we provide the following recommendations to the earth science community:

1. New particle size instruments should no longer have their results compared with those of the classical methods of sieving and pipetting, for these older techniques are imprecise and very dependent on laboratory technique and operator error. We further recommend that, where research funds are available, these classical techniques should be replaced with newer digital systems that not only are more precise and accurate, but also have an inherent ability at data handling that makes them more convenient and less open to operator error.

2. The accuracy and precision of every particle size instrument and laboratory technique should be obtained and maintained with periodic checks to monitor instrument deterioration and technician performance.

3. Once obtained, the total accuracy and precision of a given instrument should be reported within the method section of all pertinent scientific reports and manuscripts.

4. Scientific conclusions should be limited by the error associated with the particle size analysis. Far too often have researchers reached conclusions that exceed the limits of their data.

Appendix: IUGS–COS-participating instruments

Coulter Counter: A resistance pulse analyzer (RPA)

An RPA determines the number and size of particles suspended in a conductive liquid by forcing the suspension to flow through a small aperture and monitoring an electrical current that also passes through the aperture. Electrodes are immersed in the conductive liquid on opposite sides of the aperture.

As a particle passes through the aperture, it changes the resistance between the electrodes. This change in the current pulse is of short duration and has a magnitude proportional to the particle volume. The series of pulses is electronically classified by size and counted. Size results were supplied at $0.33\,\phi$ class interval in terms of equivalent spherical volume diameters. Not all the Coulter Counter users in the experiments chose the same aperture tubes. For instance, CC1 used 30- and 200-μm tubes, CC2 used 30-, 200-, 400-, and 800-μm tubes, and CC3 used 100- and 200-μm tubes.

SediGraph: X-ray attenuation

A SediGraph particle size analyzer measures the sedimentation rates of particles suspended in a liquid and automatically presents these data as a cumulative mass percent distribution in terms of the Stokesian or equivalent spherical diameter.

The instrument determines, by means of a finely collimated beam of x rays, the concentration of particles remaining at decreasing sedimentation depths as a function of time. The logarithm of the difference in transmitted x-ray intensity is electronically generated, scaled, and presented linearly as "cumulative mass percent" on the y-axis of an x–y recorder. To minimize analysis time, the position of the sedimentation cell is continually changed so that the effective sedimentation depth is inversely related to elapsed time. The cell movement is synchronized with the x-axis of the recorder to indicate the equivalent spherical diameter corresponding to the elapsed time and instantaneous sedimentation depth. Size results were supplied at $0.25\,\phi$ and $0.2\,\phi$ class intervals, respectively, for ST2 and ST1, in terms of equivalent spherical sedimentation diameters.

Malvern laser particle sizer

Laser diffraction size analysis is based on the principle that particles of a given size diffract light through a given angle, which increases with decreasing size. A narrow beam of light from a laser is passed through a suspension and the diffracted light is focused onto a detector. This senses the angular distribution of scattered light energy.

A lens placed between the illuminated sample and the detector, with the detector at its focal point, focuses the undiffracted light to a point at the centre and leaves only the surrounding diffraction pattern, which does not vary with particle movement. Thus, a stream of particles can be passed through the beam to generate a stable diffraction pattern. Using a computer, it is possible to optimize the size distribution accounting for the light-energy distribution. Size results were supplied at 0.33 ϕ class interval in terms of equivalent spherical cross-sectional area diameters.

Image analysis

A microprocessor-controlled image analysis system can measure the nominal sectional area of an individual grain, from which the nominal sectional diameter can then be calculated. A high-resolution, light-sensitive TV camera mounted on a trinocular microscope feeds an image to a high-resolution TV monitor and videodigitizer controlled by a microprocessor. The TV camera views a scene (loose grains mounted in glycerine on a glass slide and viewed with transmitted light) at a given magnification and projects that scene onto the TV monitor as a matrix of pixels (picture elements), each represented by a light-intensity value. The grain boundary is described as pixel units of relatively low light intensity when compared to the surrounding medium. The area of the particle in pixel units is then approximated from these data; a conversion factor is applied to convert the data from square-pixel units to square micrometres, from which the nominal sectional diameter is calculated. Size results were supplied as a near-continuous number frequency distribution from which a volume frequency distribution "sieved" into 0.2 ϕ class intervals of nominal diameters was produced.

Galai CIS-1 (computerized inspection system)

The CIS-1 integrates two techniques for particle sizing, which together produce a comprehensive approach to particle size and shape analysis. A laser-based optical analysis channel (which uses a finely focused He–Ne laser beam) employs the theory of "time of transition" in a photo-defined measurement. A wedge prism, rotating at a constant speed, scans the incoming laser beam circularly into a focusing objective, which then scans through the sample measurement volume. As individual particles are bisected by the laser's "flying spot," interaction signals are detected by the PIN photodiode directly across from the incoming laser beam. These signals are collected and analyzed by a personal computer in 300 discrete size intervals. Sophisticated pulse analysis algorithms are employed to reject out-of-focus and off-centre interactions. Size results were supplied at 0.1 ϕ class interval in terms of equivalent spherical cross-sectional area diameters.

Lumosed photosedimentometer

The Lumosed photosedimentometer uses the extinction of light to measure concentrations and thus to characterize particle features. This technique requires the incorporation of the substance and shape-dependent extinction coefficient in order to convert the light attenuation measured into the desired mass quantity. A powder of known particle size is used to calibrate the instrument in order to calculate the extinction coefficients for a given substance.

Three fixed beams of light at different sedimentation levels, whose extinction is measured continuously and simultaneously by sensors, detect every change in the solid's concentration. This allows for a relatively short period for the measurement of the particle distribution during the sedimentation process.

A halogen lamp produces beams of a constant white, cold light that are reflected by mirrors to pass through an accuvette at various levels measured from the suspension surface. A diaphragm bundles the beam to a light band reflected by a Plexiglas rod and projected onto the sensors. Detecting changes in light intensity, the sensors (photoelectric cells) pass these signals to an analogue–digital converter in the instrument. A personal computer controls the measurement and guides the operator. Size results were supplied at equally weighted class intervals with a resolution of $\leq 0.83 \phi$ in certain size ranges. We report the results in class intervals of equal size dimension at 0.33ϕ intervals in terms of equivalent spherical sedimentation diameters.

Settling tubes

Settling tubes measure the fall duration of particles settling in a long (> 1-m) column of water. They are based on the principle that a sample's size distribution can be obtained from measurement of the mass–velocity distribution of sand grains settling through an otherwise turbid-free liquid. A number of different types of sample introduction and detection systems can be employed (cf. Chapter 4). Below we provide a brief description of the salient features of each settling tube. Participants are given an identity label and their results described under that label.

ST1: A 2-m long, 15-cm wide settling tower. The sample is introduced using a venetian blind system, and collected on a pan suspended below a digital balance. Settling velocity is converted to size using the Gibbs et al. (1971) equation. Size results were supplied at 0.2 ϕ class intervals.

ST2: Similar to ST1, except that the tube has an inner diameter of 20.5 cm, and the sample is introduced with a surface tension plate using Kodak Photo-flo 200 surfactant on the water surface. Size results were supplied at 0.25 ϕ class intervals.

ST3: A Macrogranometer settling tube with an inner diameter of ≈18.5 cm and a particle settling distance of 1.85 m. The sample is introduced with a vibrating venetian blind system, and collected on a pan that is part of an underwater balance system. Settling velocity is converted to size using the Brezina (1979) equation. Size results were supplied at 0.02 ϕ class intervals.

SB1: A 1.5-m long, 15.25-cm wide settling tower. The sample is introduced using a dome-shaped Plexiglas shape with the sample held on by surface tension; Kodak Photo-flo 200 surfactant is used on the water surface. The trip mechanism is an electric current set up between the plate and the water. The sample is collected on a pan suspended below a digital balance. Settling velocity is converted to size using the Gibbs et al. (1971) equation. Size results were supplied at 0.25 ϕ class intervals.

ST4: A settling tube with an inner diameter of ≈15 cm and a particle settling distance of 1.5 m. The sample is introduced with a vibrating venetian blind system, and collected on a pan that is part of a sophisticated underwater balance system. Settling velocity is converted to size using the Slot (1983) equation. Size results were supplied at variable phi class intervals, from <0.01 ϕ in coarse sand range to 0.025 ϕ in the fine sand range. For plotting purposes we simplified these data using class intervals of 0.1 ϕ.

ST5: Similar to ST1. Size results were supplied at 0.2 ϕ class intervals.

Sieve analysis

Standard ASTM precision sieves, all new, were used with a Roto-Tap system tapping for a period of 20 min. Size results were supplied at 0.25 ϕ class intervals in terms of nominal sieve diameters.

Acknowledgments

We would like to pay tribute to Mr. Donald Clattenburg who, with his staff at the AGC Soft Sediment Laboratory, laboured to produce reliable standards for our experiments. A number of laboratories declined to be involved in an experiment that could potentially show their results as being wanting. We therefore thank all those who have laboured for the good of the earth science community. They include Dr. H. Pitsch, Ms. T. Forbes, Dr. J. Mazzullo, Dr. J. Halka, Dr. J. Brezina, Dr. I. N. McCave, Dr. R. E. Slot, Mr. T. Milligan, Dr. K. Kranck, Dr. D. Proudfoot, Dr. J. Coakley, Dr. A. Bale, Dr. de Boer, Dr. D. Hartmann, and Dr. B. Flemming. Supporting these individuals are a layer of technicians whom we thank deeply.

References

Allen, T. (1968). *Particle size measurement.* London: Chapman & Hall, 248 pp.

ASTM (1983). Standard practice for comparing size in the use of alternative types of particle counters. In: *1983 Annual Book of ASTM Standards.* 14.02.771–9.

Brezina, J. (1979). Particle size and settling rate distributions of sand-sized materials. *Proceedings 2nd European Symposium on Particle Characterization.* Nürnberg, Federal Republic of Germany, 21 pp. + 23 pp. of tables.

Gibbs, R. J., Matthews, M. D., & Link, D. A. (1971). Relationship between sphere size and settling velocity. *Journal of Sedimentary Petrology,* 41: 7–18.

Jones, K. P. N., McCave, I. N., & Patel, P. D. (1988). A computer-interfaced SediGraph for modal analysis of fine-grained sediment. *Sedimentology,* 35: 163–72.

Jordan, C. F., Fryer, G. E., & Hemmen, E. H. (1971). Size analysis of silt and clay by hydrophotometer. *Journal of Sedimentary Petrology,* 41: 489–96.

Komar, P. D., & Cui, B. (1984). The analysis of grain-size measurements by sieving and settling tube techniques. *Journal of Sedimentary Petrology,* 54: 603–14.

McCave, I. N., Bryant, R. J., Cook, H. F., & Coughanowr, C. A. (1986). Evaluation of a laser diffraction size analyser for use with natural sediments. *Journal of Sedimentary Petrology,* 56: 561–4.

McCave, I. N., & Jarvis, J. (1973). Use of the Model T Coulter Counter in size analysis of the fine to coarse sand. *Sedimentology,* 20: 305–15.

Muerdter, D. R., Dauphin, J. P., & Steele, G. (1981). An interactive computerized system for grain size analysis of silt using electroresistance. *Journal of Sedimentary Petrology,* 51: 647–50.

Passega, R. (1964) Grain size representation by CM patterns as a geologic tool. *Journal of Sedimentary Petrology,* 34: 830–47.

Pettijohn, F. J., Potter, P. E., & Siever, R. (1973). *Sand and sandstone*. New York: Springer–Verlag.

Schäfer, A., & Teyssen, T. (1987). Size, shape and orientation of grains in sands and sandstone – Image Analysis applied to rock thin-sections. *Sedimentary Geology*, 52: 251–71.

Singer, J. K. (1986). Results of an intercalibration experiment: An evaluation of the reproducibility of data generated from instruments used in textural analyses. Rice University Sedimentology Report, 61 pp.

Singer, J. K., Anderson, J. B., Ledbetter, M. T.,

McCave, I. N., Jones, K. P. N., & Wright, R. (1988). Assessment of analytical techniques for the size of fine-grained sediments. *Journal of Sedimentary Petrology*, 58: 534–43.

Slot, R. E. (1983). Terminal velocity formula for objects in a viscous fluid. *Journal of Hydraulic Research*, 22: 235–43.

Stein, R. (1985). Rapid grain-size analyses of clay and silt fraction by SediGraph 5000D: Comparison with Coulter Counter and Atterberg methods. *Journal of Sedimentary Petrology*, 55: 590–3.

III *In situ methods*

Sedimentologists study how particles are transported and deposited on the surface of the earth. Some use this information to make sediment transport predictions, others to interpret the depositional history of an ancient deposit. An important part of the sediment transport problem is the determination of the size distribution of grains. Grain size methods are based on the premise that the population of particles being analyzed behave as discrete entities and do not react with each other during their sizing. These same methods also assume that the sample being analyzed is a true representation of the parent population. In other words, particle–particle interactions have not been altered from the natural environment, during or subsequent to collection of the sample. Samples from some sedimentary environments, such as sand grains collected on an active beach, can provide an unbiased estimate of the size characteristics of the parent population.

Chapter 3 has already introduced readers to the concept that many sedimentary environments have grains in transport, not as single grains but in clusters of other particles. Pretreatments may alter and bias the laboratory-produced particle size distribution. Such results provide better information on the size of particles that comprise the aggregates.

For marine sedimentary environments, the act of collecting suspended particles may alter the in situ particle–particle bonding; typically, particle size is decreased. Recent progress has been made on the determination of particle size within the hydrodynamic environment. Two such techniques are highlighted in Part III. Chapter 14 describes the nature of flocculation and agglomeration of suspended particles in estuaries and coastal waters, and the scientific need for in situ values. It then demonstrates how laser diffraction (cf. Chapter 9) can be used underwater to determine the in situ size of particles in transit and undisturbed, within the natural environment.

Similarly, Chapter 15 describes how underwater stereo photography can obtain the in situ size of particles suspended in the water column. This optical technique can also be linked to an image analyzer, so that the nominal size, shape, settling velocity, sedimentation diameter, and density of each particle can be obtained.

These two methods are only the beginning. Other techniques, such as laser holography (Carder et al., 1982) and high-frequency acoustic remote sensing (Holliday, 1987), will soon enlarge the range in particle size and suspended concentration that can be measured in the natural environment.

REFERENCES

Carder, K. L., Steward, R. G., & Betzer, P. R. (1982). In situ holographic measurements of the sizes and settling rates of oceanic particulates. *Journal of Geophysical Research*, 87: 5681–5.

Holliday, D. V. (1987). Acoustic determination of suspended particulate size spectra. In: *Coastal sediments '87, vol. 1*, ed. N. C. Kraus. New York: ASCE, 260–72.

14 In situ size measurements of suspended particles in estuarine and coastal waters using laser diffraction

A. J. BALE AND A. W. MORRIS

Introduction

This chapter looks first at the sources and composition of particles in natural waters, with emphasis on estuarine and coastal environments, and highlights the importance of particles in geochemical and biogeochemical processes. It discusses the tendency of natural particles to form aggregates, and the influence this has on their transport. Recognizing the fragile nature of aggregates, the chapter then looks at the problems associated with the study of natural suspended particles using currently available techniques. Finally, it describes and assesses the performance of a recently developed laser diffraction instrument that allows particles to be sized in the water column, in real time and with minimal physical disturbance.

Sources of particles in estuarine and coastal waters

Suspended particles in estuaries and coastal waters comprise a heterogeneous mixture of mineral particles and biogenic debris as well as living organisms and their excretory products. Mineral particles originate from the weathering and erosion of crustal material and are transported mainly by rivers and, to a lesser extent, as airborne dust.

Pollutant particles, such as fly ash and dust released directly into the water system or deposited from the atmosphere, are also observed (Deuser et al., 1983). The biogenic component comprises primarily the resident phytoplanktonic assemblage, diatoms, and their exoskeletal remains plus refractory land-derived biological material. By contrast, particles in the open oceans away from land sources are predominantly autochthonous and organic (Honjo et al., 1982), though Collier & Edmond (1984) found that traces of terrigenous material were always present at the remotest oceanic sites. As well as contributing to the particle population, the biota also influence the size structure of suspended particles. Particles ingested by plankton are aggregated to varying degrees during faecal pelletization, and microbial activity contributes to degradation of organic debris.

The sizes of suspended particles in marine waters range from the submicron to macroaggregates of 3-4 mm in diameter (Eisma et al., 1983; Kranck, 1984). In open waters, away from tidal effects, particle assemblages of centimetre dimensions are often observed (Silver et al., 1978; Trent et al., 1978). Measurements of the physical characteristics of particles are necessary to understand their behaviour and transport. In recent years, however, it has become clear that the complex structure and fragility of natural particle assemblages have made analysis of their size very difficult.

In estuaries, precipitation of dissolved species may give rise to the formation of abiotic autochthonous particles. Eisma et al. (1980) reported the occurrence of 2.0-μm-diameter particles at low salinities in the Rhine Estuary that were comprised principally of calcium and sulphur, presumably as carbonate and sulphate, respectively. These particles could be generated experimentally by mixing filtered samples of Rhine river water and seawater. The precipitation of humic material with increasing salinity in estuaries (Sholkovitz, 1976) and the incorporation of significant quantities of metals within the precipitates (Sholkovitz, 1978) have also been reported. Removal of riverborne "dissolved" iron at low salinities in estuaries appears almost universal and is attributed to the coagulation of microcolloidal iron to form larger, nominally particulate (>0.45-μm-diameter) material (Duffy, 1985). In addition, suspended particles act as substrates (or surface catalysts) for the precipitation of hydrous metal oxides, for example, manganese and aluminium (Morris et al., 1982, 1986), and sites for microbiological colonization (Ellwood et al., 1982; Plummer et al., 1987).

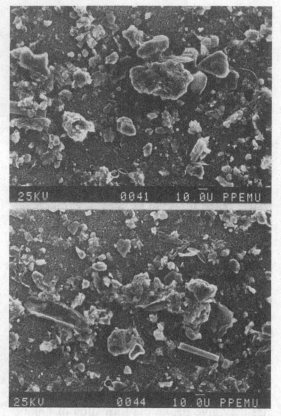

Figure 14.1. Scanning electron photomicrographs of suspended material from the Tamar Estuary trapped on a 0.45-μm pore size membrane (Millipore) demonstrate the wide range of sizes, shapes, and types of particles that comprise natural populations. A 10-μm scale bar is shown at the foot of each micrograph.

Estuarine and coastal particle characteristics

Compositional complexity is a major factor affecting the physical behaviour of particles and consequently their transport potential. The great diversity of particle types, shapes, and sizes is shown by the electron photomicrographs (Fig. 14.1) of suspended material from the Tamar Estuary, southwest England.

Studies of the charge characteristics of natural particles carried out by Neihof & Loeb (1972) and Hunter & Liss (1979, 1982) indicate that particle surfaces are invariably coated with adsorbed humic molecules that are ubiquitous in natural waters. This imparts a uniform negative surface charge to the particles irrespective of the nature of the substrate. Specific surface area measurements indicate that estuarine particles fall in the range 15–125 $m^2 g^{-1}$ and reflect the quantity and age of iron oxyhydroxides associated with the surfaces (Millward et al., 1985). Removal of the adsorbed organic matter associated with particle surfaces has been shown to increase the specific surface area (Glegg, 1987).

In estuaries, suspended particles exert a strong affinity for trace metals (see, e.g., Salomons, 1980; Millward & Moore, 1982; Davies-Colley et al., 1984; Li et al., 1984; Morris, 1986). Several workers have indicated that the exchange capacity for metals and many other species is primarily attributable to surface coatings of hydrous metal oxides (Lion et al., 1982; Tessier et al., 1985) and adsorbed humics (Hart, 1982). Keyser et al. (1978) have suggested that, as a general rule, processes at surfaces or interfaces (not just of particles) dominate the transformations and transport of the majority of pollutants in natural systems.

The efficacy of surface exchange processes combined with the physical transport of particles can lead to cycling of metals in estuaries and coastal waters over time scales that range from semidiurnal to seasonal. The cycling of lead (Elbaz-Poulichet et al., 1984), copper, zinc (Ackroyd et al., 1986), manganese (Morris et al., 1982), and aluminium (Morris et al., 1986) in estuaries is directly attributable to surface exchanges responsive to tidally dominated particle transport and seasonal changes in river flow (Bale et al., 1985). Biogeochemical processes, such as nutrient uptake by phytoplankton in estuary mouths and coastal embayments, are often strongly influenced by a combination of the physical hydrography and its effect on the suspended particles (Cloern, 1984; Pennock, 1985; Owens, 1986).

Effects of particle characteristics on transport processes

The transport of particulate material in estuaries depends basically on the settling velocity of the particles (Gibbs, 1985). Slowly or negligibly settling particles tend to follow the movements of the water, whereas faster-settling material tends to be retained at depositional sites through sedimentation. Small particles tend to be more easily mobilized than larger particles under increasingly energetic conditions. Settling velocity is a function of the density, shape, and size of a particle. Lerman et al. (1974), Komar & Reimers (1978), and McCave (1984) have thoroughly reviewed these relationships and their implications for vertical transport. Particles tend to aggregate under the conditions prevailing in natural waters, particularly at salinities in excess of a few parts per thousand (Stumm & Morgan, 1981). This has been confirmed by studies using photographic techniques and direct observation (Kranck, 1984; Eisma, 1986). It follows that the size, density, shape, and stability of these units, rather than their component particles, are the primary factors influencing natural particle transport and deposition processes (Krone, 1972; Simpson, 1982; McCave, 1984).

Settling of natural particulate material is accelerated by aggregation since it increases the size and thus settling velocity of the material, although the bulk density of the particles usually decreases. Calculations show that even when many of the primary constituents have densities typical of heavy minerals (3.0 kg dm^{-3}), the bulk density of aggregates can fall as low as 1.05 kg dm^{-3} through the inclusion of water and organic material (McCave, 1984).

In estuaries, tidal oscillations in current velocity lead to large variations in particle concentration through resuspension and deposition. In addition, tidal ebb–flood asymmetry, localized concentrations of erodable bottom sediments, and the existence of an erosion maximum in midestuary (Allen et al., 1980; Uncles et al., 1985) contribute to the generation of localized and variable concentrations of suspended solids. In these circumstances, repeated formation and disruption of temporary assemblages by the physical regime combined with the heterogeneous nature of the component materials give rise to the formation of a population of particles containing loosely bound, fragile, amorphous aggregates.

The need for in situ size determinations

Gibbs (1981) has demonstrated that aggregates are readily disrupted by passage through the pumping systems usually used for particle sampling purposes. Sampling with Niskin or similar hydrographic sampling bottles may also lead to erroneous results because large aggregates tend to sediment rapidly within the bottle and may not be withdrawn through the spigot (Gardner, 1977; Calvert & McCartney, 1979). In addition, Gibbs & Konwar (1983) have shown that passage through the narrow orifice of the water bottle spigot can fragment aggregates. Furthermore, it appears that the sample treatments associated with Coulter, light blockage, and conventional pipette analyses for particle size (Gibbs, 1982a,b; Gibbs & Konwar, 1982) also disrupt fragile aggregates. Clearly, nonintrusive, in situ measurements are the only practical approach.

Underwater photography and direct observations by scuba divers and from submersibles provided the first direct in situ information (Syvitski et al., 1983a,b). Attempts to quantify this approach are based on techniques that illuminate and photograph the particles in a specific volume of water (Eisma, 1986; Kranck, 1984). These methods have a lower limit for resolving particles of about 100 μm and, ideally, require an image analysis in order to process the photographs efficiently. Developments of the in situ photographic approach have incorporated light scattering measurements to assess the morphology and size of aggregates (Johnson & Wangersky, 1985; Richardson, 1987). Simpson et al. (1987) deployed an in situ particle collection system incorporating a modified light blockage particle counter and sizing sensor to depths of 5400 m. This significant technological achievement has produced a wealth of information on the distribution and nature of oceanic particles; unfortunately the upper size limit of 200 μm imposed by the dimensions of the optical flow cell has precluded the study of large aggregates.

In situ laser diffraction particle sizing

An in situ particle sizing instrument that avoids the disruption of fragile aggregates during sampling has been developed at the Plymouth Marine Laboratory (PML). Designed primarily for estuarine and shallow coastal work, the instrument housings were constructed and

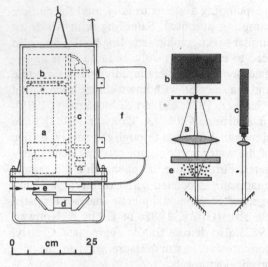

0   cm   25

Figure 14.2. A schematic layout of the main components and the optical path used in the submersible laser diffraction instrument: (a) the receiving lens and detector array; (b) electronic circuitry; (c) the He–Ne laser; (d) the light guide; (e) the measuring "cell"; and (f) the directional fin.

cabled for a maximum depth of 100 m. However, use at greater depths is limited only by engineering considerations, such as the need for stronger underwater housings and windows and more sophisticated signal and power transmission systems to overcome losses within the cables.

The instrument (shown schematically in Fig. 14.2) is based on the optical components of a Malvern Instruments Model 2200 laser diffraction particle sizer. This uses Fraunhofer diffraction theory to derive the size distribution of a particle population from the near-forward light scattering pattern generated by randomly distributed particles in a beam of monochromatic, coherent laser light. This technique is finding wide application in a range of analytical sizing problems because of its rapidity and ease of use (Weiner, 1979; Mohamed et al., 1981; Cooper et al., 1984). We have previously used a standard Malvern Instruments sizer for sedimentological studies in estuaries where the matrix-independent optical technique and the rapid, computerized data storage and analytical capabilities offered distinct advantages over other particle sizing methods (Bale et al., 1984). However, this work revealed that disruption during

sampling was leading to erroneous size distributions and prompted us to investigate the value of in situ measurements (Bale & Morris, 1987). Our present submersible instrument (shown in Fig. 14.3) is the result of several modifications that have reduced the size and weight while increasing the optical rigidity compared with the prototype described by Bale & Morris (1987). More important, the new arrangement allows the 300-mm focal length lens to be incorporated, thereby allowing size measurements throughout the range 5.8–564 μm in diameter (Table 14.1).

The main reduction in the size of the package was achieved through the use of a folded beam arrangement. The precision light guide that enabled this to be achieved is shown in Figure 14.3c,d. Great care was taken in shaping these components during construction, particularly the window retaining clamps, so that particles passing through the "cell" between the two windows experienced minimal turbulence. Likewise, the orientation of the instrument was controlled by a small fin (Fig. 14.3a) so that particles encountered minimal disturbance before and during measurements. Other engineering considerations are dealt with in detail in the description of the prototype instrument (Bale & Morris, 1987).

The laser diffraction technique

Particle sizing by laser diffraction is a relatively recent development, although the theoretical base describing the diffraction is much older (see references in the papers by Swithenbank et al. [1977]; Agrawal & Riley [1984]). This type of instrument became commercially viable with the introduction of inexpensive, low-power lasers and the advent of small computers in the mid-1970s. The facility for rapid, automated measurements with the potential for nonintrusive, on-line operation offered by this technique has several advantages over contemporary particle sizing methods. In general, the laser diffraction technique has been shown to be reliable and accurate for a range of applications (Weiner, 1979; Mohamed et al., 1981; Burkholz & Polke, 1984).

Theoretically, laser diffraction does not require calibration when the instrument is properly aligned since the derived size distribution is fundamentally related to the measured diffrac-

Figure 14.3. (a) The underside of the optical unit showing the laser and spatial filter assembly. The directional fin used to orientate the underwater unit in currents can be seen attached to the pressure case. (b) The particle sizer with the pressure case removed showing the 300-mm focal length receiving lens, the detector, and electronics package. (c) An external view of the optical unit of the PML in situ laser diffraction particle sizing apparatus. The cylinder contains the electronic and optical devices. The light guide and the measuring area can be seen at the base of the cylinder. (d) An enlarged view of the light guide assembly showing the area between the exit and entrant windows that defines the measuring "cell."

Table 14.1. *The ranges of particle size incorporated in each of the fifteen size bands recorded by the Malvern Instruments laser diffraction instrument using different focal length receiving lenses*

Band number	Particle size range (μm)		
	63 mm lens	100 mm lens	300 mm lens
1	1·2– 1·5	1·9– 2·4	5·8– 7·2
2	1·5– 1·9	2·4– 3·0	7·2– 9·0
3	1·9– 2·4	3·0– 3·8	9·0– 11·4
4	2·4– 3·0	3·8– 4·8	11·4– 14·5
5	3·0– 3·9	4·8– 6·2	14·5– 18·5
6	3·9– 5·0	6·2– 7·9	18·5– 23·7
7	5·0– 6·4	7·9– 10·1	23·7– 30·3
8	6·4– 8·2	10·1– 13·0	30·3– 39·0
9	8·2– 10·5	13·0– 16·7	39·0– 50·2
10	10·5– 13·6	16·7– 21·5	50·2– 64·6
11	13·6– 17·7	21·5– 28·1	64·6– 84·3
12	17·7– 23·7	28·1– 37·6	84·3–112·8
13	23·7– 33·7	37·6– 53·5	112·8–160·4
14	33·7– 54·9	53·5– 87·2	160·4–261·6
15	54·9–118·4	87·2–188·0	261·6–564·0

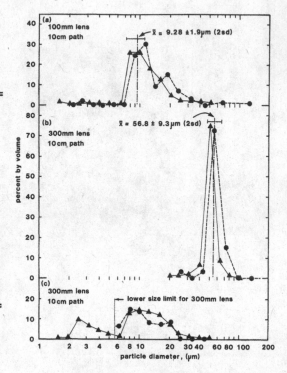

Figure 14.4. Size distribution measurements of calibration bead suspensions obtained with laser diffractometer (●) and Coulter Counter (▲). (Redrawn from Bale & Morris [1987] with the permission of Academic Press.)

tion pattern by Fraunhofer theory. Practically, it is reassuring to check the performance of the instrument using reference particles. Measurements of calibration beads suspended in seawater using our submersible instrument under laboratory conditions are compared with Coulter data in Figure 14.4. The broad size distributions and subsequent microscopic examination indicated that particle associations had occurred in the seawater matrix. Nevertheless, both techniques recorded very similar distributions; clearly, laser diffraction compares favourably with Coulter measurements when the standards fall within the working range of the lens.

However, some evaluations of laser diffraction have indicated that there are weaknesses in the technique (Evers, 1982; Cooper et al., 1984). This may in part be because different sizing techniques measure different properties of particles, although they all generally assume sphericity. This aspect is particularly problematic when dealing with extreme particle shapes and natural materials. Some inconsistencies have also been attributed to the various mathematical inversions (a "nontrivial exercise," according to Agrawal & Riley [1984]) used to derive the particle size distribution from the measured diffraction patterns (Austin & Shah, 1983; McCave et al., 1986). Other problems relate to the relative sensitivity of the individual components of the

detector array. Although our calibration results would suggest this is not always a serious problem, it appears that individual calibration of each detector element (Dodge, 1984) can significantly improve the absolute accuracy of laser diffraction measurements. The greatest inaccuracies with laser diffraction occur when there is a significant amount of material outside and close to the extremes of the measured size ranges, particularly the smallest size limit (Evers, 1982; Cooper et al., 1984; McCave et al., 1986). The algorithm used in the Malvern instrument assumes there is no material greater than the largest size band; thus, in normal use, the sample must either be prescreened to remove the larger particles, or a lens with a sufficiently large upper size limit should be used. The first of these options is not practical in the submersible mode, and we have had to accept that larger particles than can be characterized by the 300-mm lens (>564-μm diameter) are often present.

McCave et al. (1986) reported grave concern over the quality of laser diffraction measurements on samples where material smaller than the lower size limit of the lens in use was present. The Malvern instrument attempts to record the percentage of material below the lowest size limit, but it is doubtful whether it actually "sees" material of dimensions smaller than the wavelength of the light (about 0.6 μm). McCave et al. (1986) registered the greatest dissatisfaction when using the smallest focal length lens (1.2–118.4 μm), which was analyzing particles below the threshold (<7.0 μm) where Bayvel & Jones (1981) predicted that the Fraunhofer technique would break down.

Despite these limitations, it is clear that, used with an awareness of the potential problems, laser diffraction is a powerful technique that provides rapid, real-time results essential for measurements of dynamic systems. For us, awareness of the potential problems means primarily using the 300-mm focal length lens for in situ measurements. This lens is much less influenced by light scattering by colloidal sized particles than are the shorter focal length lenses, and its lower size limit more closely approximates to Bayvel & Jones's (1981) theoretical lower threshold for the diffraction technique.

In estuarine studies the laser diffraction technique is particularly advantageous in obviating the need to add electrolyte as required with impedance type instruments, and it also performs size analyses independent of the nature of the supporting matrix. Moreover, when used in an in situ mode, the technique reduces the disruption of aggregates that has led to confusion in early studies of natural particles. It is probable that future developments in the algorithms used to perform the mathematical inversion will continue to improve the accuracy of size distributions determined by laser diffraction.

Similarly, progress in the engineering field is paving the way for improved particle sizing instrumentation for incorporation in submersible apparatus. For example, the development of photon correlation spectroscopy has led to a "crop" of new instruments. This technique has a quoted working range of 0.01 μm to about 3.0 μm. Coulter Electronics has marketed an instrument of this type called the Nanosizer, and Malvern Instruments' version is the Autosizer.

However, Malvern has now combined the technique with laser diffraction in their Mastersizer. This greatly extends the measurement range over basic laser diffraction: Malvern quotes 0.1–600 μm in two ranges with sixty-four size intervals in each range. This eliminates the uncertainty associated with diffraction measurements of small particles (<7.0–μm diameter) and measurements in the presence of submicron particles. Developments in optical fibre technology offer the potential of replacing the light guide used in our submersible instrument with a much less cumbersome component. This type of technology and the advent of miniature lasers raise the possibility of constructing much smaller and therefore stronger underwater diffraction spectrometers.

Application of laser diffraction to estuarine particle sizing

Our submersible instrument has been used to demonstrate the advantage of in situ measurements during field studies in the Tamar Estuary. Particle size measurements obtained directly in the water column are compared in Figure 14.5 with measurements on discrete samples pumped from the same depth but sized using a standard bench instrument. These measurements were undertaken using the 100-mm focal length lens, which has an upper size limit of 188.0 μm; they indicate that the modal values of the size distribution measured on discrete samples fell generally in the range 10–20 μm, although there was a trend toward larger values at the lowest salinities. Measured in situ, the modal sizes fell in the largest size band (87.2–188.0 μm), indicating that even larger particles were present. These results confirm that natural aggregates are too fragile to withstand sampling by pumping and that the physical disruption induced by sampling results in smaller, more robust (10–20-μm) aggregates. The primary particles that comprise the microaggregates are, however, considerably smaller. This was demonstrated by the size measurements on the pumped samples after chemical treatment to break up any remaining associations (2 h in 30% w/v H_2O_2 followed by ultrasonic vibration for 5 min). In this case, modal values were about 10 μm, and no particles larger than 37 μm were present.

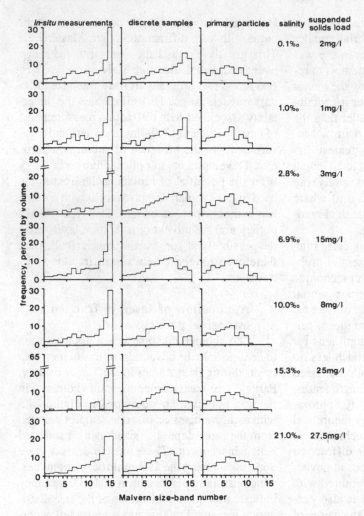

salinity | suspended solids load
0.1‰ | 2mg/l
1.0‰ | 1mg/l
2.8‰ | 3mg/l
6.9‰ | 15mg/l
10.0‰ | 8mg/l
15.3‰ | 25mg/l
21.0‰ | 27.5mg/l

Figure 14.5. A comparison of in situ measurements of suspended particles in the Tamar Estuary with measurements on discrete samples pumped from the same depth. Both instruments were fitted with 100-mm focal length lenses (see Table 14.1). The size distribution of primary particles, obtained after removal of organic material using hydrogen peroxide followed by ultrasonic treatment, is also shown. (Redrawn from Bale & Morris [1987] with the permission of Academic Press.)

In situ measurements of particle size have been taken through the water column over a tidal cycle in the upper reaches of a tidally energetic estuary. Although the in situ concentrations of suspended particles exceeded the optimum level for measurements at times of maximum current velocity, marked changes in the size structure of the particles were observed (Fig.

14.6a,b). The presence of very large aggregates, too large to be accurately sized using the largest available lens (300 mm), was registered throughout the water column during the high-water slack tide. As the current velocity maximized around midtide, the median values of the measured size distributions were much smaller (60–80 μm). This result is compatible with previous measurements (shown in Fig. 14.5) on discrete samples and with measurements on discrete samples obtained at the same time (Fig. 14.6c). This supports Eisma's (1986) conclusions that natural aggregates are easily broken into smaller subcomponents that are then relatively resistant to further breakage.

The transition in the size structure of the suspended particle population observed over the tidal cycle can be attributed to several factors.

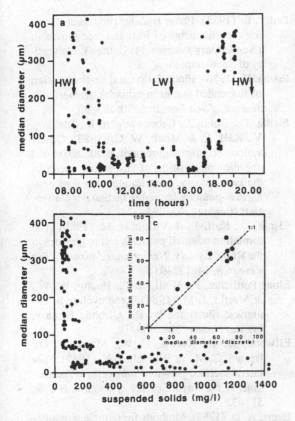

Figure 14.6. (a) Median particle diameters measured in situ over a tidal cycle at an anchor station in the Tamar Estuary. (b) Median particle diameters plotted against the corresponding concentration of suspended solids. (c) A comparison of median diameters measured in situ at the time of the maximum ebb tide current with measurements on discrete samples taken at the same time.

First, increased turbulence as the ebb and flood currents develop can physically disrupt aggregates that form at slack water. Second, the site at which these measurements were taken experienced completely fresh water at low tide. A large component of the observed increase in suspended solids would, therefore, have been due to the passage of the estuarine turbidity maximum past the moored vessel. Earlier work (Schubel, 1969; Bale et al., 1985) has demonstrated that the hydrodynamic processes that generate the turbidity maximum tend selectively to accumulate relatively small particles. Finally, the increasing currents may have mobilized discrete, relatively small mineral particles from the bed. This is consistent with the conclusions drawn by Wellershaus (1981) from organic carbon measurements in the Ems Estuary. He found that particles in the turbidity maximum comprised two populations: the first weighted by fine mineral grains that underwent periodic tidal erosion and sedimentation, and the second a more or less permanently suspended population consisting of low-density aggregates of a great variety of substances.

Conclusions

The size distributions of natural particle populations are important factors for the prediction of particle behaviour. However, the heterogeneous nature of natural particles and their tendency to form fragile aggregates makes it particularly difficult to measure their size distributions accurately using standard particle collection and sizing techniques.

In situ sizing using laser diffraction offers a practical method of studying the size distribution of natural particle populations in real time. The rapid, computerized data acquisition and nonintrusive measurements render this technique particularly suitable for the study of dynamic systems such as estuaries and coastal waters.

Field studies with an in situ instrument have been undertaken in the turbid upper reaches of a macrotidal estuary. These measurements have demonstrated the marked transition in the size structure of the particle population that occurs in response to fluctuations in tidal current strength. Even so, the present upper size limit (564 μm) did not allow the largest aggregates to be adequately characterized. Longer focal length lenses are available and need to be evaluated.

REFERENCES

Ackroyd, D. R., Bale, A. J., Howland, R. J. M., Knox, S., Millward, G. E., & Morris, A. W. (1986). Distributions and behaviour of dissolved Cu, Zn and Mn in the Tamar Estuary. *Estuarine, Coastal and Shelf Science*, 23: 621–40.

Agrawal, Y. C., & Riley, J. B. (1984). Optical particle sizing for hydrodynamics based on near forward scattering. *Society of Photo-optical Instrumentation Engineers*, 489: 68–76.

Allen, G. P., Salomon, J. C., Bassoulet, P., Du Penhoat, Y., & De Grandpre, C. (1980). Effects of tides on mixing and suspended sediment transport in estuaries. *Sedimentary Geology*, 26: 69–80.

Austin, L. G., & Shah, I. (1983). A method for inter-conversion of Microtrac sieve size distributions. *Powder Technology*, 35: 271–8.

Bale, A. J., & Morris, A. W. (1987). In-situ measurement of particle size in estuarine waters. *Estuarine, Coastal and Shelf Science*, 24: 253–63.

Bale, A. J., Morris, A. W., & Howland, R. J. M. (1984). Size distributions of suspended material in the surface waters of an estuary as measured by laser Fraunhofer diffraction. In *Transfer processes in cohesive sediment systems*, eds. W. R. Parker & D. J. J. Kinsman. New York: Plenum, pp. 75–85.

——— (1985). Seasonal sediment movement in the Tamar Estuary. *Oceanologica Acta*, 8: 1–6.

Bayvel, L. P., & Jones, A. R. (1981). *Electromagnetic scattering and its applications*. London: Applied Science.

Burkholz, A., & Polke, R. (1984). Laser diffraction spectrometers – experience in particle size analysis. *Particle Characterization*, 1: 153–60.

Calvert, S. E., & McCartney, M. J. (1979). The effect of incomplete recovery of large particles from water samplers on the chemical composition of oceanic particulate matter. *Limnology and Oceanography*, 24: 536–40.

Cloern, J. E. (1984). Temporal dynamics and ecological significance of salinity stratification in an estuary (South San Francisco Bay, USA). *Oceanologica Acta*, 7: 137–41.

Collier, R., & Edmond, J. (1984). The trace element geochemistry of marine biogenic particulate matter. *Progress in Oceanography*, 13: 113–99.

Cooper, L. R., Haverland, R. L., Hendricks, D. M., & Knisel, W. G. (1984). Microtrac particle size analyser: An alternative particle-size determination method for sediment and soils. *Soil Science*, 138: 138–46.

Davies-Colley, R. J., Nelson, P. O., & Williamson, K. J. (1984). Copper and cadmium uptake by estuarine sedimentary phases. *Environmental Science and Technology*, 18: 491–9.

Deuser, W. G., Emeis, K., Ittekkot, V., & Degens, E. T. (1983). Fly-ash particles intercepted in the deep Sargasso Sea. *Nature*, 305: 216–18.

Dodge, L. G. (1984). Calibration of the Malvern particle sizer. *Applied Optics*, 23: 2414–19.

Duffy, B. (1985). Phase transfer processes affecting the chemistry of iron and manganese in river–estuary systems. Ph.D. thesis, University of Southampton.

Eisma, D. (1986). Flocculation and de-flocculation of suspended matter in estuaries. *Netherlands Journal of Sea Research*, 20: 183–99.

Eisma, D., Boon, J., Groenewegen, R., Ittekkot, V., Kalf, J., & Mook, W. G. (1983). Observations on macroaggregates, particle size and organic composition of suspended matter in the Ems estuary. *Mitteilungen aus dem geologisch–palaontologischen Institut der Universitat Hamburg*, 55: 295–314.

Eisma, D., Kalf, J., & Veenhuis, M. (1980). The formation of small particles and aggregates in the Rhine Estuary. *Netherlands Journal of Sea Research*, 14: 172–91.

Elbaz-Poulichet, F., Holliger, P., Huang, W. W., & Martin, J.-M. (1984). Lead cycling in estuaries, illustrated by the Gironde Estuary, France. *Nature*, 308: 409–14.

Ellwood, D. C., Keevil, C. W., Marsh, P. D., Brown, C. M., & Wardell, J. N. (1982). Surface associated growth. *Philosophical Transactions of the Royal Society of London*, B 279: 517–32.

Evers, A. D. (1982). Methods for particle size analysis of flour: A collaborative test. *Laboratory Practice*, March: 215–19.

Gardner, W. D. (1977). Incomplete extraction of rapidly settling particles from water samplers. *Limnology and Oceanography*, 22: 764–8.

Gibbs, R. J. (1981). Floc breakage by pumps. *Journal of Sedimentary Petrology*, 51: 670–2.

——— (1982a). Floc stability during Coulter counter analysis. *Journal of Sedimentary Petrology*, 52: 657–70.

——— (1982b). Floc breakage during HIAC light-blocking analysis. *Environmental Science and Technology*, 16: 298–9.

——— (1985). Estuarine flocs: Their size, settling velocity and density. *Journal of Geophysical Research*, 90: 3249–51.

Gibbs, R. J., & Konwar, L. (1982). Effect of pipetting on mineral flocs. *Environmental Science and Technology*, 16: 119–21.

——— (1983). Sampling of mineral flocs using Niskin bottles. *Environmental Science and Technology*, 17: 374–5.

Glegg, G. A. (1987). Estuarine chemical reactivity at the particle water interface. Ph.D. thesis, Plymouth Polytechnic.

Hart, B. T. (1982). Uptake of trace metals by sediments and suspended particulates: A review. *Hydrobiologica*, 91: 299–313.

Honjo, S., Manganini, S. J., & Cole, J. J. (1982). Sedimentation of biogenic matter in the deep ocean. *Deep-Sea Research*, 29: 609–25.

Hunter, K. A., & Liss, P. S. (1979). The surface charge of suspended particles in estuarine and coastal waters. *Nature*, 282: 823–5.

(1982). Organic matter and the surface charge of suspended particles in estuarine waters. *Limnology and Oceanography*, 27: 322–35.

Johnson, B. D., & Wangersky, P. J. (1985). A recording backward scattering meter and camera system for examination of the distribution and morphology of macroaggregates. *Deep-Sea Research*, 32: 1143–50.

Keyser, T. R., Natusch, D. F. S., Evans, C. A., & Linton, R. W. (1978). Characterizing the surface of environmental particles. *Environmental Science and Technology*, 12: 768–73.

Komar, P. D., & Reimers, C. E. (1978). Grain shape effects on settling rates. *Journal of Geology*, 86: 193–209.

Kranck, K. (1984). The role of flocculation in the filtering of particulate matter in estuaries. In: *The estuary as a filter*, ed. V. S. Kennedy. New York: Academic Press, pp. 159–79.

Krone, R. B. (1972). A field study of flocculation as a factor in estuarial shoaling processes. Technical Bulletin no. 19. Committee on Tidal Hydraulics, U.S. Army Corps of Engineers.

Lerman, A., Lal, D., & Dacy, M. F. (1974). Stokes' settling and chemical reactivity of suspended particles in natural waters. In: *Suspended solids in water*, ed. R. J. Gibbs. New York: Plenum Press, pp. 17–49.

Li, Y.-H., Burkhardt, L., & Teraoka, H. (1984). Desorption and coagulation of trace elements during estuarine mixing. *Geochimica et Cosmochimica Acta*, 48: 1879–84.

Lion, L. W., Altmann, R. S., & Leckie, J. O. (1982). Trace-metal adsorption characteristics of estuarine particulate matter: Evaluation of contributions of Fe/Mn oxide and organic surface coatings. *Environmental Science and Technology*, 16: 660–6.

McCave, I. N. (1984). Size spectra and aggregation of suspended particles in the deep ocean. *Deep-Sea Research*, 31: 329–52.

McCave, I. N., Bryant, R. J., Cook, H. F., & Coughanowr, C. A. (1986). Evaluation of a laser diffraction size analyser for use with natural sediments. *Journal of Sedimentary Petrology*, 56: 561–4.

Millward, G. E., Glegg, G. A., Glasson, D. R., Morris, A. W., & Bale, A. J. (1985). The microstructures of estuarine particles and their role in heterogeneous chemical reactivity. *ACS Symposium Series, Division of Environmental Chemistry*, 25: 418–20.

Millward, G. E., & Moore, R. M. (1982). The adsorption of Cu, Mn and Zn by iron oxyhydroxide in model estuarine solutions. *Water Research*, 16: 981–5.

Mohamed, N., Fry, R. C., & Wetzel, D. L. (1981). Laser Fraunhofer diffraction studies of aerosol droplet size in atomic spectrochemical analysis. *Analytical Chemistry*, 53: 639–44.

Morris, A. W. (1986). Removal of trace metals in the very low salinity region of the Tamar Estuary, England. *The Science of the Total Environment*, 49: 297–304.

Morris, A. W., Bale, A. J., & Howland, R. J. M. (1982). The dynamics of estuarine manganese cycling. *Estuarine, Coastal and Shelf Science*, 14: 175–92.

Morris, A. W., Howland, R. J. M., & Bale, A. J. (1986). Dissolved aluminium in the Tamar Estuary, southwest England. *Geochimica et Cosmochimica Acta*, 50: 189–97.

Neihof, R. A., & Loeb, G. I. (1972). The surface charge of particulate matter in seawater. *Limnology and Oceanography*, 17: 7–16.

Owens, N. J. P. (1986). Estuarine nitrification: A naturally occurring fluidized bed reaction? *Estuarine, Coastal and Shelf Science*, 22: 31–44.

Pennock, J. R. (1985). Chlorophyll distributions in the Delaware estuary: Regulation by light-limitation. *Estuarine, Coastal and Shelf Science*, 21: 711–25.

Plummer, D. H., Owens, N. J. P., & Herbert, R. A. (1987). Bacteria–particle associations in turbid estuarine environments. *Continental Shelf Research*, 7: 1429–35.

Richardson, M. J. (1987). Particle size, light scattering and composition of suspended particulate matter in the North Atlantic. *Deep-Sea Research*, 34: 1301–29.

Salomons, W. (1980). Adsorption processes and hydrodynamic conditions in estuaries. *Environmental Technology Letters*, 1: 356–65.

Schubel, J. R. (1969). Size distributions of the suspended particles of the Chesapeake Bay turbidity maximum. *Netherlands Journal of Sea Research*, 4: 283–309.

Sholkovitz, E. R. (1976). Flocculation of dissolved organic and inorganic matter during the mixing of river water and seawater. *Geochimica et Cosmochimica Acta*, 40: 831–45.

— (1978). The flocculation of dissolved Fe, Mn, Al, Cu, Ni, Co and Cd during estuarine mixing. *Earth and Planetary Science Letters*, 41: 77–86.

Silver, M. W., Shanks, A. L., & Trent, J. D. (1978). Marine snow: Microplankton habitat and source of small-scale patchiness in pelagic populations. *Science*, 201: 371–3.

Simpson, W. R. (1982). Particulate matter in the oceans – sampling methods, concentrations, size distribution and particle dynamics. *Oceanography and Marine Biology Annual Review*, 20: 119–72.

Simpson, W. R., Gwilliam, T. J. P., Lawford, V. A., Fasham, M. J. R., & Lewis, A. R. (1987). In situ deep water particle sampler and real-time sensor package with data from the Madeira Abyssal Plain. *Deep-Sea Research*, 34: 1477–97.

Stumm, W., & Morgan, J. J. (1981). The solid–solution interface. In: *Aquatic chemistry*, eds. W. Stumm & J. J. Morgan. New York: Wiley-Interscience, 445–514.

Swithenbank, J., Beer, J. M., Taylor, D. S., Abbot, D., & McCreath, G. C. (1977). A laser diagnostic technique for the measurement of droplet and particle size distribution. In: *Experimental diagnostics in gas phase combustion systems*, ed. B. J. Zinn. New York: American Institute of Aeronautics and Astronautics, 53: 421–47.

Syvitski, J. P. M., Fader, G. B., Josenhans, H. W., MacLean, B., & Piper, D. J. W. (1983a). Seabed investigations of the Canadian East Coast and Arctic using Pisces IV. *Geoscience Canada*, 10: 59–68.

Syvitski, J. P. M., Silverberg, N., Ouellet, G., & Asprey, K. W. (1983b). First observations of benthos and seston from a submersible in the lower St. Lawrence Estuary. *Geographie Physique et Quaternaire*, 37: 227–40.

Tessier, A., Rapin, F., & Carignan, R. (1985). Trace metals in oxic lake sediments: Possible adsorption onto iron oxyhydroxides. *Geochimica et Cosmochimica Acta*, 49: 183–94.

Trent, J. D., Shanks, A. L., & Silver, M. W. (1978). In situ and laboratory measurements on macroscopic aggregates in Monterey Bay, California. *Limnology and Oceanography*, 23: 388–91.

Uncles, R. J., Elliott, R. C. A., & Weston, S. A. (1985). Observed fluxes of water, salt and suspended sediment in a partially mixed estuary. *Estuarine, Coastal and Shelf Science*, 20: 147–67.

Weiner, B. B. (1979). Particle and spray sizing using laser diffraction. SPIE (Optics in quality assurance II). *Society of Photo-optical Instrumentation Engineers*, 170: 53–62.

Wellershaus, S. (1981). Dredged coastal plain estuaries: Organic carbon in the turbidity maximum. *Environmental Technology Letters*, 2: 153–60.

15 The Floc Camera Assembly

DAVID E. HEFFLER,
JAMES P. M. SYVITSKI,
AND KENNETH W. ASPREY

Introduction

Visual observations of suspended sediment in coastal marine waters and in the deep ocean have confirmed that the formation of flocculated or agglomerated sediment, commonly identified as marine snow (Suzuki & Kato, 1953), is an important mechanism in the transport of sediment to the seafloor (Nishizawa et al., 1954; Shanks & Trent, 1980; Farrow et al., 1983). This is particularly true for fjords (Syvitski et al., 1985) and deep estuaries (Eisma et al., 1978), but is common to all marine environments (Kranck, 1984). Primary sediment particles in the ocean are too small to be seen with the unaided eye. The particles observed visually from submersibles are therefore in the form of floccules. The most important effect of flocculation and related processes is in controlling the net vertical flux of particles through the water column. This in turn has important implications to the fill of sedimentary basins (Syvitski et al., 1988), and in controlling the fate of pollutants (Eisma, 1981). Sampling individual flocs, or obtaining measurements on the settling rate of individual particles, however, has proved very difficult (Gibbs, 1982; Kranck, 1984).

All water sampling techniques break up the in situ structure of flocs, leading to gross errors in estimating the flux of sediment to the ocean floor (McCave, 1975). For instance, water samplers and submerged pumps can alter the characteristics of flocculated particles through mechanical interference (Gibbs, 1981). The pressure wave generated during a water bottle closure, or the multilayered baffles and filters used in submerged pumps, may destroy the larger and fragile particles of marine snow. Also the coarser, or faster-settling particles tend to be undersampled by water bottle samplers (Gardner, 1977; Syvitski & Murray, 1981). Furthermore, the concentration of suspended particles increases through sedimentation in the bottom of water samplers before withdrawal, and this can cause suspended particulate matter to interact in ways other than those found in the unrestricted ocean (Gibbs et al., 1983).

Little information is presently available on the instantaneous settling velocity of the various types of sediment particles suspended in the water. Appropriate theory is almost nonexistent (see Lal, 1977). Settling rates inferred from laboratory measured particle size distributions, obtained from disaggregated suspended matter, could be in error by several orders of magnitude (Syvitski et al., 1985). Additionally, short-duration visual observation of flocs, say, from submersibles, cannot provide a reliable estimate of floc velocity, even when grounded on the seafloor. Observations must be made with an instrument that creates little flow disturbance within the ambient water mass and is situated on a platform stationary to the sedimenting suspended particles.

Parameters not obtained using conventional sampling or profiling techniques include:

1. the number of particles of marine snow suspended in a given volume of water,

2. the spacing between these particles,

3. the in situ size and shape of these suspended particles,

4. the relative or absolute settling velocity of the different suspended particle types, and

5. the variation with time or depth of these properties.

Knowing the size/shape and settling velocity of an individual particle of marine snow, one can back-calculate floc density. The in situ size distribution of suspended particles is known to vary with depth, sediment influx mechanisms, and various oceanographic variables (Syvitski et al., 1985). Sedimentation of suspended marine particles can be controlled by biogeochemical interactions (Lal, 1977), biological communities (Alldredge, 1977; Silver & Alldredge, 1981; Lewis & Syvitski, 1983), and microbial interactions (Alldredge & Youngbluth, 1985; Alldredge et al., 1986).

To provide quantitative information on these parameters and interactions, researchers turned to underwater cameras. One such camera, designed by Johnson & Wangersky (1985), photographs individual flocs as small as 50 μm. The camera is part of a conductivity–temperature–depth profiler and has an optical sensor to detect a floc before a photograph is taken. Honjo et al. (1984) used a Benthos Model 372 deep-sea camera to record the size (>200 μm) and shape of marine snow within the water column. Asper (1987) recently incorporated the same Benthos camera within his sediment traps to photograph the sinking of marine snow (particles >500 μm in diameter). Kranck (1984) and Kranck & Milligan (1988) use a Benthos–Edgerton Model 373 Plankton Camera that collects silhouette photographs of particles >90 μm.

In situ methods for the analysis of the size of marine snow need not be photographic. Bale and coworkers (Bale et al., 1984; Bale & Morris, 1987; see also Chapter 14, this volume) collect in situ size spectra of suspended matter using a submerged laser Fraunhofer diffraction system. The technique can cope with a wide range of particle sizes (5.8–564 μm), and data handling is superior to any other in situ method. Limitations on the upper size limit, however, could prove a drawback for many marine environments. In situ instrument motion may also affect the breakup of delicate particles. Additionally the instrument may introduce population modes into the size frequency distribution where none may exist (similar to laboratory models: see Chapter 9). Carder et al. (1982) developed an in situ holographic method to determine the size and settling rate of suspended sediment within a small volume (about 1–4 cm^3). Although an improvement in the fine end of the size spectrum, at least compared to photographic techniques, the technique requires a stilling tank to reduce horizontal motion and thus cannot be considered useful for depth profiling.

Floc Camera Assembly (FCA) design

Our immediate scientific concerns were in understanding particle dynamics in fjords and deep coastal waters, such as in the Gulf of St. Lawrence and Arctic fjords. Our design parameters for a photographic package included an instrument that could reach a depth of 800 m and be operated from small oceanographic vessels

without the need for specialized handling equipment. To photograph marine snow one needs a waterproof camera and a light source that can be lowered into the sea. Most cameras have a narrow depth of field, but by only illuminating within this field one reduces the effect of out-of-focus suspended particles in both the foreground and background. Particles between the lens and the illuminated volume may cause some masking but only at suspended concentrations higher than we normally encounter (<20 g m^{-3}).

To obtain accurate size and shape information, our camera assembly uses three cameras for "redundant stereo." Two cameras with known geometry will accurately determine the three-dimensional coordinates of all points clearly photographed by both. Use of three cameras provides an internal estimate of precision. When photographing marine snow, the camera package should not disturb the ambient water. Our camera assembly has its light source directed downward with the cameras looking out horizontally into the illuminated area (Fig. 15.1). The assembly is lowered slowly enough so that the "bow" wave does not propagate into the illuminated volume. Photographs taken during recovery would be suspect.

The Atlantic Geoscience Centre (AGC) Floc Camera Assembly (FCA) consists of a frame supporting three cameras, a shrouded flash, and a computer for control and data logging. The assembly operates autonomously so it can be lowered on a simple rope or left on the seabed. A pinger (acoustic beacon) is also included for shipboard monitoring of the photo sequencing with depth. The camera can also be configured for settling velocity measurements by mounting it upside down on a large tripod and fitting it with a stilling tank (Fig. 15.2).

The camera

To keep costs to a minimum, and because of the desire for three cameras, we chose standard 35-mm SLR cameras that would fit into simple pressure cases: the Olympus OM-2 camera with 50-mm lens and motor winder. We also made provision for a "long back" film holder that carries 250-frame film, but have routinely used a "data-back" that limits us to thirty-six frames per drop. The pressure case (tested to

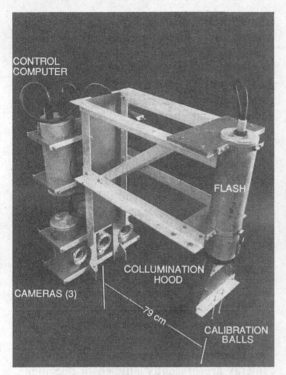

Figure 15.1. The AGC Floc Camera Assembly in profiling configuration (courtesy of BIO photography).

Figure 15.2. The AGC Floc Camera Assembly in a configuration for measuring the settling velocity of suspended particles (courtesy of BIO photography).

800-m depth) is made from aluminum pipe with a window flange welded in place (Fig. 15.3). The window is 12-mm-thick flat Plexiglas. The removable endcap has two O-ring seals and an electrical connector.

The Olympus lenses are designed for air, not water. Looking through a flat window and then through 70 cm of water might be expected to distort the field of view. Because the geometry of the three cameras and light plane are already complicated, we decided to calibrate out any distortion by photographing known objects. However, underwater calibrations using an open cube and graph paper did not show any obvious distortions. The lenses have a shallow depth of focus at $f8$, but all suspended particles that are illuminated appear in sharp focus.

The flash

The flash is a standard flash unit (Pentax AF240Z) in another aluminum pipe. The bottom end has a flat 12-mm Plexiglas port of 8-cm diameter. The Pentax flash has a linear flash

Figure 15.3. Photograph of one of the FCA cameras removed from the pressure case (courtesy of BIO photography).

tube mounted in a cylindrical parabolic reflector with a diffuser. A sliding fresnel lens is provided to focus the flash. We removed the diffuser and positioned the fresnel lens for focus at infinity (roughly). On the outside of the flash case is a collimation shroud made of polyvinyl chloride (PVC). The shroud is a tapered box with inside thickness 2 cm and 15 cm long. It flares from 8 cm at the flash port to 28 cm at its mouth. The azimuth of the flash tube is aligned with the shroud to make a volume of illumination that is a thin plain. No accurate measurements have been made of the illumination distribution, but flashing it at a wall over a metre away clearly illuminates a line about 3 cm wide. The Pentax AF240Z has a flash capacity of 17 J.

The cameras have focal plane shutters that, at 1/60 s, are only fully opened for a short time. A flash sync contact closure is provided to trigger an electronic flash (an almost instantaneous light source) at the exact instant the shutter is fully open. As the delay between tripping a camera electrically and this sync pulse is variable between cameras, we could not use one of the cameras to trigger the flash and hope that the other two would be fully open at that time. Instead, the three cameras are all set for 1/4-s exposure and are tripped simultaneously. There is about 100-ms variability between the response of the cameras, but they are all open after 200 ms when the computer triggers the flash. This method has proved reliable but makes ambient light more of a problem since the lens aperture is relatively large (*f*8), film speed is fast (ASA 400), and the shutter speed is slow (250 ms). Ambient light has not been a problem in midday operations below 30 m. Night operations are employed where information from the surface water is critical.

The computer

The three cameras and flash are controlled by a Tattletale IV computer made by Onset (North Falmouth, Massachusetts, USA). This is a small, low-power computer that is easy to program in BASIC. The computer keeps time, trips the cameras, triggers the flash, senses the output of a pressure gauge, and logs data for each frame and profile (cruise number, station number, location coordinates) in its memory. A PC is connected before the deployment to load the program and preset parameters such as depth or time intervals for photos and number of frames to expose. The PC is then disconnected and the Floc Camera Assembly operates autonomously so it can be lowered on a simple rope or even left on the seafloor. Upon recovery, the data are down-loaded from the computer memory, film is removed, the cameras reloaded, and battery strength checked.

The mounting frame

The three cameras, flash, and computer are mounted on a PVC and aluminum frame with the flash looking down and the cameras close together looking horizontally (Fig. 15.1). The distance from the camera lens to the illuminated plane is 79 cm. Initially the cameras were mounted to image the illuminated volume at angles of 60° to each other: Two were horizontal and one was mounted above, looking down. However, the photographs obtained from these wide angles were so different that the human eye could not decide which flocs on one photo corresponded to which on the other photos. The cameras are presently aligned in one plane, ~16.5 cm apart and angled toward a common point. Two 15-mm balls, mounted on the flash shroud, appear near the top of each frame. These serve as a geometric calibration for the system on all drops because the repeatability of camera placement is not perfect.

FCA profile configuration

Once loaded with film and programmed, the FCA is slowly lowered through the water column (~0.25 m/s). The descent of the FCA is traced on a 12-kHz-sounder graphic recorder at a 1-s sweep. Each time a photograph is taken, the pinger generates an additional four pings per second for 10 s, thus providing a surface indication of the photo sequence on the graphic recorder. The pressure transducer used is accurate to within ±1 m. The present flash unit needs 5 s between photographs to recharge. Our normal mode of operation is to take photo profiles based on a linear or exponential depth interval, followed by a series of photos near the seafloor triggered by a time interval.

Settling velocity configuration

The FCA settling velocity configuration has the camera frame turned upside down and mounted on a large aluminum tripod (Fig. 15.2); the illuminated volume is about 2 m above the seabed. The tripod is lowered to the seabed with a floating rope, cast off (with a buoy attached), and left for some period of time. For experiments in Bedford Basin (Nova Scotia), at a depth of 65 m and to avoid daytime ambient light, the computer was programmed to turn the system on at midnight and proceed to take a series of photos every 5 s for 3 min. Expected settling velocities were of the order of 1 mm/s (Syvitski et al., 1985), much smaller than horizontal currents near the seabed even in this sheltered basin. Therefore a large transparent acrylic box was positioned to surround the FCA-illuminated volume (but not the cameras). The aim was to eliminate horizontal currents but, after a few hours to equilibrate, allow flocs to enter the box and settle vertically through the field of view of the cameras. Particles would be photographed sequentially during their descent. Without the stilling tank, suspended particles could exit the illuminated slice of water column before the next picture was taken.

In our first test, where the top and bottom of the stilling tank remained opened, particles were seen to travel upward. This suggested that horizontal currents were causing eddies and turbulence within the box. In all subsequent tests, the bottom of the box has been blocked off and a 10-cm-high baffle having a 10-cm-square grid placed on the open top. Two spring-loaded flapper valves were mounted in the bottom to allow flooding and emptying during launch and recovery.

The data and data reduction

The photo images record a field view of 26 cm × 17 cm. The illuminated "plane" is approximately 25 mm thick. A variety of film has been tested, and recent experiments (and data provided in this report) have utilized 400 ASA colour print Kodacolor and 200 ASA colour slide Ektachrome film. The film is developed to produce uncut rolls of both negative and positive prints. The negatives from the floc camera film are analyzed on a Lietz TAS Plus automated image analysis system controlled by an LSI-11/2 microcomputer. A special floc camera program (FLOK) is used to analyze the negatives. Program FLOK scans the negatives by controlling the microscope stage recording the position coordinates (centroid), size (minimum and maximum diameters, particle area), shape, and orientation of each particle captured on film. FLOK is set up to recognize the sediment particles by edge detection (based on noting sudden changes in the grey-scale value between adjacent points).

The TAS system utilizes a programmable mechanical stage that moves new fields of view under the microscope and computer-linked television camera. Each 35 × 23 mm negative consists of 117 fields. The number of fields analyzed depends on the density of particles within the water column. Data generated for this report came from twenty to twenty-five digitized fields. For a twenty-field analysis, a 13.5 × 10.2 mm area of the slide, corresponding to 100 × 77 mm of the water column, is digitized. The thickness of the collimated flash is ~25 mm; thus the real volume of water digitized is 25 × 100 × 77 mm or ~0.2 l. The lower limit of detection of a particle suspended in the water, based on the film we presently use, is 2ϕ or 250 μm. Each negative takes about 30 min to analyze, recording information on hundreds or thousands of particles.

Data from the TAS FLOK program are eventually transferred to the BIO mainframe (Cyber 840) computer where the data are archived and analyzed by various programs including READY (Hackett et al., 1986), where the size frequency distribution of the suspended particles from each camera frame is ascertained. Where the FCA has been fixed rigidly within the water column, absolute settling speed may be determined from changes in the vertical coordinates of each particle between successive photographs. Figure 15.4 is a stereo pair from adjacent cameras. The 3-D image of the suspended particles may be seen by the reader using a standard stereoscopic viewer.

Comparison of FCA data to conventional analysis

Water was sampled synchronously with seven FCA profiles collected within the NW Gulf of St. Lawrence using 5-l Niskin samplers (see Syvitski [1988] for location and sampling details). The concentration of suspended partic-

Figure 15.4. Stereo photo pair from cameras 1 and 2 of particles suspended in the water at station 2 (48° 24.43′ N, 69° 09.61′ W) from the Gulf of St. Lawrence in Eastern Canada. The photo depth is 75 m below sea level and 175 m above the seafloor. The photos were taken on May 2, 1988. The balls seen in the FCA photograph have a real diameter of 15 mm.

ulate matter from the water samples was determined gravimetrically by filtration onto pre-weighed Nucleopore filters having a 0.45-μm nominal pore size. The number of marine snow particles (in no./l) as determined with the FCA/TAS method can then be compared to the more conventional method for determining suspended particulate matter concentration (in mg/l). We find, for the Gulf of St. Lawrence, that the number of particles of marine snow per volume of water increases with the suspended particulate matter concentration (Fig. 15.5A). The few exceptions include measurements within:

1. the surface layer of river plumes, where the suspended particulate matter concentration is high but the suspended particles have yet to reach the stage of large floccules and agglomerates; and

2. the turbulent layer near the seafloor, where suspended particles are finely divided from resuspension off the seafloor and/or from the breakup of the marine snow in response to seafloor turbulence (cf. Figs. 15.6 and 15.7).

Particles from the filtered water samples were also removed from the filter papers via initial soaking followed by low-intensity sonification in a bath. All samples were treated the same way. The primary particles composing the marine snow were dispersed in a sodium hexametaphosphate electrolyte solution. These disaggregated particles were then analyzed for their size frequency distribution using a TA II computerized Coulter Counter involving different aperture tubes so as to cover the size range of 0.63–100 μm. No correlation was found between the primary particles (i.e., the individual components that make up the marine snow) and the in situ size of the marine snow (Fig.

A

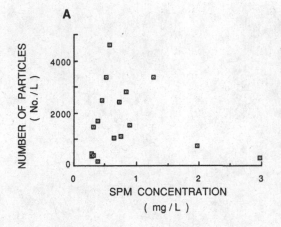

B

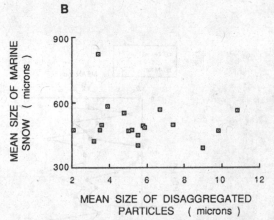

Figure 15.5. (A) Comparison of total particle concentration (in mg l⁻¹ or g m⁻³) in the water, collected by 5-l Niskin bottles, with the number of particles >250 μm (in no./l) as observed using the FCA/TAS method. (B) Comparison of mean grain size of the deflocculated suspended particles with the mean grain size of the marine snow observed using the FCA/TAS system. Water samples were collected at the same time, depth, and station as were the FCA data. All samples and data were collected on BIO cruise DA88-008 to the N.W. Gulf of St. Lawrence, Eastern Canada (cf. Syvitski, 1988).

15.5B). This is expected from theory, where floc size is dependent on a particle's residence time, biogeochemical interactions, and the distribution of turbulent shear within the water column (Syvitski, 1991).

Station FCA profile
One station in the St. Lawrence Estuary from a recent cruise (BIO DA88-008) demon-

strates the depth variations of the in situ properties of marine snow through a column of sea water (Figs. 15.6 and 15.7). Background information on the oceanography and sediment dynamics of this environment can be found in Syvitski et al. (1983). Figure 15.6 provides five representative FCA photos of the station profile. The surface layer (Fig. 15.7) is dominated by finer flocs that increase in size with depth, especially just under the halocline. Below 150 m the marine snow size is relatively constant until within the bottom mixed layer, where the particle size decreases and the particle number increases. Four layers containing high numbers of large particles appear equally spaced within the water column; these layers are considered to represent tidal influence on the sediment delivery from the St. Lawrence River to the Estuarine basin (Syvitski et al., in prep.; Fig. 15.7F).

Size frequency distributions
Analysis of the particles observed using the FCA/TAS system reveals that particles suspended in the marine environment can be very large (i.e., in the 0.5–2 mm size range). In the St. Lawrence Estuary, three types of size frequency distributions (SFD) of marine snow were recognized. A monotonically increasing SFD (see, e.g., Fig. 15.8B) typifies *surface layer* and suggests that much of the suspended particle population is of a size below the detection limit of the FCA/TAS. In other words, not all suspended particles are yet in the form of large agglomerates that normally need a longer residence time to form. A relatively flat SFD typifies samples from the *intermediate layer* (cf. Fig. 15.7) and suggests most of the suspended particulate matter load is in the form of marine snow but in a variety of sizes (see, e.g., Fig. 15.8A,B). Finally, a Gaussian-shaped SFD (see, e.g., Fig. 15.8B) typifies samples from the *bottom mixed layer* (cf. Fig. 15.7), whereby some of the marine snow particles are decreasing in their size due to the effect of turbulent shear within the layer, or from the addition of finer particles resuspended from the seafloor. The change in the shape of the SFD with water depth is obvious in the FCA curves; the Coulter Counter–produced curves of the disaggregated fraction do not show this trend (cf. Figs. 15.8B and 15.8C).

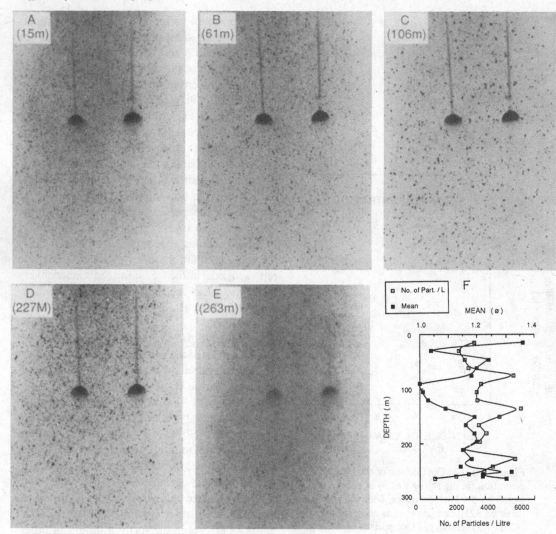

Figure 15.6. An example of an FCA profile with photographs and data from station 1 (BIO cruise DA88-008) situated at 48° 24.25′ N 69° 06.70′ W in the St. Lawrence Estuary, Eastern Canada. FCA photographs are from the middle camera taken at depths of: (A) 15 m, representing the surface layer; (B) 61 m, just below the surface layer; (C) 106 m, representing the intermediate cold water layer; (D) 227 m, representing the surface of the bottom mixed layer; and (E) 263 m, collected 2 m above the seafloor. (F) Vertical profile of the mean diameter (in phi units) and particle density (in no./l) of marine zone photographed. These data may be compared with salinity–temperature profiles and submersible observations shown on Figure 15.7.

Figure 15.9 provides a more systematic view of the change in floc size with depth at another location in the St. Lawrence Estuary,

through a series of SFD. The suspended particles change from a modal size of $1.75\,\phi$ (300 μm) at 16 m to $1.25\,\phi$ (420 μm) by 31 m. Between 226 m and the seafloor at 256 m at this location, the floc modal size became coarser once again ($0.75\,\phi = 600\ \mu$m).

Settling velocity data

Using the configuration of the FCA for collecting photographs of suspended particles settling within a stilling tank, absolute fall speeds of various particles were measured using sequential photos taken 10 s apart. The photos were collected in Bedford Basin, part of the metropolitan harbour of Halifax, Nova Scotia, Canada, at a depth of 65 m below the sea surface and 2 m above the seafloor. Figure 15.10

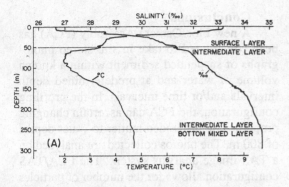

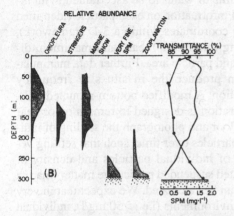

Figure 15.7. (A) Typical STD profile of the lower St. Lawrence Estuary with three water layers identified. (B) Summary of seston observations from the PISCES IV submersible (after Syvitski et al., 1983).

shows two photographs taken 20 s apart. On each of eight photos, the descent of eight particles in relation to the calibration balls was analyzed. Over 1.5–2 min of elapsed time, these particles were found to have settling rates between 0.6–1.3 mm/s. Suprisingly the largest particles did not always have the largest settling velocity (Table 15.1). The mean settling velocity of all particles analyzed was 1.05 mm/s or 91 m/day. Using more complex sampling schemes and sediment traps, Syvitski et al. (1985) found that, at similar depths in a British Columbian fjord, particles settled at 100 m/day. Their results were not based on individual particle settling rates, but rather reflected the mean settling rates for each size fraction. Table 15.1 also provides estimates on the in situ particle density of these large suspended particles. Particle density was very low, suggesting an organic nature and/

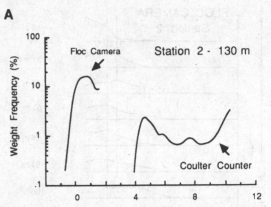

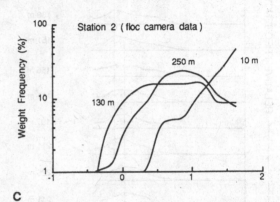

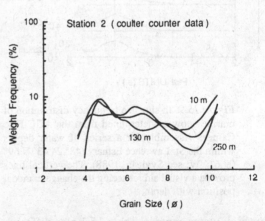

Figure 15.8. A comparison of Coulter Counter and FCA size frequency spectra from three depths representing the surface layer, the intermediate layer, and the bottom mixed layer of the St. Lawrence Estuary (cf. Figs. 15.6 and 15.7). Weight frequency data have been calculated from volume frequency data by holding particle density constant. (A) Spectra from the intermediate layer at Station 2, 130 m. (B) Variation of in situ size frequency distributions with depth. (C) Relatively invariant disaggregated or component size frequency distributions.

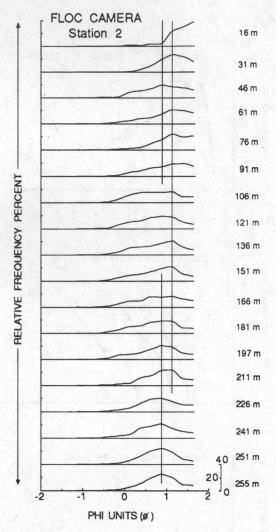

Figure 15.9. In situ size frequency distribution of marine snow as determined from the AGC Floc Camera Assembly for a series of water depths within the St. Lawrence Estuary (48° 24.43′ N 69° 09.61′ W; see Syvitski, 1988). The vertical lines provide a visual aid for seeing the change in modal position with depth.

or high water content. Table 15.1 also differentiates between individual spherical particles and those settling as strings of particles held together by some form of organic filament. Figure 15.11 provides a scanning electron view of these two distinct suspended morphologies. Again these sometimes very large strings (particle G on Fig. 15.10 is 7.6 mm long) do not necessarily settle faster than individual particles.

Conclusions

A new Floc Camera Assembly (FCA) has been constructed to take in situ stereo photographs of suspended sediment within a known volume of water and at predetermined depth intervals and/or time intervals. In the profiling configuration, the FCA can ascertain changing properties of suspended matter to water depths of 800 m. The photos collected are analyzed on a TAS image analysis system. The FCA/TAS configuration allows for the number of particles per volume of water to be ascertained, with individual information on each particle (identified through coordinates within a 3-D framework), including minimum, mean, and maximum diameter and particle area. Further data manipulation can produce the in situ size frequency distribution. A modified bottom-mounted FCA configuration is designed to remain on or near the seafloor and photograph the settling of individual particles over time, such that settling velocities of individual particles and density of flocculated suspended particulate matter (marine snow) can be calculated. We expect that in very turbid environments (i.e., >50 mg/l), individual particle identification will be difficult under the present FCA/TAS system and telephoto lenses should be used.

The FCA combined with the TAS provides a viable means to obtain quantitative data on the in situ properties of suspended particulate matter in most marine environments. Initial results have indicated little correlation between data obtained from conventional water sampling and those determined from our in situ measurements, suggesting that extreme caution must be exercised when interpreting particle dynamics from conventional methods. FCA profile data from the 270-m-deep St. Lawrence Estuary clearly show the regimes of floc production, equilibrium settling, and near-bottom disaggregation. Settling velocity experiments within the 65-m-deep Bedford Basin determined that particles in the 400–1,200-μm size range settled at speeds of 0.6–1.3 mm/s, but there was no simple correlation between floc size and settling velocity.

Future advances to our Floc Camera Assembly will include the simultaneous collection

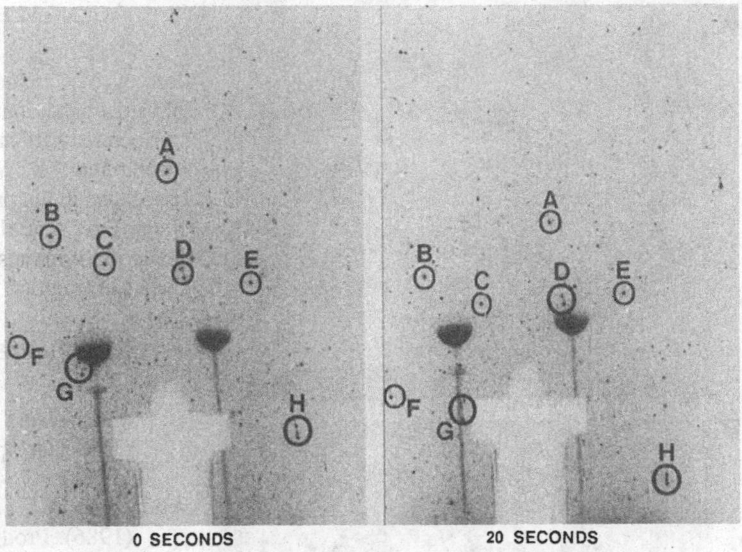

Figure 15.10. Sequential FCA photos taken 20 s apart in the bottom-mounted configuration. On each photo, eight of the same particles are identified, but at different positions with respect to the calibration balls.

Table 15.1. *Analysis of Figure 15.10 for particle dimensions, settling velocity, and density*

Particle ID[a] (Fig. 15.10)	Equiv. diam.[b] (μm)	Max. diam. (μm)	Stringer length (μm)	Settling velocity (m/day)	Density[c] (g/cm3)	ESSD[d] (μm)
A	720 ± 20	880 ± 20		119 ± 2	1.033 ± 2	55
B	740	880		97	1.031	49
C	1,240	1,490		102	1.027	52
D′1	1,010	1,100			1.027	
2	400	480				
			4,800 ± 20	76		38
E	990	1,100		54	1.026	28
F	700	740		113	1.033	54
G′1	1,060	1,230			1.028	
2	490	540				
3	400	480				
			7,580	108		53
H′1	1,050	1,220			1.028	
2	510	610				
3	490	610				
4	820	1,200				
			5,880	118		55

[a]A prime (′) indicates that the measured particle was in the form of a string of particles held together by an organic filament. The first particle measured is situated at the bottom of the string; the last particle measured is the last particle in the string. The maximum diameter is the length of the entire string. The calculated density is for the largest particle in the string, given the measured settling velocity of the entire string.
[b]Equivalent spherical diameter.
[c]Assumes a Stokes's settling velocity vs. particle density relationship, that is, $\rho_s = (18 \eta \omega / g d^2) + \rho_1$, where ρ_s is the in situ particle density, η is the molecular viscosity (=0.0188 g/cm s) at 0 °C and 30 ppt salinity, ω is the in situ particle settling velocity (calculated above, but in cgs units), g is the acceleration due to gravity (980 cm/s^2), and ρ_1 is the density of water (=1.0241 g/cm^3).
[d]Equivalent spherical sedimentation diameter of a grain of quartz (ρ_s = 2.65 g/cm^3) based on Stokes's Law.

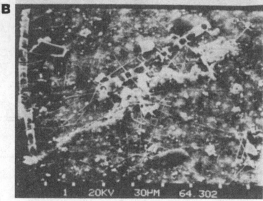

Figure 15.11. Scanning electron photomicrographs of suspended particles. (A) Organic coating surrounding marine snow is responsible for many of the large individual marine snow particles photographed by the FCA. (B) Stringer of suspended sediment with larger grouping of particles on the upper end. Scale in micrometres between white bars is given below each picture.

of attenuance data (at $\lambda = 660$ nm), so that information on the background "fines" can be ascertained. One of the cameras is being converted to handle a 200-mm lens to resolve smaller particles. We plan to attach a tilt sensor to the FCA so that we may assess any directionality of the sometimes aligned stringers. Also if the seafloor is not flat during the settling velocity configuration, the cameras and flash would not be aligned parallel to the pull of gravity, and particles would settle across rather than along the plane of the light.

ACKNOWLEDGMENTS

This project was supported by the Geological Survey of Canada under project 420-8141, "The physical behaviour of suspended particulate matter (suspended particulate matter) in natural aqueous environments." K. W. G. LeBlanc provided invaluable assistance in the preparation of this manuscript. We thank Drs. A. Bale and K. Kranck for their comments. This manuscript is Geological Survey of Canada Contribution no. 35789.

REFERENCES

Alldredge, A. L. (1977). House morphology and mechanisms of feeding in the Oikopleuridae (Tunicata, Appendicularia). *Journal of Zoology,* 181: 175–88.

Alldredge, A. L., Cole, J. J., & Caron, D. A. (1986). Production of heterotrophic bacteria inhabiting macroscopic organic aggregates (marine snow) from surface waters. *Limnology and Oceanography,* 31: 68–78.

Alldredge, A. L., & Youngbluth, M. J. (1985). The significance of macroscopic aggregates (marine snow) as sites for heterotrophic bacterial production in the mesopelagic zone of the subtropical Atlantic. *Deep-Sea Research,* 32: 1445–6.

Asper, V. L. (1987). Measuring the flux and sinking speed of marine snow aggregates. *Deep-Sea Research,* 34: 1–17.

Bale, A. J., & Morris, A. W. (1987). In situ measurement of particle size in estuarine waters. *Estuarine, Coastal and Shelf Science,* 24: 253–63.

Bale, A. J., Morris, A. W., & Howland, R. J. M. (1984). Size distribution of suspended material in the surface waters of an estuary as measured by laser Fraunhofer diffraction. In: *Transfer processes in cohesive sediment systems,* eds. W. R. Parker & D. J. J. Kinsman. New York: Plenum, pp. 75–85.

Carder, K. L., Steward, R. G., & Betzer, P. R. (1982). In situ holographic measurements of the sizes and settling rates of oceanic particulates. *Journal of Geophysical Research,* 87(C8): 5681–5.

Eisma, D. (1981). Suspended matter as a carrier for pollutants in estuaries and the sea. In: *Marine environmental pollution, vol. 2. Mining and dumping,* ed. R. A. Geyer. Amsterdam: Elsevier, pp. 281–95.

Eisma, D., Kalf, J., & Van der Gaast, S. J. (1978). Suspended matter in the Zaire estuary and the adjacent Atlantic Ocean. *Netherland Journal of Sea Research*, 12: 382–406.

Farrow, G. E., Syvitski, J. P. M., & Tunnicliffe, V. (1983). Suspended particulate loading on the macrobenthos in a highly turbid fjord; Knight Inlet, British Columbia. *Canadian Journal of Fisheries & Aquatic Sciences*, 40: 100–16.

Gardner, W. D. (1977). Incomplete extraction of rapidly settling particles from water samplers. *Limnology & Oceanography*, 22: 764–8.

Gibbs, R. J. (1981). Floc breakage by pumps. *Journal of Sedimentary Petrology*, 51: 670–2.

(1982). Floc breakage during HIAC light-blocking analysis. *Environmental Sciences and Technology*, 16: 298–9.

Gibbs, R. J., Konwar, L., & Terchunian, A. (1983). Size of flocs suspended in Delaware Bay. *Canadian Journal of Fisheries & Aquatic Sciences*, 40: 102–4.

Hackett, D. W., Syvitski, J. P. M., Prime, W., & Sherin, A. G. (1986). *Sediment size analysis system user guide*. Geological Survey of Canada Open File Report no. 1240, 25 pp.

Honjo, S., Doherty, K. W., Agrawal, Y. C., & Asper, V. L. (1984). Direct optical assessment of large amorphous aggregates (marine snow) in the deep ocean. *Deep-Sea Research*, 31: 67–76.

Johnson, B. D., & Wangersky, P. J. (1985). A recording backward scattering meter and camera system for examination of the distribution and morphology of macroaggregates. *Deep-Sea Research*, 32: 1143–50.

Kranck, K. (1984). The role of flocculation in the filtering of particulate matter in estuaries. In: *The estuary as a filter*, ed. V. S. Kennedy. New York: Academic Press, pp. 159–75.

Kranck, K., & Milligan, T. G. (1988). Macroflocs from diatoms: In situ photography of particles in Bedford Basin, Nova Scotia. *Marine Ecology – Progress Series*, 44: 183–9.

Lal, D. (1977). The oceanic microcosm of particles. *Science*, 198: 997–1009.

Lewis, A. G., & Syvitski, J. P. M. (1983). Interaction of plankton and suspended sediment in fjords. *Sedimentary Geology*, 36: 81–92.

McCave, I. N. (1975). Vertical flux of particles in the ocean. *Deep-Sea Research*, 22: 491–502.

Nishizawa, S., Fukuda, M., & Inoue, N. (1954). Photographic study of suspended water and plankton in the sea. *Hakkaido University Faculty of Fisheries Bulletin*, 5: 36–40.

Shanks, A. L., & Trent, J. D. (1980). Marine snow: Sinking rates and potential role in vertical flux. *Deep-Sea Research*, 27: 137–44.

Silver, M. W., & Alldredge, A. L. (1981). Bathypelagic marine snow: Deep-sea algal and detrital community. *Journal of Marine Research*, 39: 501–30.

Suziki, N., & Kato, K. (1953). Studies on suspended materials – marine snow in sea. Part I. Sources of marine snow. *Hakkaido University Faculty of Fisheries Bulletin*, 4: 132–5.

Syvitski, J. P. M. (1988). *DAWSON 88-008 technical cruise summary, May 01–May 17, 1988*. Geological Survey of Canada Open File Report no. 1920, 61 pp.

(1991). The changing microfabric of suspended particulate matter – the fluvial to marine transition: Flocculation, agglomeration and pelletization. In: *The microstructure of fine-grained sediments: From muds to shale*, eds. R. H. Bennett, W. R. Bryant, & M. H. Hulbert. *Frontiers in Sedimentary Geology*. New York: Springer–Verlag, pp. 131–7.

Syvitski, J. P. M., Asprey, K. W., Clattenburg, D. A., & Hodge, G. D. (1985). The prodelta environment of a fjord: Suspended particle dynamics. *Sedimentology*, 32: 83–107.

Syvitski, J. P. M., & Murray, J. W. (1981). Particle interaction in fjord suspended sediment. *Marine Geology*, 39: 215–42.

Syvitski, J. P. M., Silverberg, N., Ouellet, G., & Asprey, K. W. (1983). First observations of benthos and seston from a submersible in the Lower St. Lawrence Estuary. *Géographie Physique et Quaternaire*, 37: 227–40.

Syvitski, J. P. M., Smith, J. N., Calabrese, E. A., & Boudreau, B. P. (1988). Basin sedimentation and the growth of prograding deltas. *Journal of Geophysical Research*, 93: 6895–908.

IV Data interpretation and manipulation

Grain size analyzers provide the size frequency spectrum in terms of grain number, or the weight or volume percent within specified size intervals. If the sample is comprised of sedimentary particles, the distribution can range from well-sorted (i.e., having a narrow size range) to poorly sorted and polymodal. The distribution may be numerically represented, such as with statistical indices (Friedman, 1962; Folk, 1966; Doeglas, 1968; Davis, 1970; Buller & McManus, 1972; Roy & Biswas, 1975; McLaren, 1981), modal frequencies (Clark, 1976; Syvitski & Macdonald, 1982), or graphs (Passega, 1964; Pejrup, 1988), such as a plot of relative frequency and particle size (Burger, 1976). Size frequency distributions can be discussed on their own merits (Visher, 1969) or within larger matrices containing size information on other samples (Glaister & Nelson, 1974).

The four chapters that comprise Part IV deal with the science of grain size data interpretation and manipulation. Chapter 16 reviews how statistical parameters of particle size can describe a sedimentary environment. The author suggests that this information should come from the average of many samples of the parent population, rather than from a single sample (Hails & Hoyt, 1969). When plotted against each other, these new parameters are capable of separating the samples into the appropriate sedimentary environments. Such diagrams show aspects relating to the physics of sediment transport that may allow for a more refined paleogeographic interpretation.

There has been a long-standing debate on the form of the natural probability distribution that sedimentary samples reflect, in terms of their size spectra (cf. Middleton, 1962). Chapter 17 describes the hyperbolic distribution and shows how it can be applied to size spectral data. The distribution has properties that are quite appealing, as both the Gaussian (normal) and exponential distributions are represented as end members. The relationships between hydrodynamic parameters and the statistical measures that define the hyperbolic distribution are discussed briefly (cf. Chapter 19).

Although eigenvalues and eigenvectors are commonly used in the earth sciences, they are poorly understood, for they task the scientist to move from the comfortable four-dimensional world of space and time into the multidimensional world of multivariate statistics. Factor analysis is a part of that world. Chapter 18 looks at the utility of Q-mode factor analysis to help discern end-member size spectra from a large set of size spectral data. Simulated size frequency distributions are used to examine how this otherwise mysterious multidimensional method can sort out the similarities within such large data sets.

Finally, Chapter 19 uses both experimental and theoretical hydrodynamic considerations to interpret size spectral data. The chapter defines ways to compute suspension loads directly from a riverbed's grain size distribution, and to estimate the range in flow parameters for lognormally distributed suspended loads.

REFERENCES

Buller, A. T., & McManus, J. (1972). Simple metric sedimentary statistics used to recognize different environments. *Sedimentology*, 18: 1–100.

Burger, H. (1976). Log-normal interpolation in grain size analysis. *Sedimentology*, 23: 395–405.

Clark, M. W. (1976). Some methods for statistical analysis of multimodal distributions and their application to grain size data. *Journal of the International Association of Mathematical Geology*, 8: 267–82.

Davis, J. C. (1970). Information contained in sediment-size analysis. *Journal of the International Association of Mathematical Geology*, 2: 105–12.

Doeglas, D. J. (1968). Grain-size indices, classification and environment. *Sedimentology*, 10: 83–100.

Folk, R. L. (1966). A review of grain-size parameters. *Sedimentology*, 6: 73–93.

Friedman, G. M. (1962). On sorting, sorting coefficients, and the log-normality of the grain-size distribution of sandstones. *Journal of Geology*, 70: 737–53.

Glaister, R. P., & Nelson, H. W. (1974). Grain-size distributions, an aid in facies identification. *Bulletin of Canadian Petroleum Geology*, 22: 203–40.

Hails, J. R., & Hoyt, J. H. (1969). The significance and limitations of statistical parameters for distinguishing ancient and modern sedimentary environments of the lower Georgia Coastal Plain. *Journal of Sedimentary Petrology*, 39: 559–80.

McLaren, P. (1981). An interpretation of trends in grain size measures. *Journal of Sedimentary Petrology*, 51: 611–24.

Middleton, G. V. (1962). On sorting, sorting coefficients, and the log-normality of the grain-size distribution of sandstones: A discussion. *Journal of Geology*, 70: 737–53.

Passega, R. (1964). Grain size representation by CM patterns as a geological tool. *Journal of Sedimentary Petrology*, 34: 830–47.

Pejrup, M. (1988). The triangular diagram used for classification of estuarine sediments: A new approach. In: *Tide-influenced sedimentary environments and facies*, eds. P. L. de Boer et al. Dordrecht: D. Reidel, pp. 289–300.

Roy, R. N., & Biswas, A. B. (1975). Use of grain-size parameters for identification of depositional processes and environments of sediments. *Indian Journal of Earth Science*, 2: 154–62.

Syvitski, J. P. M., & MacDonald, R. D. (1982). Sediment character and provenance in a complex fjord; Howe Sound, British Columbia. *Canadian Journal of Earth Science*, 19: 1025–44.

Visher, G. S. (1969). Grain size distributions and depositional processes. *Journal of Sedimentary Petrology*, 39: 1074–106.

16 Suite statistics: The hydrodynamic evolution of the sediment pool

WILLIAM F. TANNER

Introduction

For a century or so the purpose of making grain size measurements was to determine the diameter of a representative particle. This is useful when one is studying reduction in grain size along a river (e.g. Sternberg, 1875). But it is a simplistic approach, and one is entitled to ask: Is the mean diameter the only information that we wish to get? Or does the simplicity of this first step make us think that we have now described the sand pool?

When we measure grain size, what do we really want to know? This does not refer to whether we measure the long axis or the short axis of a nonspherical particle, or whether we approximate the diameter by measuring a surrogate (such as fall velocity). Rather, we ask this question in order to get a glimpse of how far research has come in understanding transport agencies or conditions of deposition, and of the degree to which we might reasonably expect to improve our methods of environmental discrimination.

Can a set of parameters that describe a size spectrum for one sample permit us to compare this sample with some other, perhaps from a different transport or depositional environment? A positive response implies that a single sample may be adequate to describe the parent sand body.

Do we want to know about variability within the sand body, thereby requiring a suite of samples? This might suggest that there is only one kind of variability (hence a single parameter will do), or it might mean that we will have to explore many kinds of variability.

Do we wish to see if there is more than one sand pool contributing to the sediment body? "Sand pool" needs an improved definition, but presently includes:

(a) material presently being added (but not necessarily deposited) at the site;

(b) material located "upstream" of the site and in transport toward it; and

(c) material already present.

This concept is more easily applied to river sands than to a wide, shallow, near-shore depositional area.

Do we wish to identify any characteristics of the transporting agency? Moving from "simple diameter of one sample" to "transport agency" is a big step, and it cannot be done by using simple theory. If we are going to undertake to answer questions such as these, we must make determinations beyond simply specifying a "representative diameter."

A good deal of work has been undertaken on trying to answer such questions, much of it at Florida State University. Many environments have so far been sampled, with thousands of samples. The basic environmental identities have been as follows: dune, mature beach, river, tidal flat, settling (or closed basin: lake, lagoon, estuary, interior seaway, etc.), offshore wave, and glaciofluvial. Not all have been treated equally; for instance, glaciofluvial materials and offshore wave sediments are poorly represented. Both modern and ancient environments have been studied and sampled in the field.

Most of the terms used here are familiar ones, such as mean and standard deviation (but as calculated numerically with the method of moments); two, however, require a note of caution (Hoel, 1954; Blatt, Middleton & Murray 1980). "Skewness" and "kurtosis" refer to some form of the third- and fourth-moment measures, respectively. The actual words have been taken, in part of the literature, to identify specific geometric features of a plot of the distribution; this is *not* the intent here. For present purposes, skewness and kurtosis are convenient labels for certain moment − *not* graphic − parameters. They do not require, but may correlate with, a specific curve shape. An additional point should be made regarding skewness. This term is used to mean (in general) an asymmetry of the distribution. Positive and negative skewness cannot be defined from a priori considerations. But Folk (1974, p. 52) and Blatt et al. (1980, p. 46) used "positive skewness" to identify a sample with many classes in the fine tail, and this usage has been followed here. Phi measure has been used throughout, unless indicated otherwise.

Control factors
Air versus water
The mass density of a single grain in air is quite different from that in water. The practical effect is that realistic local variations of air velocity, by a stated amount, are more important in moving a specified grain (such as quartz sand) than are the same variations in water velocity. A unidirectional flow of water typically transports a greater range of grain sizes (provided that they are available) than does a unidirectional flow of air. The standard deviation of the grain size tends to be numerically larger in unidirectional water transport than in unidirectional air movement. One cannot take a simple statement of the standard deviation of a single sample and translate that number to "transport agency" because some of the other influences mentioned here complicate the system too much. On the other hand, with suitable treatment, the standard deviation can be the basis for a powerful tool for discriminating among certain environments. One should note that in a few instances (e.g., on a beach), back-and-forth motions over a long period result in much better sorting than can be obtained in ordinary water or air currents.

New supply from upstream; through-flow; trapping; winnowing
A river carrying a large sand and/or coarse silt load may have an abundant new supply so that winnowing cannot be very effective. On the other hand, a more-or-less isolated sand shoal perched on a wide shallow shelf may have no new supply of sand, so winnowing can be very important.

Through-flow has to do with the ability of a transport system to carry certain sizes more or less continuously (time sense), such as the wash load concept dealing with fine silt sizes in an energetic stream: They are "washed" for long distances rather than resting on the bottom for lengthy time intervals on their way along the channel.

Trapping clearly can take place in ponds, lakes, and lagoons, but likewise is important in any kind of dead-end setting: a category that can be identified as the *closed basin* (perhaps even an interior or geosynclinal seaway). Settling may be the main hydrodynamic process. A simple example is the site where a high-energy, fine-sand-laden river flows into a coastal zone having very low wave-energy density. A large supply gives roughly the same results as a closed basin: probable dominance of the finer sizes, up to the limits of availability. Winnowing, on the other hand, tends to reduce the proportion of fines, as is well known.

Back-and-forth shuffling (BAFS) and net unidirectional sediment transport (NUST)
The basic concepts of back-and-forth shuffling (BAFS) and net unidirectional sediment transport (NUST) lead to an unexpected observation confirmed in both field and laboratory. Transport in an ordinary river is clearly characterized by NUST, and one can observe the result: Sand introduced upriver is carried downstream. Under a wave field, however, the primary effect is BAFS. There is commonly an asymmetry in transport effectiveness of water motion near the bottom, so the BAFS phenomenon leads to movement of grains of the same mineral (two different sizes) in opposite directions (see May, 1973). On a low-to-moderate-energy beach without a new supply, BAFS may be so efficient in and near the swash zone that the standard deviation of the grain size may be reduced to (or close to) what may be the minimum value for quartz: ~0.26. This is even lower than is found in most dune environments, where the mass density difference between air and sand (rather than BAFS) produces very good sorting.

Multistory (multitier/multilevel) turbulence structure
A single grain of quartz sand, settling in water, generates an eddy system with dimensions controlled largely by the size of the sand grain (Tanner, 1983). Eddies of a different scale may be formed adjacent to bedforms such as ripple marks and/or giant ripple marks. Sandbars and sandbanks are typically responsible for a third scale, and river bends create turbulence

at a fourth level. It is not known how many stories may exist, but the number could reach six or more. In certain environments, the structure generally has only one or two levels. The turbulence structure is much more complicated (of higher numerical order) in a large, energetic river than on a low-energy beach. The standard deviation of the grain size distribution tends to be numerically smaller as the number of stories decreases (within the limits of availability). A high-order system (many stories) typically produces greater mixing of sizes at any one locality (if such sizes are available).

Grain–grain interactions

Because the eddy system generated by a single grain may occupy a volume orders of magnitude larger than the grain itself, interactions between grains may be frequent and important (Tanner, 1983). These include trapping, tailgating, and ejection, and involve the motion of one grain fairly close to another, perhaps of dissimilar size; the two may then be deposited together. The result is a significant numerical increase in the standard deviation compared with what would have been obtained had there been no such interactions.

Transport directionality

Unlike BAFS (under one wave train), this has to do with differences arising from varying wave approach angles and the mixing of two (or more) sand pools that may result. Such mixing tends to increase the standard deviation numerically, and generally modifies other parameters as well.

Bivariate plots

Sedimentologists have long tried to make environmental sense out of bivariate plots of parameters that describe a sample size spectrum. The investigator selects two parameters (such as mean size and standard deviation), and plots them against each other. There are many examples in the literature (see, e.g., Friedman & Sanders [1978, pp. 78–81], where skewness is plotted against sorting).

The results have been less than enchanting, for several reasons (Socci & Tanner, 1980):

1. *sampling technique* (taking too large a sample; cutting across many adjacent laminae);

2. *lab procedure* (using whole- or half-phi screens instead of quarter-phi sieves; making more than one split, thus escalating the splitting error; using too short a sieving time; putting too large or too small a sample on the coarsest screen; and so on);

3. *graphic or other low-precision parameters* (e.g., the sorting coefficient instead of the standard deviation);

4. *overlimited options* (e.g., beach versus river, as if dunes and other settings can be ignored).

In Friedman & Sanders (1978, p. 78), beach samples are compared with river samples by plotting "simple skewness measure" against "sorting measure" (their terms for graphic measures). The dividing line that was drawn between the two number fields is far from straight, does not emphasize anything like a natural division, and for some twenty-seven samples gives the wrong answer. Use of moment measures (p. 81) produces an improvement (perhaps nineteen misidentifications), but the dividing line is still complex, many samples of both types cluster on or near it, and the number field for beaches is centered within that for rivers. One understands why many workers have given up on the procedure. However, the basis for this kind of diagram is hydrodynamic (BAFS for beaches vs. NUST for rivers); it lacks only a small modification to become a valuable tool.

An obvious way to improve the graph is to define sample suites, by field or subsurface study, and plot only suite-mean values for each parameter. This greatly reduces the scatter, minimizes (perhaps eliminates) overlap, and pinpoints the center of gravity of the measurements for any one suite. Aberrant or anomalous values, still available to the investigator, no longer appear as clutter.

Ideally a suite contains perhaps fifty or more samples, but this is commonly impractical. Experience has shown that, for sand and coarse silt sizes, twenty samples generally make a stable suite, and in many instances twelve to fifteen may be enough; this can be verified statistically.

Advantages of the suite statistics approach include the following:

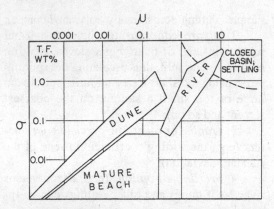

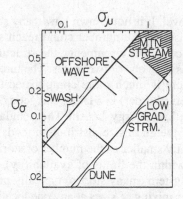

Figure 16.1. The tail-of-fines diagram. The means and standard deviations of the weight percents on the 4 ϕ screen and finer are shown here. Four fairly distinct number fields appear, as labeled above, with relatively little overlap. Many suites plot neatly in a single field. In certain other cases the apparent ambiguity may be useful; for example, a point at a mean of 0.01 and a standard deviation of 0.017 might indicate either dune or mature beach, and not formed in a closed basin. This diagram commonly gives a "river" position when in fact the river was the "last-previous" agency, but not the final one.

1. The number field* is smaller.
2. Anomalous points (although still present) do not clutter the plot.
3. Distinctions between transport agencies are more obvious.
4. Transition suites or mixtures may be easy to identify.
5. Three or more sedimentary environments may be represented conveniently on one graph.

Doeglas (1946) observed that the tails of the distribution provide much of the important information that is available to us; so we are less likely to be able to make fine distinctions by using the mean (surrogate for the first moment) than by using higher moments (such as the third and fourth). Therefore, we can construct one or more bivariate plots of tail data, using weight percents in the fine tail (material on the 4 ϕ screen and finer). For a suite of samples, one can plot the mean weight percent of the fine tail (4 ϕ and finer) against the standard deviation of the same fine-tail data (Fig. 16.1).

Editor's note: This is a numerical matrix of distribution parameters.

Figure 16.2. The variability diagram, showing the suite standard deviation of the sample means and of the sample standard deviations. Except for the extremes, the plotted position indicates two possible agencies (such as swash or dune). The decision between these two can be made, in most instances, by consulting other plots (such as Fig. 16.1). This diagram considers specifically the variability, within the suite, from one sample to the others.

This diagram has four reasonably distinct number fields: mature beach, dune, river, and closed basin (settling). The tail-of-fines plot works well because it depends on hydrodynamic factors: The suite mean tends to separate "large new supply" (river or closed basin) from winnowing (beach and dune), and the suite standard deviation tends to separate BAFS (mature beach and near-shore) and "large mass–density difference" (dune) from settling and poor winnowing. Furthermore, mature dunes (large mass–density difference) are generally readily distinguished from mature beaches (BAFS) because the number of transport events per year on the beach may be 10^5 or 10^6 larger than in a dune field.

A plot (not shown here) of standard deviation (S.D.) versus kurtosis K is commonly useful because a river, dune, or beach suite may appear within a small distinctive area, whereas tidal flat or other settling suites may plot in a long narrow band showing a very closely controlled relationship between the two parameters (form: $\ln K = a * \exp(-b * \text{S.D.})$, where a and b are numerical values to be determined for any given suite of samples; R^2 about 0.92–0.99). The standard deviation need not be small for this relationship to hold. On one well-studied tidal flat, settling effects produced a straight line of data points (Tanner & Demirpolat, 1988).

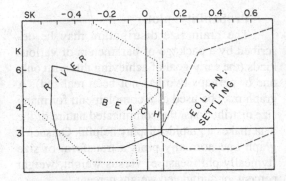

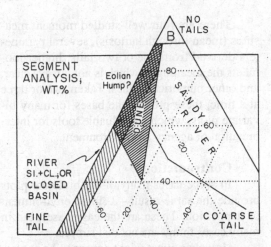

Figure 16.3. Skewness vs. kurtosis. The suite means of these two parameters are used. Positive skewness, as used here, identifies a geometrically distinctive fine tail; if there is also a distinctive coarse tail, it is the smaller (weight percent) of the two. The closed basin (settling) environment typically produces an obvious fine tail, much more so than beach or river sands. Eolian sands commonly have, instead of a well-developed fine tail, a feature called the *eolian hump* (cf. Fig. 16.5), which the skewness indicates in the same way as it does a distinctive fine tail. Therefore the two tend to plot together. Negative skewness identifies a distinctive coarse tail, either fluvial coarse tail (large *K*) or surf "break" (=kink in the probability plot; *K* in the range of 3–5 or so). Many river and beach suites appear in the same part of the diagram but are ordinarily easy to identify by using this figure first and then the tail-of-fines diagram (Fig. 16.1).

The variability diagram (S.D. of means vs. S.D. of standard deviations) typically places any given suite in one or perhaps two categories (e.g., dune or beach). Figure 16.2 identifies variability from sample to sample in a band ranging from "very small variations" (dune) to "very large variations" (high-energy stream), perhaps because it identifies the multistory turbulence level.

Skewness indicates the balance between the two tails. If the coarse tail is well developed and the fine tail small or nonexistent, a negative skewness (as defined here) results. If, on the other hand, the fine tail is dominant, skewness is positive. Experience to date is that beach sands (surf break on the probability plot) and river sands (coarse fluvial tail) tend to show skewness $Sk < 0.1$. Settling or closed basin deposits, with a well-developed fine tail but little or no coarse tail, typically show $Sk > 0.1$. Mature dune sands commonly have the *eolian hump*

Figure 16.4. The segment analysis triangle. The procedure for picking segments and obtaining the necessary numbers is outlined in the text. The apex is characterized by very small or negligible distinctive tails ("no tails"), and the base (not shown) connects distinctive coarse tail (to the right) with distinctive fine tail (to the left). Four different environments are distinguished reasonably clearly, except for one area of overlap; in this area, one examines the probability plots for the eolian hump in order to see which of the two is indicated.

(a convex-up inflection, coarser than 50%, on the probability plot). Because the skewness parameter may see the main part of the sample as a large fine tail, it generally exceeds 0.1 for these sands. Therefore skewness can be combined to good advantage with some other parameter – perhaps kurtosis, as shown here, to make a useful diagram (Fig. 16.3).

The probability curve (see next section) can be divided into three convenient parts: A central segment (straight line) is identified first, and whatever is left over is then assigned to either the coarse tail or the fine tail. The percentages of these three parts can be averaged for the suite, and the resulting point can be plotted on a three-dimensional (triangular) "segment analysis" diagram, which separates mature sandy beaches (minimal tails) from rivers and dunes. Sand-and-gravel rivers are likewise separated from dunes and silt-and-clay rivers because the former tend to have larger coarse tails, and the latter, larger fine tails (like settling). Mature dunes can be identified provided the eolian hump is present on some of the probability plots (Fig. 16.4).

There are four well-studied moment measures (mean through kurtosis), several parameters derived from one or two tails, higher moments the meaning of which is not always clear, and other numerical values. Taken two or three at a time, these provide the bases for many bivariate plots, which are valuable tools for interpretation of agency and environment.

Contradictions?

In some cases many of the bivariate plots provide the same result – a single environment of deposition. These are the easy cases; but in many cases things are not that simple.

However, what might appear to be contradictory is commonly correct throughout, and helpful as well, because not all sand masses were moved by only one transport agency. Samples from the inner continental shelf (see Arthur et al., 1986) may give a river or fluvial identification, despite the marine location, because a sand mass delivered by one agency to another may not lose the granulometric fingerprint of the first until some considerable time has passed. Dune sand may be blown into a river or creek, and be reworked by running water. Suite analysis of the creek sands might then give a primary dune indication, yet also show that there have been complications. The analyst should not insist on selecting a single agency, but instead be aware that there may be two or more. Analysis of beach ridge sands commonly indicates both beach and dune origins, since sand beach ridges are typically built by both agencies (swash, wind). A mix of dune and beach indices might be formed in several ways, but unless there is other evidence for dune buildup, one should pick the beach as the preferred site.

Wind deposits filtered down from above, without any dune migration, have been identified correctly (see Tanner & Demirpolat, 1988) by suite methods as eolian plus settling, although beach and river components were suggested (also correctly) by the analysis.

There are many kinds of transitions and mixing, and the different suite parameters provide different kinds of clues to this – the hydrodynamic basis for any one parameter is not necessarily the same as for the others. Rarely, an environmental decision cannot be made.

Probability plot

The grain size distribution may be described by a package of parameters of various kinds (the early goal of achieving this with only one or perhaps two has not been realized). A graph may be used as a complement; for many size distributions, the complicated nature of the data makes a graphic display helpful. On such a graph one normally plots a transform of size (typically phi measure) versus weight, weight percent, or cumulated weight percent.

Histograms and cumulated S-curves fall into this category but suffer from a defect: They either display a slope reversal (which is confusing in highly detailed work) or exhibit marked curvature. In well-sorted sediments having few data points, these curves can be drawn in various ways without violating the points, and thus may be works of art rather than tools for study.

The purpose of a plot of a single sample is to provide information or an impression not readily gotten from the various numerical parameters. One therefore seeks a procedure that shows neither modal peaks nor artistic curve-fitting, and several are available. Well-known examples are the probability plot (Otto, 1938; Pettijohn, 1975), the Rosin-law plot ("crushed particle" distribution applicable to milled or ground-up material [Rosin & Rammler, 1934; Irani & Callis, 1963; Pettijohn, 1975]), and a method based on the log-exponential concept (Bagnold, 1941; Barndorff-Nielsen, 1977). Any additional variety can be created simply by noting that one must pick one or two transforms for the data (e.g., the probability plot uses a log transform for size [phi] and the Gaussian transform for cumulated weight percent). The Rosin-law plot has not been exploited much: Crushed materials follow it, but transported sediments do not. The Bagnold procedure has not been popular, perhaps because the computations required are time consuming, even on a computer.

There are certain advantages to using the probability plot:
Several varieties of suitable paper are already available commercially.
It is easy to make.
Modes and saddles are relatively easy to spot.
Certain parameters unavailable via ordinary statistical techniques can be read directly.

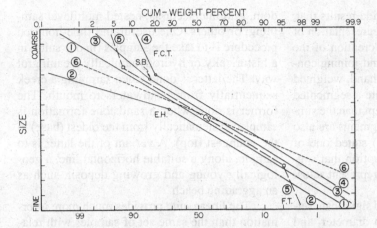

Figure 16.5. Diagrammatic probability plots. (1) The Gaussian, rare among sands. (2) The distinctive eolian hump (E.H.) is common, but not universal, in dune sands, and so far has not been observed in other sands that did not have any previous eolian history. (3) The surf break (S.B.) has been demonstrated to form in the surf zone, as the sorting improves. (4) The fluvial coarse tail is geometrically distinctive, but cannot be distinguished in every case from the surf break. (5) This curve has both a fluvial coarse tail and a fluvial fine tail; the central segment (C.S.) is the line between the two small squares. However, it is not the modal swarm (see text). (6) The modal swarm (a grain size concept, not a graphic one) obtained by subtraction from the original distribution; it shows the actual size distribution of the central segment (graphic device) of line 5. Lines of these kinds help one visualize the effects summarized in the bivariate plots.

Interpretation is fairly easy once the operator is used to its distinctive properties.

The main disadvantage (shared with some other methods) is that the learning period may be long.

The rationale for the probability plot may have been the idea that many (if not most) sands and coarse silts should be essentially Gaussian; therefore, one would quickly learn to distinguish "standard" sediments (straight line on probability paper) from anomalous ones. In fact, very few samples are perfectly Gaussian, but segments of the grain size distribution curve indeed are. Therefore, a better approach might be to study the probability plot in order to identify these particular segments, with an eye to providing an improved interpretation (as long as one remembers that they are segments of the plot, not components of the population) (Fig. 16.5).

Moss (1962–3) identified three common segments on such a plot and designated them by capital letters (A, B, C). Two of these are tails, and one is the central part of the curve. Rather than letters, the present writer prefers descriptive terms: *coarse tail* (segment), *central segment*, and *fine tail* (segment). The central segment commonly (but not in every case) crosses

the 50% line, and generally shows numerically smaller (i.e., better) standard deviation (sorting) than the other two. Distinctive kinks, or inflections, mark their mutual boundaries. (The fine tail [segment] does not have the same definition or use as the tail of fines discussed earlier; here, it is set off by a kink in the curve, rather than by a stated size.)

The central segment must be distinguished from the *modal swarm*, which is a sample component (not segment) having a purely Gaussian form on probability paper. This component can be separated, in most instances, by simple subtraction. When combined with the pertinent tail components, the modal swarm yields the complete curve, which now shows segments (Tanner, 1983). Components, made of actual grains, are generally not visible on the plot. Segments are visible on the plot, but represent the effects of combining two or more components in one sample, and do not show quantities of various sizes actually present in any one component.

There is a widely used procedure based on the assumption that each straight-line segment on probability paper is also a component that has combined with its neighbors via butt-end joining (e.g., Visher, 1969). Because grain–

grain interactions in water provide results that, in turn, include significant misrepresentation of true grain diameters as part of creation of the perceived size spread, the butt-end-joining concept is erroneous. On the other hand, weighted addition of components, to create a segmented curve, is a better analytical concept than the simplistic and mistaken idea that segments are also the components. LeRoy (1981) stated one of several objections to the assumption that segments (rather than components) represent actual sand grain clusters in transport.

Bergmann (1982) examined significant settling effects on perceived grain diameter, and Tanner (1983) studied the hydrodynamic processes (capture, tailgating, ejection, and others). Their results show that butt-end joining is not possible, but that addition of overlapping components is not only possible but common. The resulting curve is, indeed, made up of two or more segments, but these segments are not the same as the actual components.

A few grain size curves are almost Gaussian on probability paper – mostly mature beach or dune sands, where long-term winnowing has been effective without the addition of new supplies. Other curves show the adjustment from one Gaussian form to another (Stapor & Tanner, 1975). Most curves have at least two segments and many have three, four, or more. (The number of segments does not specify the number of components; combining two well-sorted sands having considerable disparity in mean size may well produce a three-segment curve.)

The probability plot can be used:

to identify distinctive inflections (e.g., eolian hump, surf break),

to separate components where mixing is suspected,

to permit quick estimation of internal sorting,

to allow easy reading of certain useful parameters, and

to permit (where a linear suite is taken along the travel path) direct analysis, on the plot, of important hydrodynamic changes.

The linear suite

A suite of samples may be taken in any pattern preferred by the investigator: areally random, grid intersections, nested multilevel sampling, or others. One useful and time-honored procedure is to take the samples (in one suite) in a historically or hydrodynamically meaningful way. The latter is done when sampling a creek sequentially from headwaters to mouth. The former is done when a sandstone formation is sampled systematically from the oldest (base) to the youngest (top). A variant of the latter is to sample, along a suitable horizontal line, a geologically young and growing deposit, such as an aggrading beach.

The linear suite provides much more information than the same set of samples with relative positions in time or space unknown. A long stream without tributaries should yield a linear suite in which various size parameters (e.g., the mean) change in a predictable way from one end to the other. If tributaries are sampled, sediment mixing processes can be studied to good advantage in such a suite. If it is not known to be linear, this kind of work cannot be done.

Linear suites, unless the line is too short, commonly show changes. Down the river profile, tributaries may make important contributions. The changes in key parameters should correlate with the fact that not even the river profile itself is smoothly curving from one end to the other (Tanner, 1974). In time-dependent linear suites, historical events may be evident in the data.

The presence of changes or contributions, as stated in the previous paragraphs, may well preclude the use of simple regression models. One does not get high assurance by trying to force a nonlinear curve with several peaks into a simple linear model. If the nature of the change can be identified or assumed in some acceptable way, then detailed study can be undertaken on the separate segments of the line. Again, a good example is a linear suite along a stream profile: Peaks (or troughs) on the curve may correlate closely with tributaries. If so, the analyst should recognize this fact and adjust the treatment accordingly.

Demirpolat, Tanner, & Clark (1986) studied a line of samples across a beach ridge plain that has had a history of adding one new sand ridge every 20–25 years (roughly 180 ridges). In addition to areal sampling, they focused on a single transverse line, along which they sampled each available ridge (on that line). Because the ridges were clustered in visibly and measurably distinct sets, each containing some eight to twenty ridges, these workers were able to treat their samples in separate suites. One also can look at the whole sequence as a single linear suite and examine quasi-cyclical changes with time (oldest to youngest). Changes along the sample line are clear when plotted (sample by sample) in terms of standard deviation and/or kurtosis. Each of these two parameters showed striking changes having a quasi-regular periodicity of some centuries (two or more), and each change is associated with a change in ridge-set altitude. Because each set was deposited over an interval of centuries, these changes cannot reflect storms (or even stormy seasons); therefore these two parameters appear to identify small changes of sea level (of a meter or two, from plane-table profiling).

This work also produced other interesting results. On the tail-of-fines diagram (Fig. 16.1), successive sets of ridges plot as follows:

D,   river (perhaps closed basin);
E1,   mature river sediment;
E2,   more mature river;
F,   even more mature river;
G,   beyond the river number field, to the edge of the "mature beach" number field.

Because this plot typically provides information about the "next to last" agency, the conclusion is drawn that set D reflects early reworking of river (probably deltaic) sands, and that by the time set G was being deposited, waves had come pretty close to producing mature beach or nearshore sand. This, of course, is correct, but could not have been deduced from the samples had they not been handled in linear fashion.

Storm data

Rizk (1985) studied beach sands along a large spit in the Florida Panhandle and determined grain size parameters along five transects taken at right angles to the beach, for both the high- and low-energy seasons of the year. His work provided a baseline data bank representing almost ten years since the area had last been struck by a hurricane (in 1975, by Eloise).

Shortly after he completed his analysis,

three hurricanes (Elena, Juan, and Kate) affected the coast at intervals of only weeks (all in 1985). He therefore resampled his earlier transects at more or less regular time intervals: once shortly after Elena, once shortly after Juan, and twice within a month after Kate (Rizk & Demirpolat, 1986). Juan was the mildest of the three and did not produce any significant changes. This left six sample dates: two following a hurricane-free period of nearly a decade, two between Elena and Kate, and two after Kate.

The data for each traverse and date were examined in terms of the range for each moment measure; for example, for Traverse A, immediately after Kate, the range was calculated for the size mean, for the size standard deviation, and so on. The smaller the range, the less heterogeneity in that traverse on that date. The range of the mean size (in phi units) on Traverse A dropped from 1.0 prior to Elena to slightly less than 0.3 immediately after Kate, then climbed back to about 0.4 a month later. The standard deviation behaved in the same fashion (smaller values immediately after the storm). Changes were minor or not clear in the other parameters.

From their data the conclusion should be drawn that high storm energy resulted in mixing (homogenization), in contrast to the well-defined areal banding that existed after nearly a decade of relative calm and that was being reestablished a month or so after the third hurricane. For linear suites, this suggests that storm effects on this kind of coast are minimized in a matter of weeks to months, and therefore should not be expected to show up clearly in samples taken at much greater intervals than this. If major storms were common in the sample area (e.g., weekly intervals), then one would not expect the kind of recovery seen by Rizk & Demirpolat, and there would be no evidence for occasional violent storm activity.

Rizk & Demirpolat (1986) noted that whatever is represented (locally) by fair weather provides important long-term influences on the grain size distribution. The before- and afterstorm data were also studied in terms of other suite parameters. The effect was greater suite uniformity after each storm. This is the same observation made from comparing traverses, but the presentation is different and the conclusions appear to be more general when seen in suite form.

Evolution of the sand pool

A suite may be taken in areal, linear, or some other fashion. An areal suite should approximate a fairly narrow time slice, spread over an area. If the area is large, local suites should be taken at various places (perhaps only a few meters to some hundreds of meters across). The suite analysis should provide roughly the same kind of information produced by isoplething each parameter, such as mean size, but with less variability and more different kinds of information. An *isopleth map* of mean size (or of coarsest median diameter, or of some similar measure) commonly pinpoints the part of the study area closest to the source. If this is all one wants to know, and if the simple isopleth map is clear, then suite methods are not indicated. On the other hand, suite procedures for a full house of parameters provide a great deal more information.

The linear suite may be geographical (e.g., along a river), as stated earlier, or it may have a time dimension. If the sample line goes, say, from oldest to youngest, then one has the opportunity of studying the evolution of the sand pool. One can plot various parameters – for individual samples or for small suites – along the time line, and observe the changes. If small suites (e.g., eight to twenty closely spaced samples each) are used, suite methods provide that there is very little noise, and one worries about neither the validity of statistical "lumping" techniques nor the exponential proliferation of numbers of samples in any useful nested-sampling design.

Beach ridge sets on a central traverse on St. Vincent Island have suite means and standard deviations of mean sizes (phi units) as follows:

Set				
D	E1	E2	F	G
Mean 2.2867	2.3812	2.4750	2.3045	2.1750
S.D. 0.0925	0.1014	0.0682	0.1324	0.0787

This produces a relatively smooth history of fining, followed by coarsening, from oldest (D) to youngest (G). The changes are real, as can be

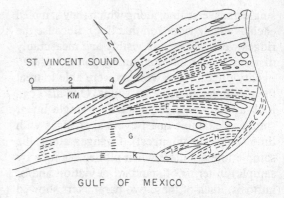

Figure 16.6. Sketch map of St. Vincent Island, Florida, showing twelve beach ridge sets easy to recognize on air photos or by plane tabling. Some of these sets, each consisting of ten to fifteen ridges and representing a few centuries, stand higher than others. Grain size parameters identify correctly those sets that stand high (or low).

seen by inspecting the standard deviations: The mean for E2 is four standard deviations larger than the mean for G. The history of the offshore sand pool, for these ridges, includes an early fining sequence (the model of May [1973]), followed by coarser sediments after much of the fine-sand population had been depleted (not included, but implied, in the model). The time is ~2,000 years (Figs. 16.1, 16.6).

Sets D–G are almost parallel with each other in map view (although set boundaries are distinct); however, they do not cover the full history of the island. Younger than G are sets H–L. Sets H, I, and J are located at the eastern end of the island, have distinctive map patterns including very short ridges, and do not parallel anything older; therefore, they do not belong in the same history as D–G. In fact, the suite-mean mean size for H–J is 2.3, an apparent fining after the previous coarsening. The offshore sand pool has been isolated for longer than the history of St. Vincent Island; hence it is unlikely that this is merely a "new wave" of fine materials, but from the same source as older ridges.

Alternatively we can see if sets H–J represent introduction of finer material from the east (St. George Island). The suite mean of H–J is clearly within the range of D–G, so it is not finer than the offshore sand pool may have been at an earlier date. It is finer than set G.

Other parameters may be helpful. The standard deviation of H–J is numerically larger (mean value 0.414, with a suite standard deviation of only 0.025) than any set in D–G (maximum, 0.39). The kurtosis of H–J is larger (3.936) than anything in D–F, and equally large with G. For low-to-moderate wave energy, these are distinctive results.

The map patterns show that these short, almost isolated ridges were supplied from the east; but the granulometry alone, with linear methods, indicates that a new source of sand became important in the study area between G and H. The history of this one island is really not the topic here. The key item is the chain of trends shown by sequential suites (or samples) in a linear system. Such a study yields more information than can be obtained by sampling in a nonlinear fashion.

Conclusions

The grain size distribution of a sand contains much information about transport, deposition, or both. Analysis of a single sample of that sand may not provide very much of the information since variability from one sample to another may be an important facet of the nature of the sand body. The joint study of ten or more samples from the same sand is much more useful.

Several effects or controls operate so that the suite of samples shows distinctive characteristics. Among these effects are the following:

1. the mass–density contrast between mineral in air and mineral in water;
2. supply rates, through-flow, trapping, and winnowing;
3. back-and-forth shuffling (BAFS) and net unidirectional sediment transport (NUST);
4. multistory turbulence structure;
5. grain–grain interactions; and
6. transport directionality.

New procedures have been developed based on the concept of the *sample suite:* a set of closely related samples taken from a single transport and/or depositional system. Suite data include the means and standard deviations of the usual statistical parameters, plus additional measures such as the mean and standard deviation of the weight percent in the tails-of-fines (4ϕ and finer). These can be plotted to good advantage on bivariate diagrams, where interpre-

tation is reasonably clear. These diagrams are particularly useful because they show certain aspects of hydrodynamics, sediment supply and resupply, trapping (settling; closed basin), and similar items. The result is a markedly improved analysis.

Some of the advantages over traditional plots, which show all of the data points, are as follows:

1. The number of plotted points is now much smaller and easier to visualize.
2. Dubious or anomalous points, due perhaps to sampling problems, do not clutter the diagram.
3. The effects of various transport agencies are easier to differentiate.
4. Transitions (in a historical sense) from one agency to another are easier to identify.
5. Four or more environments are conveniently placed on any one plot, without confusion.

The use of five or six properly selected bivariate plots commonly provides additional information, such as "last previous agency," and may also give strong clues as to the reliability of the analysis. The use of linear suites, where appropriate, can pay off in terms of important geographic or historical information: for example, the maturing of a sediment pool, the change from one agency to another, changes in energy level or in sea level, or the advent of short-term contributions from outside the system.

REFERENCES

Arthur, J., Applegate, J., Melkote, S., & Scott, T. (1986). Heavy mineral reconnaissance off the coast of the Apalachicola River delta, Northwest Florida. *Florida Geological Survey, Report of Investigation no. 95,* Tallahassee, 61 pp.

Bagnold, R. A. (1941). *The physics of blown sand and desert dunes.* London: Methuen and Co., 265 pp.

Barndorff-Nielson, O. (1977). Exponentially decreasing distributions for the logarithm of particle size. *Proceedings Royal Society of London, Ser. A,* 353: 401–19.

Bergmann, P. C. (1982). Comparison of sieving, settling and microscope determination of sand grain size. Unpublished M.S. thesis, Florida State Univ., Tallahassee, 178 pp.

Blatt, H., Middleton, G., & Murray, R. (1980). *Origin of sedimentary rocks*. Englewood Cliffs, N.J. : Prentice–Hall, 782 pp.

Demirpolat, S., Tanner, W. F., & Clark, D. (1986). Subtle mean sea level changes and sand grain size data. In: *Proceedings 7th Symposium on Coastal Sedimentology*, ed. W. F. Tanner. Geology Department, Florida State Univ., Tallahassee, 113–28.

Doeglas, D. J. (1946). Interpretation of the results of mechanical analyses. *Journal of Sedimentary Petrology*, 16: 19–40.

Folk, R. L. (1974). *Petrology of sedimentary rocks*. Austin, Texas: Hemphill, 182 pp.

Friedman, G., & Sanders, J. (1978). *Principles of Sedimentology*, New York John Wiley, 792 pp.

Hoel, P. G. (1954). *Introduction to mathematical statistics*. New York: John Wiley, 318 pp.

Irani, R. R., & Callis, C. F. (1963). *Particle size: Measurement, interpretation and application*. New York: John Wiley, 165 pp.

LeRoy, S. D. (1981). Grain-size and moment measures: A new look at Karl Pearson's ideas on distributions. *Journal of Sedimentary Petrology*, 51: 625–30.

May, James P. (1973). Selective transport of heavy minerals by shoaling waves. *Sedimentology*, 20: 203–12.

Moss, A. J. (1962–3). The physical nature of common sandy and pebbly deposits. *American Journal of Science*, 260: 337–73; 261: 297–343.

Otto, G. H. (1938). The sedimentation unit and its use in field sampling. *Journal of Geology*, 46: 569–82.

Pettijohn, F. (1975). *Sedimentary rocks* (3rd ed.). New York: Harper and Row, 628 pp.

Rizk, F. (1985). Sedimentological studies at Alligator Spit, Franklin County, Florida. Unpublished M.S. thesis, Florida State Univ., Tallahassee, 171 pp.

Rizk, F., & Demirpolat, S. (1986). Pre-hurricane vs. post-hurricane beach sand. In: *Proceedings 7th Symposium on Coastal Sedimentology*, ed. W. F. Tanner. Geology Department, Florida State Univ., Tallahassee, 129–42.

Rosin, P. O., & Rammler, E. (1934). Die Kornzusammensetzung des Mahlgutes im Lichte der Wahrscheinlichkeitslehre. *Kolloid Zeitschrift*, 67: 16–26.

Socci, A., & Tanner, W. F. (1980). Little-known but important papers on grain-size analysis. *Sedimentology*, 27: 231–2.

Stapor, F. W., & Tanner, W. F. (1975). Hydrodynamic implications of beach, beach ridge and dune grain size studies. *Journal of Sedimentary Petrology*, 45: 926–31.

Sternberg, H. (1875). Untersuchungen über Längen- und Quer-profil geschiebeführende Flüsse. *Zeitschrift Bauwesen*, 25: 483–506.

Tanner, W. F. (1974). Bed-load transportation in a chain of river segments. *Shale Shaker, Oklahoma City, Okla., USA*, 14: 128–34.

(1983). Hydrodynamic origin of the Gaussian size distribution. In: *Nearshore sedimentology: Proceedings 6th Symposium on Coastal Sedimentology*, ed. W. F. Tanner. Geology Department, Florida State Univ., Tallahassee, pp. 12–34.

Tanner, W. F., & Demirpolat, S. (1988). New beach ridge type: Severely limited fetch, very shallow water. *Transactions, Gulf Coast Association of Geological Societies*, 38: 367–73.

Visher, G. S. (1969). Grain size distributions and depositional process. *Journal of Sedimentary Petrology*, 39: 1074–106.

17 The hyperbolic distribution

CHRISTIAN CHRISTIANSEN AND
DANIEL HARTMANN

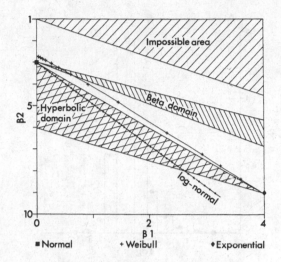

Figure 17.1. Locations in (β_1, β_2) plane for various distributions.

Introduction

The question of the relations between grain size parameters and the dynamical and geological factors that govern the populations sampled has been a long-standing and often controversial issue in the sedimentological literature. In spite of that a few conclusions of broad applicability have emerged (Viard & Breyer, 1979; Sedimentation Seminar, 1981; Ehrlich, 1983). A number of authors (esp. Bagnold, 1979) have recognized that this line of study may give significant results if close attention is paid to the detailed features of the size frequency distributions. The normal (Gaussian) distribution, commonly used in sedimentology (Leroy, 1981), is only one of a number of styles of frequency distributions. Therefore, we suggest the more encompassing hyperbolic distribution for descriptive needs of sedimentology.

The log-hyperbolic distribution was introduced by Barndorff-Nielsen (1977) to describe the mass–size distributions of eolian sand (see also Bagnold, 1941). The distribution has wide application, not only to other kinds of sediments (Bagnold & Barndorff-Nielsen, 1980; Christiansen et al., 1984; Barndorff-Nielsen & Christiansen, 1988; Christiansen & Kristensen, 1988), but also to other types of frequency data: size distributions of droplets and aerosols (Durst & Macagno, 1986), turbulence (Barndorff-Nielsen, 1979), wind shear (Barndorff-Nielsen et al., 1988), and in astronomy, biology, and economics (Barndorff-Nielsen & Blæsild, 1981, 1983; Barndorff-Nielsen et al., 1985). The hyperbolic distribution is also related to the principle of self-similarity in fractal geometry (Mandelbrot, 1977).

The hyperbolic distribution is defined by its log probability function being a hyperbola, just as that of the normal distribution is a parabola.

Size distributions

For many studies it is sufficient to represent grain size data by a histogram or a frequency distribution. There are, however, situations in which it is desirable to represent data by a theoretical distribution. Reasons include (1) objectivity, (2) sophistication and flexibility, and (3) automated data analysis.

A method for evaluating which distribution could give the most useful information about a data set is the use of moments to make a bivariate plot of β_1 and β_2, where β_1 is standardized skewness squared and β_2 is standardized kurtosis. Figure 17.1 shows the possible regions for normal, β (uniform = special case), γ (exponential = special case), lognormal, Weibull, and hyperbolic distributions.

For normal distributions $\beta_1 = 0$ and $\beta_2 = 3$. Therefore these distributions are represented in Figure 17.1 by a *single point,* as are the exponential and the uniform distributions. These distributions involve only a single shape and have no shape parameters. They therefore do not provide the degree of generality that is frequently desirable. The γ and lognormal distributions are *lines* in Figure 17.1 and provide for some more generality. The β and hyperbolic distributions have two shape parameters and occupy *regions* in Figure 17.1, and thus provide greater generality than any of the other distributions.

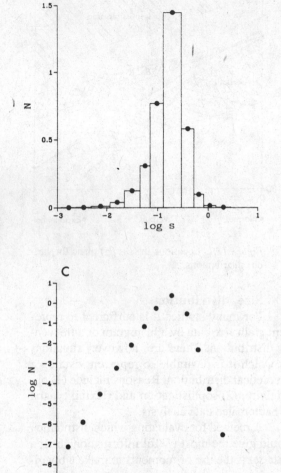

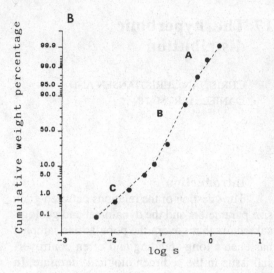

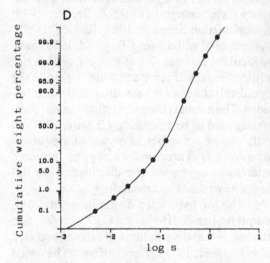

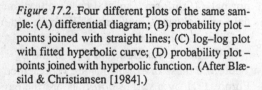

Figure 17.2. Four different plots of the same sample: (A) differential diagram; (B) probability plot – points joined with straight lines; (C) log–log plot with fitted hyperbolic curve; (D) probability plot – points joined with hyperbolic function. (After Blæsild & Christiansen [1984].)

Historical background

The pioneering work on the hyperbolic distribution was carried out by Bagnold (1937, 1941) in a search for the optimal way to represent grain size data. First he showed that plotting size data on a differential diagram (Fig. 17.2A, where N represents percentage weight divided by sieve interval) allowed the predominant diameter to be seen at a glance. However, the differential diagram suffers from the same

disadvantage as a traditional block diagram: The proportions of extremely fine and extremely big grains in the samples were too small to be plotted at all. In order to give all values equal prominence, Bagnold (op. cit.) plotted the data in a log diagram where the ordinate was $\log N$ (Fig. 17.2C). Plotting on log diagrams led Bagnold to conclude:

It is now for the first time apparent that outside a definite central zone the grades to right and left of the peak fall off each at its own constant rate, this means that we are again confronted with the same logarithmic law of distribution which runs through the whole subject. (Bagnold, 1941, p. 114)

Bagnold's approach did not receive much attention in the geological community for a long time. Until the 1970s only a few authors used the log diagram (Simonett, 1960; McKinney & Friedman, 1970). The reason probably was that following Krumbein's (1936) introduction of moment measures and lognormal transformations in sedimentology, a large number of authors assumed that their data could be described by the lognormal distribution. Also, plotting of weight percent cumulative curves on Gaussian probability paper became and is still popular among sedimentologists. Normal distributions plot as straight lines on this type of paper, but the main observation is that most samples plot as line segments or broad arcs (Grace et al., 1978) – a common explanation being that these segments result from lognormal subpopulations that are either truncated or overlapping (Visher, 1969; Middleton, 1976). Graphical measures of mean, sorting, skewness, and kurtosis became increasingly popular, especially those of Folk & Ward (1957).

What was really employed was not the lognormal distribution itself, but *deviations* from it, either in the form of the segmented shape on Gaussian probability paper (Visher, 1969; Glaister & Nelson, 1974) or in the use of skewness (a normal distribution has skewness 0) as an important grain size parameter (Sly et al., 1983). As a logical consequence, Schleyer (1987) even suggested a measure of deviation from lognormality as a new grain size parameter. Hartmann (1988b) strongly argued against this latter approach.

In the 1970s a number of authors (Reed et al., 1975; Jackson, 1977; Viard & Breyer, 1979; Sedimentation Seminar, 1981) questioned the above approach. At the same time, doubts on the traditional interpretation of the segmented shape on Gaussian probability paper started cooperation between geologists and mathematicians at the University of Aarhus (Barndorff-Nielsen, 1986).

The first paper from the group was the introduction of the mathematical background for the hyperbolic distribution (Barndorff-Nielsen, 1977). Christiansen et al. (1984) showed that when a hyperbolic distribution is plotted on Gaussian paper (as in Fig. 17.2D) its shape can be misinterpreted as consisting of line segments (Fig. 17.2B).

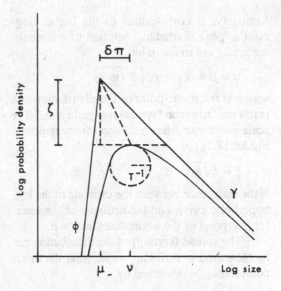

Figure 17.3. The geometrical interpretation of the parameters of the hyperbolic distribution.

The distribution and its parameters

The hyperbolic distribution needs four parameters for its specification, two of which define the position and scale of the hyperbola and two which define the "shape" of the hyperbola. We give here one of the parametrizations of the model function:

$$p(x; \mu, \delta, \varphi, \gamma) = a(\delta, \varphi, \gamma) \exp(-1/2(\varphi h_+ + \varphi h_-)) \quad (17.1)$$

where x indicates the observed variable, $\mu, \delta, \varphi,$ and γ are parameters,

$$h_\pm = \sqrt{\{\delta^2 + (x - \mu)^2\}} \pm (x - \mu)$$

and

$$a(\delta, \varphi, \gamma) = \sqrt{\varphi \gamma} / [\delta(\varphi + \gamma) K_1(\delta \sqrt{\varphi \gamma})]$$

K_1 being a Bessel function. For fixed values of $\mu, \delta, \varphi,$ and γ, equation (17.1) determines a probability (density) function. A distribution model for phi size can be found in McArthur (1987).

Figure 17.3 shows the geometrical interpretations of the parameters and some of their useful combinations. The parameters φ and γ are simply the slopes of the two linear asymptotes of the hyperbolic log probability function. They therefore correspond to Bagnold's (1941) "small grade" and "coarse grade," respectively.

Similarly, μ corresponds to the log of Bagnold's "peak diameter." Applied to sediments, we prefer not to use μ but

$$\upsilon = \mu + \delta(\varphi - \gamma)/\left\{2\sqrt{\varphi\,\gamma}\right\}$$

which is the mode point of the distribution and is referred to as the "typical log-grain size." The scale parameter δ has no direct interpretation in Figure 17.1; but

$$\zeta = \delta\sqrt{\varphi\,\gamma}$$

is the difference between the ordinate of the log-hyperbolic curve and the ordinate of the intersection point of the asymptotes, at $x = \mu$.

The spread (sorting) of the distribution can be measured in different ways. Near the mode point it may be described by

$$\tau^2 = \zeta\delta^{-2}(1 + \pi^2)^{-1}$$

where

$$\pi = 1/2(\varphi - \gamma)/\sqrt{\varphi + \gamma}$$

(Note that this is not the π of classical mathematics.)

The parameter τ^2 represents the curvature of the hyperbola at that point. Parameters δ, ζ, and $\kappa = (\varphi\gamma)^{1/2}$ are also measures of spread (see Fig. 17.3).

The skewness and the kurtosis of the log-hyperbolic distribution, as traditionally defined in statistics, are very complicated functions of φ, γ, and δ (Barndorff-Nielsen & Blæsild, 1981), but in most cases ($\zeta > 1$ and $|\pi| < 0.5$).

We may approximate the kurtosis by

$$\xi = \left(1 + \delta\sqrt{\varphi\,\gamma}\right)^{-1/2} \tag{17.2}$$

and the skewness by

$$\chi = ((\varphi - \gamma)/(\varphi + \gamma))\xi \tag{17.3}$$

The domain of variation of χ and ξ is a triangle referred to as the *log-hyperbolic shape triangle* (Fig. 17.4). Values of χ and ξ falling outside the triangle respresent other distributions.

From the probability density function of the hyperbolic distribution (17.1) it can, for example, be seen that for $\xi = 0$ and $\chi = 0$ (see also (17.2) and (17.3)) we obtain the normal distribution, and that for $\xi = 1$ and $-1 < \chi < 1$ we obtain the symmetrical and skew log-Laplace distributions as two limiting distributions of the

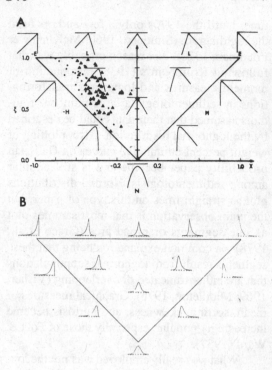

Figure 17.4. The hyperbolic shape triangle. The letters at the boundaries indicate limit distributions: N, normal; L, Laplace; E, exponential; H$^+$ and H$^-$, positive and negative hyperbolic (for details, see text). (A) Log probability functions corresponding to selected (χ, ξ) values. (B) Representative probability functions corresponding to (A).

log-hyperbolic distribution. Negatively skewed distributions (= positively skewed on the phi scale) plot to the left of the central line and positively skewed distributions to the right; more peaked distributions plot higher in the triangle. Figure 17.4 thus shows that the samples plotted in the triangle are far from being normally distributed. The samples plotted in Figure 17.4 are littoral samples from the Mediterranean coast of Israel. A few of the samples have a Laplace distribution. Fieller et al. (1984), with considerable success, used this latter distribution for environmental discrimination.

For any observed mass–size distribution, determined by sieving or settling behaviour, there is the problem that the sample size in the form of the number of single particles is unknown. An idea of the precision of the estimates is therefore not readily available. Even if the actual number of particles were known, this would give an exaggerated impression of the precision because there are added uncertainties stemming from the sampling and laboratory processes and possibly also from other sources. It is, however, possible to define a "quasi–sample size" N based on the assumption that the data follow the log-hyperbolic model (Barndorff-Nielsen et al., 1985). This enables one to describe the relative uncertainties of the estimates of the hyperbolic distribution by maximizing the function

$$\sum r_i \ln p_i \ (\mu, \delta, \varphi, \gamma)$$

or, equivalently, by minimizing the function

$$I = I(\hat{\mu}, \hat{\delta}, \hat{\varphi}, \hat{\gamma}) = \sum r_i \ln \left(r_i / p_i(\mu, \delta, \varphi, \gamma) \right)$$

where r_i is the relative weight of the grains in the ith sieving interval, p_i is the corresponding probability mass of the hyperbolic distribution in the same interval, and

$$\hat{\mu} = \text{estimate of } \mu$$

The function I can be considered as being proportional to a log-likelihood function l, where the constant of proportionality is minus the reciprocal of the quasi–sample size (Barndorff-Nielsen et al., 1985). If N and l were real and N large, then

$$2N\hat{l} = 2NI(\hat{\mu}, \hat{\delta}, \hat{\varphi}, \hat{\gamma})$$

would follow, approximately, a χ^2 distribution on $k-1-4$ degrees of freedom, where k is the number of sieving intervals. This suggests the quasi–sample size may be defined as

$$N = (k-3)(2I)^{-1}$$

where $(k-3)^{-1}$ is the mode point of the distribution of the reciprocal of a χ^2 variate on $k-5$ degrees of freedom (see Barndorff-Nielsen et al. [1985] for further details). Using the modified profile likelihood (Barndorff-Nielsen & Blæsild, 1983) it is then possible to plot contours of (approximate) likelihood or confidence regions in the hyperbolic shape triangle (Fig. 17.5). The quasi–sample size for the 35-g backshore sample shown in Figure 17.5 is 10,715 grains. This method for estimating N perhaps yields a better approximation than those suggested by Meland et al. (1964) or Jones (1969). The latter obtained $N = 500$ for a 100-g sample.

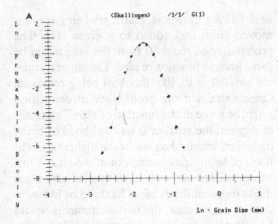

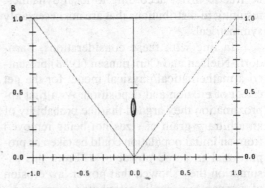

Figure 17.5. (A) Log–log plot of backshore sample with fitted hyperbolic curve. Quasi-$N = 10,715$. (B) Hyperbolic shape triangle plot of the same sample as in (A). Contours are 3 and 5 log-likelihood, respectively, 95% and 99% confidence.

Estimation of the hyperbolic parameters is rather complicated. Christiansen & Hartmann (1988a; see also Jensen, 1988) documented a package of PC programs for estimating the hyperbolic parameters as well as the standard moments. The package includes plotting routines of a log–log plot with histogram data and fitted hyperbolic function, as well as a plot of the hyperbolic shape triangle with confidence contours. McArthur (1986) presented mathematical calculations and FORTRAN subroutines for analysis of the hyperbolic function.

Generation of hyperbolic shape

Bagnold & Barndorff-Nielsen (1980) briefly presented a dynamical model for the generation of hyperbolic curves. Their explanation was further discussed by McArthur (1987). The explanations were based on the following assump-

tion. During transport, grains are being both removed from and added to a given area. The
probability of removal from the area might be
some inverse function of size: The larger a grain,
the smaller is its likelihood of being removed.
Once entrained the probability of deposition
might be some direct function of size: The smaller a grain, the smaller is its likelihood of being
deposited within the area, or the higher its likelihood of being transported out of the area. Thus,
some median size would dominate, and since
the two probabilities seem likely to be independent of each other, the two decrements would
be free to differ according to local dynamics,
resulting in distributions that are not necessarily
symmetrical.

In line with these considerations, Barndorff-Nielsen and Christiansen (1988) presented a mathematical/physical model for the net
effect of erosion and deposition. As a first approximation they argued that the probability of
an arbitrary grain of sizes not being removed
from an initial population could be taken as proportional to some power of s. From this assumption they showed that power law erosion
of a given hyperbolic distributed source sediment would move its (χ, ξ) position toward the
right-hand part of the shape triangle. Similarly,
power law deposition onto a hyperbolic distributed sediment causes a shift toward the left-hand
part of the triangle. Both types of shift take place
along specific curves that satisfy

$$\chi^2 = \xi^2 - \left(\delta \frac{\varphi + \gamma}{2}\right)^{-2} (\xi - \chi^{-1})^2$$

Instances of such hummocky curves, with equi-
distant

$$\left(\delta \frac{\varphi + \gamma}{2}\right)$$

are shown in Figure 17.6A. Deigaard & Fredsøe (1978) showed that their transport model
with logarithmic normal distributed source material resulted in a logarithmic hyperbolic distributed sediment.

In the above reasoning, no account is tak-
en of the influence of the difference in size between the grains in the sediment. This difference causes deviations from the pure power law
(=e in their notation), erosion/deposition, for in-

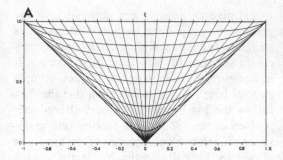

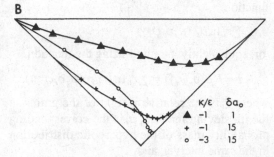

Figure 17.6. (A) Hyperbolic shape triangle with
erosion/deposition curves. The "hammock" curves
represent ε erosion/deposition and the straight lines
κ erosion/deposition. The curves are drawn equidis-
tant from each other. (B). Erosion/deposition curves
showing instances of combined ε and κ erosion/
deposition.

stance, so that both grains that are small (sheltered) and grains that are large, compared with
the typical size, have a relatively higher probability of not being removed from the sand
bed. Barndorff-Nielsen & Christiansen (1988)
showed that, if the influence of difference in
size depended on the relative occurrence of fine
grains compared to coarse grains, and if $\varepsilon = 0$,
then this effect can be modelled as corresponding to one of the shifts along the straight lines
$\sigma = \chi/\xi$ in Figure 17.6A. Barndorff-Nielsen &
Christiansen (1988) referred to this case as
"pure" κ erosion/deposition.

The combined net effect of ε and κ ero-
sion/deposition corresponds to the curves in
Figure 17.6B satisfying $\chi = \sigma \xi$ and

$$\xi = \left(1 + \delta \alpha [\exp - (\kappa/\varepsilon) \sigma] \sqrt{1 - \sigma^2}\right)^{-1/2}$$

For a given population on the bed, deposition
will lead toward the left and upward part of the
shape triangle, whereas erosion will lead toward
the right-hand part of the triangle. An attractive
element in the model is that erosion and deposi-

tion act over a continuum of skewness (χ) values, both positive and negative. This is in accordance with McArthur (1987) but contradicts other models (Duane, 1964) where a change in skewness sign implies a change in the process producing asymmetry. Barndorff-Nielsen & Christiansen (1988) found good agreement between the model and the observations of erosion and deposition on a microtidal flat in Christiansen & Kristensen (1986, 1988).

Basically, the above model describes the changes with time of the size distribution at a single location. Barndorff-Nielsen & Sørensen (1989) showed how the model may be related to differences between size distributions at different locations, but at a fixed time point. Applying this enlarged model, Barndorff-Nielsen & Sørensen (1989) showed that, due to converging streamlines on the upwind part of a simple dune form, grain size became coarser and sorting improved toward the dune crest. Such results are in accordance with observations from simple linear dunes in nature (Folk, 1971; Lancaster, 1986). Furthermore, the enlarged model corroborated results on sediment sorting in alluvial rivers modelled and observed by Rana et al. (1973) and Deigaard & Fredsøe (1978).

Applications in sedimentology

Dalsgaard et al. (see Chapter 5) have devised a method of calibrating sieves by means of empirical grain size distributions of hyperbolic shape. Nielsen (1985) used log-hyperbolic distributed sand samples to estimate a mean grain shape factor for each sieve fraction. A number of recent papers have shown that hyperbolic parameters provide for better environmental information than the traditional moments (Vincent, 1986; McArthur, 1987; Hartmannn & Christiansen, 1988a) or their Folk & Ward (1957) equivalents (Christiansen, 1984; Christiansen & Kristensen, 1988). Wyrwoll & Smyth (1985) presented a case showing that there was no apparent gain in using log-hyperbolic parameters instead of lognormal parameters. Their findings have been discussed by Christiansen & Hartmann (1988b).

Which of the hyperbolic parameters is most relevant in a given study depends on the circumstances and the level of resolution. To some extent it is also an empirical question. This is particularly so at the present stage of development, using hyperbolic parameters for either environmental discrimination or to infer dynamics from grain size data. Christiansen (1984) and McArthur (1987) could use almost all the hyperbolic parameters in distinguishing between dune subenvironments. From their studies it appears that the sorting parameter δ has a high discriminative power when the size range is small. For a bigger size range Christiansen & Kristensen (1988) had success with the sorting parameter κ. Vincent (1986) showed that π was very efficient in distinguishing between beach and dune sands. From a dynamical point of view there can be little doubt that, in addition to the typical grain size υ (the invariant parameter), χ and ξ, and their ratio σ, provide the most useful information (Barndorff-Nielsen & Christiansen, 1988; Hartmann, 1988a; Hartmann & Christiansen, 1988a).

A common procedure for discriminating between sedimentary environments is to plot grain size parameters in bivariate plots and look for differences between the environments (Friedman, 1967). Samples from closely connected (sub)environments – for example, beach subenvironments – often show a considerable variation and overlap. This has caused some authors (Krauss & Nakashima, 1986; Pranzini, 1983) to describe the beach as chaotic or even anarchistic. The use of multivariate tests of differences – as, for example, Mahalanobi's D^2 or Hotelling's T^2 – might show that two groups of samples that do not differ significantly on the basis of any single parameter do differ when all parameters are considered simultaneously (Greenwood, 1969; McArthur, 1987). However, most such tests demand that the parameters used in the procedure must be normally distributed and show homoscedasticity (equal variance–covariance matrices). A glance at the many bivariate plots of parameters in the literature suggests that this condition is very seldom fulfilled. Very little has been published on the validity of an assumption of normally distributed parameters. Hartmann (1988a) showed that distributions of moments and hyperbolic parameters from eight littoral subenvironments could be nicely fitted with β distributions. In this sec-

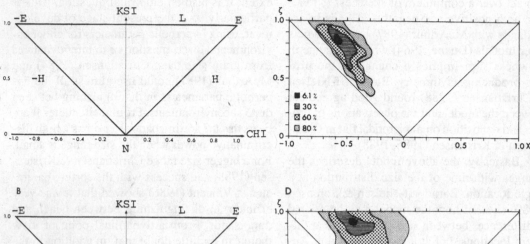

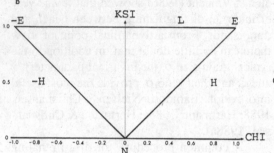
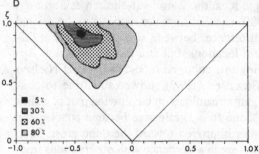

Figure 17.7. Hyperbolic shape triangles: (A) shape positions of coastal eolian sediments; (B) shape positions of swash zone sediments; (C) coastal eolian sediments contoured according to percent density; (D) swash zone sediments contoured according to percent density.

tion we highlight methods that are independent of normality in the involved parameters.

We adapt a "population concept" assuming that all samples taken in a given environment reflect the dynamical conditions in that environment. The samples are stochastically independent, and together they form a population that typifies the environment. We use two sets of samples to illustrate the methods:

1. 247 samples taken in the mid–swash zone along a 54-km stretch of the Mediterranean coast of Israel; and

2. 70 coastal eolian samples taken from the same stretch.

The question to be answered by these methods is this: Do the different dynamical conditions in these two environments produce samples that can be said to represent two statistically different populations?

Hartmann & Christiansen (1988b) suggested one method that builds on the two hyper-

bolic parameters χ and ξ. We find it important to note that, as χ and ξ are location and scale invariant (Barndorff-Nielsen & Christiansen, 1988), the method is independent of traditional size and sorting in the samples that build the two populations. As the method uses the population concept, it demands several samples – more than twenty – to build a population.

The method consists of both a graphical and a numerical approach that are performed in a number of steps.

Graphical approach

(1) Estimate the χ and ξ values for all the samples in the two environments and plot their shape position in the shape triangles. This can be done by the SAHARA program (Christiansen & Hartmann, 1988a). It is evident from Figure 17.7A,B that there is a considerable overlap between the shape positions of the two environments (mid–swash zone and coastal dunes) plotted in the triangle.

(2) Divide the hyperbolic shape triangle into 200 cells by plotting equally spaced ε and κ curves and straight lines in the triangle (Fig. 17.6A). We use this nonlinear (in χ, ξ space) subdivision instead of cells based on equally

spaced χ and ξ because these ε, κ subdivisions allow one to attach some dynamical interpretation to the plot according to the model in Barndorff-Nielsen & Christiansen (1988).

(3) Count the number of shape positions from each of the environments for every cell. Depending on the level of resolution and the number of samples, the number of cells can be reduced – for instance, by combining four cells into one bigger cell. Similar procedures are used for contouring stereo nets.

(4) Because the χ, ξ position of each single sample has an associated area of uncertainty (see Christiansen & Hartmann [1988a] for details), combine the counts for blocks consisting of four neighbouring cells and write the resulting number at each grid line intersection. The next block consists of two cells from the last block and two new cells. This procedure has a smoothing effect. At the margin of the shape triangle the blocks consist of only two cells. The procedure is followed until a contour value has been obtained for each grid line intersection.

(5) Normalize the above contour values to percent values and contour the shape triangle with lines of equal percent density (Fig. 17.7C, D). The dark shaded area in Figure 17.7C,D represents the modal shape of the distributions from the two environments.

Although most of the above methods employ a manual procedure, they could easily be programmed into a computer-plotter system.

Numerical approach

Having performed abovementioned steps 1–3, we now proceed by summing the number of observations from each environment in blocks consisting of four cells without any overlap. The number of observations n_i and k_i ($i = 1, 2, 3, \ldots, p$) in each block is recorded in a table. The expected number in each block should preferably be >5 (Crow et al., 1960).

To test, at the significant level β, the null hypothesis that the two groups of samples were drawn from the same population, we use a chi-square test and note that if

$$X^2 > \chi^2_{\beta(p-1)}$$

where $p - 1$ is the number of degrees of freedom, then we can conclude that the samples were taken from two statistically different environments. The two populations shown in Figure 17.7A,B were different at 99.99% probability. For further results using the method we refer to Hartmann (1988a).

Another method for environmental assignment using hyperbolic parameters that is not dependent on normality in parameter distributions was introduced by Vincent (1986), who used a logistic regression approach.

Conclusions

We suggest geologists go beyond the sole use of the (log-)normal distribution as a model for grain size distributions. In general we consider the hyperbolic distribution to be a much more useful tool. However, having run >5,000 samples on the hyperbolic estimation program SAHARA (Christiansen & Hartmann, 1988a), the result is that about 78% of them could be estimated and therefore regarded as having a hyperbolic distribution. Many of the remaining 22% were bimodal; others belonged to some other distribution. Thus, in line with LeRoy (1981), we want to stress that there is a larger number of frequency distributions. One of these, the β distribution, might turn out to be another useful tool in geology (Hartmann, 1988a; Barndorff-Nielsen et al., in prep.).

ACKNOWLEDGMENTS

We wish to thank Prof. O. E. Barndorff-Nielsen for critically reading the manuscript and B. Gilbert and R. W. Dalrymple for their competent reviews of it.

LIST OF SYMBOLS

E	Exponential distribution
H^+	Positive hyperbolic distribution
H^-	Negative hyperbolic distribution
K_1	Bessel function
L	Laplace distribution
N	Normal distribution (Fig. 17.4)
N	Quasi-sample size (after Barndorff-Nielsen et al., 1985)
N	Weight percentage divided by sieve interval (after Bagnold, 1941)
lc	Number of sieving intervals
l	Log-likelihood function

p	Degrees of freedom
p_i	Hyperbolic probability mass in the ith sieve
r_i	Relative weight of the sample in the ith sieve
α	$= \varphi + \gamma/2$
α_0	Value of a for a symmetrical distribution
β	Significant level
β_1	Standardized skewness
β_2	Standardized kurtosis
γ	Slope of coarse tail
δ	Measure of roundness
ε	Power law erosion/deposition
ζ	Sorting parameter
κ	Sorting parameter
μ	Peak diameter
υ	Typical grain size
ξ	Approximate of kurtosis
π	Measure of asymmetry
σ	$= K/\xi$
τ^{-1}	Radius of circle inside hyperbolic curve
φ	Slope of fine tail
χ	Approximate of skewness
$\hat{\mu}$	Estimate of μ

REFERENCES

Bagnold, R. A. (1937). The size-grading by wind. *Proceedings of the Royal Society (London)*, 163: 250–64.

— (1941). *The physics of blown sand and desert dunes*. London: Methuen, 265 pp.

— (1979). Acceptance of Sorby medal. *Sedimentology*, 26: 189–90.

Bagnold, R. A., & Barndorff-Nielsen, O. E. (1980). The pattern of natural size distributions. *Sedimentology*, 27: 199–207.

Barndorff-Nielsen, O. E. (1977). Exponentially decreasing distributions for the logarithm of particle size. *Proceedings of the Royal Society (London)*, 353: 401–19.

— (1979). Models for non-Gaussian variation: With application to turbulence. *Proceedings of the Royal Society (London)*, 368: 501–30.

— (1986). Sand wind and statistics: Some recent investigations. *Acta Mechanica*, 64: 1–18.

Barndorff-Nielsen, O. E., & Blæsild, P. (1981). Log-hyperbolic distributions and ramifications: Contributions to theory and applications. In: *Statistical distributions in scientific work*, eds. C. Taillic, G. P. Patil, & B. A. Baldessari. Dordrecht: Reidel, vol. 4, pp. 19–44.

— (1983). Hyperbolic distributions. *Encyclopedia of statistical sciences*. New York: Wiley, vol. 3, pp. 700–7.

Barndorff-Nielsen, O. E., Blæsild, P., Jensen, J. L., & Sørensen, M. (1985) The fascination of sand. In: *A celebration of statistics*, eds. A. C. Atkinson & S. E. Feinberg. New York: Springer–Verlag, pp. 57–87

Barndorff-Nielsen, O. E., & Christiansen, C. (1988). Erosion, deposition and size distributions of sand. *Proceedings of the Royal Society (London)*, 417:335–52.

Barndorff-Nielsen O. E., Christiansen, C., & Hartmann, D. (in prep.). Distributional shape triangles with applications in sedimentology.

Barndorff-Nielsen, O. E., Jensen, J. L., & Sørensen, M. (1988). Wind shear and hyperbolic distributions. Research Report no. 145, Department of Theoretical Statistics, University of Aarhus.

Barndorff-Nielsen, O. E., & Sørensen, M. (1989). On the spatial variation of sediment size distributions. Research Report no. 194, Department of Theoretical Statistics, University of Aarhus.

Blæsild, P., & Christiansen, C. (1984). Sand statistik. *Statistiske Interna. no. 40*, University of Aarhus, 11 pp.

Christiansen, C. (1984). A comparison of sediment parameters from log–probability plots and log-log plots of the same sediments. University of Aarhus, *Geoskrifter*, 20: 10–30.

Christiansen, C., Blæsild, P., & Dalsgaard, K. (1984). Re-interpreting 'segmented' grain-size distributions. *Geological Magazine*, 121: 47–51.

Christiansen, C., & Hartmann, D. (1988a). SAHARA – a package of PC-computer programmes for estimating both hyperbolic grain-size parameters and standard moments. *Computers and Geoscience*, 14: 557–625.

— (1988b). On using the log–hyperbolic distribution to describe the textural characteristics of eolian sediments – Discussion. *Journal of Sedimentary Petrology*, 58: 159–60.

Christiansen, C., & Kristensen, S. D. (1986). Two time scales of erosion and accumulation on a micro-tidal flat. 2: Topographic chanes. In: *Twenty-five years of geology in Aarhus*, ed. J. T. Møller, University of Aarhus. *Geoskrifter* 24: 103–12.

— (1988). Two time scales of micro-tidal flat erosion and accumulation. 1: Sediment changes. *Geografiska Annaler*, 70: 47–58.

Crow, E. L., Davis, F. A., & Maxfield, M. W. (1960). *Statistics manual*. New York: Dover, 288 pp.

Deigaard, R., & Fredsøe, J. (1978). Longitudinal grain sorting by current in alluvial streams. *Nordic Hydrology*, 9: 7–16.

Duane, B. (1964). Significance of skewness in recent sediments, Western Pamlico Sound, North Carolina. *Journal of Sedimentary Petrology*, 34: 864–74.

Durst, F., & Macagno, M. (1986). Experimental particle size distributions and their representation by log hyperbolic functions. *Powder Technology*, 45: 223–44.

Ehrlich, R. (1983). Size analysis wears no clothes, or have moments come and gone. *Journal of Sedimentary Petrology*, 53: 1.

Fieller, N. R. J., Gilbertson, D. D., & Olbricht, W. (1984). A new method for environmental analysis of particle size distribution data from shoreline sediments. *Nature*, 311: 648–51.

Folk, R.L. (1971). Longitudinal dunes of the northwestern edge of the Simpson Desert, Northern Territory, Australia. I. Geomorphology and grain-size relationships. *Sedimentology*, 16: 5–54.

Folk, R. L., & Ward, W. C. (1957). Brazos River bar, a study in the significance of grain-size parameters. *Journal of Sedimentary Petrology*, 27: 3–27.

Friedman, G. M. (1967). Dynamic processes and statistical parameters compared for size-frequency distributions of beach and river sands. *Journal of Sedimentary Petrology*, 37: 327–54.

Glaister, R. P., & Nelson, H. W. (1974). Grain size distributions: An aid in facies identification. *Bulletin of Canadian Petroleum Geology*, 22: 203–40.

Grace, J. T., Grothaus, B. T., & Ehrlich, R. (1978). Size frequency distributions taken from within sand laminae. *Journal of Sedimentary Petrology*, 48: 1193–202.

Greenwood, B. (1969). Sediment parameters and environment discrimination: An application of multivariate statistics. *Canadian Journal of Earth Science*, 6: 1347–58.

Hartmann, D. (1988a). Coastal sands of the southern part of the Mediterranean coast of Israel: Reflections of dynamic sorting processes. Ph.D. thesis, Geological Institute, University of Aarhus.

(1988b). The goodness-of-fit to ideal Gauss and Rosin distributions: a new grain-size parameter – Discussion. *Journal of Sedimentary Petrology*, 58: 913–17.

Hartmann, D., & Christiansen, C. (1988a). Settling velocity distributions and sorting processes on a longitudinal dune: A case study. *Earth Surface Processes and Landforms*, 13: 649–56.

(1988b). The hyperbolic shape triangle as a tool for discrimination of populations of sediment samples from closely connected origins (unpublished).

Jackson, R. G. (1977). Mechanisms and hydrodynamic factors of sediment transport in alluvial streams In: *Research in fluvial geomorphology (Proceedings of the 5th Guelph Symposium on Geomorphology)*, eds. R. Davidson-Arnott & W. Nickling. Norwich: Geo Books, pp. 9–44.

Jensen, J. L. (1988). Maximum likelihood estimation of the log-hyperbolic parameters from grouped observations. *Computers and Geoscience*, 14: 389–408.

Jones, T. A. (1969). Determination of *n* in weight frequency data. *Journal of Sedimentary Petrology*, 39: 1473–6.

Krauss, N. C., & Nakashima, L. (1986). Field methods for determining rapidly the dry weight of wet sand samples. *Journal of Sedimentary Petrology*, 56: 550–1.

Krumbein, W. C. (1936). Application of logarithmic moments to size frequency distributions of sediments. *Journal of Sedimentary Petrology*, 6: 35–47.

Lancaster, N. (1986). Grain-size characteristics of linear dunes in the southwestern Kalahari. *Journal of Sedimentary Petrology*, 56: 395–400.

LeRoy, S. D. (1981). Grain-size and moment measures: A new look at Karl Person's ideas on distributions. *Journal of Sedimentary Petrology*, 51: 625–30.

McArthur, D. S. (1986). Derivatives and FORTRAN subroutines for a least-squares analysis of the hyperbolic function. *Mathematical Geology*, 18: 441–50.

(1987). Distinction between grain-size distributions of accretion and encroachment deposits in an inland dune. *Sedimentary Geology*, 54: 147–63.

McKinney, T. F., & Friedman, G. M. (1970). Continental shelf sediments of Long Island, New York. *Journal of Sedimentary Petrology*, 40: 1–23.

Mandelbrot, B. B. (1977). *Fractals: Form, chance and dimension*. San Francisco: Freeman.

Meland, N., Ferm, J. C., & Norrman, J. O. (1964).

On the problem of *n* in weight frequency data. *Journal of Sedimentary Petrology,* 35: 984–5.

Middleton, G. V. (1976). Hydraulic interpretation of sand size distributions. *Journal of Geology,* 84:405–26.

Nielsen, H. L. (1985). Shapes of sand grains estimated from grain mass and sieve size. Proceedings of international workshop on the physics of blown sand. *Aarhus,* 3: 677–88.

Pranzini, E. (1983). Random changes in beach sand grain-size. *Bolletino della Societa Geologica Italiana,* 102: 177–89.

Rana, S. A., Simons, D. B., & Mahmood, K. (1973). Analysis of sediment sorting in alluvial channels. *Journal of the Hydraulics Division, ASCE,* 99: 1967–80.

Reed, W. E., LeFevre, R., & Moir, G. (1975). Depositional environment interpretation from settling velocity (Psi) distributions. *Bulletin of tthe Geological Society of America,* 86: 1321–8.

Schleyer, R. (1987). The goodness-of-fit to ideal Gauss and Rosin distributions: A new grain size parameter. *Journal of Sedimentary Petrology,* 57: 871–80.

Sedimentation Seminar (1981). Comparison of methods of size analysis for sands of the Amazon–Solimoes Rivers, Brazil and Peru. *Sedimentology,* 28: 123–8.

Simonett, D. S. (1960). Development and grading of dunes in Western Kansas. *Annals of the Association of American Geographers,* 50: 216–41.

Sly, P. G., Thomas, R. L., & Pelletier, J. A. (1983). Interpretation of moment measures from waterlain sediments. *Sedimentology,* 30: 219–33.

Viard, J. P., & Breyer, J. A. (1979). Description and hydraulic interpretation of grain-size cumulative curves from the Platte River system. *Sedimentology,* 26: 427–39.

Vincent, P. (1986). Differentiation of modern beach and coastal dune sands – A logistic regression approach using the parameters of the log-hyperbolic function. *Sedimentary Geology,* 49: 167–76.

Visher, G. S. (1969). Grain-size distribution and depositional processes. *Journal of Sedimentary Petrology,* 39: 1074–106.

Wyrwoll, K.–H., & Smyth, G. K. (1985). On using the log-hyperbolic distribution to describe the textural characteristics of eolian sediments. *Journal of Sedimentary Petrology,* 55: 471–8.

18 Factor analysis of size frequency distributions: Significance of factor solutions based on simulation experiments

JAMES P.M. SYVITSKI

Factor analysis is often considered a part of statistical methodology. But in view of the fervor shown by its advocates as well as by its detractors, it is suggested that factor analysis might better be classified as a religion. (Wallis, 1968)

Introduction

Many factors influence the size frequency distribution (SFD) of natural sediment populations. Solohub & Klovan (1970) list:

1. distribution of the source material;
2. mineralogy and texture of the source material;
3. the type and amount of energy at a depositional site;
4. the rate of sediment supply;
5. the possibility of sediment removal;
6. the mixing, sorting, and remixing of separate populations; and
7. digenetic alterations.

The scale and nature of sampling can also be very important (Grace et al., 1978). Analysis of the SFD of sedimentary particles remains an important tool for understanding sedimentary processes despite the varying degrees of success achieved by its application. Methods include interpreting bivariate plots of "statistical" parameters derived from SFD, or the shape of individual SFD, to more complex multivariate methods, such as factor or discriminate analyses.

Despite the potential pitfalls of the method (Ehrenberg, 1962; Matalas & Reiher, 1967; Glaister & Nelson, 1974; Temple, 1978), Q-mode factor analysis has provided useful geologic information (Klovan, 1966; Yorath, 1967; Beall, 1970; Allen et al., 1971; Clague, 1976; Dal Cin, 1976; Chambers & Upchurch, 1979). Its application, as with most of the other techniques involving size frequency data, has been based on empirical evidence; the results seem to make "sense." Exactly how the method achieves these results from size frequency data was first reported by Syvitski (1984).

Mathematical analysis and simulation are used in this paper to determine relationships between characteristics of SFD and the results of Q-mode factor analysis. Some fundamental questions are raised:

1. How are differences in means of SFD resolved?
2. How are differences in sorting resolved?
3. What patterns emerge from the mixing together of end-member SFD having distinct mean and sorting (among other statistical) values?

In short, the approach is to determine what Q-mode factor analysis will produce given some known size frequency distributions to start with.

Q-mode factor analysis described

Davis (1973) writes:

Factor analysis is commonly regarded as a deep and mysterious methodology of great complexity ... Nevertheless, it is one of the most widely used multivariate procedures and extends a beguiling promise to experiments faced with a welter of complex data and little insight into the structure of the data. (p. 475)

For readers familiar with dimensional analysis used widely in hydrodynamics, Q-mode factor analysis is similar in its objective – to discern regularity and order in a multivariate data set.

The most common forms of factor analysis used in natural sciences are R-mode and Q-mode. *R-mode factor analysis* investigates interrelations in a matrix of correlations between variables. The factors that are created are new variables having the form of linear combinations of the original variables. In *Q-mode factor analysis* the role of samples and variables is reversed. Q-mode is concerned with interrelationships between samples. The objective of Q-mode analysis is much the same as the objective of cluster analysis – to arrange a suite of samples into a meaningful order so that the relationships between one sample and another may be deduced.

The general principles of Q-mode factor analysis have been described adequately in the literature (see Armstrong, 1967; Rummel, 1967; Castaing, 1973; Jøreskog et al., 1976). Discussed below are those aspects of special importance to its use in the analysis of grain size data.

The data

Klovan (1966) was the first to use Q-mode factor analysis to interpret size frequency data from a suite of sediment samples. He argued that each class interval is a unique attribute of the particular sediment sample. His data set comprised ten whole-phi intervals (as the variables) that covered the entire size frequency spectrum (i.e., the cumulative weight frequency of a given sample totalled 100%). As more automated size analytical methods developed, the number of variables increased: Clark (1978), for example, used twenty-one quarter-phi intervals.

Butler (1976) noted that principal component analysis (an important phase within factor analysis) induces negative correlations within the similarity matrix when the original data set is composed of closed arrays (i.e., size frequency distributions that total 100%). As a consequence, the correlation matrix is overdetermined. Chayes & Trochimczyk (1978) noted that transformation of the data provides an elegant escape from closure correlation if the problem can be restated entirely in terms of component scores (transformed variables). Chambers & Upchurch (1979), in an effort to avoid closure problems, eliminated one class interval and thus opened up the data matrix. More recently, Perillo & Marone (1986) demonstrated the advantages of transforming SFD using the maximum entropy concept such that a uniform distribution is produced by altering the width of class intervals. They further suggested an optimal number of class intervals be used.

In my experiment, each sediment sample is considered to be totally described in terms of its size frequency distribution by the weight percent of sediment contained in each of p size classes. Geometrically, each sample can be visualized as a vector whose coordinates on p mutually orthogonal axes are the weight percentages. Because the sum of the weight percentages is 100 for all samples, the ends of the vectors are constrained to be on a $p-1$-dimensional hyperplane in p-dimensional space. Occasionally, the original data set was transformed before analysis through range transformation, which essentially eliminates the variance along the variables (class intervals).

Similarity

Of the variety of measures that can be used to describe mathematically the degree of similarity between pairs of samples, the cos θ index of Imbrie & Purdy (1962) provides several advantages. The index between samples i and j, assuming p size classes, is computed from:

$$\cos \theta_{ij} = \left(\sum_{k=1}^{p} X_{ik} X_{jk} \right) \left(\sum_{k=1}^{p} X_{ik}^2 \sum_{k=1}^{p} X_{jk}^2 \right)^{-0.5}$$

(18.1)

where X_{ik} is the amount of sediment in the kth class (size interval) for sample i, and likewise for sample j.

As is evident from equation (18.1), cos θ simply measures the cosine of the angle separating any two sample vectors in p space. Two aspects of this index are worth emphasizing:

1. The denominator of the equation, in effect, normalizes the vectors so that they are each of unit length.

2. The angular separation is determined solely by the proportion of sediment in each size class, and therefore absolute amounts of sediment are ignored.
A cos θ value of 1.0 signifies perfect similarity; a value of 0.0 signifies perfect dissimilarity.

In practice, all pairs of samples are substituted into equation (18.1) and an $N \times N$ matrix of cos θ values is computed (where N is the total number of samples present in the study). This matrix contains all the information concerning the mutual similarities (and dissimilarities) between all of the samples, and is the starting point of the Q-mode analysis itself.

Q-mode analysis

Although mathematically complex, the Q-mode procedure as outlined by Klovan & Imbrie (1971), Klovan & Miesch (1976), and Miesch (1976) relies on the rather simple notion

Table 18.1. *Some helpful definitions for Q-mode factor analysis terminology*

Communality: Proportion of variation that each sample is involved in the factor patterns; equal to the sum of squared factor loadings; indicates to what degree a sample is unrelated to others.

Eigenvalue: Sum of squared factor loadings of a particular factor. May be thought of as a length of an elliptical axis in p-dimensional space that, together with other eigenvalues, confine the data set.

Factor loadings: Measure of which samples are involved with which factor pattern and to what degree. They can be interpreted like correlation coefficients.

Factor patterns: Separate (meaningful) patterns between samples that are independent of (uncorrelated to) one another.

Factor score: Variable score weighted proportionally to its involvement in a factor. Compositional scores decribe, for example, an end-member distribution.

Factor variance: Percent of the total variance accounted for by a particular factor; is determined by summing the column of squared loadings for a factor, dividing by the number of vectors, and multiplying by 100.

Oblique rotation: The individual rotation of factors to fit distinct sample clusters; by definition, there will be some correlation between the factors.

Principal component analysis (PCA): Involves the production of the unrotated factor matrix from the solution of eigenvalues and eigenvectors. Specifically, the matrix is equal to the eigenvectors times the reciprocal square root of their associated eigenvalues.

Varimax: Orthogonal rotation such that each unrotated factor has been rotated until it defines a distinct cluster of interrelated samples. Thus the number of samples loaded highly on a factor is minimized.

---

of *principal components*. The matrix of similarity coefficients is decomposed into two matrices. One, the matrix of *factor loadings*, provides the coordinates of the samples in a space of reduced dimensionality. The other, the matrix of *factor scores*, provides the position of the axes used to determine the reduced space in terms of their coordinates on the original p variables that define the data space. (Table 18.1 provides a simplified glossary of factor analysis terms.)

Miesch (1976) developed a method that permits the use of the most divergent samples as reference axes and gives their composition in terms of the p weight percent classes used to describe the samples originally. For grain size analyses, then, the Miesch method attempts:

1. to find the most different types of samples, based on their total SFD;

2. to express all other samples as mixtures of these end-member distributions; and

3. to describe the composition of the end members in terms of their total grain size characteristics.

Although we shall examine the fundamentals of Q-mode factor analysis using the Miesch approach, Clarke (1978) has argued that an oblique factor analysis solution is best for the analysis of mixtures. Also, as discussed later, more advanced algorithms may provide solutions to some of the problems encountered using the Miesch approach (Full et al., 1981, 1982).

Analytical approach
On understanding cos θ

Because the $\cos\theta$ matrix of similarities contains all the information on which Q-mode method works, it is important to establish exactly how the $\cos\theta$ index describes similarity between grain size distributions. Initially, these distributions are considered lognormal in character; if grain size intervals are measured in phi units, we may consider Gaussian distributions.

Our first case is to compute $\cos\theta$ between two perfect Gaussian distributions whose properties are solely determined by $\mu_1, \mu_2, \sigma_1, \sigma_2$, the respective means and standard deviations. The equation for a Gaussian distribution is

$$y_i = (2\pi\sigma)^{-0.5}\exp[-0.5((x_i-\mu)/\sigma)^2] \quad (18.2)$$

where (in terms of grain size distributions) y_i is the weight of sediment in the ith phi interval and less, x_i is the θ value, μ is the mean value, and σ is the standard deviation in phi units. The $\cos\theta$ index can be considered as a correlation function between two such distributions.

Denoting the first distribution as $f_1(x)$ and the second as $f_2(x)$, it is seen that

$$\cos\theta_{12} = \left[\int_{-\infty}^{\infty} f_1(x)f_2(x)\,dx\right]\left[\int_{-\infty}^{\infty}[f_1(x)]^2\,dx\right]^{-0.5}$$
$$\cdot\left[\int_{-\infty}^{\infty}[f_2(x)]^2\,dx\right]^{-0.5} \quad\quad (18.3)$$

and upon integration we obtain

$$\cos\theta_{12} = \left[\frac{(2\sigma_1\sigma_2)}{(\sigma_1{}^2+\sigma_2{}^2)}\right]^{0.5}\exp-\left[\frac{(\mu-\mu_2)^2}{2(\sigma_1{}^2+\sigma_2{}^2)}\right]$$
$$(18.4)$$

It is thus apparent that $\cos\theta$ is a complex, non-linear function of the two basic parameters of the Gaussian distributions being compared.

For the special case where the two grain size distributions (samples) have the same standard deviation (sorting), the relationship becomes

$$\cos\theta_{12} = \exp-((\mu_1-\mu_2)/2\sigma)^2 \quad (18.5)$$

Curves generated from this equation (Fig. 18.1) show that, for a given standard deviation, $\cos\theta$ decreases as differences in mean value increase. Also evident is the fact that a pair of poorly sorted samples are designated as being more similar than a pair of well-sorted samples having the same difference in mean values. For very well-sorted samples, small differences in the mean value drastically reduce $\cos\theta$.

For the special case where the distributions have zero mean difference, the relationship becomes:

$$\cos\theta_{12} = (2\sigma_1\sigma_2/(\sigma_1{}^2+\sigma_2{}^2))^{0.5} \quad (18.6)$$

Figure 18.2, generated from this equation, shows that $\cos\theta$ is less sensitive to contrasts in sorting between two poorly sorted distributions than it is to two well-sorted distributions.

A set of ideal, lognormally distributed, grain size samples was mathematically generated for our second case, with means ranging from $0\,\phi$ to $4\,\phi$ in half-phi increments and standard deviations from $0.5\,\phi$ to $2\,\phi$ in increments of 0.5 standard deviations. There are thus thirty-six samples in the set. The distributions were "sieved" into twenty intervals between $-8\,\phi$ and $11\,\phi$. Even with this degree of separation, con-

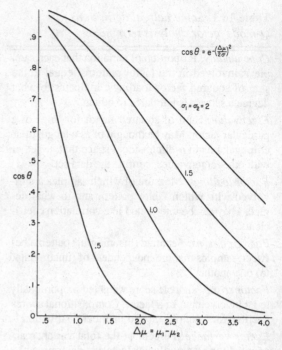

Figure 18.1. Relationship between $\cos\theta$ and difference in mean value ($\Delta\mu$) between two Gaussian distributions having the same standard deviation σ.

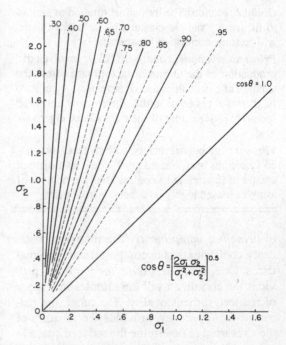

Figure 18.2. Relationship between $\cos\theta$ and standard deviation of two Gaussian distributions with zero mean difference.

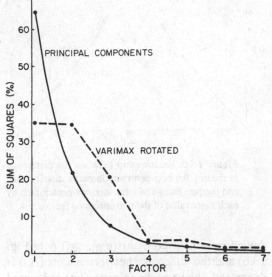

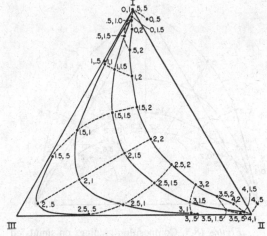

Figure 18.3. Percent sums of squares extracted by successive factors for both principal components and varimax solutions. A three-factor solution is indicated as being sufficient.

Figure 18.4. Normalized varimax factor components on simulated size frequency distributions. Sample points are identified by a two-member code: The first describes the mean value; the second, the standard deviation. Heavy lines join samples with equal standard deviations; broken lines join samples with equal means.

siderable deviation from an ideal distribution has been introduced, particularly for those samples whose means fall on a class boundary. Nevertheless, cos θ values computed from the simulated discrete distributions differ only in the third decimal from the theoretical cos θ values derived from equation (18.2).

On understanding the Q-mode factor analytical solution

The thirty-six discrete, lognormal distributions generated above, each with a different mean and/or standard deviation, were analyzed by Q-mode. The percent sums of squares extracted by successive factors for both the principal components and varimax cases (Fig. 18.3) suggest that a three-factor solution would be optimal. Only the phi classes between −1 and 5 were found to contribute to the analysis. Communalities for a three-factor solution range from 0.7026 to 0.9990. Thus, even with errorless input data, considerable distortion is present in the factor analysis results.

A plot of the normalized varimax factor components for the three-factor case is shown

in Figure 18.4. The typical "horseshoe" pattern caused by a closed data set is evident. Factor I contains high loadings for samples with means of 0.0 and 0.5; Factor II contains high loadings for samples with means of 3.5 and 4.0; and Factor III has a single high loading for the sample with a mean of 2.0 and a standard deviation of 0.5.

The most divergent samples are determined to be (0, 0.5), (2, 0.5), and (4, 0.5), with the first number referring to the mean, the second to the standard deviation. A plot using these samples as reference axes is shown in Figure 18.5. The "horseshoe" shape is again evident. Samples with the same standard deviation are arranged along arcs subparallel to the line joining (0, 0.5) to (4, 0.5). Samples with the same mean are arranged on arcs subparallel to the perpendicular of the line joining (0, 0.5) to (4, 0.5), which extends to (2, 0.5).

Although each sample used in this analysis represents a perfect lognormally distributed sediment, the usual interpretation of this diagram would consider them to be mixtures of three end-member samples: (0, 0.5), (2, 0.5),

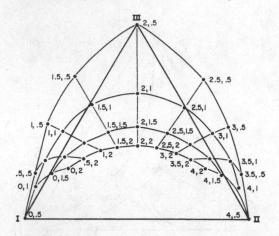

Figure 18.5. Composition loadings on simulated size frequency distributions. Symbols as in Figure 18.4.

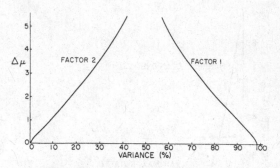

Figure 18.6. Relationship between the difference in mean ($\Delta\mu$) between two Gaussian distributions and the percentage of total variance contributed by each eigenvalue of the two unrotated factors.

and (4, 0.5) (see Klovan, 1966). Because many samples plot outside the positive triangle defined by these three samples (due to their negative loadings), it is apparent that a simple additive mixing model is not appropriate. For example, sample (1, 0.5) is a mixture of 72.47% of sample (0, 0.5), 45.74% of sample (2, 0.5), and −18.60% of sample (4, 0.5).

The compositional scores of the three end members for the three-factor solution have their mean and standard deviation preserved but are distorted with numerous negative values of composition. Clearly, a three-factor model is not a satisfactory solution; but experiments with more factors do not materially improve the situation, and there are no good a priori reasons for using a higher-order factor solution.

Let us next compare two Gaussian distributions predetermined to be different by altering the mean and/or standard deviations. Twenty nearly identical distributions were generated by the Cornish–Fisher expansion equation for each of these two distributions (see Kendall & Stuart, 1969, pp. 165–6; Swan et al., 1978). Therefore, the data input for factor analysis was a total of forty samples, but only two distributions.

The unrotated factor matrix results in two factors accounting for all variance in the data set (Fig. 18.6). The second factor separates the distributions by the loading sign. When standard

deviation is held constant (e.g., at 1 ϕ in Fig. 18.6) and the difference between means ($\Delta\mu$) is decreased, the variance (eigenvalue) accounted for by the unrotated factor 1 increases; factor 2 decreases proportionally. At $\Delta\mu < 0.15$, factor 1 accounts for nearly 100% of the data variance (Fig. 18.6). Similarly, when the mean is held constant and the difference between standard deviations ($\Delta\sigma$) is decreased, the variance accounted for by factor 1 increases in the unrotated factor matrix. In both cases, the rotated factor matrix separates the distributions by the loading values. When $\Delta\sigma$ or $\Delta\mu$ decreases, these loading values approach one another – factor analysis cannot distinguish these distributions when $\Delta\sigma < 0.25$ or $\Delta\mu < 0.5$.

Figure 18.7a gives the SFD of two distributions having the same mean but different standard deviation, and shows how the factor scores mirror the variance along any class interval; that is,

$$V_k = \left(\sum_{k=1}^{N} (X_i - \bar{X}_j)^2 \right) N^{-1} \qquad (18.7)$$

where V is the variance in the kth class, X_i is the weight of the ith sample in the kth class, $\bar{X}$ is the mean weight of the kth class, and N is the total number of samples. In an attempt to circumvent this situation, the above experiment was repeated on range-transformed data. The total variance in the data set accounted for by the rotated factor matrix is considerably lower than that of the nontransformed data (Fig. 18.7b).

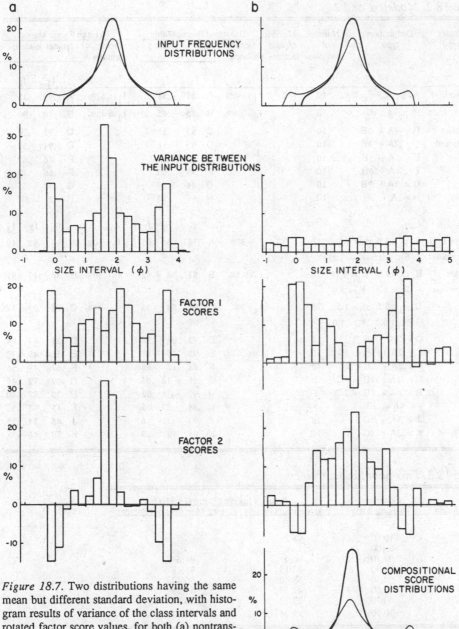

Figure 18.7. Two distributions having the same mean but different standard deviation, with histogram results of variance of the class intervals and rotated factor score values, for both (a) nontransformed and (b) range-transformed data. Compositional scores from the range-transformed data indicate the two original distributions.

However, the loadings produce a better separation of the two distributions. Though the rotated factor scores are hard to interpret, the compositional scores mimic the original distributions (Fig. 18.7b).

Q-mode factor analysis of distribution mixtures

To clarify the mixing model approach, a series of grain size distributions were constructed by mixing together various proportions of particular end members (see Table 18.2). The

Table 18.2. *Models 1 and 2*

Model type	Distribution type	Number of samples	Number of real factors	Unrotated Factor Matrix factor variance	Unrotated typical loading f_1	f_2		Rotated Factor Matrix factor variance	Rotated typical loading f_1	f_2	
1: Two End	A, μ_1=4.5 σ_1=0.5	10	2	f_1 = 80%	A .81	−.58		f_1 = 52%	A .99	.13	
Members	B, μ_2=6.5 σ_2=1.0	10	2	f_2 = 19%	B .75	.65		f_2 = 48%	B .10	.99	
plus	C = .9A + .1B	10			C .85	−.51			C .98	.20	
mixtures	D = .7A + .3B	10			D .95	−.31			D .91	.39	
	E = .5A + .5B	10			E .99	−.05			E .76	.63	
	F = .4A + .6B	10			F .99	.05			F .66	.73	
	G = .3A + .7B	10			G .96	.24			G .52	.85	
	H = 1A + .9B	10			H .99	.52			H .23	.97	

Model type	Distribution type	Number of samples	Number of real factors	Unrotated Factor Matrix factor variance	Unrotated typical loading f_1	f_2	f_3	Rotated Factor Matrix factor variance	Rotated typical loading f_1	f_2	f_3
2: Three End	A, μ_1 = 3.5 σ_1 = 0.5	10	3	f_1= 80%	A .76	−.65	.01	f_1= 30%	A .10	.98	.19
Members	Sk_1= 0.5 K_1= 5.5										
plus	B, μ_2= 4.5 σ_2= 0.75	10		f_2= 13%	B .82	.38	−.42	f_2= 30%	B .36	.17	.92
mixtures	Sk_2= 0.0 K_2= 3.0										
	C, μ_3= 5.5 σ_3= 1.5	10		f_3= 7%	C .76	.54	.35	f_3= 31%	C .92	.07	.37
	Sk_3= 0.0 K_3= 3.0										
	D = .2A + .8B	10			D .92	.08	−.38		D .28	.47	.84
	E = .2A + .5B + .3C	10			E .98	.15	−.14		E .51	.47	.72
	F = .2A + .2B + .6C	10			F .98	.15	.08		F .67	.49	.56
	G = .4A + .2B + .4C	5			G .99	−.14	.03		G .49	.72	.49
	H = .6A + .2B + .2C	5			H .92	−.38	−.02		H .30	.87	.40
	I = .8A + .2B	5			I .84	−.53	−.06		I .15	.93	.32
	J = .5A + .5C	10			J .93	−.33	.16		J .45	.84	.29
	K = .2A + .8C	10			K .94	.18	.29		K .80	.46	.38

Table 18.3. *Two-population problem*

Population	Generated Mixing Ratio	Factor Loading Ratio (l^2 x100) End Members Present	End Members not Present
A	100:0	98:2	-
B	0:100	1:99	-
C	90:10	96:4	94:6
D	70:30	83:15	81:18
E	50:50	58:40	58:42
F	40:60	43:54	41:58
G	30:70	27:72	26:74
H	10:90	5:94	4:94

first mixture model used two distributions, (4.5, 0.5) and (6.5, 1.0), as end members (Table 18.2, Model 1). Two factors account for 99% of the variance in the data set. The factor scores of the rotated matrix approximate the original end-member populations. The squared factor loadings of the rotated factor matrix (×100 for percentage values) give the approximate mixing ratios (Table 18.3; the loadings as indicators of mixing ratios are improved upon oblique rotation [Clarke, 1978]). When the experiment was repeated with all six mixtures but no end members, the rotated factor loadings were nearly the same (Table 18.3), and the compositional scores gave nearly the same end-member distributions.

The second mixture model generated and mixed three distributions (Table 18.2, Model 2;

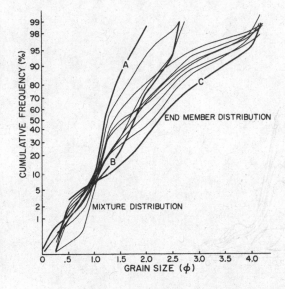

Figure 18.8. Log probability plot of typical distributions from Model 2 (Table 18.2), the three mixed distributions.

Fig. 18.8). Three factors account for 100% of the data variance. The three rotated factors each account for approximately 33% of the data variance, and the varimax scores indicate that each factor corresponds to one of the three end-member distributions – again, whether or not the original data set included end-member data. Mixing ratios calculated from the varimax loadings indicate that an orthogonal rotation is no longer satisfactory for systems as complex as three end-member mixtures (Syvitski, 1984). Figure 18.9 indicates part of the problem. The factor model appears only to designate an end-member distribution from its central mode. Thus, these modal size intervals are overcompensated in the compositional scores and cause a reduction in score values over the same size intervals in the remaining end members.

The fact that only the central grain size classes carry much weight in factor analysis supports the notion that information on the character of the tails of the distribution does not contribute significantly in the computation of similarity. Given the composition of the end members and mixing ratios, the mean and standard deviation of these mixtures can be perfectly predicted in terms of their position on the factor plot. However, it is not a unique solution: One point on the factor plot can represent

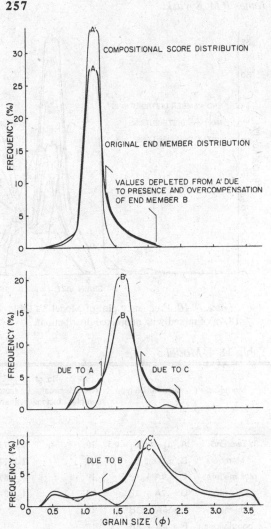

Figure 18.9. Frequency plots of each of the end members of Model 2 (Table 18.2), compared to the compositional scores from the rotated factor matrix. Compositional score A′ is affected by B′; B′ is affected by A′; and C′ shows the effect of B′.

samples with quite different grain size characteristics, as long as the central modes are similar. For instance, a lognormal sample (μ, σ = 3.5, 2.0) was found to fall very close to a $1:4:5$ mixture of $\mu, \sigma = 0, 0.5$; $2, 0.5$; $4, 0.5$ on a three-end-member factor plot (Syvitski, 1984). The attempt to circumvent the situation through use of range transformation was not successful in that the same inconsistencies emerged as in the untransformed three-factor solution.

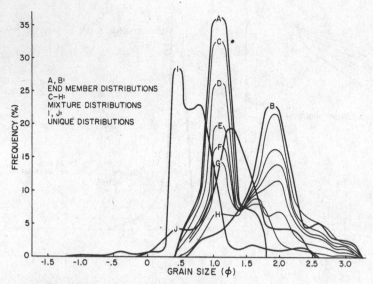

Figure 18.10. Frequency plot of Model 3 (Table 18.4), of mixed with nonmixed distributions.

Table 18.4. *Model 3*

Model type	Distribution type	No. of samples	No. of real factors	Unrotated Factor Matrix factor variance	Unrotated typical loading				Rotated Factor Matrix factor variance	Rotated typical loading			
					f_1	f_2	f_3	f_4		f_1	f_2	f_3	f_4
3: Two End	A, $\mu_1 = 4.5$ $\sigma_1 = 0.5$	10	4	$f_1 = 73\%$	A .83	−.51	−.19	−.08	$f_1 = 45\%$	A .98	.12	.15	.08
Members	B, $\mu_2 = 6.5$ $\sigma_2 = 1.0$	10		$f_2 = 17\%$	B .72	.68	.11	−.00	$f_2 = 39\%$	B .09	.99	.04	.08
plus mixtures	C = .9A + .1B	10		$f_3 = 8\%$	C .87	−.45	−.18	−.08	$f_3 = 11\%$	C .96	.20	.14	.09
plus two	D = .7A + .3B	10		$f_4 = 2\%$	D .95	−.26	−.14	−.07	$f_4 = 4\%$	D .90	.40	.14	.09
non-mixed	E = .5A + .5B	10			E .99	.00	−.08	−.06		E .75	.63	.12	.10
populations.	F = .4A + .6B	10			F .98	.16	−.04	−.05		F .64	.75	.11	.10
	G = .3A + .7B	10			G .94	.32	−.00	−.04		G .50	.85	.09	.10
	H = .1A + .9B	10			H .79	.60	.08	−.02		H .21	.97	.05	.08
	I, $\mu_3 = 3.5$ $\sigma_3 = 0.75$ $sk_3 = 1.0$ $k_3 = 10.5$	10			I .41	−.38	.83	−.02		I .24	.06	.97	.03
	J, $\mu_4 = 5.0$ $\sigma_4 = 1.0$	10			J .90	−.20	−.05	.38		J .73	.39	.20	.53

Data sets containing both mixed and nonmixed distributions

Data input to Model 1 (Table 18.2) was used with the addition of two unrelated distributions (Fig. 18.10; Table 18.4, Model 3). Four factors resulted, one for every type of distribution excluding those that are mixtures of the two end-member distributions. There are some notable similarities and differences between Model 1 (Table 18.2) and Model 3 (Table 18.4). The end-member factors of the rotated factor matrix

account for less variance in Model 3; the factor loadings, however, are nearly identical. Nonmixed distribution *I* (of Model 3) has significantly higher rotated factor loadings than nonmixed distribution *J,* the numerical difference being due to distribution *I* having more easily recognized marker variables (class intervals). Distribution *J* is nearly hidden within the mixed distributions (Fig. 18.10).

When additional nonmixed distributions were added to Model 3 factors, the original factors had smaller eigenvalues (or factor variance), which decreases the chance of getting true compositional scores. Factors that have accountable variance of <5% have unrealistic compositional scores.

Summary of simulation experiments

1. Cos θ is a complex, nonlinear function of the means and standard deviations of grain size distributions (eqs. 18.4–18.6). Cos θ is more affected by differences that occur between well-sorted samples than those between poorly sorted samples.

2. The Q-mode factor solution can be used to separate distributions with $\Delta\sigma > 0.25\,\phi$ and for $\Delta\mu > 0.5\,\phi$. It concentrates on similarities and differences involving only the central portions of the size distributions and thus effectively classifies samples on the basis of mean and standard deviation only, rather than on the higher-moment measures of skewness and kurtosis.

3. Q-mode factor analysis provides a mixing model for the interpretation of grain size distributions whether or not such a solution is applicable.

(a) When not applicable, the composition factor scores reflect only the standard deviation along the class intervals between the different distributions, and not the original sample compositions.

(b) When applicable, the factor scores of the rotated matrix approximate the original end-member distributions, even if the original data set only contained mixed distributions. If the original data set is complex (i.e., it contains mixtures of three or more end members), the compositional scores begin to show marked deviations from the actual end-member distributions.

4. Consequently, Q-mode factor analysis can be an effective method of dissecting mixtures of sediment sources. The squared factor loadings of the rotated factor matrix give the approximate mixing ratios, but the accuracy decreases with increasing complexity: For example, when three or more end members or extraneous distributions are included in the data set.

5. When a data set contains a majority of sample distributions that are a result of the mixture of end-member distributions, extraneous distributions not part of the mixing model do not interfere with the appropriate demixing or end-member solution to the mixed distributions. Other factors, accounting for little variance in the data, appear to relate to these nonmixed sediment distributions. However, when the accountable variance per factor is <5%, these secondary factors produce unrealistic compositional scores.

6. There is no unique place on a factor plot for a given size frequency distribution. Grain size distributions that are similar only in the central portions of their frequency distributions will be interpreted as being the same, even if one is generated as a mixture of end-member distributions and another is a perfect Gaussian distribution.

Implications for real data sets

The above results indicate that Q-mode factor analysis can be applied to the interpretation of depositonal processes and environments only to the extent that

1. the central portions of sediment distributions (as given by mean and standard deviation) reflect these processes and environments, and

2. the data sets are confined to samples that delimit simple physical processes or environments.

Below we examine two previously published case studies where Q-mode factor analysis of size frequency data was employed and where the above two conditions were met.

Klovan (1966) analyzed sixty-nine sediment samples collected from Barataria Bay on the Mississippi Delta using data from their sieved distributions. Employing Q-mode factor analysis, a three-factor solution was obtained (Fig. 18.11) with the first two factors accounting for 92% of the sample variance, and the third for another 6%. In his interpretation, Klovan found that each sediment sample was derived from the mixing of two fundamental source populations (i.e., sand and mud) through the action of three distinct processes: Factor I, current energy; Factor II, gravitational energy;

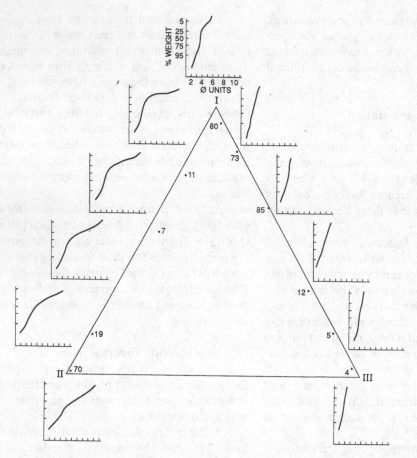

Figure 18.11. Normalized factor components and associated grain size distributions. (From Klovan [1966].)

and Factor III, surf energy. I would suggest that a more reasonable interpretation, in the light of the above simulated experiments, is that there are only two real factors (his factors II & III on Fig. 18.11), and that these represent end-member distributions within a mixing model. (Note that the order of factors is without significance after varimax rotation.) Compositional score related to his Factor I appear intermediate between the other two end-member distributions. Employing a two-factor solution, these two new factors might represent end-member samples respectively derived from the settling out of suspended sediment (his original factor II: Fig. 18.11) and samples representing bed-load transport and deposition (his original factor III: Fig. 18.11).

The second case study concerns the analysis of 125 sediment samples from various geomorphic environments within a drainage basin (East, 1985, 1987). Each sample was subdivided into seventeen size classes (whole-phi intervals), and this data set was analyzed by Q-mode factor analysis. Three factors accounted for 94.5% of the total variance. They were interpreted as representing idealized end members of the sample suite, and their particle size distributions were considered to represent distinct processes of weathering and surface transport: Factor I, suspended sediment transport and deposition; Factor II, weathering associated with groundwater transport; and Factor III, bedload transport and deposition. Six textural groups were found to fit into four geomorphic environments as shown in Figure 18.12. Although I question whether there are three or two real factors (the third factor only accounts for 6% of the total variance), East's interpretation makes conceptual sense (Fig. 18.12).

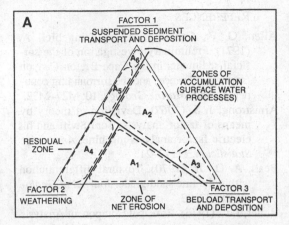

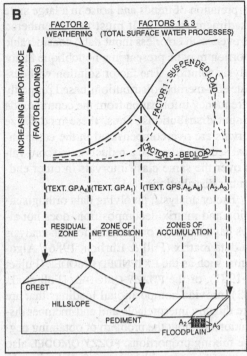

Figure 18.12. Process–response model of four geomorphic environments: (A) graphical representation of inferred process–particle size relationships; (B) diagrammatic representation of the spatial significance of relationships shown in (A). (After East [1987].)

In evaluating both the Klovan (1966) and the East (1987) studies, the concept of mixing needs to be addressed. Mixing, in the sedimentological sense, may be defined as two sediment populations having unique size frequency distributions that are mixed in various proportions to provide a suite of sediment samples with attributes falling between these two end-member "source" distributions. For example, aeolian sediment may be added to an ambient suspended sediment load found in an ocean or lake. The final lake- or seafloor deposits could be considered true mixtures of different sediment sources that are transported by independent sedimentary transport mechanisms. Another example might be the mixing of suspended sediment loads discharged into a sedimentary basin from two or more rivers, each associated with unique fluvial-sediment regimes.

However, end-member distributions may also represent the beginning and end of a singular depositional process. For instance, suspended sediment discharged from a river will systematically change in its size characteristics with distance from the river mouth (see Syvitski et al., 1985). Subsequently, basin sediments will also reflect this transition in their size frequency

distributions. In such cases, the interpretation of factors identified after a Q-mode factor analysis should not be one of a suite of samples developed from the mixture of two sediment sources, but rather as the transition resulting in changing energy conditions related to a single process. This could prove a complication within closed systems, such as a sedimentary basin. For example, there could exist a continuum between proximal and distal prodelta sediments based on sedimentation from under the seaward flowing river plume, *and* a reverse continuum related to the seaward transport of coarse delta front sediments transported by turbidity currents. Neither involves true mixing of source populations. In such a system should one expect a Q-mode solution with two, three, or four factors, especially since there is a communality of the same proximal sediment source, yet two independent transport mechanisms?

Summary

This paper discovers how factor analysis operates using Q-mode on size frequency data. Much more work is necessary, starting with real data from simple sedimentary systems, to see if its full potential as a method will aid in the interpretation of trends and noise in a large suite of sediment samples. It must be remembered that even with errorless input data, considerable distortion may be present in a nonoblique factor analysis solution. The factor solution will designate end-member distributions based primarily on frequency information from the central mode of input distributions. Thus, these modal size intervals are overcompensated in the compositional scores and cause a reduction in score values over the same class intervals in other end-member distributions.

Factor analysis, involving data orthogonalization and matrix decomposition, does not always have a benign or neutral effect on analysis of a data matrix (Full & Ehrlich, 1986). Algorithms, such as the EXTENDED QMODEL (Full et al., 1981), or the FUZZY QMODEL (Full et al., 1982), provide compositional loadings that are more directly interpretable and end-member distributions without the problem of obtaining negative mixing proportions. FUZZY QMODEL also provides the user with a means for filtering aberrant data samples.

A fundamental problem is that Q-mode factor analysis provides a mixing model solution for the interpretation of grain size distributions whether or not such a solution is applicable. Also, there is no unique place on a factor plot for a given size frequency distribution. Grain size distributions similar only in the central portions of their frequency distributions will be interpreted as being the same, even if one is generated as a mixture of end-member distributions and another is a perfect Gaussian distribution.

ACKNOWLEDGMENTS

Our study was supported by grants from the Natural Science and Engineering Research Council of Canada and the Geological Survey of Canada. I thank Ed Krebes for his development of equation (18.2), and Ed Klovan and Bob Dalrymple for their encouragement and detailed discussions. Drs. Peta Mudie and Charles Schafer reviewed an earlier version of this manuscript – many thanks. This manuscript comprises Geological Survey of Canada contribution no. 34589.

REFERENCES

Allen, G. P., Castaing, P., & Klingebiel, A. (1971). Preliminary investigation of the surficial sediments in the Cape Breton Canyon (southwest France) and the surrounding continental shelf. *Marine Geology,* 10: M27–M32.

Armstrong, J. S. (1967). Deviation of theory by means of factor analysis: Tom Swift and his electric factor analysis machine. *American Statistician,* 21: 17–21.

Beall, A. O., Jr. (1970). Textural differentiation within the fine sand grade. *Journal of Geology,* 78: 77–93.

Butler, J. C. (1976). Principal Component Analysis using the hypothetical closed array. *Mathematical Geology,* 8: 25–36.

Castaing, P. (1973). Remarques sur l'utilization de l'analyse factorielle en sédimentologie. *Bulletin Institute Geologie Bassin Aquitaine,* 13: 53–85.

Chambers, R. L., & Upchurch, S. B. (1979). Multivariate analysis of sedimentary environments using grain-size frequency distributions. *Mathematical Geology,* 11: 27–43.

Chayes, F., & Trochimczyk, J. (1978). An effect of closure on the structure of principal components. *Mathematical Geology,* 10: 323–33.

Clague, J. J. (1976). Surficial sediments of the Northern Strait of Georgia, British Columbia. *Geological Survey of Canada,* paper no. 75-1, part B: 151–6.

Clarke, T. L. (1978). An oblique factor analysis solution for the analysis of mixtures. *Mathematical Geology,* 10: 225–41.

Dal Cin, R. (1976). The use of factor analysis in determining beach erosion and accretion from grain size data. *Marine Geology,* 20: 95–116.

Davis, J. C. (1973). *Statistics and data analysis in geology.* New York: John Wiley & Sons, 550 pp.

East, T. J. (1985). A factor analytical approach to the identification of geomorphic processes from soil particle size characteristics. *Earth Surface Processes and Landforms,* 10(5): 441–63.

(1987). A multivariate analysis of the particle size characteristics of regolith in a catchment on the Darling Downs, Australia. *Catena,* 14: 101–18.

Ehrenberg, A. S. C. (1962). Some questions about factor analysis. *Statistician,* 12: 191–208.

Full, W. E., & Ehrlich, R. (1986). Fundamental problems associated with "eigenshape analysis" and similar "factor" analysis procedures. *Mathematical Geology,* 18: 451–63.

Full, W. E., Ehrlich, R., & Bezdek, J. C. (1982). FUZZY QMODEL – A new approach for linear unmixing. *Mathematical Geology*, 14: 259–70.

Full, W. E., Ehrlich, R., & Klovan, J. E. (1981). EXTENDED QMODEL – Objective definition of external end members in the analysis of mixtures. *Mathematical Geology*, 13: 331–44.

Glaister, R. P., & Nelson, H. W. (1974). Grain-size distributions an aid in facies identification. *Bulletin of Canadian Petroleum Geology*, 22: 203–40.

Grace, J. T., Grothaus, B. T., & Ehrlich, R. (1978). Non-normal size frequency distributions taken from within sand laminae. *Journal of Sedimentary Petrology*, 48: 1193–202.

Imbrie, J., & Purdy, E. G. (1962). Classification of modern Bahamian carbonate sediments. *American Association of Petroleum Geologists*, Memoir 1: 252–72.

Jøreskog, K. G., Klovan, J. E., & Reyment, R. A. (1976). *Geological factor analysis*. Amsterdam: Elsevier, 178 pp.

Kendall, M. G., & Stuart, A. (1969). *The advanced theory of statistics, vol. 1: Distribution theory* (3rd ed.). New York: Hafner Publishing, 439 pp.

Klovan, J. E. (1966). The use of factor analysis in determining depositional environments from grain-size distributions. *Journal of Sedimentary Petrology*, 36: 115–25.

Klovan, J. E., & Imbrie, J. (1971). An algorithm and FORTRAN IV program for large-scale Q-mode factor analysis and calculation of factor score. *Mathematical Geology*, 3: 61–7.

Klovan, J. E., & Miesch, A. T. (1976). Extended CABFAC and QMODEL computer programs for Q-mode factor analysis of compositional data. *Computers and Geosciences*, 1: 161–78.

Matalas, N. C., & Reiher, B. J. (1967). Some comments on the use of factor analyses. *Water Resources Research*, 3: 213–23.

Miesch, A. T. (1976). Q-mode factor analysis of geochemical and petrologic data matrices with constant row-sums. *U.S. Geological Survey*, prof. paper no. 574 G.

Perillo, G. M. E., & Marone, E. (1986). Application of maximum entropy and optimal numbers of class interval concepts: Two examples. *Mathematical Geology*, 18: 465–77.

Rummel, R. J. (1967). Understanding factor analysis. *Journal of Conflict Resolution*, 11: 444–80.

Solohub, J. T., & Klovan, J. E. (1970). Evaluation of grain-size parameters in lacustrine environments. *Journal of Sedimentary Petrology*, 40: 81–101.

Swan, D., Clague, J. J., & Luternauer, J. L. (1978). Grain-size statistics: Evaluation of the Folk and Ward graphic measures. *Journal of Sedimentary Petrology*, 48: 863–78.

Syvitski, J. P. M. (1984). Q-mode Factor Analysis of grain size distributions. Geological Survey of Canada, Open File Report no. 965, 37 pp.

Syvitski, J. P. M., Asprey, K. W., Clattenburg, D. C., & Hodge, G. D. (1985). The prodelta environment of a fjord: Suspended particle dynamics. *Sedimentology*, 32: 83–107.

Temple, J. T. (1978). The use of factor analysis in Geology. *Mathematical Geology*, 10: 379–87.

Wallis, J. R. (1968). Factor analysis in hydrology: An agnostic view. *Water Resources Research*, 4: 521–7.

Yorath, C. J. (1967). Determination of sediment dispersal patterns by statistical and factor analysis northestern Scotian Shelf. Ph.D. Thesis, Queen's University, Ontario, Canada.

19 Experimental–theoretical approach to interpretation of grain size frequency distributions

Supriya Sengupta, J. K. Ghosh,
and B. S. Mazumder

Introduction

The conventional method of interpretation of grain size data, popular among sedimentologists since the late 1950s, is based on the method of $\log(\phi)$ probability plots. Grain size distributions of sediment populations transported in nature generally show up as segmented curves when the cumulative frequencies are plotted against logarithms of grain sizes. Phi values of grain sizes ($\phi = -\log_2 D$, where D is grain diameter in mm [Krumbein, 1936]) are generally used for this purpose. A linear phi-probability plot being indicative of a lognormal distribution, the segmented cumulative curves are interpreted as mixtures of several (truncated) lognormal grain size distributions, each corresponding to a particular mode of transportation. It is further assumed that the segments having coarse, medium, and fine modal sizes were transported as bed-, saltation, and suspension loads, respectively. This method of interpretation has developed out of the studies by Inman (1949), Sindowski (1957), Moss (1963), Spencer (1963), and Visher (1965, 1969). The history of the development of the ideas leading to this popular technique of grain size interpretation has been summarized by Visher (1969).

Of late, the validity of this method of interpretation has raised questions in some minds. Cumulative grain size distributions of samples collected wholly from bed or suspension loads, for example, have been shown to plot as segmented curves on probability paper (Sengupta, 1975a,b). It has also been recognized that one-to-one correlation between modes of sediment transportation and segments of log probability plots are based on simplistic assumptions (Ko-

mar, 1986). In some cases moreover, grain size distributions of natural sediments are found to fit log-hyperbolic rather than lognormal models (Bagnold & Barndorff-Nielsen, 1980; Christiansen et al., 1984).

The purpose of the present work is to re-examine the validity of the conventional method of grain size interpretation based on log probability plots, and also to investigate the relationship between the lognormal and log-hyperbolic models. For this purpose, samples collected on earlier occasions from natural as well as artificial (experimental) channels have been used. Mathematical techniques are presented:

1. for direct computation of suspension loads from bed's grain size distribution, and

2. for estimation of range of flow parameters for lognormally distributed suspended loads.

This review presents work done by the authors on grain size problems. Review of the experimental–theoretical studies conducted by other authors is beyond the scope of this presentation.

Observations on ancient and modern fluviatile sediments

Grain size distributions of the Kamthi sandstone samples belonging to the Permo-Triassic, fluviatile Gondwana deposits of India were obtained. Phi probability plots of these size spectra show segmented patterns. Following the conventional approach, the coarse and fine fractions of each of these phi probability plots were respectively interpreted as the bed- and suspension loads of the ancient Kamthi River (Sengupta, 1967, 1970). The plot of a representative Kamthi sample together with the interpreted segments is shown in Figure 19.1.

Grain size distributions of sand samples collected from a present-day river bed, the Usri River in Bihar, India, were examined at various distances from the source. In this case phi probability plots of the source material were irregular (Fig. 19.2), but linearity in these plots developed with transportation. In the samples collected about 44 km downstream a straight line could be fitted through the plots (Fig. 19.3). It was assumed that sediments derived from a nonlognormal source may develop lognormality

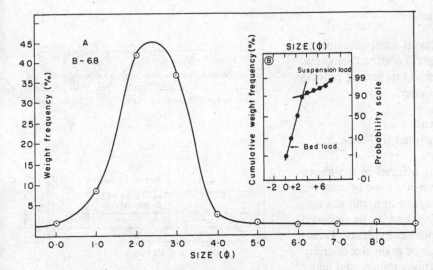

Figure 19.1. (A) Grain size frequency distribution of a representative sandstone sample from the Kamthi Formation, Godavari valley, India. (B) Log(φ) probability plot of cumulative grain size frequency distribution of the same sample. (Reproduced from Sengupta [1977, Fig. 1, p. 96] with permission.)

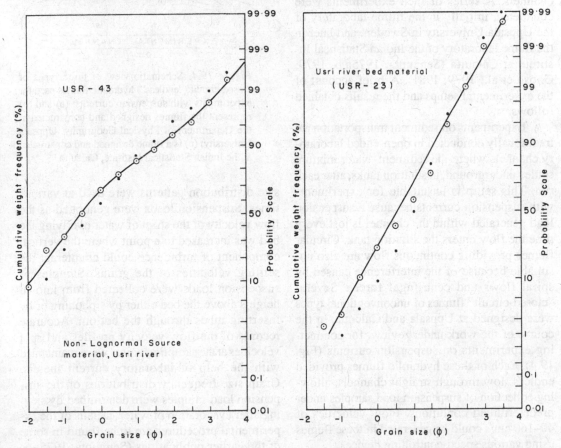

Figure 19.2. Cumulative grain size frequency distribution of the Usri River bed sample near the source area, Bihar, India: ○, log(φ) probability plot; •, log-hyperbolic plot.

Figure 19.3. Cumulative grain size frequency distribution of the bed sample collected at a distance of 44 km from the source area of the Usri River, Bihar, India: ○, log(φ) probability plot; •, log-hyperbolic plot.

under conditions of fluvial transportation by means of a process of grain sorting (Sengupta, 1975b). The mechanism of this sorting process is presently being investigated.

Flume experiments
Experimentally generated grain size distributions

The interpretations offered in the above studies were essentially speculative because the mode of sediment transportation and the mechanisms of size sorting had to be deduced from various lines of indirect evidence. Direct information on the nature of grain size distribution of sediment populations transported under predetermined conditions could be obtained only through controlled experiments in laboratory channels. A series of such experiments were conducted, initially in the flume laboratory of the Uppsala University in Sweden, and later in the flume laboratory of the Indian Statistical Institute at Calcutta (Sengupta, 1975a,b, 1979; Ghosh et al., 1979, 1981). A brief account of the experimental setups and the results obtained follows.

Experiments on sediment transportation are traditionally conducted in open-ended laboratory channels where the sediment–water mixture settles underground, in siltation tanks, after each run. This setup is unsuitable for experiments with suspension currents because a suspension load generated within the channel is lost every time the flow enters the siltation tank. Circular flumes providing continuous flow are also unsuitable because of the interference caused by spiral flows and centrifugal forces. Several "closed-circuit" flumes of unconventional types were designed at Uppsala and Calcutta, in the course of the work under review, for conducting experiments on suspension currents (Fig. 19.4). Each of these hydraulic flumes provided endless flow through straight channels, allowing collection of suspension load samples under predetermined conditions. Flow velocities of 40–160 cm/s could be generated in these flumes using various speed-controlling devices.

Experiments on suspension transportation were performed by placing sand mixtures of known grain size distributions on the beds of the flumes. Bed materials of six different grain

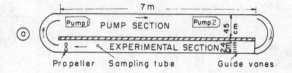

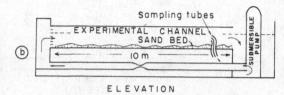

Figure 19.4. Schematic view of three types of closed-circuit ("endless") hydraulic flumes used for experiments with suspension currents: (a) and (b) represent the flumes designed and constructed at the Department of Physical Geography, Uppsala University; (c) is a flume designed and constructed at the Indian Statistical Institute, Calcutta.

size distribution patterns were used at various times. Suspension loads were generated as the flow velocity of the sheet of water overlying the bed was increased to a point where the vertical component of turbulence could counteract the settling velocities of the grains. Samples of suspension loads were collected from known heights above the bed either by siphoning or by inserting tubes through the bottom. Accurate records of the flow-velocity profiles and spot velocities at the sampling points were maintained with the help of laboratory current meters. Grain size frequency distributions of the suspension load samples were determined by sieving after evaporation of water. Details of the experimental procedures are to be found in some of the earlier publications (Sengupta, 1975a,b, 1979). Following is a summary of the major experimental results.

Grain size distributions of four of the sand beds used for experiments, together with the

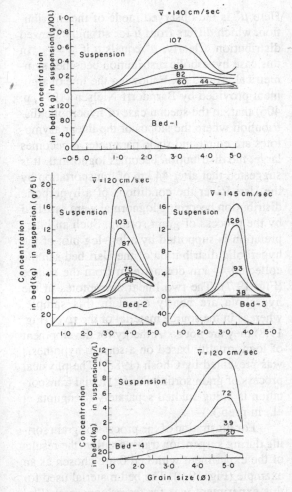

Figure 19.5. Grain size frequency distributions of the suspended loads at a height of about 20 cm for different flow velocities, above bed materials of four different grain size distributions. (Redrawn from Sengupta [1979] with permission from *Sedimentology*.)

corresponding suspension loads at various flow velocities, are shown in Figure 19.5. To facilitate comparison, grain size distributions of the samples collected from comparable heights above the bed are represented in this figure. With increase of flow velocity, as also with decrease of the bed's average grain size, the total amount of materials in suspension markedly increased. Still, irrespective of the actual grain size distribution of the bed, the modal size of the suspended load in each case ranged between medium sand and coarse silt. Clearly, the coarsest as well as the finest fractions were eliminat-

ed from suspension, compared to the medium size fraction. In no case was any visual resemblance between the size distribution patterns of the bed material and the suspended load noticed.

The chances of "armouring" (fine grains hiding within intergranular space) are eliminated because at high flow velocities the whole bed migrated downstream in the form of ripples, thereby exposing each grain to the water flow above. Because of their weight, the coarsest grains available in the bed were not eroded; the finest ones resisted erosion due to their homogeneity. The proportion of bed materials that went into suspension steadily increased with decrease of grain size, but the concentration of fines in suspension hardly ever reached 100% of what was available in the bed. The negatively skewed grain size distribution that developed in suspension out of this process of grain sorting (Fig. 19.6), became symmetrical after logarithmic (phi) transformation, giving the impression of lognormality.

The results of these experimental studies, discussed in detail elsewhere (Sengupta, 1979), led to the conclusion that, for a given bed material, the suspension load's grain size distribution is related to flow velocity and height of suspension above the bed. Lognormality of grain size distribution in suspension develops only when the combination of bed material, flow velocity, and height of suspension above the bed are appropriate.

Lognormal versus log-hyperbolic models

Grain size frequency distributions of the source materials in recent rivers, as exemplified by the samples from the Usri River (Fig. 19.2), are neither lognormal nor log-hyperbolic. Lognormality in the Usri River bed samples developed only after prolonged downstream transportation, as shown by a fairly good linear fit in Figure 19.3.

The "hyperbolic distribution" refers to a distribution for which a plot of the probability density function with a logarithmic scale on the ordinate axis is a hyperbola. Thus, denoting the size variable by ϕ and the density function by $S_{k_b}(\phi)$, the equation of the curve is of the form (Barndorff-Nielsen, 1977, p. 406):

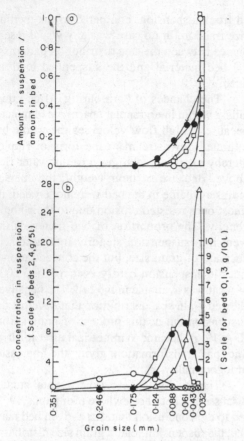

Figure 19.6. (a) Proportion of bed materials of different grain sizes in suspension at a height of 20 cm and at flow velocities varying between 72 and 93 cm/s. (b) Suspension concentration above the five different beds used for experiments. Grain sizes are plotted in arithmetic scale in both the diagrams. Explanation of symbols: ○, bed 0; X, bed 1; ●, bed 2; □, bed 3; △, bed 4. (Reprinted from Sengupta [1979] with permission from *Sedimentology*.)

$$\log S_{ks}(\phi) = v - \tfrac{1}{2}(\gamma_1 + \gamma_2)\sqrt{\delta^2 + (\phi - \mu)^2}$$
$$+ \tfrac{1}{2}(\gamma_1 - \gamma_2)(\phi - \mu) \qquad (19.1)$$

where γ_1 and γ_2 are the slopes of two linear asymptotes of hyperbola, μ is the abscissa of the intersecting point of these two asymptotes, the ordinate v equals the logarithm of the norming constant, which is adjusted to agree with the observed frequency, and scale parameter δ (>0) can be expressed as

$$\delta = \frac{2(\mu^* - \mu)\sqrt{\gamma_1\,\gamma_2}}{\gamma_1 - \gamma_2} \qquad (19.2)$$

Here μ^* is the observed mode of the distribution, which differs from μ for strongly skewed distribution. Clearly, μ^* equals μ if $\gamma_1 = \gamma_2$. In this case hyperbolic distribution becomes symmetrical around μ^*. Following the line of argument provided by Barndorff-Nielsen (1977, p. 406) that, in the special case of hyperbolic distribution where the slopes of the linear asymptotes are equal and scale parameter δ becomes large, the distribution becomes lognormal. It is suggested that after 44 km of transportation by the Usri River the conditions of a hyperbolic distribution becoming lognormal were satisfied by the process of grain sorting. Such an interpretation is supported by a log–log plot of the hyperbolic distribution of the Usri bed sample collected 44 km downstream from the source (Fig. 19.7). The two linear asymptotes of the hyperbola are nearly equal in this sample, whereas in the source material of the river (Fig. 19.8) they are not so. A theory for development of lognormality based on a sorting hypothesis was presented by Ghosh (1988). The physical process of grain sorting during fluvial transportation is being studied separately (Sengupta et al., in prep.).

For illustration of the process of grain sorting during suspension transportation, the results of the experiment with bed 0 are chosen as an example (Fig. 19.9). The bed material used for this experiment was far from lognormal (Fig. 19.10), but the suspension load – particularly at a flow velocity of 146 cm/s – was nearly perfectly lognormal. The log(ϕ) probability plot for the suspension data is linear, excepting only the finest fraction (Fig. 19.11). Following Barndorff-Nielsen's line of approach it is suggested that lognormality developed in suspension because two asymptotes of the hyperbola became nearly equal ($\gamma_1 = 4.2$, $\gamma_2 = 3.8$; see Table 19.1) when the flow velocity was high. Elimination of the very coarse and fine end members during suspension transportation by the process discussed earlier led to the development of symmetrical hyperbolic distribution.

Relationship between flow parameters and grain size distributions

Cumulative grain size distributions of the suspended load samples, particularly those col-

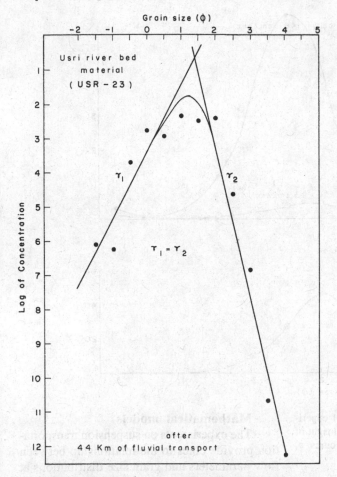

Figure 19.7. Log–log plot of the hyperbolic distribution of the Usri River bed sample collected at a distance of 44 km from the source area (data same as in Fig. 19.3).

lected at low flow velocities, have segmented appearance in log probability plots. A good example of this is the suspension sample above bed 0 at 93 cm/s (open circles in Fig. 19.12). The distribution at this stage, however, seems to be log-hyperbolic because a straight line can be fitted through all the hyperbolic plots (filled circles in Fig. 19.12). With increase of flow velocity, as larger quantities of the medium-sized particles went into suspension and the coarse and fine end members were eliminated, the dis-

Figure 19.8. Log–log plot of the hyperbolic distribution of the Usri River bed sample (USR 43) collected from the source area (data same as in Fig. 19.2). Note: –log2 has been used on both the coordinates in all diagrams of hyperbolic plots for the sake of uniformity.

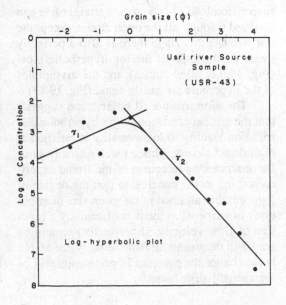

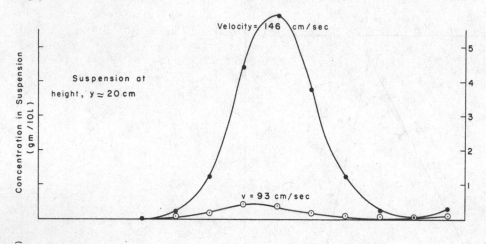

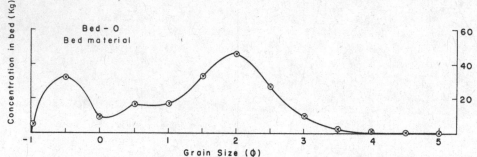

Figure 19.9. Grain size distributions of experimental bed 0 and suspension loads at a height of about 20 cm, at two different flow velocities: ○, 93 cm/s; •, 146 cm/s.

tribution became nearly perfectly lognormal. In Figure 19.11, for example, which represents the suspension load at 146 cm/s, a straight line can be fitted through all the open circles except the last two. At this stage there is also a perfectly linear fit through all the log-hyperbolic plots (Fig. 19.11, filled circles), and the asymptotes of the hyperbola are nearly equal (Fig. 19.13).

The aforementioned observation suggests that the critical condition of log-hyperbolic distribution leading to lognormality is satisfied at high flow velocity. This, as well as many similar observations in course of the flume experiments, led to the conclusion that the degree of lognormality attained in the grain size distributions of suspension loads is essentially a function of flow velocity. Other factors remaining constant, the greater the flow velocity, the higher is the chance the particles in suspension will be lognormally distributed.

Mathematical models

The experiments on suspension transportation provide clues to the relationship between flow parameters and grain size distributions at known heights above given bed materials. Following these clues, mathematical models have been developed for predicting the grain size distributions in suspension at a given height when the flow parameters and the bed's grain size composition are known. This also affords estimation of palaeoflow velocity for ancient sedimentary deposits under certain assumptions, provided some knowledge of the bed is available. The methods developed have been discussed in detail in some of the earlier publications (Ghosh et al., 1979, 1981, 1986; Ghosh & Mazumder, 1981). A summary of these methods and their applications is given below.

A method for direct computation of suspension load

A mathematical model based on the diffusion equations of sediment and water with some modifications of Hunt's (1954) method is developed for estimating the suspended sediment

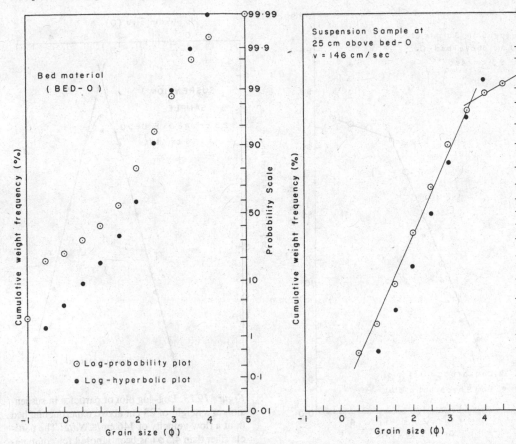

Figure 19.10. Cumulative grain size frequency distribution of experimental bed 0: ○, log(φ) probability plot; ●, log-hyperbolic plot.

Figure 19.11. Cumulative grain size frequency distribution of suspension load at 25 cm above experimental bed 0, at a flow velocity of 146 cm/s: ○, log(φ) probability plot; ●, log-hyperbolic plot.

concentration for given bed concentration and flow parameters. In a uniform flow where the concentration varies only in the vertical coordinate y, and the diffusion coefficients of sediment and water are assumed to be equal (i.e., $\varepsilon_s = \varepsilon_m$), the formula for computing the suspended sediment concentration (S_y) is given by Ghosh et al. (1979, 1981) as

$$\log_e\left[\frac{S_y(1 - S_{k_s})}{S_{k_s}(1 - S_y)}\right] = K_1 \log_e\left[\frac{(d - y_1)}{(d - k_s)}\right]$$

$$+ K_2 \log_e\left[\frac{\dfrac{(1 - y/d)^{1/2}}{(y_1/d)^{1/2}}[B - (1 - y_1/d)^{1/2}]}{B - (1 - y/d)^{1/2}}\right] \quad (19.3)$$

where

$$K_1 = \frac{c(\phi)du_{y_1}}{u*^2(y_1 - k_s)}, \qquad K_2 = \frac{c(\phi)}{B\chi u*} \quad (19.4)$$

(see the "List of symbols" for explanation), and hence the relative concentration S'_y is found out by

$$S'_y = \frac{S_y(\phi)}{\sum_\phi S_y(\phi)} \quad (19.5)$$

With the help of equation (19.5), it is possible to calculate the suspension concentration at any height $y \geq y_1$ above the bed if the relative concentration S_{k_s} of sediment of the bed is known. It is also possible to compute the average concentration of sediment of different sizes in the bed layer by inserting $y = y_1$ in equation (19.5).

Applications

The efficiency of the theoretical model developed above has been verified by comparing the observed and computed suspended load con-

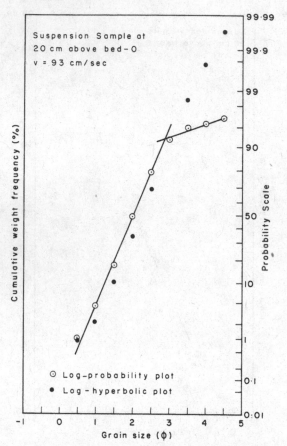

Figure 19.12. Cumulative grain size frequency distribution of suspension load at 20 cm above experimental bed 0, at a flow velocity of 93 cm/s: ○, log(ϕ) probability plot; •, log-hyperbolic plot.

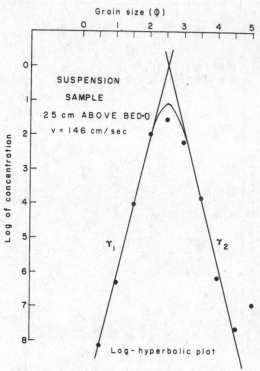

Figure 19.13. Log–log plot of particles in suspension at a height of 25 cm above experimental bed 0, at a flow velocity of 146 cm/s. *Note:* The particle finer than 4.5 ϕ has been ignored for computational purposes.

Table 19.1. *Parameters of hyperbolic density function of experimental bed 0 and its suspension loads*

Sample	γ_1	γ_2	δ	μ
Bed 0	1.1484	4.3151	0.1487	2.3058
Suspension at:				
$v = 93$ cm/s				
(II-20-48)	3.3242	4.0809	0.8722	2.3396
$v = 146$ cm/s				
(III-25-1)	4.1729	3.8194	0.8109	2.4641

centrations over three different sand beds (beds 2, 3, and 5). The suspended load data were obtained from experiments conducted at Uppsala and Calcutta in hydraulic flumes of two different designs. The results were also compared to

the values obtained with the help of the well-known equation proposed by Rouse (1938). The result of one such comparison conducted for the suspension load above bed 3 is presented in Figure 19.14. It has been shown elsewhere (Ghosh et al., 1981) that the method developed is as good as any other method for computation of suspended load. The present method, moreover, has a distinct advantage over others available because, unlike the others, it allows direct computation of suspended load from the bed without going through an intermediate stage – namely, computation of the bedload.

The applicability of this technique of calculation of suspended load has been further tested in a natural stream (suspension load data of the Niobrara River, Nebraska [Colby & Hembree, 1955, p. 134, table 6]). The result of comparison of the computed and observed suspended load grain size distribution is presented in Figure 19.15. The distribution patterns, particularly

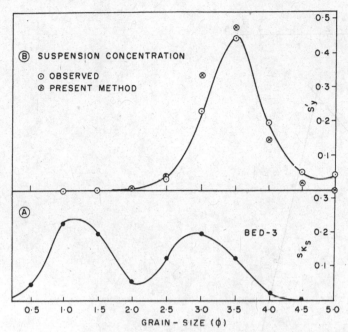

Figure 19.14. (A) Grain size distribution (relative concentration, S_{k_s}) in experimental bed 3. (B). Grain size distribution in suspension S'_y (at y = 17.5 cm). ⊙, observed; ⊗, computed by a method developed by Ghosh et al. (1981). (Relevant portions reproduced from Ghosh et al. [1981, Fig. 3, p. 784] with permission from *Sedimentology.*)

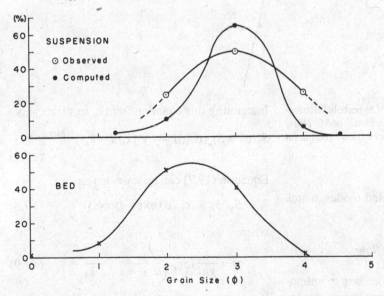

Figure 19.15. Grain size frequency distributions in the bed and in the suspension load in the Niobrara River, Nebraska, USA, at sample station no. 22, on June 19, 1952 (data from Colby & Hembree [1955, table 6]): ×, observed concentration in the bed; ○, observed concentration in suspension at y = 33.53 cm, v = 151.5 cm/s; ●, suspension concentration computed by a method developed by Ghosh et al. (1981).

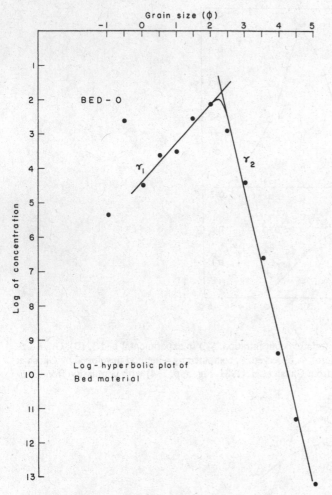

Figure 19.16. Log–log plot of hyperbolic distribution of particles in the experimental bed 0. *Note:* Particles coarser than 0ϕ have been ignored for computational purposes.

the observed and the calculated modes, match well.

A method for estimating flow parameters

Theory: The conditions leading to unimodality, symmetry, and lognormality in suspension load for a given hyperbolic bed distribution have been deduced by Ghosh & Mazumder (1981) from the well-known Rouse equation.

Under equilibrium conditions, the conventional Rouse (1938) equation is given by

$$\varepsilon(y)(dS_y/dy) + c(\phi)S_y = 0 \qquad (19.6)$$

Integrating this equation from k_s to y, one gets

$$S_y(\phi) = S_{k_s}(\phi)[(d - y/y)(k_s/d - k_s)]^{c(\phi)/\chi\mu_*} \qquad (19.7)$$

Equation (19.7) can be rewritten as

$$S_y(\phi) = S_{k_s}(\phi) \exp(-\eta c(\phi)) \qquad (19.8)$$

where

$$\eta = \frac{1}{\chi u_*} \log_e \left(\frac{y}{d - y} \cdot \frac{d - k_s}{k_s}\right) \qquad (19.9)$$

is a parameter summarizing the effect of flow. Hence the expression S'_η for relative suspension concentration distribution at any height η (varying with y) is given by

$$S'_\eta(\phi) = \frac{S_\eta(\phi)}{\sum_\phi S_\eta(\phi)} = S_{k_s}(\phi)\, e^{-\eta c(\phi) + g(\eta)} \qquad (19.10)$$

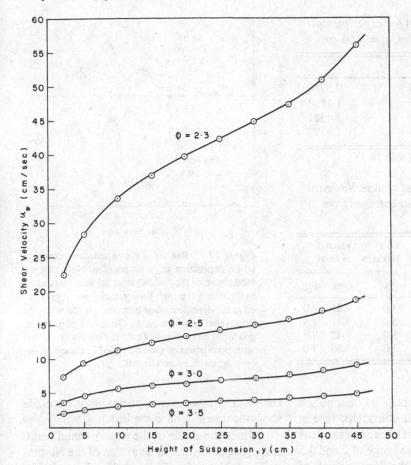

Figure 19.17. Relationship between shear velocity u_* and height of suspension y above experimental bed 0. Phi values on the curves indicate modal grain sizes of the suspension loads.

where

$$e^{-g(\eta)} = \sum_\phi S_{k_s}(\phi)\, e^{-\eta c(\phi)} \qquad (19.11)$$

While studying the question of unimodality of $S'_\eta(\phi)$, it is noted that $\log_e c(\phi)$ is approximately linear in ϕ; that is, $\log_e c(\phi) \approx a + b\phi$, $1 \le \phi \le 5$, where $a = 3.5758$ and $b = -1.1840$.

Let $\psi_\eta(\phi) = \log_e S'_\eta(\phi)$ and $\psi_{k_s}(\phi) = \log_e S_{k_s}(\phi)$; equation (19.10) can be written as

$$\psi_\eta(\phi) = \psi_{k_s}(\phi) - \eta c(\phi) + g(\eta) \qquad (19.12)$$

Now $\psi_\eta(\phi)$ may be expected to be unimodal if one can find a unique solution to the equation for the mode $\hat\phi$. Then, from equation (19.12),

$$(d\psi_\eta/d\phi)\big|_{\hat\phi} = (d\psi_{k_s}/d\phi)\big|_{\hat\phi} - \eta b e^{a+b\phi} = 0 \qquad (19.13)$$

Since

(I) $b < 0$, equation (19.13) can have a solution only when $d\psi_{k_s}/d\phi < 0$ at $\phi = \hat\phi$, and

(II)   the solution of equation (19.13) is unique if $d^2\psi_\eta/d\phi^2 \le 0$ at $\hat\phi$ for $2 \le \hat\phi \le 4$, the range within which the solution is sought.

It is clear that a sufficient condition for (II) is $d^2\psi_{k_s}/d\phi^2 \le 0$ at $\hat\phi$ for $2 \le \hat\phi \le 4$.

The condition holds if S_{k_s} is normal or hyperbolic (Barndorff-Nielsen, 1977) with mode to the left of 2. For such $S_{k_s}(\phi)$ the set of values of η for which equation (19.13) has a solution will determine the region of unimodality. From the hyperbolic plot of bed 0 (Fig. 19.16), it follows from the condition (I) that equation (19.13) can have a solution only to the right of $\phi = 2$ and the condition (II) also holds for $\phi \ge 2.3$. Whenever equation (19.13) has a solution, it is unique, so that the corresponding $S'_\eta(\phi)$ is unimodal. Further details are provided below.

Table 19.2. ϕ versus η for experimental bed 0, where $d = 50$ cm, $k_s = 0.06$ cm

	ϕ	η
1.	2.3	0.3980
2.	2.5	1.1863
3.	3.0	2.4321
4.	3.5	4.4373

Note: $\phi = 2.3 = \mu$ (see Table 19.1).

Table 19.3. *Observed values of u_* in Niobrara River compared to the values obtained from Figure 19.15*

Tab. no.[a]	Sta. no.	Sampling date	Obs. height y (cm)	Obs. mode in suspens. (ϕ)	Value of u_* from Obs.	Value of u_* from Fig.
6	135	15 Oct. '49	36.58	3.0	9.21	8.0
6	22	19 June '52	33.53	3.0	9.21	7.5
28	44	20 May '53	33.53	3.0	9.42	7.5
28	102	20 May '53	27.43	3.0	7.86	7.0

[a]Colby & Hembree (1955).

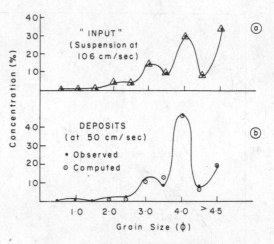

Figure 19.18. Results of the experiments conducted on deposition from suspension. Note that the modal size of the "input" injected into suspension in the flume at a high flow velocity was represented in the deposit formed from suspension when the flow velocity was reduced. (Relevant portions reproduced from Ghosh et al. [1986, Fig. 6, p. 61] with permission of the Society of Economic Paleontologists & Mineralogists.)

Applications: The suspension distribution above experimental bed 0 is considered here in detail as an illustration of these ideas. This study is restricted to the size range of 2 and 3.5 ϕ, because it is clear from the derivation (I) of equation (19.13) and the bed's grain size distribution that equation (19.13) can have a solution only for $\phi \geq 2$, and the observed data with $\phi \geq 4$ is likely to be scarce.

For fixed ϕ in the range 2–3.5 for bed 0, equation (19.13) is solved for η. The results are shown in Table 19.2. The plots of y versus u_* for suspension above bed 0 are shown in Figure 19.17. A family of curves relating y and u_* have been derived by inserting various values of η in equation (19.9). These curves give the values of y and u_* for unimodal suspension distributions with peak at some $2 \leq \hat{\phi} \leq 3.5$. Similar curves for experimental beds 2 and 3 were derived earlier, and it was also shown that these unimodal distributions will be approximately lognormal (Ghosh & Mazumder, 1981). With the help of these curves it is possible to work out the values of u_*, and hence the flow velocity, for given y's when the mode of lognormal grain size distribution in suspension is

known and some assumptions may be made about the distribution in the bed.

Since the mode of the experimental bed 0 is approximately the same as that of the Niobrara River, Nebraska (Colby & Hembree, 1955), and the nature of the distributions are also similar, an attempt was made to test the utility of u_* versus y curves (Fig. 19.17) for predicting the values of u_* from the modal size of the suspension load for given heights. The predicted values are shown against the observed u_* (obtained from $u_* = \sqrt{g\,dJ}$) in the Niobrara River in Table 19.3. The observed and the calculated values match with reasonable degree of accuracy.

For estimation of flow parameters from grain size distributions of ancient sediments by equation (19.13), it is imperative that the grain sizes taken into suspension by running water be also preserved in the deposits formed from suspension at a reduced flow velocity. That this is so has been demonstrated through a series of experiments conducted at Uppsala and Calcutta. It has been shown that all grain sizes present in suspension at a high flow velocity are involved in the process of formation of a bed when the flow velocity is decreased. Figure 19.18, based on the results of the experiments conducted at Calcutta, for example, shows that the modal sizes of the suspension load at a high velocity (106 cm/s) and that of the deposit formed from suspension at a lower velocity (50 cm/s) are the same. A mathematical model for computing the deposition from suspension for a wide range of grain sizes was also developed (Ghosh et al., 1986). With the help of this model it may be possible to work out palaeoflow velocities of suspension currents from grain size parameters of ancient deposits.

Summary and conclusions

It is customary for some sedimentologists to interpret grain size distribution patterns in terms of log probability plots. A straight-line fit through such plots is indicative of lognormality. Cumulative frequency patterns of naturally transported sediments show as segmented curves in log probability plots, and have been interpreted as mixtures of several lognormally distributed populations, each transported by a particular mechanism. It is further assumed that the segments belonging to coarse, medium, and fine modal sizes in log probability plots represent fractions transported as the bed-, saltation, and suspension loads, respectively.

The validity of this technique of grain size interpretation, although popular among sedimentologists, needs restudy. Sediment populations known to have been transported by a single mechanism (either as bed- or suspended load) also show segmented patterns. Segmented nature of the curves vanishes, and the phi probability plots become linear, only after prolonged transportation as bedloads in rivers or as suspended loads in artificial channels under high flow velocity. At this stage the grain size distribution is nearly perfectly lognormal.

Theoretical studies by Barndorff-Nielsen (1977) showed that lognormality may develop as a special case of hyperbolic distribution when the slopes of the linear asymptotes of the hyperbola are equal and a scale parameter δ becomes large. Controlled experiments in laboratory channels indicated that this stage is attained in the suspension loads at a high flow velocity by a process of grain sorting eliminating both coarse and fine fractions from suspension. The mechanism of grain sorting in river channels leading to a similar situation is being studied.

Controlled experiments with suspension currents in specially designed hydraulic flumes have shown that:

1. for a given bed, the suspension load's grain size distribution is related to flow velocity and height of suspension above the bed; and

2. the degree of lognormality attained in the grain size distribution of the suspended load is a function of flow velocity.

Other factors remaining constant, the greater the flow velocity, the higher is the chance the suspended particles will be lognormally distributed. At a lower flow velocity, when the sorting is incomplete, the cumulative grain size distribution may show up as a segmented curve on the log probability plot.

Following the clues obtained from experiments, a mathematical model based on the diffusion equations of sediment and water has been developed for estimating the suspended sediment concentration for a given bed concentration and flow parameter. Following this model it is now possible to calculate the suspension concentration at any height above the bed if the relative concentration of the particle at the bed is known. It is also possible to compute the average concentration of sediment of different sizes in the bed layer. This technique, unlike others, allows direct computation of suspended load from the bed without going through an intermediate (bed-layer) stage. The experimental results seem to indicate that for the purpose of predicting suspension distribution from the distribution in the bed, the sorting at the intermediate bed-layer stage does not play a very important role. Moreover, use of any available models for the bed layer does not improve the prediction at the level of suspension. Much finer models are needed to achieve the improvement. The method allowed computation of heterogeneous suspension loads in natural and artificial channels with a reasonable degree of accuracy.

Theoretical studies on the conditions leading to unimodality, symmetry, and lognormality have led to the construction of a family of curves relating height of suspension y and shear velocity u_* for unimodal suspension distributions, with known peaks, which are approximately lognormal. With the help of these curves it is possible to work out the shear velocities for given heights when the mode of lognormal grain size distribution in suspension is known and some assumptions can be made about the grain size distribution in the bed.

A separate series of experiments, conducted with different types of hydraulic flumes, has shown that the modal sizes of the suspension loads at a high velocity, and that of the deposits formed from suspension at a lower velocity, are the same. Following this clue, and using the set of y versus u_* curves mentioned above, it may be possible to estimate palaeoflow velocities from grain size distributions of deposits preserved in geological record.

ACKNOWLEDGMENTS

The experiments outlined here were initiated by the first author (SS) with the flume facilities provided by the Department of Physical Geography, University of Uppsala. These experiments and theoretical studies were continued at Calcutta with the facilities provided by the Indian Statistical Institute. A portion of the work was done by the second author (JKG) while visiting the University of Purdue, and by the third author (BSM) while visiting the Illinois State Water Survey, Champaign, Illinois. We record our sincere thanks to all these institutions and to Becky Howard for typing the final version of the manuscript.

LIST OF SYMBOLS

B Integrating constant
c Settling velocity of a particle
d Initial depth of water as measured from flume base
D Grain diameter in mm
g Acceleration due to gravity
J Slope or energy gradient of stream
k_s Bed roughness
S_y Concentration of material per unit volume
S_{k_s} Relative concentration distribution at k_s
$S'_y(\phi) = S_y(\phi) / \sum_\phi S_y(\phi)$, relative concentration distribution at height y above the bed
v Time average velocity of water in the flow direction
u_* Shear velocity $\sqrt{\tau_0/\rho}$
u_{y_1} Extrapolated velocity at $y = y_1 = 25$ cm above bed
y Vertical height above the bed

$\gamma_1, \gamma_2, \mu, \delta$ Parameters of hyperbolic distribution
ε_m Water diffusion coefficient
ε_s Sediment diffusion coefficient
μ^* Observed mode of the distribution
ρ Density of water
τ_0 Bottom shear stress
ϕ $-\log_2 D$, where D is size in mm
χ von Kármán constant (0.4)

REFERENCES

Bagnold, R. A., & Barndorff-Nielsen, O. (1980). The pattern of natural size distributions. *Sedimentology*, 27: 199–207.

Barndorff-Nielsen, O. (1977). Exponentially decreasing distributions for the logarithm of particle size. *Proceedings of the Royal Society, London* A353: 401–19.

Christiansen, C., Blæsild, F., & Dalsgaard, K. (1984). Re-interpreting "segmented" grain-size curves. *Geological Magazine*, 121: 47–51.

Colby, B. R., & Hembree, C. H. (1955). Computation of total sediment discharge, Niobrara River near Cody, Nebraska. Geological Survey Water-Supply Paper no. 1357, Washington, D.C.: USGPO, 187 pp.

Ghosh, J. K. (1988). The sorting hypothesis and new mathematical models for changes in size distribution of sand grains. *Indian Journal of Geology*, 60: 1–10.

Ghosh, J. K., & Mazumder, B. S. (1981). Size distribution of suspended particles – unimodality, symmetry, and lognormality. In: *Statistical distributions in scientific work*, vol. 6, eds. C. Taillie et al. Dordrecht: Reidel, pp. 21–32.

Ghosh, J. K., Mazumder, B. S., Saha, M. R., & Sengupta, S. (1986). Deposition of sand by suspension currents: Experimental and theoretical studies. *Journal of Sedimentary Petrology*, 56: 57–66.

Ghosh, J. K., Mazumder, B. S., & Sengupta, S. (1979). Methods of computation of suspended load from bed materials and flow parameters. Flume Project Technical Report no. 1/79. Calcutta: Indian Statistical Institute, 34 pp.

(1981). Methods of computation of suspended load from bed materials and flow parameters. *Sedimentology*, 28: 781–91.

Hunt, J. N. (1954). The turbulent transport of suspended sediment in open channels. *Proceedings of the Royal Society, London*, A224: 322–35.

Inman, D. L. (1949). Sorting of sediment in light of fluvial mechanics. *Journal of Sedimentary Petrology*, 19: 51–70.

Komar, P. D. (1986). "Breaks" in grain size distributions and application of the suspension criterion to turbidities: Reply to a discussion by J. V. Tassell. *Sedimentology*, 33: 437–40.

Krumbein, W. C. (1936). Application of logarithmic moments to size frequency distributions of sediments. *Journal of Sedimentary Petrology*, 6: 35–47.

Moss, A. J. (1963). The physical nature of common sandy and pebbly deposits, part II. *American Journal of Science*, 261: 297–343.

Rouse, H. (1938). *Fluid mechanics for hydraulic engineers* (Dover ed., 1961). New York: Dover, 422 pp.

Sengupta, S. (1967). Grain-size frequency distribution as indicator of depositional environment in some Gondwana rocks. *7th International Sedimentological Congress, England*, Preprint, 4 pp.

(1970). Gondwana sedimentation around Bheemaram (Bhimaram), Pranhita–Godavari valley, India. *Journal of Sedimentary Petrology*, 40: 140–70.

(1975a). Size-sorting during suspension transportation – lognormality and other characteristics. *Sedimentology*, 22: 257–73.

(1975b). Sorting processes during transportation of suspended sediments – An experimental–

theoretical study. UNGI Report no. 40, Uppsala University, Department of Physical Geography, 49 pp.

(1977). Experimental approach to grain size data interpretation. *Indian Journal of Earth Sciences*, S. Ray volume, pp. 95–102.

(1979). Grain-size distribution of suspended load in relation to bed materials and flow velocity. *Sedimentology*, 26: 63–82.

Sengupta, S., Maji, A. K., Mukhopadhyay, A. K., & Sarkar, J. (in prep.), Downstream changes in grain size and roundness in Dwarakeswar River, India, and their implications.

Sindowski, K. H. (1957). Die synoptische Methods des Kornkurben – Vergleiches zur Ausdeutung fossiler Sedimentationsraume. *Geologische Jahrsbuch*, 73: 235–57.

Spencer, D. W. (1963). The interpretation of grain size distribution curves of clastic sediments. *Journal of Sedimentary Petrology*, 33: 180–90.

Visher, G. S. (1965). Fluvial processes as interpreted from ancient and recent fluvial deposits. In: *Primary sedimentary structures and their hydrodynamic interpretations*, ed. G. V. Middleton. Society of Economic Paleontologists and Mineralogists, Special Publication no. 12, pp. 116–32.

(1969). Grain size distributions and depositional processes. *Journal of Sedimentary Petrology*, 39: 1074–106.

V *Applications*

There are myriad ways that particle size data are used daily by the earth science and oceanographic community. The data may be used to predict where flooding will occur on the banks of rivers, or how oil-contaminated sediment will move along a beach. It has been used to understand how toxic chemicals dumped into estuaries and fjords are deposited and trapped (Syvitski et al., 1987). Particle size data provide important clues to the deposition of precious heavy minerals, such as gold and titanium (ilmenite, rutile), along our rivers and beaches (Slingerland & Smith, 1986), and are commonly used in the production of geological maps and stratigraphic sections (Pickering et al., 1990). Particle size information has been used to define benthic habitats of bottom dwellers (Pearson & Rosenberg, 1978), and the positioning of anchorages for large ships. It can also play an important role in submarine warfare.

Within the many applications for particle size data, we highlight five central themes: stratigraphy, glacial geology, geochemistry, oceanography, and geotechnical studies. Chapter 20 applies the methodology of suite statistics (defined in Chapter 16) to the interpretation of coastal stratigraphy and sea-level fluctuations, in both modern and ancient geological settings. The method is used to identify both the dominant and previous transport agency or sedimentary environment, as well as evidence for small changes in sea level.

Chapter 21 uses size sequence data to understand the glacial and paraglacial transport and partitioning of sediment. A sedimentary sequence refers to the variable deposits that can be related by their stratigraphic position and generalized sedimentary environment. A mass balance approach is used to understand how sediments are cycled from land to the sea in a glacial fjord environment.

Chapter 22 reviews how marine chemists use particle size data to understand the fluxes, cycles, budgets, sources, and sinks of chemical elements in nature. Particle size distributions are an intimate reflection of, or controlling factor in, most of these geochemical processes. Four case histories are used to detail the effects of mineralogy, surface adsorption effects, provenance, and diagenesis. The chapter covers most of the marine environments, including a glacial fjord, a mixed estuary, the continental shelf, and the deep ocean.

Many geological and biological processes in the sea are concerned with the behavior and fate of particles. Chapter 23 outlines how these nonfluid components of the ocean are linked by complex physical processes. The role of flocculation is highlighted and described in terms of a floc-grain settling model.

Finally, Chapter 24 provides an eloquent description of the mass physical properties of seafloor sediments and how grain size data play a dominant role in our understanding of the marine geotechnical environment. Particle size is discussed as an influence on sedimentary fabric and mass physical properties, including consolidation and shear strength of sediment.

REFERENCES

Pearson, T. H., & Rosenberg, R. (1978). Macrobenthic succession in relation to organic enrichment and pollution in the marine environment. *Marine Biological Annual Review,* 16: 229–311.

Pickering, K. T., Hiscott, R. N., & Hein, F. J. (1990). *Deep marine environments.* London: Unwin Hyman, 416 pp.

Slingerland, R., & Smith, N. D. (1986). Occurrence and formation of water-laid placers. *Annual Review of Earth and Planetary Science,* 14: 113–47.

Syvitski, J. P. M., Burrell, D. C., & Skei, J. M. (1987). *Fjords: Process and products.* New York, Springer–Verlag, 379 pp.

20 Application of suite statistics to stratigraphy and sea-level changes

WILLIAM F. TANNER

Introduction

The methodology underlying the work described in this chapter is given by Tanner in Chapter 16. Development of the basic procedures has at all times been coupled with field and laboratory work on modern sediments, in a variety of environments. Over the course of almost fifteen years, the data bank has grown greatly. The plots and other devices that have evolved have been based on that data bank. These methods are considered to represent an improvement over traditional procedures (interpreting one sample at a time). Because the suite of samples contains more useful information than can a single sample, the results obtained by using suite parameters should provide better, more detailed answers.

Certain standards have been observed: laminar sampling (where possible), precision sieving, ≤ 100 g of field sample, not more than one split, ≤ 50 g for sieving using 0.25ϕ screens, and 30-min shaking time (Socci & Tanner, 1980). All of the work has been focused on quartz-rich clastics in the coarse silt–fine gravel size range; sediments made of chemically mobile materials are not discussed here.

Modern environments

Many modern environments have been studied using suite statistics: inner continental shelves, beaches, beach ridges, coastal dunes, interior dunes, river channels, tidal flats, delta fronts, aeolian nondune deposits, and others. A few deposits are problematical, even though they accumulated in late Holocene time and have not been altered very much since deposition. Except for the latter group, all of the modern examples are of known (observed) origin; thus the results of using suite methods are necessarily correct in the simplest, most obvious cases

(criteria were based on these cases). Only one easy example will be given, followed by a presentation of a few marginal and perhaps contradictory examples.

Cape San Blas

Cape San Blas is located on the Gulf of Mexico coast of Florida southwest of the city of Tallahassee and southwest of the village of Apalachicola. All samples were taken from the mid-tide position on the modern (medium-energy) sand beach. The suite means (and standard deviations) of sample means and standard deviations were 2.517ϕ (0.175ϕ) and 0.368ϕ (0.046ϕ). This is a well-sorted and homogeneous suite.

Suite statistics parameters are summarized in Table 20.1. If this were a set of samples from a lithified rock exposure, one should conclude without hesitation that the environment of deposition was open beach (not lake, lagoon, estuary, or any other small closed basin). One might also wish to infer that there was a river influence of some kind, perhaps earlier in the history of the sand pool; but it is clear from Table 20.1 that these sands do not represent a river deposit, and there is no specific unique evidence for a river source. The single dune item should not be given much weight either, even though wind work is common on beaches: The variability diagram merely does not distinguish (in many instances) between beach and dune.

Medano Creek

Medano Creek flows along the eastern edge of the modern sand dune field, in Great Sand Dunes National Monument, near Alamosa, Colorado, at an altitude of >2.3 km. Dune sand is blown by the wind along what is generally a west-to-east path. Therefore the creek has two main sources of sediment: the dunes to the west and the mountain range to the north and east. The creek bed was sampled at 100-m intervals, producing fifteen samples in what was presumed to be a single suite. The basic measures for the suite (suite mean and standard deviation) follow: diameter 2.195ϕ (0.218ϕ), standard deviation 0.396ϕ (0.053ϕ), skewness 0.1 (0.25), kurtosis 6.882 (3.952), and tail of fines $(\leq 4\phi)$ 0.001 (0.0005). The standard deviations of the

Table 20.1. *Summary of suite statistics parameters for Cape San Blas*

Mean of the skewness (two choices):	beach	river
Variability (of mean, of S.D.; two):	beach	dune
Relative dispersions (mean, S.D.):	beach	
Skewness and mean of tail of fines:	beach	
Skewness and S.D. of tail of fines:	beach	
Tail-of-fines diagram:	beach	
Comparison of tails:	beach	

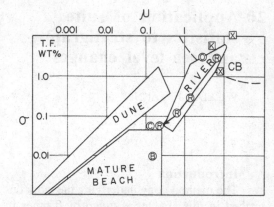

Figure 20.1. Tail-of-fines diagram: means versus standard deviations of sample values. The tail of fines is taken as the weight percent resting on the 4ϕ screen, and finer. The closed basin (CB, settling), river, dune, and mature beach number fields are defined. °, modern environments; □, ancient environments. Abbreviations: B, Cape San Blas beach; C, Medano creek; D, Great Sand Dunes; K, Oklahoma sandstones; O, Florida offshore; R, river; X, New Mexico sandstones.

means (0.218 ϕ) and standard deviations (0.053 ϕ) indicated that this is a uniform suite. Nevertheless the analysis was run with fifteen samples, then with thirteen, and finally only eleven samples. Discarding was done on the basis of anomalously high fifth and sixth moments. The mean and the standard deviation of the tail of fines (Fig. 20.1), coupled with the skewness, indicate that these are dune sands, if only two treatments are used ($n = 13$ or $n = 11$ samples).

Likewise, the large standard deviations of skewness and kurtosis indicate a dune origin. Two other plots show a settling process; but settling, as used here, covers several environments, including aeolian, small lake, lagoon, or other closed basin setting – hence this finding is not helpful. The variability diagram (Fig. 20.2) gives a choice of dune or mature beach. Probability plots show a faint but sharply developed coarse fluvial tail, not at all representative of mature sand beaches or dunes. There is a single aeolian hump on the probability plots; this (by itself) is modest evidence for dunes. The analysis indicates that these sands are probably aeolian (dune and/or settling), but might have formed on a beach. The probability plots provide the additional clue that there is a small (but not dominant) influence of stream type. The beach evidence vanishes when the suite is reduced to $n = 13$, and so is rejected.

A suite of twenty-one samples is available from the dunes proper. This suite is characterized by great homogeneity. Suite parameters clearly indicate dune or settling, and many probability plots show the aeolian hump. Two obvious differences can be seen on the probability plots: Aeolian humps are common in the dune sands, and a faint but sharply developed fluvial coarse tail appears in the creek samples.

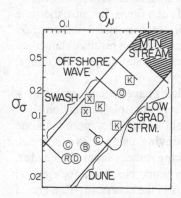

Figure 20.2. The variability diagram: standard deviations of sample means and of sample standard deviations. High energy is in the upper right, low energy in the lower left. Except for some dunes and high-energy streams, there are commonly two possibilities for any one point; this ambiguity can be resolved in Figure 20.1, Figure 20.3, or with some of the other methods given in text. See Figure 20.1 for key.

If the creek sands were lithified and exposed in an ordinary stratigraphic section, the environmental interpretation would probably be dune, but there would be a necessary note that there had been a minor fluvial influence of some kind. The analyst might consider the suggestion of a mature beach (hence the ocean coast), and might wish to opt for coastal dunes (a serious error), but a little attention to suite homogeneity would avoid this mistake.

Florida shelf

Arthur et al. (1986) studied surficial continental shelf sediments south of the panhandle (extreme northwestern part) of the state of Florida in the Gulf of Mexico. Their study area extended from a point south of Tallahassee, westward almost to the the westernmost boundary of the state. The east–west extent of the area is roughly 300 km, and the north–south width of that band is about 18 km, extending from 5.5 km offshore to almost 24 km offshore. Water depths were mostly in the range 3–30 m. There were 32 north–south sample lines, spaced about 9.5 km apart in the east–west direction, with 250 samples collected from the entire area.

The published report contains:

I. loran coordinates, water depth, and longitude and latitude for each sample;

II. the first four grain size moments, the tail of fines, and the median diameter for each sample;

III. running four-point averages, for each transect, showing smoothed mean, standard deviation, and weight-percent heavy minerals in each of three size classes;

IV. occurrence data for each of twelve heavy minerals in each sample; and

V. modal analyses for each of twelve minerals, in one size fraction (3–4 ϕ), for thirty-one selected samples.

In general, mean grain size coarsens to the west, but with large excursions, nearly as great as the entire range of values. The coarsest sizes were found opposite estuaries, and appear to reflect river channel positions when sea level was somewhat lower than it is now. The finest sizes, at the eastern end of the study strip, were found in an area where mean wave-energy density is extraordinarily low. Breaker heights are typically only 3 or 4 cm, despite the fact that this is an open sea coast (Tanner, 1960).

Standard deviations (running averages) of these size spectra decrease from about 0.94 at the eastern end to close to 0.67 at the western end, but with large excursions. Sediment was best sorted on a large shoal off of Cape St.

George, where wave refraction and energy convergence are important.

A plot of skewness versus standard deviation from individual size spectra indicates samples in a beach or river regime, but fails to separate the two. The kurtosis versus skewness plot of Friedman & Sanders (1978) also did not make a clear distinction between these environments.

In this particular study, the large area clearly contains two or more sample suites, but the exact boundaries of the latter are not known. Therefore small subareas were treated, arbitrarily, as suites. Several results are apparent. The variability diagram of Arthur et al. (1986: cf. Fig. 20.2) placed the grand mean in the offshore wave or coastal plain stream category, but subareas also extended into the swash regime. The area is presently being reworked by offshore waves; it was subject to surf action in middle-to-late Holocene time; and the sands were indeed delivered to the region by one or more rivers.

The tail-of-fines scatter plot placed each subarea within the fluvial regime, and showed that these sediments are much more nearly like various rivers than they are like typical beach sands. The two environments, on the diagram, are separated by a factor of 10. This approach largely eliminates the uncertainty between river and wave (or beach). The location of the aeolian area on this scatter plot is well known, and these samples are in no sense aeolian.

Probability plots of representative samples show no aeolian hump, most having the obvious tail of fines (which eliminates the mature beach category) and a distinctive coarse tail (more typical of rivers than wave-dominated environments).

These samples were all obtained from the continental shelf, yet scatter plots of their suite statistics identify them as beach or river sands that still exhibit grain size characteristics of river sands. The apparent environmental interpretation is that these stream-deposited sediments date largely from a lower stand of sea level, and that there has not been enough time since they were laid down, and the water has been too deep, for the local waves to rework them to any great degree. Thus we should not expect that a sand mass, delivered from agency *A* to *B*, should instantly acquire the characteristics of the latter environment, which in due time could be imposed, especially as transport processes of the latter environment cannot operate effectively.

Further study of their data with several suite methods indicates the settling category for some of the subareas. One of several interpretations is that of open water, where storm waves can stir the bottom but fair-weather waves cannot – hence the stirred sediment settles back to the bottom, without evidence of wave action.

If this material were lithified and then examined in outcrops, one could immediately conclude from the lab work that these are river sands. Because channels (although present) are rare, one should also conclude that there was a second agency that spread the sediment in thin sheets over a large area without producing floodplain characteristics. If any of the many shells in the area were fossilized, and/or if numerous wave-type ripple marks were preserved, it would be obvious that a large water body was involved. If no such field evidence were found, and if exposures were adequate, it could be seen that various parts of the area gave different results: None favored an aeolian or beach origin, but fluvial, offshore wave, and settling were indicated. Perhaps these could be combined correctly. Lagoon, lake, or delta, fed by a small stream, do not seem to be reasonable choices.

Beach Ridge Plain

St. Vincent Island, Florida, has been studied extensively by many people because it is made up of some 180 well-defined sand beach ridges in a relatively simple pattern, showing a clear succession from the oldest (about 4,000 yr BP) to the present. The area is located southwest of the coastal town of Apalachicola, Florida, and is one of a chain of barrier islands and spits that rims the delta of the Apalachicola River (Chattahoochee River plus Flint River).

Demirpolat et al. (1986) summarize much of the earlier work and also add more granulometric information. The ridges occur in sets, clearly visible as such from the air or on the ground because of differences in soil gray scale, ridge height, soil moisture, and small changes in map orientation at set boundaries. There are perhaps twelve sets, with roughly fifteen ridges per set.

A centrally sited orthogonal traverse, about 4 km long, was sampled by two different investigators, on two different occasions, using standardized field techniques. Each worker did his own lab processing. For comparison, the two projects yielded suite mean sizes of 2.34ϕ and 2.32ϕ, and suite mean standard deviations of 0.385ϕ and 0.39ϕ. Sample mean diameters were within 10% of the suite mean. Sample standard deviation differed from the suite value by less than 14%. Therefore the grain size data represent a homogeneous suite from a central strip across the island.

For present purposes, fifty-nine ridges in the central strip were sampled, each ridge on the seaward face at half the mean altitude. Because of the great uniformity of the sand, these fifty-nine samples were taken to constitute one suite, but with the knowledge that several ridge sets were included. The numerically low suite value for the standard deviation indicates dune or mature beach, but not settling, river, glacial meltwater, or tidal flat. The suite skewness (–0.123) points to river or mature beach, but the first of these is no longer an attractive possibility. The variability diagram (suite standard deviations of sample means and of sample standard deviations; Fig. 20.2) suggests dune or mature beach, but the skewness value (Fig. 20.3) largely precludes the former. Taken together, the evidence identifies mature beach as the environment of deposition.

The tail-of-fines procedure (mean and standard deviation of weight percents on the 4ϕ screen and finer; Fig. 20.1) commonly gives last previous (penultimate) agency (as in the preceding section on continental shelf suites). For St. Vincent Island beach ridges, set D plotted with river or settling; at the time of set D, the ridges were being built more or less directly from river-mouth sediments. Sets E-1, E-2, and F plotted with rivers, but the three sets lie in a succession, each farther from the settling area than its predecessors; and set G appeared on the edge of the mature beach number field on this plot. That is, with time (from D through G), the sand became less and less like delta deposits and more and more like open beach sands. These

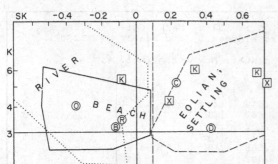

Figure 20.3. Suite skewness versus suite kurtosis. The primary dividing line is at a skewness value of 0.1, separating aeolian or settling sediments (to the right) from river or beach sediments (to the left). The skewness provides a comparison of the two tails, and the kurtosis indicates their size. In some instances, river and beach sediments can be distinguished on the basis of kurtosis. The latter (K) also provides a general clue to energy levels: small K suggests high energy, and large K, low energy. On the right-hand side, dune sands typically have $K \geq 3$, and closed basin sediments may have $K < 3$. See Figure 20.1 for key.

four sets span about 2,000 yr of more or less simple evolution of the sand pool.

At the southeastern tip of the island are several small ridge sets that do not belong with the main body of the island. These short ridges also indicated (with various criteria) settling, river, and mature beach, but also an aeolian influence. The combination of swash and aeolian effects is essentially standard in sand beach ridges (not necessarily in single massive coastal ridges [Tanner, 1987]). However, the primary agency was wave action (swash) with secondary wind work. This is true of most sand beach ridges that I have studied, where obvious wind deposition typically represents 5%–20% of the entire coastal ridge.

The ridges crossed by the central transect were deposited in an unambiguous sequence from oldest to youngest. This fact provides much more information than can be obtained from a suite of random samples. Several parameters are of interest here, because set boundaries are clearly visible (e.g., on air photos), and therefore ridge sets can be studied as such. Certain ridge sets stand low relative to others. Each ridge set spans a few centuries, and therefore

Table 20.2. *Mean/kurtosis of ridges*

Topographically low ridges	Ridges with low mean/kurtosis
1, 2, 3, 4, 5, 6	2, 3, 4, 5, 6
15, 16, 17, 18, 19, 20	14, 15, 16, 17, 18, 19
46, 47, 48, 49, 50, 51	48, 49, 50, 51

each has a vertical position that does not reflect storms or even stormy decades.

Where the mean/kurtosis parameter is numerically low, the ridge set was found to stand low. This is demonstrated in Table 20.2, where ridge numbers are from oldest to youngest. A critical mean/kurtosis value of 0.68 was used here; data were examined in terms of a three-point moving average (Fig. 20.4). Other statistical parameters successful in distinguishing between topographically high and low sets, using three-point moving averages, were 1/kurtosis (0.3), S.D./kurtosis (0.115 ϕ), the differences of the standard deviations (which show changes only, but not positions), and differences of kurtosis (which also show changes only). Each of these five plots identifies either changes and positions, or changes, correctly.

Set means of the means and set means of the standard deviations provide some additional information. The sand got finer and better sorted from ridge 1 to 15 (a boundary value, as seen above); then coarser and less well sorted from 15 to 32 (another boundary value); and finally finer and better sorted from 32 to 45. There were matching changes in skewness and kurtosis, also. Ridges 15 and 32 appear to mark important boundaries in terms of these size parameters.

Each of these sets covers hundreds of years (e.g., ridges 15–20, plus any ridges missing on the traverse, times roughly 20 yr each). Therefore the numerical results show sand-pool effects, regardless of whether a low set represents a slightly lower mean sea level. Seasonal or storm events are too brief to be shown, but there is evidence for slower (few-century) modifications of the sand-pool history. Wave-climate changes are also not shown since differences in set heights are too large to be due to long-term wave changes. It is concluded that suite size

```
This is LINEAR.  Data source:STVIN.4AE.  N is  37 .  Mn/Ku Cut-off: .68
Std.D/Ku Cut-off:  .125 .   Mn/Ku MovAvgs Window= 3  and mean:  .7131271 .

I      Std.D./K     Mean     Kurt.   Mn/Kurt. Mov.Avg. (incl. $)           I
Mean value of Mn/Kurt.:                                       #

 1     .096         2.318    3.928                                         1
 2     .106         2.362    3.578    .619  LOW?    $                       2
 3     .103         2.274    3.728    .667  LOW?         $                  3
 4     .12          2.346    3.158    .662  LOW?         $                  4
 5     .135    @    2.23     3.459    .677  LOW?          $                 5
 6     .101         2.381    3.658    .665  LOW?         $                  6
 7     .119         2.392    3.403    .692                $                 7
 8     .113         2.445    3.355    .729                     $            8
 9     .13     @    2.473    3.266    .765                       $          9
10     .15     @    2.338    2.861    .792                         $       10
11     .143    @    2.434    3.019    .793                         $       11
12     .136    @    2.315    3.048    .74                        $         12
13     .109         2.303    3.455    .703                     $           13
14     .123         2.297    3.33     .649  LOW?         $                 14
15     .096         2.355    3.924    .643  LOW?         $                 15
16     .12          2.293    3.542    .63   LOW?       $                   16
17     .122         2.212    3.406    .615  LOW?   $                       17
18     .11          2.14     3.844    .613  LOW?   $                       18
19     .088         2.435    3.819    .651  LOW?         $                 19
20     .138    @    2.306    2.897    .704                     $           20
21     .106         2.416    3.449    .741                  $             21
22     .156    @    2.203    2.992    .796                           $     22
23     .146    @    2.415    2.39     .813                            $     23
24     .112         2.365    3.198    .772                          $       24
25     .106         2.427    3.742    .71                 $               25
26     .112         2.454    3.255    .731                $                26
27     .116         2.518    3.121    .785                      $          27
28     .151    @    2.238    2.8      .817                            $     28
29     .126    @    2.468    2.918    .817                            $     29
30     .117         2.531    3.136    .834                             $    30
31     .135    @    2.474    2.901    .817                            $     31
32     .132    @    2.35     2.962    .807                            $     32
33     .116         2.457    3.153    .78                           $       33
34     .11          2.566    3.327    .766                         $       34
35     .151    @    2.188    2.931    .7                  $               35
36     .096         2.228    3.714    .649  LOW?         $                 36
37     .107         2.19     3.523                                         37

Mean value of Mn/Kurt.:                                       #
I      Std.D./K     Mean     Kurt.   Mn/Kurt. Mov.Avg. (incl. $)           I

This is LINEAR.  Data source:STVIN.4AE.  N is  37 .  Mn/Ku Cut-off: .68
Std.D/Ku Cut-off:  .125 .  The latter helps identify boundaries.
```

Figure 20.4. The ratio of mean to kurtosis, for each sample, along a time line across part of the St. Vincent Island beach ridge plain. The full output covers ridges 1 (oldest) through 59, but is not shown here for space reasons. The note LOW?, derived from the ratio itself, identifies topographically low ridge sets (not low individual ridges; these are three-point moving averages). In this area as on other sand ridges that have been studied, this ratio, or any of several other parameters listed in the text, identifies set vertical position changes and hence presumably small sea-level changes.

data provide useful clues to long-term wave-energy-level changes and therefore to small (1–2 m) changes in mean sea level. That this is not a geographically localized effect is shown by the fact that the same relationship between suite data and ridge height appears in data from Mesa del Gavilan, a beach ridge plain located south-east of Port Isabel, Texas (Tanner & Demirpolat, 1988).

What if this sand sheet were lithified, and then exposed at some future date? Almost completely without marine fossils, it might contain a few thin peat layers. Ripple marks are nonexistent; cross-bedding, if visible, would be largely in the 8°–20° range, dipping southward (at present, toward the open sea), but having a uniformity that might very well be puzzling. Perhaps the best information would be granulometric. One should be able to identify the sand as having been placed on the beach rather than in some other environment. River influence should be obvious, and one should even be able to point toward the river mouth (north of the oldest sands, by present coordinates).

Identifying modest sea-level changes conceivably would be possible, but the investigator would not be able to select sample sites on the basis of ridge geometry (as was actually done), and perhaps would not even be able to establish a linear suite.

Ancient environments
Oklahoma

The Vamoosa formation, of late Pennsylvanian age, was chosen for sampling. This sandstone and conglomerate unit is well exposed along several roads in Seminole County, in east central Oklahoma. It was selected because previous work (e.g., Tanner, 1953) showed that it contains (from north to south) sediments that represent seafloor, barrier island, and lagoon (or estuarine) environments of deposition. As part of a larger project, five sample suites, totaling forty-eight samples, were collected (Tanner, 1988).

The suites were taken along two lines that cross the axis of the barrier island at about 60°. One of these lines, representing the basal, or oldest, part of the formation, was 23 km long and included three suites: A on the northern (seaward) flank, B near the middle, and E on the southern flank of the barrier. There were ten samples in suite A, fifteen in B, and five in E. The other two suites (fourteen and four samples) were collected near the middle of the formation, on the northern flank of the barrier island. There were twenty-nine more samples from rock units higher in the section (slightly younger).

The three suites from the basal Vamoosa formation contained tail of fines with weight percents and standard deviations of those percents in the closed basin or settling category: quantities of sediment on the 4ϕ sieve (and finer) greater than found in most rivers. The tail-of-

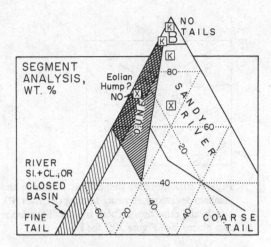

Figure 20.5. The segment analysis triangle. The raw data were taken from the probability curves for individual samples. Each curve was dissected into a straight-line central segment and two tails (whether straight lines or not). This provided three segments for each sample. Percentages were read for each segment, and the three mean values then calculated and plotted. The "B" near the apex marks mature beaches; other environments are labeled clearly. The main overlap involves dune (aeolian) sediments and clay- or silt-carrying streams; this ambiguity can be resolved, in most instances, by checking for the aeolian hump on the probability plots, and/or by comparing the mean sizes and standard deviations. See Figure 20.1 for key.

fines plot (Fig. 20.1) commonly shows the penultimate transport agency; therefore, these data may indicate a fluvial source.

The suite mean values of skewness for the three suites plot as follows (Fig. 20.3): B and A (closest to the sea), settling or aeolian; E (closest to the fluvial source to the south), river. There was no clear evidence for an aeolian origin, so this option was discounted. The standard deviations of suite means and suite standard deviations indicate beach, offshore wave, or low-gradient stream (Fig. 20.2). The lack of coarse tails in these sand units – coupled with high weight percents in the central segment on probability plots (typically ≥90) – places A and B (seaward suites) with beaches, and E (landward suite) with rivers (Fig. 20.5). The relative dispersions of means and standard deviations (Fig. 20.6) indicate a beach origin for B, a probable beach origin for A, and a probable river source for E.

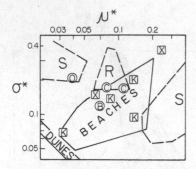

Figure 20.6. Relative dispersions of means and standard deviations, showing settling (S), river (R), beach, and dune areas. There is a small overlap at two places. See Figure 20.1 for key.

Suite E, taken within the lagoon (or estuary), appeared on bivariate plots with river environments. Suites A and B, on the other hand, showed clear beach characteristics, but also evidence of settling. This last indication is here taken to match the restricted (interior) seaway in which these sediments accumulated. The penultimate agency for all of the suites was indicated as fluvial.

Because the suite analysis provided the same interpretation as had been made on the basis of other information, the conclusion is drawn that the methodology was useful (Tanner, 1988). The suite approach did not produce as much detail as did the combination of fieldwork and paleontology, but in the absence of the latter would have reached essentially the same general conclusions.

New Mexico

The Yeso formation of Permian age is well exposed in many places in San Miguel County, in north central New Mexico. It is typically a thin-bedded sandstone, with lesser amounts of coarse siltstone and shale, and a few thin limestone beds. It lies immediately above the Sangre de Cristo formation, mostly of continental origin in this part of the state, and directly below the Glorieta sandstone, which contains wave-type ripple marks indicating fetch in the hundreds of kilometers. The Yeso itself also contains wave-type ripple marks, but all are small, ranging down to <1 cm in spacing. Based on this information, the Yeso was identified earlier as of tidal flat origin (Tanner, 1963).

Two suites (eight and twelve samples) have been processed from the Yeso; they are designated A and B. The two were collected within about a kilometer of each other, horizontally, but the former represents the middle of the formation, and the latter, the upper third.

Tail-of-fines data (Fig. 20.1) place both suites in the closed basin or settling category. This is taken to indicate the geosynclinal seaway in which Yeso sediments accumulated. The suite mean skewness value (Fig. 20.3) also indicates settling for each suite, although the two do not have similar values. The alternative is aeolian, but the fairly uniform thin bedding, the very small wave-type ripple marks, and the absence of any positive evidence eliminate this possibility.

Each of the two suites provides one clue suggesting a tidal flat depositional environment: the ratio of suite mean standard deviation to suite mean kurtosis (0.19 in each case). The relative dispersion of the means and of the standard deviations (Fig. 20.6) puts each suite in the beach category, and the variability plot (Fig. 20.2) shows each as beach or low-gradient stream. The latter, along with the wind possibility, must be rejected for lack of support; but most of the criteria indicate settling, which is consistent with the low-energy tidal flat environment. The interpretation to be drawn from the suite statistics analysis is closed basin or settling, probably beach or tidal flat, but with a possible river episode in its history. The ripple mark data indicate very small fetch values, in general only hundreds of meters or a few kilometers at the most. The Permian sea in Yeso time was actually much larger than a few kilometers, so the ripple mark measurements are taken to mean restricted ponds on the tidal flat, such as are actually observed on low-relief, low-energy, modern examples (Tanner & Demirpolat, 1988).

Conclusions

The suite statistics procedure requires that a suitable set of samples, from a given sand body, be used for analysis, and that suite parameters (such as the mean of the sample means) be employed for environmental interpretation. The results of such a study typically include one or more of the following: identification of a domi-

Table 20.3. *Summary of suite statistics results for examples in the text*

	CSB	Med	GSD	Off	StV	Okla	NewM
Skewness > 0. 1							
(2 possibilities)		D,S	D,S			D,S	D,S
Variability			D			S	S
Tail of fines						S	S
Aeolian hump			D				
Skewness < 0. 1							
(2 possibilities)	B,R			B,R	B,R	B,R	
Variability	B	B,D		B,R	B	R	
Tail of fines	B	?		S	R		
Fluvial coarse tail		R					
Segment analysis						B,R	B,R
Relative dispersion	B	B,R		S		B,R	?
Summary	B	?	D	B,R,S	B,R	B,R,S	B,R,S

Note: Only the first column shows a simple case: obvious uncomplicated modern beach. The other columns illustrate problems, commonly due to mixing of environments, or sequence of environments (such as dune-to-creek, or river-to-sea). A more detailed analysis is given in the text.

Abbreviations: CSB, modern beach; Med, modern creek in Great Sand Dunes; GSD, modern dunes; Off, Florida offshore, modern shelf; StV, modern beach ridge plain; Okla, ancient barrier island, lagoon, and estuary; NewM, ancient near-shore tidal flat; B, beach; D, dune or aeolian; R, river or creek; S, settling (closed basin, lagoon, estuary, lake, interior seaway, delta).

nant agency or environment (if there was only one), identification of a penultimate agency, evidence for combination of two agencies, details on maturing of the sand pool, and evidence for small changes in mean sea level.

Only one easy modern example has been included in this study. The other modern examples were selected to illustrate the kinds of problems commonly encountered. Although interpretation is not invariably easy, the results (Table 20.3 gives a summary) are nevertheless superior to what one gets by making single-sample statistical studies.

The ancient examples are representative of various projects that have been undertaken in that kind of work: Where other evidence has been available, suite methods have given essentially the same results. This provides some encouragement that, when other information is scarce or ambiguous, suite methods provide useful results.

However, some of the lithified sandstones (of unknown or uncertain origin) studied have not provided satisfactory answers: some because of pervasive silica cement, precluding lab analysis, and others because suite procedures yielded indicators of too many different environments. This is not a statement that the suite parameters were erroneous, but only that no clear interpretation appeared in the course of the work. Many modern sands represent mixed agencies, or transition from one agency to another (e.g., river to marine). The suite methods have been useful, in general, in these cases; the examples given are representative of the more difficult suites.

Despite problems that one encounters in using suite statistics methods, the results are much better than are generally obtainable by trying to base an analysis on data from individual samples. The method is not foolproof for various reasons, including inadequate number of samples, inadequate areal coverage, poor sampling technique, absence of a single well-defined environment of deposition, and (for ancient rocks) pervasive silica cementation. Perhaps the most important difficulty lies in the fact that certain sand bodies have been studied at a moment of transition; that is, when there is in fact no single responsible agency or environment. This may be problematic for our wish to erect sharply defined class boundaries, but it is not a problem in the method of study.

REFERENCES

Arthur, J., Applegate, J., Melkote, S., & Scott, T. (1986). Heavy mineral reconnaissance off the coast of the Apalachicola River delta, Northwest Florida. Florida Geological Survey, Report of Investigation no. 95, Tallahassee, 61 pp.

Demirpolat, S., Tanner, W. F., & Clark, D. (1986). Subtle mean sea level changes and sand grain size data. In: *Proceedings of Seventh Symposium on Coastal Sedimentology*, ed. W. F. Tanner. Geology Department, Florida State Univ., Tallahassee, pp. 113–28.

Friedman, G., & Sanders, J. (1978). *Principles of sedimentology*. New York: John Wiley, 792 pp.

Socci, A., & Tanner, W. F. (1980). Little-known but important papers on grain-size analysis. *Sedimentology*, 27: 231–2.

Tanner, W. F. (1953). Facies indicators in the Upper Pennsylvanian of Seminole County, Okla.: *Journal of Sedimentary Petrology*, 23: 220–8.

(1960). Florida coastal classification. *Transactions, Gulf Coast Association of Geological Societies*, 10: 259–66.

(1963). Permian shoreline of central New Mexico. *Bulletin of American Association of Petroleum Geologists*, 47: 1604–10.

(1987). Spatial and temporal factors controlling overtopping of coastal ridges. In: *Flood hydrology, vol. V*, ed. P. Singh. Dordrecht: Reidel, 241–8.

(1988). Paleogeographic inferences from suite statistics: Late Pennsylvanian and early Permian strata in Central Oklahoma. *The Shale Shaker (Oklahoma City, Okla.)*, 38(4): 62–6.

Tanner, W. F., & Demirpolat, S. (1988). New beach ridge type: Severely limited fetch, very shallow water. *Transactions, Gulf Coast Association of Geological Societies*, 38: 553–62.

21 Application of size sequence data to glacial–paraglacial sediment transport and sediment partitioning

JAY A. STRAVERS, JAMES P. M. SYVITSKI, AND DAN B. PRAEG

Introduction

Fjord systems are an important interconnecting link between the continental shelf and the glaciated portions of continental cratons. As effective sediment traps, fjords and fjord valleys are filled with the full spectrum of sedimentary particles derived from the chemical and physical erosion of the hinterland terrains by alpine glaciers and former continental ice sheets. Much of the sediment that accumulates within these overdeepened coastal basins is related to the glacial/proglacial deposition during retreat from the last major ice advance, or several episodes of paraglacial basin filling.

For many of the world's fjords, glaciers once converged through their heads, draining the continental ice sheets. This funneling of large volumes of ice gave rise to rapid flow velocities and, depending on the nature of the bed contact, increased power of erosion. During the deglacial cycle, the outlet glacier acted as a vast conveyor belt delivering debris to the fjord and its tributary valleys. Large volumes of clastic sediments, sometimes over 1 km in thickness (Syvitski, Burrell, & Skei, 1987), were deposited in fjords during this critical period.

As the fjord outlet glaciers retreated onshore, outwash deposits of thick sequences of glaciofluvial sediments (sandurs) aggraded onto the valley floors (Church, 1972). Contemporaneous deposition of marine sediment occurred in the deltaic environment at the fjord head and in the deep fjord basins. As the ice sheet thinned and eventually disappeared, isostatic rebound of the inner fjord region caused coastal emergence. The shoreline position is thus controlled by a dynamic balance between the rate of delta aggradation/progradation and the rate of coastal emergence/submergence.

Our objective is to examine the Holocene fractionation of sediment within a complex of sequences of terrestrial and marine sedimentary deposits found in a single arctic fjord system. We quantify the general characteristics of early to mid Holocene glacial and marine sediments, and suggest how late Holocene erosion recycles sediment to produce modern terrestrial and marine deposits. We identify the transfer processes that control the escape or capture of terrestrial and marine sediment, and determine both the temporary and ultimate storage locations of different grain size fractions.

Our field area, Cambridge Fiord of northern Baffin Island (Figs. 21.1 and 21.2), has received intensive study within SAFE, the international project investigating Canadian arctic fjords (Syvitski & Schafer, 1985). The fjord hinterland contains a diverse sequence of sedimentary deposits, including an extensive series of mid to late Holocene raised marine and terrestrial sediments, which form a thick valley fill in the Keel and Cambridge (informal) river valleys at the head of the fjord (Fig. 21.3). These deposits are actively being cannibalized (recycled) to produce sediment for the present-day alluvial, deltaic, and marine deposits.

The sedimentary sequence

We employ a concept of sedimentary sequence divisions in order to examine the partitioning of grain sizes within deposits from differing sedimentary environments. Our approach is similar to that of Andrews (1985), who analyzed size frequency trends for Quaternary sediments of many different ages, environments, and locations across Baffin Island. However, we have restricted the geographic extent of our data set, for the most part, to a single sedimentary (fjord) system (Fig. 21.2) and, in the temporal sense, to sediments that are actively being processed (eroded/transported/deposited) at the present time.

Deposits from eight sedimentary sequences have been identified, and these sequences have been further grouped into three divisions: source deposits, temporary transport deposits, and ultimate deposits. *Source deposits* consist of Holo-

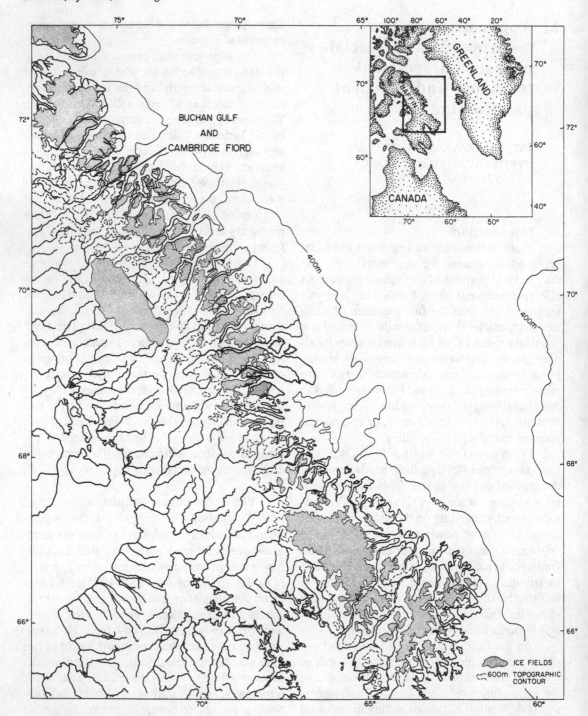

Figure 21.1. Location map of Cambridge Fiord of Northern Baffin Island.

cene deposits that are currently being eroded and thus serve as sediment sources. These include early Holocene tills, raised glaciomarine sedi-

ments, and glaciofluvial sediments, as well as mid to late Holocene raised alluvial and deltaic deposits. *Temporary transport deposits* comprise modern aeolian sands, alluvial channel sediments, and prodelta sediments. This division includes sediments periodically transported

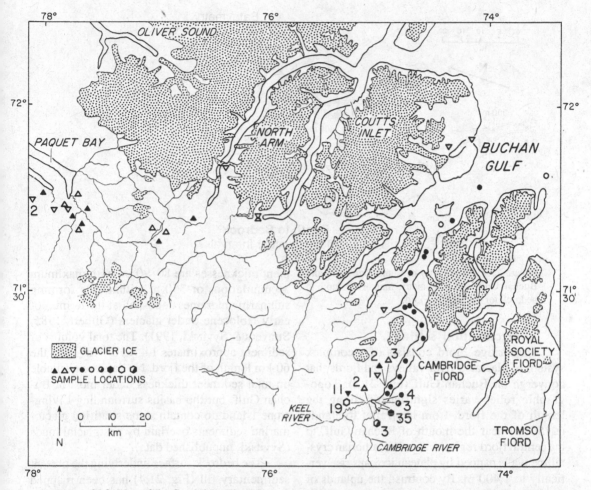

Figure 21.2. Map of the Cambridge Fiord System and adjacent regions showing sample localities, ice fields, and land drainage. Numbers beside sample stations indicate the total number of samples from near that locality.

and deposited by seasonally variable processes. The net result is transfer of sediment to the sea, but involving multiple, complex transport pathways punctuated by periods of temporary storage en route. *Ultimate deposits* consist of marine basinal sediments currently being deposited in the deeper waters of Cambridge Fiord. The fjord basin is characterized by net deposition and sediment stability on Holocene time scales. These basin deposits are viewed as the end result of modern sedimentary processes and will remain in the fjord at least until the next glacial cycle.

Figure 21.3. Oblique aerial photograph (looking NW) of the Keel and Cambridge (informal) River valleys at the head of Cambridge Fiord.

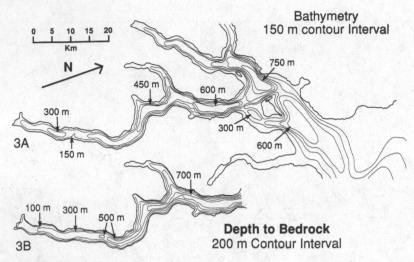

Bathymetry
150 m contour Interval

750 m

450 m   600 m

300 m

300 m

3A

600 m

150 m

700 m

100 m   300 m

500 m

Depth to Bedrock
200 m Contour Interval

3B

Figure 21.4. Bathymetry of Cambridge Fiord and Buchan Gulf, and bathymetric map of the depth to bedrock for Cambridge Fiord.

Environmental setting

Cambridge Fiord consists of a complex system of glaciated river valleys and fjords that converge on Buchan Gulf (Fig. 21.2). Topographic relief varies significantly along the length of the fjord, from low-relief forelands (≤150 m) near the mouth of Buchan Gulf, to the central fjord region where Precambrian crystalline walls capped by plateau ice caps rise vertically to 1,400 m. By contrast, the uplands of the inner fjord region consist of rounded bedrock surfaces that are ice-free and exhibit lesser relief (<900 m). Evidence for inundation of the inner fjord by the Laurentide ice sheet is ubiquitous in the form of grooved and polished bedrock and large-scale, ice-sculptured bedrock forms. Below sea level, the fjord system consists of five sedimentary basins divided by bedrock or morainic sills (Fig. 21.4). Deep basins ranging from 600 m to 750 m water depth are arrayed to the north and west of Livingstone Island in Buchan Gulf and along the central reaches of Cambridge Fiord. Shallower basins of around 300 m water depth are found to the south of Livingstone Island and landward of a submarine moraine that isolates the innermost basin of Cambridge Fiord (the Keel Basin).

Thick sequences of Quaternary glacial marine sediments underlie the seabed of Cambridge Fiord (Syvitski, 1984a). Average sedi-ment thicknesses are 80–90 m, with maximum accumulations of 350 m within sills forming submarine moraines deposited at the termini of early Holocene outlet glaciers (Gilbert, 1985; Stravers & Syvitski, 1991). The total volume of sediment approximates 1.4×10^{10} m^3 over the 60-km length of the fjord. No data are available on total sediment thickness under most of Buchan Gulf, but the basins surrounding Livingstone Island do contain some stratified glacio-marine sediments overlain by postglacial muds (Syvitski, unpublished data).

The bedrock surface underlying the present sedimentary fill (Fig. 21.4) has been mapped from air-gun seismic reflection data and shows a much less complex morphology than the present bathymetry (Stravers & Syvitski, 1991). The surface forms a prominent, flat-floored parabolic trough 700 m below sea level in outer Cambridge Fiord that shallows gradually to 100 m at the fjord head.

Glacial history

Cambridge Fiord represents a major glacial outlet trough, probably occupied many times throughout the Quaternary by Laurentide ice flowing northeastward from Foxe Basin (Miller, 1985; Andrews, 1987). During the late Foxe–early Holocene maximum (c. 10,000–8,000 yr BP), ice was restricted to a complex network of outlet glaciers flowing through Cambridge and its tributary fjords (Hodgson & Haselton, 1974; Stravers & Syvitski, 1991). The ice may have terminated as far seaward as outer Buchan Gulf, where a series of low-elevation moraines and raised morainal banks have been mapped (Stravers & Syvitski, 1991). Their exact age remains uncertain, though they may correlate to ice-contact deposits of the "Cape Hatt interval" dated at 9,500 yr BP (see Klassan, 1985). This scenario disagrees with more regional reconstructions (Feyling-Hansen, 1985; Miller, 1985).

The Holocene emergence history for inner Cambridge Fiord is characterized by continuous emergence at a rate of 12 m/ka for the period from 6,700 yr BP (marine limit in Keel Valley) to 3,000 yr BP (Stravers & Syvitski, 1991). Terrestrial and marine sedimentary deposits dating from this period are preserved in the Keel and Cambridge river valleys as raised terraces or terrace remnants. Specific ages are estimated by extrapolation of the terrace elevations to the emergence curve (Syvitski, 1985). All terraces were formed between about 6,700 and 1,500 yr BP, with the majority of terraces in the upper Keel River valley (below the marine limit) forming during the mid Holocene climatic optimum between 6,000 and 4,000 yr BP (for climatic reconstruction, see Fisher & Koerner [1981]; Short et al. [1985]).

Present climate

Sparse meteorological measurements collected since the late 1950s have resulted in a record of climate that characterizes regional rather than local climatic conditions, and may simply reflect short-term climatic variability rather than long-term or stable climatic conditions (Barry et al., 1975; Jacobs et al., 1985). Nevertheless, there appears to be a significant climatic gradient from the outer coastal regions to the mountainous highlands of the central and inner fjords. Mean annual precipitation for northern Baffin Island ranges from about 200 mm on the coast to >300 mm over the highland ice caps, with 20%–25% of the latter falling as liquid precipitation (Maxwell, 1981). Mean daily January temperatures vary from $-20°$ to $-25°C$ along the outer coast and $-23°$ to $-28°C$ in the landward highland regions. Mean daily July temperatures are very similar ($+5°C$) for the two areas; however, summer temperatures for the lower-elevation valleys at the fjord heads may be considerably warmer. Unfortunately, the meteorological data are not sufficient to quantify the elevational temperature gradient for the fjord head regions. The present distribution of ice caps reflects the local alpine climate (Andrews & Miller, 1972; Syvitski et al., 1984a) and differs markedly from the early Holocene. At that time a continental ice sheet was advancing on the fjord from the south, while alpine glaciers were of similar or diminished extent due to the partial interception of precipitation by the Laurentide ice sheet (Andrews et al., 1972; Miller, 1976).

The present hinterland that drains into Cambridge Fiord is 2,045 km^2. Based on the climate–discharge model outlined in Syvitski et al. (1984a), 58% of the annual freshwater discharge of 0.42 km^3 enters Cambridge Fiord during the spring discharge (freshet) from melting snow banks (late June–mid July) from the melting of winter snow. This is followed by lower summer flows punctuated by periodic rainstorm floods that are induced orographically and account for a further 37% of the annual discharge. Only 5% of the fluvial discharge can be related to the late summer ablation of the local ice caps.

Our data suggest that, under the present climate, the fluvial regime of the Cambridge Fiord region is ineffective at completely reworking the early Holocene deposits. The coarse fraction of the tills and ice proximal deposits are rarely transported and thus are remaining in the floodplain as coarse lag gravels derived from the erosion of adjacent terrace scarps.

Paraglacial sedimentation

The paraglacial cycle of sedimentation was first defined by Church & Ryder (1972) for terrestrial deposits and later expanded for marine sedimentation by Syvitski et al. (1987). Paraglacial sedimentation refers to the abnormally high flux of sediment from land to sea via fluvial discharge from a terrestrial ice sheet that is experiencing rapid ablation. During this period, vast quantities of glacial, proglacial, and possibly raised marine sediments are available for fluvial erosion and transport, and thus account for abnormally high rates of denudation. Generally, this type of sedimentation occurred leading into the diachronous Hypsithermal period (~8,000–6,000 yr BP in northern Baffin Island), a time of warm, dry summers associated with ice-sheet meltwater discharge over an order of magnitude greater than is presently found for modern rivers draining a glaciated hinterland. For further discussion of the magnitude and timing of late Quaternary ice melt for Baffin Island, see Quinlan (1985).

Glacial versus paraglacial denudation rates

Glacial denudation rates for the late Foxe and early Holocene glaciation of this region was estimated using an ice sheet reconstructed from three data sets:

1. morainal distribution data (Hodgson & Haselton, 1974; Stravers & Syvitski, 1991);

2. the position of the ice divide over Foxe Basin (Andrews & Miller, 1979; Dyke et al., 1982); and

3. ice flow lines (Dyke & Prest, 1987).
The ice that flowed through Cambridge Fiord apparently drained a glacial "basin" of approximately 9,000 km^2. Using the approximate volume of sediment in Cambridge Fiord and Buchan Gulf (1.4×10^{10} m^3), this gives a minimum estimate of material eroded from the 9,000 km^2; that is, it does not account for sediment that may have escaped from the system to the continental shelf and open ocean. Adjusting for sediment porosity, we get an average of ~40 cm of solid rock eroded from the area drained by the ice, most of which would have been from along the coast of Foxe Basin and the rounded highlands inland from the head of Cambridge Fiord. The area between was probably covered by cold-based ice and therefore subject to little erosion (Andrews et al., 1985). The period for this denudation of 40 cm was some time between 18,000 and 6,000 yr BP. If we assume:

1. that pre–late Foxe sediments are absent within the fjord basin,

2. that the fjord was occupied by glacial ice by 10,000 yr BP, and

3. that the ice was completely ablated from the Cambridge Fiord drainage basins by 6,000 yr BP (for details, see Stravers & Syvitski [1991]),
then ~3.5×10^6 m^{3}a^{-1} of sediment was transported and deposited into the fjord basin during this period.

Modern denudation rates were estimated using the climate–sediment transport model outlined in Syvitski et al. (1984a). This approach generates a synthetic discharge curve for individual drainage basins based on estimates of (1) summer rain runoff, (2) nival freshet runoff, and (3) glacier melt runoff. The suspended load for individual drainage basins within the hinterland of Cambridge Fiord was estimated next using appropriate sediment-discharge-rating algorithms that depended on the areal coverage and altitude of modern ice caps (Fig. 21.2). Approximately 3.0×10^4 m^{3}a^{-1} of suspended sediment is found to discharge into the fjord presently; an additional 3.0×10^5 m^{3}a^{-1} of bedload transported sand and gravel reaches the coastline in the form of prograding deltas and rivers. Approximately 65% of this total sediment load of 0.33 $\times 10^6$ m^{3}a^{-1} enters at the head of the fjord; the remainder enters laterally along the margins where most of the present glaciers are found. Thus, we find the early Holocene glacial sediment transport rate of 3.5×10^6 m^{3}a^{-1} to be one order of magnitude larger than the modern transport rates. Syvitski et al. (1984a) calculated the transport rate during the transition between these two distinct glacial/interglacial periods based on mass balance of raised marine terraces. In general, as the rate of emergence has decreased, so has the annual rate of erosion. The denudation rates are considered to have reached steady state by 4,500 yr BP.

Field methods

Reasonable sample coverage (Fig. 21.2) has been obtained for the terrestrial glacial deposits and Holocene raised marine deposits, as well as marine samples from the modern prodelta and the deep fjord basins (Clattenburg et al., 1983; Schafer et al., 1984; Syvitski et al., 1983, 1984a,b; Stravers, 1987). Offshore investigations were undertaken during cruises 82-031 and 83-028 of *CSS Hudson* (Syvitski & Blakeney, 1983; Syvitski, 1984b). Grab samples of the modern fjord basin marine sediments were collected using a 40×40 cm Van Veen or Shipek sampler. Launches deployed by *CSS Hudson* were used to collect Van Veen grab samples from the shallow, near-shore reaches of both the Keel and Cambridge river prodeltas.

Additional grab samples were collected from a Boston whaler using an Eckman dredge. Onshore investigations were briefly conducted during both of the Hudson cruises. Extended field sampling, including transect surveys and stratigraphic mapping, was undertaken with the support of a Polar Continental Shelf helicopter during the summers of 1985–7 (Syvitski & Praeg, 1987).

Laboratory methods

Size frequency distributions for 132 samples from the Cambridge Fiord system were determined at the sedimentology laboratory of the Atlantic Geoscience Centre (Bedford Institute of Oceanography). The gravel fraction was separated from the finer fraction using a standard 2-mm sieve. If the gravel fraction was >4% of the total sample and weighed >0.25 kg, that fraction was analyzed for its nominal grain diameters at 0.25ϕ intervals ($\phi = -\log_2 d$, where d is particle diameter in millimetres) using ASTM (1959) procedures. The initial sand fraction was separated from the mud using a 53-μm wet sieve, and was analyzed for its equivalent spherical sedimentation diameter at 0.2ϕ intervals using the computerized Atlantic Geoscience Centre settling tube (2 m long, 0.15 m wide). The mud fraction, when significant (>5% of the total sample), was analyzed on a computerized Sedi-Graph 5000D for the particle equivalent spherical sedimentation diameter (ESSD) at 0.2ϕ intervals over the range of 0.5–63 μm. Total sample grain size distribution plots were produced by combining data from the various methods with program MERGE (Hackett et al., 1986). The final proportion of sand and mud was based on the traditional 62.5-μm class boundary determined from the merged weight frequency distribution. In addition to the size frequency plots and moment statistics generated for individual samples, "sequence" frequency distributions were determined by averaging all samples from a given sedimentary facies.

Results

The results of the laboratory analysis of each individual sample are presented in Syvitski & Blakeney (1983), Syvitski (1984b), and Syvitski & Praeg (1987). Mean grain size frequen-cy and moment statistics for the "averaged sediment" in each of the eight sedimentary categories are given in Table 21.1. We also present this information in the form of ternary diagrams of gravel–sand–mud and sand–silt–clay (Figs. 21.5–21.7) and cumulative frequency plots (Fig. 21.8).

Source deposits
Early Holocene till

These deposits occur in prominent lateral moraines along valley walls of the inner fjord and as thin, discontinuous blankets (≤2 m) of lodgment till draping ice-scoured bedrock. The data set also includes till collected from an extended region outside of the modern Cambridge drainage basin. We chose to include these samples in order define the regional rather than local character of the till. It should be noted that all the samples appear to have been derived from the erosion of similar crystalline rock types of the northern Baffin uplands, and that the grain size frequency of the Cambridge till samples is indistinguishable from those of samples derived from outside the Cambridge drainage. All till samples included were deposited by the Laurentide ice sheet and are not related to neoglacial activity of the local ice caps. The samples were collected from the C horizons of soil pits excavated into the morainal surface (usually to a depth of >60 cm).

The tills are classified as sandy gravels, gravelly sands, or gravelly sandy muds – typically with <10% mud content (Table 21.1, Figs. 21.5A and 21.8). The till matrix contains >80% sand. Glacial erosion of igneous and metamorphic terrains commonly produces a multimodal sediment (see Schafer et al., 1989) or one containing a subequal mixture of gravels, sands, silts, and clays (Vorren et al., 1983). If this was true for tills produced in the Cambridge hinterland, then they have been effectively washed by an efficient subglacial plumbing system, such that the fines were removed before their deposition.

Andrews (1985) has suggested that the fine fraction in other Baffin Island tills is related to postdepositional processes involving in situ weathering or aeolian input. The matrix size of the tills from the Cambridge Fiord region are

Table 21.1. *Averaged grain size distributions (major fraction percentages and moment statistics) for samples in each of the major sediment categories*

Sedimentary Environment	Samples	Gravel %	Sand %	Mud %	Silt %	Clay %	Mean Ø	S.D. Ø	Sk.	Ku.
1. Till	N=11	47.63	44.60	7.77	---	---	-0.31	3.22	1.19	3.81
2. Gl. Fluvial	N=8	37.96	57.45	4.59	---	---	-0.21	2.76	1.00	4.18
3. Gl. Marine	N=9	14.77	24.76	60.46	32.95	27.51	4.84	4.60	-0.27	2.11
4. Raised Deltas:										
Foresets	N=10	0.61	77.88	21.50	17.37	4.13	2.95	2.19	1.27	5.20
Topsets	N=5	40.94	57.84	1.22	---	---	-0.67	2.22	0.63	3.81
5. Modern Alluvial	N=23	17.27	81.60	1.13	---	---	0.52	1.92	-0.19	4.81
6. Aeolian	N=3	0.30	97.33	2.37	---	---	1.85	1.32	2.11	12.22
7. Fj. Prodelta	N=35	12.65	58.05	29.30	23.89	5.41	2.71	3.14	0.02	3.05
8. Fjord Basin	N=17	1.02	24.40	74.58	42.10	32.47	6.23	3.16	-0.37	2.60
9. Modern Sediment Megasequence		9.31	58.95	31.74	20.13	11.61	3.02	3.52	0.44	2.76
10. Source Sediment Megasequence		22.48	47.22	30.31	17.7	12.61	2.38	4.24	0.51	2.48

comparatively coarser. Andrews's (1985) samples included tills of many different ages and also tills derived from the erosion of carbonate bedrock or marine sediments. Our data suggest that if in situ weathering or aeolian input is responsible for the finer grain sizes of the pan–Baffin Island tills, then those processes only affect sediments that have been exposed for longer periods than the early Holocene samples that we present here.

Early Holocene glaciofluvial sediments

These samples consist of terrestrial deposits preserved as terrace remnants of outwash trains from early Holocene ice margins. They comprise part of the raised sequence of terraces in both the Keel and Cambridge river valleys

(see Fig. 21.3), and consist primarily of coarse sandy gravels or gravelly sands (Figs. 21.5A and 21.8). Although the modal characteristics of these samples are very similar to the till source material, the mud fraction has been further reduced.

The reconstructed distribution of the early Holocene sandur surfaces and the character of the glaciofluvial sediment provide important clues to environmental conditions. They indicate that large quantities of readily transportable glacial sediment were available and were probably subjected to rapid buildup and decay of flood discharges from the ablating ice tongues. The depositional environment was similar to that described by Boulton & Eyles (1979), in which derived sediment is redeposited as planar,

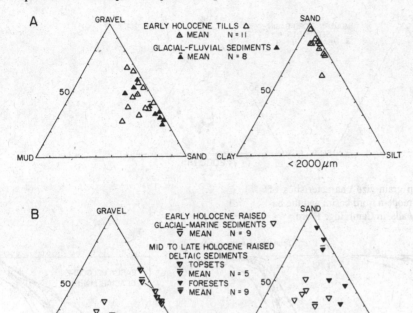

Figure 21.5. Mean grain size characteristics of the early Holocene tills, glacial–fluvial sediments and raised glacial–marine sediments, and mid to late Holocene raised deltaic sediments, from the Cambridge Fiord region.

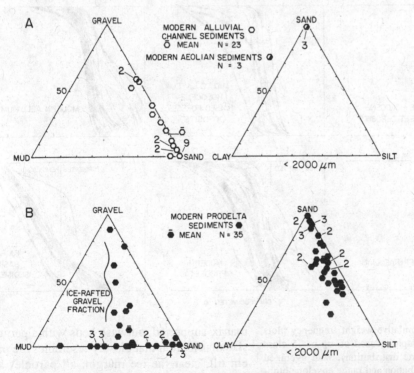

Figure 21.6. Mean grain size characteristics of modern alluvial/deltaic sediments and aeolian sands from the Keel River Valley, Cambridge Fiord, and modern prodelta sediments offshore of the Keel and Cambridge Rivers.

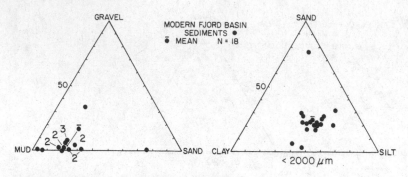

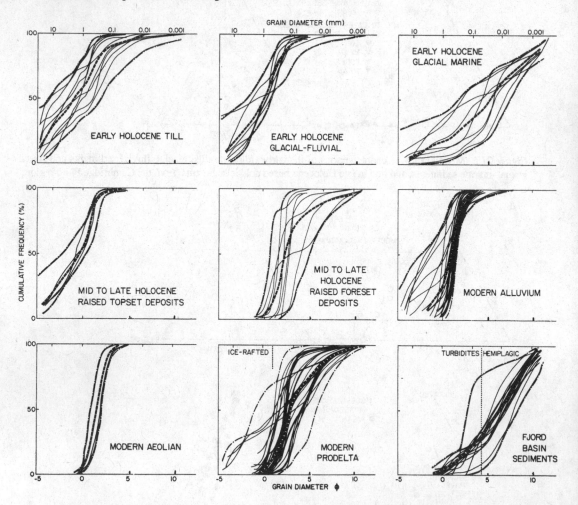

Figure 21.7. Mean grain size characteristics of sediment from the modern fjord basin, and the basin sills and steep walls, in Cambridge Fiord.

Figure 21.8. Cumulative weight frequency plots of individual samples (based on up to 75 class intervals over one distribution), with averaged "sequence" distribution and range envelope highlighted.

matrix-supported, outwash beds with a particle size distribution that closely resembles the parent till. Near the ice margin, all particles are transported and deposited simultaneously. Within a few kilometres of the ice front, however, normal fluvial hydrodynamics become important, concomitant with a decrease in grain size and increase in sorting down-valley of the glacier (Church & Gilbert, 1975). The particle size similarity between our samples of tills and glaciofluvial sediments (Fig. 21.5A) indicates that the latter represent primarily ice-proximal deposition. The finest-grained samples may simply reflect incipient sorting at sites slightly more distal to the ice margin.

Early Holocene raised glaciomarine sediments

Raised glaciomarine sediments are of limited extent within the Cambridge Fiord system (see Fig. 21.2). Therefore, we have chosen to include samples from outside the Cambridge drainage basin in order to quantify more accurately the character of the sediment. The latter samples are indistinguishable from those collected in Cambridge, and their addition does not significantly change the mean or total variance of the group.

We interpret these samples as comprising ice-contact or ice-proximal sediments deposited near a tidewater glacier in shallow water depths (<80 m). They are characteristically muddy (even clayey) with subequal amounts of gravel, sand, and mud (Table 21.1, Figs. 21.5B and 21.8), and are similar to modern equivalents from the seafloor fronting tidewater glaciers (Syvitski, 1989; Powell, in press). The finer-grained component is believed to be largely derived from glaciofluvial discharge into a marine basin, with coarser-grained contributions from iceberg dumping of supraglacial material and sea-ice rafting. Their exposure above present sea level is evidence of coastal emergence following retreat of the ice sheet.

Mid to late Holocene raised deltaic sediments

These deposits are of widespread extent in the Keel River valley and other valleys tributary to Cambridge Fiord. The sediments include relatively well-sorted foreset and more poorly sorted topset deposits being derived principally from the reworking of early Holocene till and glaciofluvial sediments. There is probably some contribution of the coarser fraction from the erosion

of raised glaciomarine deposits; yet the paucity of glaciomarine outcrops in the Keel River valley suggests that they were completely eroded during the mid to late Holocene, or were never raised above present sea level in this valley.

Throughout the inner fjord system, mid to late Holocene raised deltas are the most recently exposed coastal sediment, as the fjord head remains emergent. The topset deposits, composed of gravelly sand (Table 21.1, Figs. 21.5B and 21.8), are similar to or slightly coarser than glaciofluvial sediments from which they are principally derived, although they contain a better-sorted sand mode. The foreset deposits, largely silty sands or sandy silts, do not contain the clay fraction associated with the raised glaciomarine sediments, and are much better sorted (Table 21.1, Figs. 21.5B and 21.8).

Temporary transport deposits
Modern alluvial channel sediments

These deposits differ from their early Holocene raised counterparts (cf. Fig. 21.6A with 21.5A), in that they contain even less mud and have a better sorted sand fraction. This is to be expected from the second-cycle sorting of older glaciofluvial sediments. The gravel content is variable, decreasing toward the delta front. The modern alluvial channel deposits are also $\sim 1\,\phi$ finer than the early Holocene counterparts, reflecting the comparatively lower transport energy associated with the modern rivers and streams (Table 21.1, Figs. 21.6A and 21.8). The coarser-grained samples are derived from longitudinal midchannel bars. We believe that this coarse fraction is rarely transported under the present fluvial regime and thus represents alluvial channel lag deposits.

Modern aeolian sediments

These sands are largely derived from wind erosion of the raised fluvial and deltaic sediments and the modern alluvium covering the seasonally desiccated floodplains (McKenna-Neuman & Gilbert, 1986). Winds in excess of 25 m/s are common in Cambridge Fiord (Syvitski et al., 1984b; Syvitski & Praeg, 1987) and are capable of transporting much of the sand-sized alluvium, as little vegetation exists to protect the sediment from deflation or saltation. The

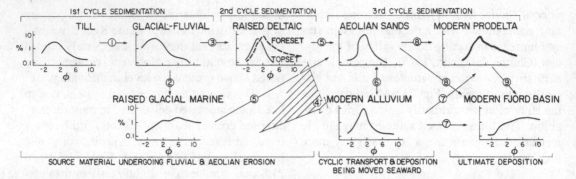

Figure 21.9. A schematic model of the erosion, transport, and ultimate deposition in the Cambridge Fiord system, with actual size frequency distributions of the eight Cambridge Fiord sedimentary sequences (see text for details on numbered sequence linkages).

aeolian sands are moderately well sorted, containing virtually no silt or gravel (Table 21.1, Figs. 21.6A and 21.8). The coarser fraction of the source deposits has been left behind as a lag, with the finer fraction being blown into the fjord basin.

Modern prodelta sediments

These deposits include the sands and silts that have been deposited from bedload along steep-angled foresets (up to 15°), or scavenged from river plume transport carrying the suspended load – the latter demonstrated by the prominent tail seen in most samples (Fig. 21.8). The early scavenging of sands and coarse silts suspended in the plume accounts for the marked fining trend observed in the sand–silt–clay ternary diagram of Figure 21.6B. During the summer and fall months, aeolian silts and sands may additionally be deposited near the delta front. During the winter, aeolian sediment may be transported greater distances seaward over the frozen fjord surface (Gilbert, 1982, 1989). The prodelta gravel fraction, although variable (0%–90%), is <1% on average (Table 21.1, Figs. 21.6B and 21.8). Although some of the gravel might be transported during exceptionally large river flood events, much of the gravel is believed to relate to ice rafting (Gilbert, 1984).

Ultimate deposits
Modern fjord basin sediments

These sediments represent the ultimate submarine deposits within a fjord sedimentary transport system. They are presently at suffi-

cient water depths to remain below wave base, regardless of glacial/interglacial fluctuations in sea level due to isostasy or eustasy. They contain subequal amounts of silt and clay (Table 21.1, Figs. 21.7A and 21.8) with minor amounts of sand and gravel. The gravel fraction is largely related to modern ice rafting, especially from sea-ice-rafted talus debris from along the fjord margins (Gilbert, 1989). A minor amount of gravel could also be the result of debris flows from failures along the steep fjord walls and even from the prodelta environment (Syvitski & Farrow, 1989).

The sand content is largely the result of deposition of turbidity currents and cohesionless debris flows that have occurred episodically within Cambridge Fiord (Syvitski & Farrow, 1989). Submarine channels, observed on sidescan sonargraphs collected off the mouth of the Cambridge River delta, could effectively transport sand into the fjord basin after bypassing the prodelta environment. One sample containing 80% sand, and collected from 500 m of water depth, is considered typical of such a deposit (Figs. 21.7A and 21.8).

Trends in sequence size frequency distributions

In order to visualize the grain size partitioning of sediments as they are recycled through time, we generated "sequence" size frequency distributions for eight sedimentary environments (Fig. 21.9). The size frequency plots were produced by combining all of the samples from that environment and recalculating to pro-

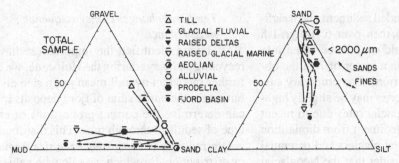

Figure 21.10. Sediment transfer pathways leading to the ultimate deposition of sediment in Cambridge Fiord.

duce a mean distribution. From this we developed a total system size fractionation model (Fig. 21.9) that distinguishes not only among source deposits, temporary transport deposits, and ultimate deposits, but also defines three major cycles of sedimentation.

The first sedimentation cycle is the deposition of the early Holocene glacial sediments that were later to serve as source sediments for mid to late Holocene fluvial erosion. Due to efficient plumbing of the early Holocene glaciers, the tills and glaciofluvial deposits, although poorly sorted, are largely lacking the mud fraction that has been washed into the glaciomarine deposits through linkages ① or ② in Figure 21.9. Once isostatically raised and/or exposed, the glaciomarine and glaciofluvial deposits are recycled during linkage ③ to comprise the mid to late Holocene raised deltaic deposits (second-cycle sedimentation). All four of these deposits have been reworked to variable extent by the modern fluvial regime (linkage ④ in Fig. 21.9), though the coarsest gravel fraction is rarely transported and remains as a lag deposit on the floor of the fjord valley. This completes the second cycle of terrestrial sorting.

The finer fractions of the mid to late Holocene deltaic deposits and early Holocene glaciomarine deposits are further susceptible to erosion by wind (linkage ⑤ in Fig. 21.9). There is a two-way linkage ⑥ between modern aeolian and alluvial deposits, which at separate times of the year rework the other deposit; that is, aeolian dunes and sheet sands are eroded by fluvial action during spring–summer, and fluvial channel bar and floodplain deposits are reworked

during fall–winter aeolian storms. The finer sediments that pass through the fjord valley are transported by river plumes (linkage ⑦) and dust clouds (linkage ⑧) into the prodelta and fjord basin. Finally, through the action of submarine slides, prodelta sediment is subjected to its third cycle of sorting during its transport into the fjord basin.

Changes in mean grain size and the sediment recycling pathways described above can also be plotted as trends on the ternary diagrams (Fig. 21.10). The finer fraction of the source sediments is progressively reworked by major streams, such as the Keel River, which act to partition the coarse glacial deposits into transportable and nontransportable fractions. Thus the sand fraction is primarily residing in the alluvial and prodelta environments, whereas the muds are partitioned into the marine basin. The final result is a very fine-grained marine sediment (Fig. 21.10). If the coarse channel lag deposits could be reliably sampled and analyzed, they would plot at essentially 100% gravel. We can, nevertheless, estimate the total weight percentage of the lag component by mass balance calculations. The following is a budget model that attempts to account for all sediments recycled during the mid to late Holocene.

Mass balance of prodelta and marine basin sediments

Suspended sediment mass balance

Based on sediment volume data from Horvath (1986) and Stravers & Syvitski (1991), between 8.5×10^7 and $1 \times 10^8 \, \text{m}^3$ of basinal sediments have been deposited in Cambridge Fiord proper (not including Buchan Gulf) since deglaciation at about 6,700 yr BP, when the glacier retreated onshore in the Keel River valley. Interestingly, if we take the estimate for modern delivery rates of suspended sediment given earlier (i.e., $3 \times 10^4 \, m^3 a^{-1}$), then, over 6,700 yr, 1.8 $\times 10^8 \, m^3$ of mud would need to be taken into account. This is within a factor of 2 of the observed volume in the fjord. Contemporary suspended sediment deliveries may be slightly higher than average postglacial rates due to recent exposure of surficial sediment from diminution of late Neoglacial outlet glaciers and perennial snowbanks. If we consider that the Neoglacial period (4,200–100 yr BP) is marked by cool, dry summers (with net ice accumulation and glacier advance [Syvitski et al., 1987]), then the modern delivery rates should be reduced for this period, and the two data sets are in reasonable agreement.

Bedload sediment mass balance

The total volume of sediment deposited within the Keel River valley since 6,700 yr BP, through delta progradation, is approximately $3.36 \times 10^8 \, m^3$ ($= 0.5 \times 80 \, m \times 12,000 \, m \times 700 \, m$) comprised of bedload-transported sediment. We initially assume that a negligible amount has been removed through submarine slides and turbidity currents. Based on the modern annual rate of bedload delivery through the Keel River of $1.6 \times 10^4 \, m^3$ (Syvitski et al., 1984a), 1.04×10^8 m^3 of sediment would be delivered in 6,700 yr. This would imply that during this period the modern bedload delivery rate underpredicts (by a factor of 3) the progradation/aggradation observed on the valley floor. The work of Church (1978) may provide the solution to this minor discrepancy. He found that half of the annual sediment yield during the Neoglacial, and for a similar Baffin Island sandur, was derived from erosion of the isostatically uplifted Hypsithermal deposits. We calculate, for the Keel River raised section of the valley (i.e., from the mouth to the marine limit about 12 km upstream; above that point there has been very little erosion of the valley fill), $1 \times 10^8 \, m^3$ was removed through fluvial cannibalism. Most of that sediment would be from early to mid Holocene deltaic and glaciofluvial sediment, and late Holocene alluvial sediments. This brings the two estimates to within a factor of 2 – that is, $1.04 \times 10^8 \, m^3$ for the sediment delivery rate compared to 2.36×10^8 m^3 for the observed volume in the valley.

Temporal changes in megasequence mass balance

Having identified the important sediment recycling pathways during the Holocene, we can further model an overall mean grain size distribution for the entire suite of fjord deposits if we can determine the correct proportions of each type of sediment through time. This distribution is referred to as the *megasequence size frequency distribution,* for which we calculate values at two different time intervals: the early to mid Holocene (first- and second-cycle sediments) and late Holocene or modern (third-cycle sediments). For the megasequence model, the data set is restricted to the mass balance of the Keel River, the Cambridge River, and the Omega Bay system, of which the Keel River is by far the most important drainage.

For the third-cycle sediments the rate of bedload delivery for the Keel River valley (i.e., the volume of sediment in the prodelta) and the suspended sediment delivery to the fjord basin (i.e., the volume of sediment going into the marine basin) are approximately in the same proportion. We take this as a clue and combine these two modern marine sediments in equal proportions in the megasequence. For the terrestrial deposits, proportions were estimated on the basis of the areal distribution of modern aeolian and alluvial deposits and their relative volumes in relation to the marine sediments. We estimate that the following are reasonable proportional values: 10% aeolian, 30% alluvium, 30% prodelta, and 30% basin sediments. Thus we get a megasequence size frequency distribution (Fig. 21.11) having a mean size of 3ϕ and dominated by sand (~60%) with <10% gravel (Table 21.1).

We can similarly generate a megasequence distribution from the first- and second-cycle sediments (Fig. 21.11). The following proportions are based on areal distribution data and relative volume estimates for each of the deposits:
10% till (based on the estimated volume of moraines in the Keel Valley that date from about 7,200–6,000 yr BP);
20% glaciofluvial (based on the reconstructed outwash sandurs of similar age);
40% glaciomarine (based on volume estimates from Stravers & Syvitski [1991]); and

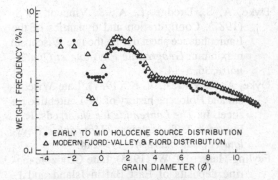

Figure 21.11. Comparison of averaged size frequency "megasequence" distribution comprised of modern fjord valley and fjord sedimentary deposits and compared to a similar megasequence distribution representing the first- and second-cycle source glacial and paraglacial deposits (see text for details).

30% raised deltaic (based on the extent of reconstructed delta surfaces.

We find this "source" megasequence is coarser (2.4ϕ) with less sand (47%) and more gravel (22%: Table 21.1). This may confirm our field observations of incomplete recycling of the coarsest fraction and its storage as immobile gravel lags at interchannel bars.

Conclusions

The interface between the marine and terrestrial environments of Arctic fjords represents a dynamic interplay among glacial isostasy, climate, and sediment supply. The dramatic changes in these three factors through the Holocene are responsible for changes in the partitioning of various grain sizes within the fjord system. Under the present climate regime, the coarse glaciofluvial sediment that was stored in the terrestrial system during the early Holocene is only partially reworked by modern streams of lesser competency. Thus, grain size characteristics of the fjord sediment are responding directly to the change in Holocene climate and a marked decrease in glacier-melt discharge.

The coarse nature of the terrestrial glacial deposits ensures that they are actively transported only during the glacial–deglacial regime, when the drainage basin is naturally much enlarged by the presence of the Laurentide ice sheet. In effect, mid to late Holocene precipitation over the given surface area of the Cambridge drainage basins is insufficient to provide enough runoff to transport the entire suite of valley floor sediments. Thus, the present fluvial energy is in disequilibrium with the fluvial sediments occupying the valley floor. The late Holocene evolution of the fjord drainage represents only minor resculpturing of a landscape, dominated by ice-sheet-scale drainage basins. Thus, the problem of discharge–sediment size disequilibrium can be viewed as a larger-scale problem of basin area and water storage disequilibrium.

Acknowledgments

We would like to thank the officers and crew of *CSS Hudson,* who were of great help during the data collection phase of this research. Onshore field investigations were also supported by the Polar Continental Shelf Project. All of the samples presented in this study were analyzed by Donald Clattenburg and Ken Asprey, and we thank them for the quality of the data. Paul Gottler and Wayne Prime also assisted in data reduction. Drs. John Anderson and John Andrews provided many useful comments on an earlier draft. This manuscript comprises Geological Survey of Canada Contribution no. 35489.

References

Andrews, J. T. (1985). Grain size characteristics of Quaternary sediments, Baffin Island region. In: *Quaternary environments: Eastern Canadian Arctic, Baffin Bay, and Western Greenland,* ed. J. T. Andrews. Boston: Allen and Unwin, pp. 124–53.

(1987). Late Wisconsin glaciation and deglaciation of the Laurentide ice sheet. In: *North America and adjacent oceans during the last deglaciation,* eds. W. F. Ruddiman & H. E. Wright, Jr. The Geology of North America, vol. K-3. Boulder, Co.: Geological Society of America, pp. 13–38.

Andrews, J. T., Barry, R. G., Bradley, R. S., Miller, G. H., & Williams, L. D. (1972). Past and present glaciological responses to climate in eastern Baffin Island. *Quaternary Research,* 2(3): 303–14.

Andrews, J. T., Clark, P., & Stravers, J. (1985). The patterns of glacial erosion across the eastern Canadian Arctic. In: *Quaternary environments: Eastern Canadian Arctic, Baffin Bay, and Western Greenland,* ed. J. T. Andrews. Boston: Allen and Unwin, pp. 69–92.

Andrews, J. T., & Miller, G. H. (1972). Quaternary history of northern Cumberland Peninsula, Baffin Island, N.W.T., Canada: Maps of the present glaciation limits and lowest equilibrium line altitude for northern and southern Baffin Island. *Arctic and Alpine Research,* 4(1): 45–59

——— (1979). Glacial erosion and ice sheet divides, northeastern Laurentide ice sheet, on the basis of the distribution of limestone erratics. *Geology,* 7: 592–6.

ASTM (1959). Symposium on particle size measurement. American Society for Testing Materials Special Technical Publication 234, 303 pp.

Barry, R. G., Bradley, R. S., & Jacobs, J. D. (1975). Synoptic climatological studies of the Baffin Island area. In: *Climate of the Arctic,* eds. G. Weller & S. Bowling. Geophysical Institute, University of Alaska, pp. 82–90.

Boulton, G. S., & Eyles, N. (1979). Sedimentation by valley glaciers: A model and genetic classification. In: *Moraines and varves,* ed. C. Schliichter (Proceedings of a 1978 INQUA Symposium on Genesis and Lithology of Quaternary Deposits), pp. 11–25.

Church, M. (1972). Baffin Island sandurs: A study of Arctic fluvial processes. *Geological Survey of Canada, Bulletin no. 216,* 208 pp.

——— (1978). Palaeohydrological reconstructions from a Holocene valley fill. In: *Fluvial sedimentology,* ed. A. D. Miall, Canadian Society of Petroleum Geologists Memoir, vol. 5. Calgary: CSPG, pp. 743–72.

Church, M., & Gilbert, R. (1975). Proglacial fluvial and lacustrine environments. In: *Glaciofluvial and glaciolacustrine sedimentation,* eds. A. V. Jopling & B. C. McDonald, Society of Economic Paleontologists and Mineralogists Special Publication no. 23. Oklahoma: SEPM, pp. 22–100.

Church, M., & Ryder, R. M. (1972). Paraglacial sedimentation: A consideration of fluvial processes conditioned by glaciation. *Geological Society of America Bulletin,* 83: 3059–72.

Clattenburg, D., Cole, F., Kelly, B., LeBlanc, W., Bishop, P., Rashid, M., Schafer, C. T., & Syvitski, J. P. M. (1983). SAFE: 1982 bottom grab samples. In: *Sedimentology of Arctic Fjords Experiment: Hu83-028 Data Report, vol. 1,* eds. J. P. M. Syvitski & C. P. Blakeney. Canadian Data Report of Hydrography and Ocean Sciences, no. 12, pp. 8-1–8-94.

Dyke, A. S., Dredge, L. A., & Vincent, J. S. (1982). Configuration and dynamics of the Laurentide ice sheet during the late Wisconsin maximum. *Géographie Physique et Quaternaire,* 36, 5–14.

Dyke, A. S., & Prest, V. K. (1987). Late Wisconsin and Holocene history of the Laurentide ice sheet. In: *The Laurentide ice sheet,* eds. R. J. Fulton & J. T. Andrews. *Géographie Physique et Quaternaire,* 41: 199–214.

Feyling-Hansen, R. W. (1985). Late Cenozoic marine deposits of East Baffin Island and E. Greenland: Microbiostratigraphy, correlation, and age. In: *Quaternary environments: Eastern Canadian Arctic, Baffin Bay, and Western Greenland,* ed. J. T. Andrews. Boston: Allen and Unwin, pp. 354–93.

Fisher, D. A., & Koerner, R. M. (1981). Some aspects of climatic change in the High Arctic during the Holocene as deduced from ice cores. In: *Quaternary paleoclimate,* ed. W. C. Mahaney. Norwich, England: GeoAbstracts, pp. 249–72.

Gilbert, R. (1982). Contemporary sedimentary environments on Baffin Island, N.W.T., Canada: Glaciomarine processes in fjords of eastern Cumberland Peninsula. *Arctic and Alpine Research* 14: 1–12.

——— (1984). Coarse particles in the sediments of Cambridge, McBeth and Itirbilung Fiords. In: *Sedimentology of Arctic Fjords Experiment: Hu83-028 Data Report, vol. 2,* ed. J. P. M. Syvitski. Canadian Data Report of Hydrography and Ocean Sciences, no. 28, pp. 9-1–9-25.

——— (1985). Quaternary glaciomarine sedimentation interpreted from seismic surveys of fiords on Baffin Island, N.W.T. *Arctic,* 38(4): 271–80.

——— (1989). Rafting in glaciomarine environments. *Abstract of the Geological Society (London) Symposia on Glaciomarine Environments: Processes and Sediments, March 16–18, 1989.*

Hackett, D. W., Syvitski, J. P. M., Prime, W., & Sherin, A. G. (1986). Sediment size analysis system user guide. Geological Survey of Canada, Open File Report no. 1240, 25 pp.

Hodgson, D. A., & Haselton, G. M. (1974). Reconnaissance glacial geology, northeastern Baffin Island. Geological Survey of Canada Paper no. 74-20, 10 pp.

Horvath, V. V. (1986). Glacimarine sedimentation in Cambridge Fiord, Baffin Island, N.W.T., Canada. Unpublished M.Sc. thesis. Kingston, Ontario: Queen's University, 226 pp.

Jacobs, J. D., Andrews, J. T., & Funder, S. (1985). Environmental background. In: *Quaternary environments: Eastern Canadian Arctic, Baffin Bay, and Western Greenland*, ed. J. T. Andrews. Boston: Allen and Unwin, pp. 26–68.

Klassen, R. A. (1985). An outline of the glacial history of Bylot Island, District of Franklin, N.W.T. In: *Quaternary environments: Eastern Canadian Arctic, Baffin Bay, and Western Greenland*, ed. J. T. Andrews. Boston: Allen and Unwin, pp. 428–60.

McKenna-Neuman, C., & Gilbert, R. (1986). Aeolian processes and landforms in glaciofluvial environments of southeastern Baffin Island, N.W.T., Canada. In: *Aeolian geomorphology*, ed. W. G. Nickling. Norwich: Geo Books, pp. 213–35.

Maxwell, J. B. (1981). *The climate of the Canadian Arctic Islands and adjacent waters. Climatological Studies, vol. 30*. Toronto: Atmospheric Environment Service, .532 pp.

Miller, G. H. (1976). Anomalous local glacier activity, Baffin Island, Canada: Paleoclimatic implications. *Geology*, 4, 502–4

(1985). Aminostratigraphy of Baffin Island shell-bearing deposits. In: *Quaternary environments: Eastern Canadian Arctic, Baffin Bay, and Western Greenland*, ed. J. T. Andrews. Boston: Allen and Unwin, pp. 394–427.

Powell, R. D. (in press). Processes at grounding-lines fans and their growth to ice-contact deltas. *Geological Society Special Publication*.

Quinlan, G. (1985). A numerical model of postglacial relative sea level change near Baffin Island. In: *Quaternary environments: Eastern Canadian Arctic, Baffin Bay, and Western Greenland*, ed. J. T. Andrews. Boston: Allen and Unwin, pp. 560–85.

Schafer, C. T., Clattenburg, D., & Cole, F. (1984). SAFE: 1983 Hudson bottom grab samples. In: *Sedimentology of Arctic Fjords Experiment: Hu83-028 Data Report, vol. 2*, ed. J. P. M. Syvitski. Canadian Data Report of Hydrography and Ocean Sciences, no. 28, pp. 7-1–7-73.

Schafer, C. T., Cole, F. E., & Syvitski, J. P. M. (1989). Bio- and Lithofacies of modern sediments in Knight and Butte Inlets, British Columbia. *Palaios*, 4, 107–26.

Short, S. K., Mode, W. N., & Davis, P. T. (1985). The Holocene record from Baffin Island: Modern and fossil pollen studies. In: *Quaternary environments: Eastern Canadian Arctic, Baffin Bay, and Western Greenland*, ed. J. T. Andrews. Boston: Allen and Unwin, pp. 608–42.

Stravers, J. A. (1987). Late Quaternary glacial and raised marine stratigraphy of northern Baffin Island fjords. In: *Sedimentology of Arctic Fjords Experiment: Data Report, vol. 3*, eds. J. P. M. Syvitski and D. B. Praeg. Canadian Data Report of Hydrography and Ocean Sciences, no. 54, pp. 2-1–2-57.

Stravers, J. A., & Syvitski, J. P. M. (1991). Early Holocene land–sea correlations and glacial reconstruction of Cambridge Fjord, Northern Baffin Island. *Quaternary Research*, 35: 72–90.

Syvitski, J. P. M. (1984a). 1983 geophysical investigations: In: *Sedimentology of Arctic Fjords Experiment: Hu83-028 Data Report, vol. 2*, ed. J. P. M. Syvitski. Canadian Data Report of Hydrography and Ocean Sciences, no. 28, pp. 16-1–16-26.

(ed.) (1984b). *Sedimentology of Arctic fjords experiment: Hu83-028 Data Report, vol. 2*. Canadian Data Report of Hydrography and Ocean Sciences, no. 28, 1130 pp.

(1985). The influence of sea level fluctuations, discharge variations, and sea conditions on Arctic delta formation: Examples from Baffin Island. In: *14th Annual Arctic Workshop Abstracts; Arctic Land–Sea Interactions*. Dartmouth, Nova Scotia: Bedford Institute of Oceanography, pp. 145–9.

(1989). On the deposition of sediment within glacier influenced fjords: Oceanographic controls. *Marine Geology*, 85, 301–29.

Syvitski, J. P. M., Asprey, K. W., Blakeney, C. P., & Clattenburg, D. (1983). SAFE: 1982 delta report. In: *Sedimentology of Arctic fjords experiment: Hu83-028 Data Report, vol. 1*, eds. J. P. M. Syvitski and C. P. Blakeney. Canadian Data Report of Hydrography and Ocean Sciences, no. 12, pp. 18-1–18-41.

Syvitski, J. P. M., & Blakeney, C. P. (eds.) (1983). *Sedimentology of Arctic fjords experiment: Hu83-028 Data Report, vol. 1*. Canadian Data Report of Hydrography and Ocean Sciences, no. 12, 960 pp.

Syvitski, J. P. M., Burrell, D. C., & Skei, J. M. (1987). *Fjords: Processes and products*. New York: Springer–Verlag, 379 pp.

Syvitski, J. P. M., & Farrow, G. E. (1989). Fjord sedimentation as an analogue for small hydrocarbon-bearing fan deltas. In: *Deltas: Sites and traps for fossil fuels*, eds. M. K. G.

Whateley and K. T. Pickering. Geological Society Special Publ. no. 41, pp. 21–43.

Syvitski, J. P. M., Farrow, G. E., Taylor, R., Gilbert, R., & Emory-Moore, M. (1984a). SAFE: 1983 delta survey report. In: *Sedimentology of Arctic Fjords Experiment: Hu83-028 Data Report, vol. 2*, ed. J. P. M. Syvitski. Canadian Data Report of Hydrography and Ocean Sciences, no. 28, pp. 18-1–18-91.

Syvitski, J. P. M., Hay, A. E., Schafer, C. T., & Asprey, K. W. (1984b). SAFE: 1983 bay head pro-delta investigations. In: *Sedimentology of Arctic Fjords Experiment: Hu83-028 Data Report, vol. 2*, ed. J. P. M. Syvitski. Canadian Data Report of Hydrography and Ocean Sciences, no. 28, pp. 17-1–17-62.

Syvitski, J. P. M., & Praeg, D. B. (1987). *Sedimentology of Arctic Fjords Experiment: Hu83-028 Data Report, vol. 3*. Canadian Data Report of Hydrography and Ocean Sciences, no. 54, 468 pp.

Syvitski, J. P. M., & Schafer, C. T. (1985). Sedimentology of Arctic Fjords Experiment (SAFE): Project introduction. *Arctic*, 38: 264–70.

Vorren, T. O., Hald, M., Edvardsen, M., & Lind-Hansen, O.-W. (1983). Glacigenic sediments and sedimentary environments on continental shelves: General principles with a case study from the Norwegian shelf. In: *Glacial deposits in north-east Europe*, ed. J. Ehlers. Rotterdam: Balkema Publ., pp. 61–73.

22 The use of grain size information in marine geochemistry*

DALE E. BUCKLEY AND
RAY E. CRANSTON

Introduction

In recent years there has been renewed interest in environmental studies that combine knowledge gained from several specialized sciences. Particularly noteworthy examples of such studies are those that combine analyses of particle size distributions with chemical analyses of the composition of sediments. Geochemical studies lead to understanding cycles, fluxes, budgets, sources, and sinks of chemical elements in nature. Particle size distributions are an intimate reflection of, or a controlling factor in, most of these geochemical processes.

It has been well known that fine-grained sediments tend to have relatively high metal contents, due in part to the high specific surface area of the smaller particles. This enrichment is mainly due to surface adsorption and ionic attraction (Balistrieri et al., 1981; Li, 1981; McCave, 1984; Horowitz & Elrick, 1987). Trace elements can also be concentrated in iron and manganese oxyhydroxides, which tend to be associated with fine-grained sediments (Tessier et al., 1985). Finally, coatings of organic matter are prevalent in fine-grained sediments, and these coatings bind a variety of trace elements (Wangersky, 1986).

Geochemical characteristics of a sediment can be used to infer the provenance and transport history of the sediment. This knowledge has been used to determine mineral source areas (Baldi & Bargagli, 1982; Tessier et al., 1982), and to determine the source of pollution (Oliver, 1973; Forstner & Salomons, 1980). Such knowledge is often essential in dealing with problems of disposal of drilling, mining, or dredging wastes. In the marine environment the

*Geological Survey of Canada Contribution No. 12689.

highly dispersed fine-grained sediments often pose the greatest threat to marine living resources because of their high mobility and potential higher toxicity.

Fine suspended particulate matter often flocculates in estuaries where mixing of fresh and marine waters takes place. This process concentrates dissolved organic compounds and metals in the flocs (Cranston & Buckley, 1972; Rashid et al., 1972; Cranston, 1976; Sholkovitz, 1978).

Physical properties and chemical characteristics of sediments are also intimately related. Permeability and porosity of sediments determine both the diffusive and advective transport fluxes of chemical constituents through a sediment column. Under certain conditions of diffusion and advection, chemical precipitation may be enhanced. This is accentuated as the contact time between pore water and associated sediments is increased and as the concentration gradient through a sediment layer increases.

When assessing the significance of relationships between sediment particle sizes and the chemical data, analytical techniques and methods must be evaluated. Some of the analytical techniques used to prepare sediment samples for analyses may produce an artifact that distorts the results. For example, freezing sediment samples for storage or drying purposes can cause disaggregation of floccules and may cause precipitation of insoluble salt coatings on particle surfaces. Storage of sediment samples at ambient temperatures can result in bacterial degradation of organic matter and precipitation of oxide coatings. The use of chemical dispersants during particle size analysis may artificially increase the population of fine particles and distort the chemical results. Total elemental analyses are often difficult to interpret because major constituents, such as quartz or biogenic particles, obscure significant variations in minor components. Also, the change in mineralogy with particle size may not be easily detected in bulk geochemical analyses (Fulghum et al., 1988). To overcome some of these difficulties, selective chemical leaches of specific size fractions have been used, although these techniques can alter the surface area and particle size distribution (Horowitz & Elrick, 1987). In some studies the geochemical analyses have been performed on discrete size fractions (Ackerman, 1980; Forstner & Salomons, 1980; Ackerman et al., 1983); however, this latter approach is time consuming and expensive. Loring (1988) has suggested that certain normalization techniques may be effective in discerning the effects of grain size or variations in mineralogy; one such technique compares ratios of metals, such as metal/Al or metal/Li, to discern differences between sediments or reference materials that may have similar mineralogy but different metal contents in different size fractions.

In order to illustrate some of the important relationships that exist between particle size information and geochemical characteristics, we present four case histories of studies that have been carried out from our laboratory. The examples are not intended to be inclusive of all the problems that may be encountered, but they do illustrate several different types of problem that may be tackled by sedimentologists and geochemists working together. By using our data with known confidence limits, we draw conclusions with regard to correlations between particle size measurements and geochemical values.

Effects of mineralogy on chemistry of different size classes: Geochemistry of suspended and bottom sediments in a glacial fjord

A study of an Alaskan fjord was undertaken to examine geochemical processes active during weathering, transport, and deposition of erosion products in the marine environment. The sediments were from weathering and glacial erosion of acid igneous rocks in the southeastern coastal range of Alaska (Buckley & Loder, 1968). Sampling stations were selected from the toe of an alpine glacier, along the glacial meltwater stream, across the fjord delta, and along the axis of the marine fjord. The sampling consisted of collection of suspended particulate matter by centrifugation of large volumes of water, and collection of bottom sediments from the same locations. The suspended particulate matter was later separated, by a combination of elutriation and centrifugation techniques, into four size fractions (Buckley, 1972): >20 μm, 4–20 μm, 0.5–4 μm, and <0.5 μm. Bottom sediments were separated as follows: >4 μm, 2–4 μm, 0.5–2 μm, and <0.5 μm. Each size fraction (<20 μm) was studied by a number of analytical techniques, including total elemental analysis by atomic absorption spectroscopy (Buckley & Cranston, 1971), mineralogical analyses by means of x-ray diffraction (Buckley, 1972) with verification by scanning electron microscopy, and optical microscopy.

Compilation of a very extensive data base on the chemical and mineralogical nature of sediment size fractions from a relatively small geographic area allows one to examine some important relationships between geochemistry and particle size distributions. Table 22.1 shows that the mineral composition changes quite significantly from one size fraction to the other. The phyllosilicates, chlorite, and biotite account for more than half of the minerals in the fine size fractions. The remaining silicates, feldspars, amphiboles, and quartz are most abundant in the coarse fractions. There is some difference in the mineral composition of the suspended sediments as compared with the same size fraction of the bottom sediments. For example, the feldspar content in the finest fraction of suspended sediments is about 6%, whereas the same size fraction in the bottom sediments contains about 16% feldspars. An opposite trend is seen in the amphibole content, where nearly three times more amphiboles are found in the fine fraction of the suspended sediments as compared with the equivalent fraction in the bottom sediments.

It can also be demonstrated that the chemical composition of the minerals varies from one size fraction to another. This was accomplished in this study by carrying out a detailed x-ray diffraction study of the phyllosilicates in each of the size fractions, in both the suspended sediments and the bottom sediments. The composition of the biotite was deduced from an examination of the relative intensity ratios of the (003) and (005) basal reflection planes and comparison with data in Grim (1968). These data showed that the overall general formula for the biotite in the bottom sediments was

$$K(Mg_{0.4}Fe_{0.6})_3 (AlSi_3O_{10}) (OH)_2$$

Table 22.1. *Summary data of relative percentages of minerals in fine-grained suspended and bottom sediments from an Alaskan fjord*

Mineral	Bottom Sediments				Suspended Sediments		
	> 4 μm	2 - 4 μm	0.5 - 2 μm	< 0.5 μm	4 - 20 μm	0.5 - 4 μm	< 0.5 μm
Biotite	35 ± 10	38 ± 10	45 ± 4	55 ± 11	31 ± 11	38 ± 13	51 ± 9
Chlorite	18 ± 6	20 ± 6	23 ± 2	22 ± 12	16 ± 5	16 ± 5	25 ± 3
Feldspars	20 ± 5	19 ± 3	17 ± 6	16 ± 9	18 ± 4	18 ± 4	8 ± 6
Amphiboles	19 ± 13	18 ± 13	13 ± 1	6 ± 5	31 ± 14	31 ± 14	16 ± 7
Quartz	8 ± 3	6 ± 3	2 ± 2	1 ± 3	5 ± 4	5 ± 4	<1

Notes: Mineralogy was determined semiquantitatively by x-ray diffraction analyses. Relative percentages are mean values from nineteen samples of suspended sediments and six samples of bottom sediments. Variation about the mean is indicated by the standard deviation.

whereas the formula for the suspended sediments was

$$K(Mg_{0.5}Fe_{0.5})_3 \, (AlSi_3O_{10}) \, (OH)_2$$

A similar type of x-ray diffraction analysis was carried out for the chlorite minerals in the sediments. In this case the degree of isomorphous substitution of Mg and Fe and of Si and Al were determined by measuring the (060) reflection plane and the (001) basal reflection dimension, according to the relationships described by Brindley (1961). These measurements allowed estimates to be made of the degree of alteration to the general composition formula for chlorites:

$$(Mg,Fe,Al)_6 \, (Al,Si)_4O_{10}(OH)_8$$

The results indicated that the Mg/Fe ratio changed from 1.45 in the largest size fraction of chlorites in the bottom sediments to 1.31 in the finest size fraction. Also the Si/Al ratio for the same compared mineral fractions changed from 4.5 in the largest size fraction to 1.5 in the finest size fraction. The same type of comparison for the suspended sediments yielded Mg/Fe ratios of 0.88 for the coarse fraction and 0.78 for the finest fraction, and Si/Al ratios of 1.5 for the coarse fraction and 1.17 for the fine fraction. These data indicate that the sediment dispersal processes lead to different mineral assemblages and different composition within mineral groups being segregated in the various size fractions of suspended sediments and bottom sediments.

Using information gained from the detailed mineral analyses in this study, it is possible to assign elemental compositions to the mixed assemblage of minerals of each size fraction. This was accomplished by combining the relative abundance of each mineral with the appropriate chemical composition formula for the minerals. This calculated total elemental composition was then compared with the analytically determined total bulk chemical composition (Cranston & Buckley, 1971) of the sediment in each size fraction (Tables 22.2 and 22.3). This comparison demonstrates that bulk chemical data do reflect mineralogical variations: Note the high degree of correlation between the two sets of data. These data also demonstrate the significant variation in composition with each size fraction. Both sets of data show:

1. a decreasing concentration of Na and Ca with size, attributable to decreasing amounts of feldspars in the finer size fractions; and

2. an increase in the concentration of K and Mg with decreasing size, which corresponds to the increased abundance of the phyllosilicates in the finer size fractions.

There is a considerable difference in the quantity of Fe calculated to be present on the basis of mineral formulas and the amount actually measured by the analytical technique. This difference appears to be due to excessive assignment

Table 22.2. *Major element contribution to total bulk chemical composition of bottom sediments by the five most abundant minerals*

Relative Percent Cation Contribution

Size Fraction	Composition Determined From:	Na	K	Ca	Mg	Fe	Mn	Al	Si	Ti	Correlation
> 4 μm	Mineral Abundance	1.4	3.6	2.3	4.9	13.5	0.9	6.8	20.2	0.7	r = 0.857
	Chemical Analyses	2.3	1.8	3.9	1.6	4.1	0.09	8.6	26.2	0.6	p = 99.9
2 - 4 μm	Mineral Abundance	1.4	3.9	2.1	5.2	14.4	1.0	7.1	19.9	0.7	r = 0.903
	Chemical Analyses	2.1	3.1	3.1	3.1	7.1	0.14	9.4	23.5	0.8	p = 99.9
0.5 - 2 μm	Mineral Abundance	1.1	4.3	1.7	5.7	15.4	1.2	7.3	18.9	0.5	r = 0.909
	Chemical Analyses	1.7	3.7	2.5	3.7	8.2	0.18	9.4	20.7	0.8	p = 99.9
< 0.5 μm	Mineral Abundance	0.9	5.2	1.0	5.8	16.4	1.1	7.3	18.7	0.2	r = 0.908
	Chemical Analyses	1.6	3.7	2.1	3.8	8.8	0.19	9.3	19.6	0.9	p = 99.9

Notes: Data are calculated from formula weight percentages and the average abundance of minerals in each size fraction. Analytical results are averages of bulk chemical analyses by AAS of each size fraction.

Table 22.3. *Major element contribution to total bulk chemical composition of suspended sediments by the five most abundant minerals*

Relative Percent Cation Contribution

Size Fraction	Composition Determined From:	Na	K	Ca	Mg	Fe	Mn	Al	Si	Ti	Correlation
4 - 20 μm	Mineral Abundance	1.6	3.2	3.3	5.2	14.0	0.7	6.6	20.0	1.2	r = 0.916
	Chemical Analyses	2.2	2.8	3.6	3.3	7.1	0.13	8.9	22.5	0.7	p = 99.9
0.5 - 4 μm	Mineral Abundance	1.3	3.7	2.7	5.8	15.1	0.9	6.9	18.9	1.0	r = 0.916
	Chemical Analyses	1.8	3.3	3.1	3.9	8.5	0.15	9.0	21.5	0.9	p = 99.9
< 0.5 μm	Mineral Abundance	-0.8	4.6	1.7	6.7	17.0	1.1	6.8	17.9	0.6	r = 0.879
	Chemical Analyses	1.6	3.4	2.6	4.0	8.8	0.14	8.6	20.6	0.9	p = 99.9

Notes: Data are calculated from formula weight percentages and the average abundance of minerals in each size fraction. Analytical results are averages of bulk chemical analyses by AAS of each size fraction.

Table 22.4. *Correlation of major cation composition as determined by atomic absorption spectroscopy, with mineralogy of suspended sediments (n = 43)*

Major Cations	Minerals				
	Biotite	Chlorite	Amphiboles	Feldspars	Quartz
Na	-0.415	-0.419	0.233	0.475	0.552
K	0.658	0.698	-0.686	-0.555	-0.264
Ca	-0.587	-0.554	0.465	0.585	0.519
Mg	0.691	0.690	-0.549	-0.695	-0.614
Fe	0.709	0.708	-0.574	-0.627	-0.696
Mn	0.461	0.524	-0.439	-0.449	-0.270
Al	0.011	0.006	0.059	-0.157	0.211
Si	-0.625	-0.582	0.605	0.532	0.323
Ti	0.718	0.705	-0.647	-0.675	-0.470

Note: Percent probability for the significance of r is < 95% for $r < 0.301$; 95% for $0.30 \leq r < 0.389$; 99% for $0.389 \leq r < 0.485$; and 99.9% for $r \geq 0.485$.

of Fe to the biotite and amphibole minerals. However, both sets of data indicate that Fe is strongly associated with the phyllosilicates.

To further demonstrate these chemical and mineral affinities, we have compiled a correlation table that compares the variations in analytically determined major chemical elements with the measured abundance of minerals in all the size fractions (Table 22.4). These data show very clearly the mineral associations of most of the major elements. Note, however, that Al is not strongly correlated with any of the major minerals, because this element is a major constituent in both the feldspars and the phyllosilicates. Also note that Ti is strongly correlated with the abundance of the phyllosilicates. These observations are important when one considers elemental normalization techniques to demonstrate the influence of specific mineral abundances in certain size fractions. In this case, normalization with Al would be a poor choice if one assumed that Al was representative of the abundance of phyllosilicates. However, normalization with Ti might be ideal to show the relative influence of the phyllosilicates.

The association of minor elements with certain size fractions and with specific minerals

was also evaluated with these samples. Table 22.5 shows the amount of minor elements found in each of the size fractions from the suspended sediments and the bottom sediments. Only the elements Cu, Zn, Li, and Sr show any clear trend with changing size in both the suspended sediments and the bottom sediments. Cu, Zn, and Li are more concentrated in the finer size fractions, whereas Sr is more concentrated in the coarser fractions. There is little clear indication that any element is preferentially concentrated in either the bottom sediments or the suspended sediments, with the exception of Cu, which may be more concentrated in the suspended sediments.

When the variations in minor element concentrations were correlated with the variations in mineral abundance it was found that Co, Cu, and Zn had probable associations with both biotite and chlorite, whereas Li was strongly associated with these two phyllosilicates (Table 22.6). Ni, V, and Cr did not appear to have any particular association with any of the major minerals. Feldspars are known to have minor amounts of Sr in the lattice structure, so it is not surprising that there is a strong association of Sr and the feldspars. Because of the highly sig-

Table 22.5. *Summary of minor element concentrations in various size fractions of bottom and suspended sediments*

Mineral Element	Bottom Sediments				Suspended Sediments		
	> 4 µm	2 - 4 µm	0.5 - 2 µm	< 0.5 µm	4 - 20 µm	0.5 - 4 µm	< 0.5 µm
Ni	92 ± 29	126 ± 84	115 ± 28	108 ± 51	92 ± 27	101 ± 31	91 ± 25
Co	64 ± 21	61 ± 23	69 ± 7	89 ± 39	50 ± 8	50 ± 9	53 ± 15
Cu	20 ± 6	45 ± 15	52 ± 9	60 ± 8	58 ± 16	93 ± 23	112 ± 27
Zn	98 ± 20	180 ± 27	236 ± 24	309 ± 50	203 ± 53	294 ± 80	435 ± 312
Pb	10 ± 1	10 ± 3	9 ± 3	10 ± 0	145 ± 90	180 ± 112	281 ± 233
V	214 ± 41	236 ± 66	283 ± 48	274 ± 43	150 ± 35	164 ± 29	155 ± 23
Cr	130 ± 95	202 ± 119	226 ± 137	266 ± 83	162 ± 96	162 ± 70	150 ± 62
Li	10 ± 6	34 ± 10	58 ± 8	68 ± 12	38 ± 7	50 ± 9	55 ± 8
Sr	482 ± 28	435 ± 43	392 ± 47	350 ± 28	344 ± 32	307 ± 28	258 ± 32

Notes: Mean concentration is in ppm (µg/g) of sediment. Standard deviation is calculated from variation about the mean for each size fraction.

Table 22.6. *Correlation of minor element concentration with mineralogy of suspended sediments (n = 43)*

Minor Elements	Most Abundant Minerals			
	Biotite	Chlorite	Amphiboles	Feldspars
Ni	-0.058	-0.092	0.118	0.004
Co	0.370	0.282	-0.164	-0.513
Cu	0.380	0.408	-0.274	-0.353
Zn	0.344	0.360	-0.359	-0.181
Pb	0.110	0.113	-0.081	-0.084
V	0.087	0.090	0.076	-0.127
Cr	0.158	0.117	-0.107	-0.176
Li	0.672	0.711	-0.658	-0.582
Sr	-0.529	-0.496	0.322	0.555

Note: Percent probability for the significance of r is < 95% for $r < 0.301$; 95% for $0.30 \leq r < 0.389$; 99% for $0.389 \leq r < 0.485$; and 99.9% for $r \geq 0.485$.

nificant correlation of Li with the two phyllosilicates, these data suggest that the element Li may be a good candidate for elemental normalization, when it is desired to show the influence of phyl-losilicates or clay minerals on the chemical composition of sediments.

The clear trends in associations of some of the major and minor elements with both the min-

Table 22.7. *La Have River/Estuary characteristics*

Location	Salinity (ppt)	pH	Suspended Load (mg/L)	Bottom Sediments	
				Organic Carbon (%)	< 16 µm (%)
River	0.8 ± 0.4 (n = 6)	6.4 ± 0.2 (n = 6)	0.6 ± 0.1 (n = 6)	0.3 ± 0.04 (n = 3)	1.0 ± 0.0 (n = 3)
Estuary	4.0 ± 0.9 (n = 4)	7.2 ± 0.2 (n = 4)	0.5 ± 1.0 (n = 4)	4.0 ± 1.0 (n = 4)	13.0 ± 1.0 (n = 4)
Ocean	27.0 ± 1.0 (n = 6)	8.1 ± 0.1 (n = 6)	0.2 ± 0.4 (n = 6)	0.5 ± 0.3 (n = 3)	2.0 ± 0.2 (n = 6)

eralogy and size fractions of these sediments from the glacial fjord environment were easily detected because of the relatively simple mineral composition of the sediments. Another factor that made these associations relatively evident was the lack of a strong influence by organic matter. In research on these same samples by Loder (1971), it was found that the percentage of organic carbon in the suspended sediment was only rarely above 1.0%, and averaged 0.67%. The organic carbon content of the bottom sediments was usually <0.1%, with an average for all samples of 0.074%. This very significant difference in organic carbon content between the suspended sediments and bottom sediments is matched by the dramatic difference in lead content in these respective sediments (Table 22.5). An independent check of correlation of lead content in the suspended sediments with organic carbon content showed that the relationship is definitely significant, with $r = 0.705$.

Summary: We have shown that chemical variability in sediment analyses can be related to mineralogy and sediment texture. Concentration of specific minerals in some size fractions may result in correlations of chemical elements with particle size. This might suggest surface area concentration factors; however, in this case these effects can be directly related to mineralogical factors. The significant correlation of Ti with phyllosilicates and of Li with mica-type clay minerals suggests that these elements may be useful indicators of mineralogical variations in some studies. In such cases they can be used in normalization techniques, in which the variability of other elements relative to Ti and Li can

be used to discern the influence of particle size and adsorption effects.

Surface adsorption effects: Dissolved and particulate metal interactions in a mixed estuary

Evaluation of the fate of metals dispersed from a river into a marine estuary is of great importance in assessing the potential impact of urban and industrial development on coastal marine areas.

One such study conducted on the La Have River and Estuary system on the Atlantic coast of Nova Scotia was designed to examine the processes that might lead to the concentration of potential metal contaminants in estuarine sediments. Interactions between suspended particulate matter and dissolved metals carried to the estuary would be very important in determining the eventual fate of most potential contaminants.

The La Have Estuary reaches 24 km inland from the Atlantic coast of Nova Scotia. Saline water reaches the head of the estuary. At the time of the study of this system in 1970–1, there were no major industrial developments along the river and upper estuary. However, domestic sewage from the town of Bridgewater was discharged into the upper estuary.

Analyses of chemical constituents in the river and estuarine system (Cranston & Buckley, 1972; Cranston, 1976) were used to characterize the processes that might be responsible for the transport of potential contaminants through the system. Table 22.7 contains a summary of some of these constituents in surface water and bottom sediments. From these data it is evident that the mean salinity and pH increased seaward reflecting the mixing of marine water with river water in the estuarine region, defined by salinities between 2 and 20 ppt. These areas also contained elevated levels of suspended particulate matter, with the estuary containing especially high concentrations of organic carbon. The increased content of fine-grained mud in the estuarine sediment was deduced to be the result of processes of flocculation. As the fine-grained colloids, with a net negative charge, were discharged from the river into the estuary, the charge was neutralized in the saline waters and flocculation occurred (Boyle et al., 1977). At the same time, the increase in pH would be sufficient to cause the precipitation of dissolved iron as an oxyhydroxide phase (Stumm & Morgan, 1970).

An examination of trace metal analyses of the water, suspended particulate matter and bottom sediments (Table 22.8) makes it possible to identify various processes responsible for the dispersion or accumulation of metals throughout the estuarine system. The rapid decrease in dissolved Fe concentration by >100 ppb between the river and estuary, coincident with the increase of particulate Fe by >100 ppb, suggests that iron is being transformed from a dissolved or colloidal phase into a particulate phase as the salinity increases. This increase in particulate iron is also reflected in bulk analyses of the total bottom sediment, and the separated fine fraction (<16 μm) of the bottom sediments. Similar trends are also observed for Cu, Zn, and Pb analyses, where dissolved concentrations in the estuarine waters are lower than those found in the river water.

Also, the concentration of these metals in the suspended particulate matter and the bottom sediment from the estuary are considerably higher than in the river or marine areas. The likely mechanism for removal of the dissolved metals from the fresh or brackish water is by coprecipitation with the Fe oxyhydroxides or by adsorption on the flocculating particulate matter.

The partition and dispersion of mercury appear to differ somewhat from those of the other metals. The main difference is that the concentration of dissolved Hg in the estuarine water is greater than in any other part of the system. Cranston & Buckley (1972) attributed this to the addition of dissolved Hg from the nearby sewage disposal outfalls at the town of Bridgewater. Mercury adsorbed on the organic rich floccules in the estuary causes considerable enrichment of Hg in both the suspended particulate matter and the bottom sediments. The especially high concentration factors for the fine fraction of the bottom sediments indicates that this metal is adsorbed in proportion to the specific surface area of the particles (Cranston, 1976; Baldi & Bargagli, 1982).

The distribution of Mn throughout the river and estuarine system is markedly different than any of the other metals. There is a decrease in both the dissolved and particulate phases of the metal as river water mixes with the marine water through the estuary to the open coastal marine area. The concentration of Mn in the bottom sediments increases in the seaward direction, reaching the highest concentrations in the coastal sediments. Unlike most of the other trace elements, Mn is not enriched in the fine fraction of the bottom muds. This observation leads to the conclusion that Mn is not removed from solution by adsorption on the finest particles, as is the case for the other trace metals. These observations are similar to those made in other estuarine studies (Buckley & Winters, 1983). One possible explanation of this behaviour is that Mn reactions with particulate matter occur rather slowly, especially if Mn removal from solution is due to bacterial oxidation (Kepkay, 1985).

Removal of metals from solution is often thought to be related to the size and specific surface area of the associated suspended particles in the water. Also inorganic coatings of oxyhydroxides or adsorbed organic coatings on particles may enhance removal of metals from solution. Data from this study were used in an effort to evaluate the effect of particle interaction in the estuary. The metal and organic carbon content of the bulk mud and the fine silt–clay fraction (<16 μm) of the muds were determined from a series of bottom sediment samples from the estuary. The results of these analyses were then compared in a multiple correlation analysis (Table 22.9). The metals Fe, Cu, Zn, Pb, and Hg are very significantly positively correlated, but Mn is negatively correlated with Fe. This

Table 22.8. *Metal distribution in the La Have Estuary*

	Location	Bulk Sediment	Fine Fraction	Dissolved	Particulate
Fe		(Fe, %)	(Fe, %)	(Fe, ppb)	(Fe, ppb)
	River	3.7 ± 0.5 (n = 3)	- -	140.0 ± 30.0 (n = 6)	44.0 ± 5.0 (n = 8)
	Estuary	4.8 ± 0.2 (n = 4)	5.6 ± 0.1 (n = 18)	34.0 ± 9.0 (n = 4)	160.0 ± 12.0 (n = 4)
	Marine	2.1 ± 0.3 (n = 5)	4.4 ± 0.2 (n = 5)	5.0 ± 2.0 (n = 6)	60.0 ± 4.0 (n = 19)
Cu		(Cu, ppm)	(Cu, ppm)	(Cu, ppb)	(Cu, ppb)
	River	11.0 ± 0.3 (n = 3)	-	1.2 ± 0.2 (n = 6)	0.04 ± 0.1 (n = 8)
	Estuary	38.0 ± 10.0 (n = 4)	26.0 ± 1.0 (n = 18)	0.8 ± 0.1 (n = 4)	0.23 ± 0.04 (n = 17)
	Marine	18.0 ± 2.0 (n = 5)	20.0 ± 2.0 (n = 5)	0.5 ± 0.1 (n = 6)	0.12 ± 19.0 (n = 19)
Zn		(Zn, ppm)	(Zn, ppm)	(Zn, ppb)	(Zn, ppb)
	River	33.0 ± 18.0 (n = 3)	-	5.3 ± 0.8 (n = 6)	0.16 ± 0.04 (n = 8)
	Estuary	130.0 ± 10.0 (n = 4)	150.0 ± 11.0 (n = 18)	4.7 ± 0.7 (n = 4)	0.50 ± 0.05 (n = 17)
	Marine	40.0 ± 10.0 (n = 5)	130.0 ± 20.0 (n = 5)	2.5 ± 1.0 (n = 6)	0.20 ± 0.02 (n = 19)
Pb		(Pb, ppm)	(Pb, ppm)	(Pb, ppb)	(Pb, ppb)
	River	- -	- -	1.3 ± 0.4 (n = 6)	0.24 ± 0.5 (n = 8)
	Estuary	45.0 ± 5.0 (n = 4)	120.0 ± 10.0 (n = 18)	0.45 ± 0.1 (n = 4)	0.68 ± 0.13 (n = 17)
	Marine	55.0 ± 15.0 (n = 5)	60.0 ± 10.0 (n = 5)	0.16 ± 0.05 (n = 6)	0.13 ± 0.01 (n = 19)
Hg		(Hg, ppm)	(Hg, ppm)	(Hg, ppb)	(Hg, ppb)
	River	0.55 ± 0.1 (n = 3)	-	0.07 ± 0.03 (n = 6)	0.006 ± 0.001 (n = 8)
	Estuary	1.2 ± 0.5 (n = 4)	6.2 ± 1.2 (n = 18)	0.18 ± 0.15 (n = 4)	0.020 ± 0.01 (n = 17)
	Marine	0.44 ± 0.11 (n = 5)	4.1 ± 0.7 (n = 5)	0.06 ± 0.01 (n = 6)	0.008 ± 0.001 (n = 19)
Mn		(Mn, ppm)	(Mn, ppm)	(Mn, ppb)	(Mn, ppb)
	River	550.0 ± 160.0 (n = 3)	-	29.0 ± 3.0 (n = 6)	2.2 ± 0.5 (n = 8)
	Estuary	720.0 ± 30.0 (n = 4)	580.0 ± 7.0 (n = 18)	19.0 ± 4.0 (n = 4)	1.6 ± 0.2 (n = 17)
	Marine	740.0 ± 130.0 (n = 5)	600.0 ± 9.0 (n = 5)	3.0 ± 1.0 (n = 6)	0.8 ± 0.04 (n = 19)

agrees with the earlier discussion of data from Table 22.8. None of the metals was found to correlate with the organic carbon content, thus suggesting that organic complexing or coatings were not collecting a significant amount of met-al, even though organic matter plays a predominant role in flocculation (Eisma, 1986).

Summary: The results from these experiments verified that the finer muds are more effective in concentrating some metals such as Fe,

Table 22.9. *Correlation coefficients for trace elements and organic carbon in total mud fraction sediments from La Have Estuary samples*

	Fe	Cu	Zn	Hg	Pb	Mn
Fe	-	0.80	0.83	0.75	0.69	-0.83
Cu	0.80	-	0.74	0.81	0.76	-
Zn	0.83	0.74	-	0.73	-	-
Hg	0.75	0.81	0.73	-	0.69	-
Pb	0.69	0.76	-	0.69	-	-
Mn	-0.83	-	-	-	-	-
Org. C.	-	-	-	-	-	-
< 16 μm*	0.64	-	-	-	-	-0.70

Note: Coefficients are significant at a 99% probability level, $n = 23$.
* Percentage of total mud in the <16-μm size fraction.

Cu, Zn, and Pb from solution. The apparent mechanism of removal of these metals from solution is by coprecipitation with Fe oxyhydroxide coatings. Mercury is also enriched in the finer particulate matter, but the mechanism is likely by adsorption with organic matter. Manganese is not enriched in the finer sediment particles, and therefore is negatively correlated with the other trace elements.

Textural and geochemical properties as provenance indicators: Distal turbidites in the Madeira Abyssal Plain

The Madeira Abyssal Plain is located in the northeast Atlantic Ocean between the Madeira Islands on the east and the abyssal hills leading to the Mid-Atlantic Ridge on the west. It is adjacent to the northwest African Continental Margin, which feeds turbidites to the abyssal plain (Weaver & Kuijpers, 1983). Long piston cores to subseafloor depths of 35 m were collected at three locations in the area. Detailed discussions of the occurrence and origin of eighteen turbidite units have been presented by DeLange et al. (1989) and by Weaver et al. (1989). Geochemical data for the three cores have been discussed by Buckley & Cranston (1988), DeLange et al. (1989), and Weaver et al. (1989). However, detailed examination of the size analyses of sediments by the Coulter Counter technique have

not previously been presented. This case history will deal with turbidite characteristics based on results of the Coulter Counter analyses and associated sedimentary geochemistry. The complete data set on which these discussions are based is available from Buckley et al. (1989).

Most of the eighteen turbidites identified in the long cores were >1 m thick and could be correlated between cores spaced up to 100 km apart. Only 12 subsamples out of 289 from these cores had mean particle sizes >4 μm, and these were found at or near the base of the turbidites; the remaining 277 samples had mean sizes in the clay size class (<4 μm). No significant correlation exists between the mean particle size and depth in the turbidite, leading to the conclusion that sorting of the clays did not occur during transport of the turbidite material to the abyssal plain. This conclusion agrees with the discussion by McCave (1988), in which he concluded that the turbidity currents "freeze" during transport and the high concentration of particles in the clay fraction are prevented from differential settling by interparticle forces.

DeLange et al. (1989) studied the geochemical characteristics of the major turbidites in an effort to classify the turbidites into categories based on source and composition. One class of turbidites, consisting of only three turbidites, was identified as carbonate-rich. These relatively thin turbidites (average thickness 100 cm)

Table 22.10. *Particle size and geochemical data for major turbidite types, Madeira Abyssal Plain*

Parameter	Organic-Rich	Volcanic-Rich	Carbonate-Rich
No. Observations	148	119	22
Mean Size (μm)	3.1 ± 0.6	3.4 ± 1.0	3.3 ± 2.9
Sand (%)	0.0 ± 0.0	0.0 ± 0.0	1.0 ± 7.0
Silt (%)	39.0 ± 7.0	43.0 ± 10.0	28.0 ± 8.0
Clay (%)	61.0 ± 7.0	57.0 ± 10.0	70.0 ± 12.0
Organic Carbon (%)	0.77 ± 0.44	0.12 ± 0.06	0.06 ± 0.02
Fe/Al Ratio	0.64 ± 0.11	0.89 ± 0.19	0.55 ± 0.11
CaCO$_3$ (%)	50.0 ± 6.0	56.0 ± 6.0	82.0 ± 3.0

make up only 10% of the sediment column. The average carbonate content of these turbidites was 82%.

A more common turbidite type was identified as volcanic-rich and accounted for 30% of the sediment column. This type of turbidite tended to be somewhat thicker (250 cm) on average, and contained an average of only 0.12% organic carbon. The volcanic-rich turbidites contained significantly more Fe, normalized to total Al. DeLange et al. (1989) also found that the volcanic-rich turbidites were enriched in Ti and Zr as well as Fe, and that these turbidites contained volcanic glass and pumice. They suggested that the source of these turbidites was the volcanic islands to the east of the abyssal plain (Madeira) and to the north (Azores).

DeLange et al. (1989) concluded that half of the sediment column consisted of organic-rich sediment that was transported to the abyssal plain from the northwest African Continental Margin. The ten turbidites in this class had average thickness of 150 cm. The average organic carbon content of these turbidites was 0.77%.

In the subsequent discussion of the turbidite characteristics three long cores are considered. Two of these cores were collected on the eastern side of the Madeira Abyssal Plain (cores 24 and 37) and one core on the western side of the abyssal plain (core 10). Table 22.10 contains a summary of textural data and selected geochemical parameters for the three types of

turbidites described above. From these data it may be seen that the three types of turbidites show a significant difference in the silt content (the means are different at the 99% probability level). Also, the mean particle size of the organic- and volcanic-rich turbidites are significantly different when examined by a *t*-test. Each type of turbidite is examined below in some detail to determine if particle size data and geochemical characteristics can be used together to identify the origin of the turbidites.

Carbonate-rich turbidites

Individual carbonate-rich turbidites were found in each of the three cores. Table 22.11 contains a partial summary of particle size and geochemical data for these turbidites. Data from the eastern cores can be compared with equivalent data from the western core. The means of the various characteristics have been *t*-tested to determine whether they are significantly different. The probability that any means are different is given in the probability column, if it is equal to or greater than 99%.

The size data appear to show that carbonate-rich turbidites from the western side of the abyssal plain have slightly more silt and less clay than do samples from the same turbidites in the eastern part, suggesting that the coarser sediment originated from the west, although higher-resolution sampling would be required to prove this conclusion at a 99% confidence level. Based

Table 22.11. *Particle size and geochemical data for carbonate-rich turbidites in the Madeira Abyssal Plain*

Parameter	Eastern Side of Plain	Western Side of Plain	Probability * Means are Different (%)
No. Observations	11	11	
Mean Size (μm)	2.4 ± 0.1	4.1 ± 4.1	-
Sand (%)	0.0 ± 0.0	3.0 ± 9.0	-
Silt (%)	25.0 ± 2.0	32.0 ± 9.0	-
Clay (%)	75.0 ± 2.0	66.0 ± 15.0	-
Organic Carbon (%)	0.06 ± 0.01	0.06 ± 0.02	-
Fe/Al Ratio	0.47 ± 0.06	0.62 ± 0.11	99.0
$CaCO_3$ (%)	82.0 ± 3.0	82.0 ± 4.0	-

* Only probabilities of ≥99.0 are reported.

in part on paleontological evidence, DeLange et al. (1989) also concluded that these carbonate-rich turbidites originated in the abyssal hills to the west of the Madeira Abyssal Plain.

From a chemical point of view, the sediments taken in the western core (core 10) are more enriched in Fe relative to Al as compared with the sediments obtained from the eastern cores (cores 24 and 37). This chemical difference is probably due to the additional small amounts of clay minerals and volcanic minerals present in the eastern samples (DeLange et al. 1989; Weaver et al. 1989). There is no difference in the amounts of $CaCO_3$ or organic carbon found in the eastern and western core samples of the carbonate-rich turbidites.

Volcanic-rich turbidites

When the size and chemical data for the volcanic-rich turbidites are compared on an eastern versus western basis, there appears to be no difference at the >99% significance level. This result may be somewhat surprising in that Weaver & Kuijpers (1983), DeLange et al. (1989), and Weaver et al. (1989) all report that the bases of the turbidites are progressively more fine grained toward the west. This trend supported the conclusion that the source of most of the turbidites was to the east. In addition, DeLange et al. (1989) found that the volcanic-

rich turbidites contained significant quantities of heavy minerals.

Geochemical analyses of the volcanic-rich turbidites in the three cores compared in this study do not show any highly significant difference in the heavy mineral content in either the eastern or western side of the Madeira Abyssal Plain. These results support the above conclusion, that the fine fraction of distal turbidites is often prevented from differential settling and sorting.

Organic-rich turbidites

As is the case with the volcanic-rich turbidites, there is no evidence that the organic-rich turbidites can be differentiated between eastern and western types, based on chemical or textural properties; however, organic-rich turbidites display a particular feature that allows them to be subdivided on a vertical basis. Buckley & Cranston (1988) showed that the organic-rich turbidites had been partially oxidized at the top in what they called a "paleo-oxidized zone." The diagenetic process responsible for this feature resulted from penetration of bottom-water oxygen into the surface of freshly deposited turbidites, resulting in the oxidation of most of the organic matter contained in the upper few centimetres of the turbidite. This process also resulted in the dissolution of some of the carbonates

Table 22.12. *Particle size and geochemical data for oxidized and reduced portions of organic-rich sediments*

Parameter	Oxidized Portion		Reduced Portion		Probability* Means are Different (%)
No. Observations	27		109		
Mean Size (μm)	2.9	± 0.2	3.0	± 0.2	-
Sand (%)	0.0	± 0.0	0.0	± 0.0	-
Silt (%)	37.0	± 4.0	39.0	± 4.0	-
Clay (%)	63	± 4.0	61.0	± 4.0	-
Organic Carbon (%)	0.24	± 0.14	0.93	± 0.36	99.9
Fe/Al Ratio	0.67	± 0.12	0.63	± 0.11	-
$CaCO_3$ (%)	46.0	± 6.0	50.0	± 4.0	99.9
Water (% wet wt.)	47.0	± 3.0	51.0	± 4.0	99.9

* Only probabilities of ≥ 99.0 are reported.

in the oxidation zone. With the exclusion of four basal turbidite silt layers, there was no significant difference in the particle size in the unoxidized portion of the individual turbidites as compared with the paleo-oxidized zones at the top of the turbidites. There was, however, a significant difference in the water content of the two portions of each of the turbidites. These results are summarized in Table 22.12.

In order to understand more about the subtle differences between the paleo-oxidized zone and the underlying unoxidized portion of the turbidites, more than 100 textural and geochemical variables were correlated. The most significant of these results are shown in Table 22.13. The nature of the paleo-oxidation zone can be depicted somewhat by the oxidation potential (pE), in which the more oxidized sediments have higher pE values. The fact that present-day measurements of pE still reflect relative differences in the redox potential, in spite of the fact that some of these turbidites were deposited several hundred thousands of years ago (Weaver et al., 1989), indicates that some biogeochemical processes may still be active. Thus, the negative correlation between pE and $CaCO_3$ and water content confirms the statistical difference shown in Table 22.12, which showed that the relative-

ly more oxidized paleo-oxidized zone had lower water contents and lower $CaCO_3$. The negative correlation between the mean particle size and $CaCO_3$ throughout the turbidite units suggests that finer-grained turbidites contain more carbonate material. The positive correlation of mean size with SiO_2 suggests that slightly coarser-grained turbidites may contain more quartz or opaline silica. This variability is probably a reflection of the source of the individual turbidites, in which some have finer particle sizes but more carbonate content and less silica.

The negative correlations between water content and the variables pE, Mn, Fe, and SiO_2 require special attention. The negative correlation with SiO_2, and positive correlation with $CaCO_3$, is logical, since we had already concluded that the paleo-oxidized zones with less water also contained less carbonate due to oxidation reactions. The relative depletion in carbonates would be offset by a relative increase in silica. The negative correlation between water content and the two metal variables provides a clue to the processes responsible for the lower water contents in the paleo-oxidized zones. The Mn and Fe variables are the quantity of leachable metal that could be extracted from the sediment by a hot reducing leach (Buckley & Cran-

Table 22.13. *Correlation of analytical results for oxidized, organic-rich turbidites, Madeira Abyssal Plain*

	Depth	Size	Org. C	pE	CaCO$_3$	Water	Mn*	Fe*	SiO$_2$
Depth (cm)	1.00								
Size (μm)	-	1.00							
Org. C (%)	-	-	1.00						
pE	-	-	-	1.00					
CaCO$_3$	-	-0.51	-	-0.46	1.00				
Water (%)	-	-	-	-0.48	0.56	1.00			
Mn (ppm)*	-	-	-	-	-	-0.58	1.00		
Fe (ppm)*	-	-	-	-	-0.53	-0.55	0.81	1.00	
SiO$_2$ (%)	-	0.57	-	-	-0.85	-0.46	0.48	0.52	1.00

Note: Only those correlation coefficients that are significant at $p > 99\%$ are listed.
* Reducible metal fraction only (HA).

Table 22.14. *Geochemical and sedimentological variables in Emerald Basin core*

	(Mean ± SD)	Sand (%)	Silt (%)	Clay (%)	Mud (%)	Mean (phi)	Mean (μm)	SD (phi)	Kurt	Skew
Sand (%)	(0.7 ± 0.5)	1.000			-1.000					
Silt (%)	(70.4 ± 8.2)		1.000							0.963
Clay (%)	(29.2 ± 8.7)		-0.999	1.000						-0.967
Mud (%)	(99.4 ± 0.5)				1.000					
Mean (phi)	(6.74 ± 0.8)	-1.000	-0.963	0.973	-0/593	1.000			-0.626	-0.986
Mean (μm)	(8.6 ± 2.7)	0.593	0.893	-0.906		-0.971	1.000		0.768	0.969
SD (phi)	(1.65 ± 0.07)							1.000		
Kurt	(2.29 ± 0.40)					-0.626	0.768		1.000	0.681
Skew	(0.40 ± 0.31)		0.963	-0.967		-0.986	0.969		0.681	1.000

Fe_{wA} (ug·g^{-1})	(2794 ± 1517)	-0.611	0.616	0.685	-0.709		-0.726
Mn_{wA} (ug·g^{-1})	(147 ± 39)	-0.695	0.706	0.807	-0.845	-0.602	-0.828
Cu_{wA} (µg·g^{-1})	(0.35 ± 0.34)						
Fe_{HA} (µg·g^{-1})	(528 ± 222)	-0.696	0.701	0.782	-0.805		-0.812
Mn_{HA} (µg·g^{-1})	(13.4 ± 2.7)	-0.636	0.650	0.757	-0.828	-0.687	-0.770
Fe_{HHA} (µg·g^{-1})	(5615 ± 686)				-0.624		
Mn_{HHA} (µg·g^{-1})	(26 ± 6)						
Cu_{HHA} (µg·g^{-1})	(1.0 ± 0.4)						
Fe_R (µg·g^{-1})	(24953 ± 5446)			0.620	-0.672		-0.633
Mn_R (µg·g^{-1})	(247 ± 62)						
Cu_R (µg·g^{-1})	(15 ± 3.7)			0.661	-0.712		-0.666
Fe_{TR} (µg·g^{-1})	(26800 ± 8090)			0.623	-0.675		-0.638
Mn_{TR} (µg·g^{-1})	(273 ± 69)						
Cu_{TR} (µg·g^{-1})	(16 ± 4)			0.665	-0.716		-0.672
Fe_{SUM} (µg·g^{-1})	(33892 ± 7418)			0.668	-0.717		-0.686
Mn_{SUM} (µg·g^{-1})	(484 ± 74)						
Cu_{SUM} (µg·g^{-1})	(16 ± 3.7)			0.657	-0.707		-0.651
Fe_T (µg·g^{-1})	(34600 ± 7809)			0.640	-0.693		-0.663
Mn_T (µg·g^{-1})	(459 ± 103)						
Cu_T (µg·g^{-1})	(14 ± 2.6)					-0.607	
OC (%)	(0.48 ± 0.12)	-0.668	0.672	0.736	-0.764	-0.583	-0.764
$CaCO_3$ (%)	(5.1 ± 1.0)				-0.622		-0.622
pE (-log M)	(4.21 ± 0.63)						
SiO_2 pw (mM)	(0.424 ± 0.117)						
SO_4 pw (mM)	(17.7 ± 4.3)	0.673	-0.678	-0.761	0.756		0.776
NH_4 pw (mM)	(1.31 ± 0.45)	-0.680	0.692	0.783	-0.797		-0.797
Depth (cm)	(317 ± 177)	-0.710	0.715	0.786	-0.795		-0.795

Note: Correlation coefficients are shown where the significance is > 99.9% probability. Partial correlation matrix for 26 samples.

ston, 1988). Metals extracted by this technique are thought to originate from relatively stable oxide coatings or cements. Thus as the amount of these coatings increases and fills pore spaces the water content would decrease. This would not necessarily lead to a change in particle size distribution, as measured by the Coulter Counter technique, because during the preparation stages for the analysis the coatings might be removed and the initial particles could be disaggregated.

Summary: Grain size information and geochemistry can be a useful means of recognizing and classifying different types of turbidites. However, in this case study particle size information and geochemical characteristics of fine-grained (clay-size) turbiditic sediments could not be used to distinguish source and sorting processes of some carbonate- and volcanic-rich turbidites. On the other hand, organic-rich turbidites derived from continental margin sources retained both textural and geochemical (mineralogical) imprints of their source areas. The imprints of diagenetic processes, such as postdepositional oxidation, are clearly recognizable in geochemical characteristics of some turbidites. However, these diagenetic processes do not appear to affect the initial grain size characteristics of the turbidite.

Discrimination between detrital characteristics and chemical processes: Depositional facies and diagenesis in a continental shelf basin

An important interpretation task, and often the most difficult, is to decipher the relative influences of depositional processes and diagenetic processes. This case history is typical of this type of problem, and illustrates how a combination of sedimentological and geochemical data can be used to determine the relative importance of these two influences.

A 6.5-m-long piston core was collected from the southeastern margin of the Emerald Basin on the Scotian Shelf off the east coast of Canada (Piper, 1988). This core intersected a sequence of postglacial silty mud representing deposition extending back as far as about 14 ka (Gipp & Piper, 1989). Sediments below the

depth of 2.9 m in the core are described as grey-brown mud with occasional dispersed gravel and dropstones, and with abundant bioturbation traces. The sediments above this depth are olive grey muddy silt with some laminae of sand or silt.

Sedimentological and geochemical analyses included Coulter Counter analyses of particle size distributions, total silicate analyses (Buckley & Cranston, 1971), and a series of sequential leach tests to extract labile metals under varying chemical conditions (for details and complete data set, see Fitzgerald et al. [1987, 1989]). Our selected data include twenty-eight chemical and sedimentological variables from twenty-seven subsamples.

Chemical data used in this study include results from weak acetic acid leaching of the sediments for the metals Fe, Mn, and Cu. These data are identified by the subscript WA in Table 22.14. Results from analyses of the easily reducible form of Fe and Mn were obtained from a hydroxylamine leach using $1\ M$ NH_2OH–HCl, and are identified by the subscript HA in Table 22.14. Analyses of the chemical fraction associated with more stable oxides were obtained from a heated hydroxylamine leach at $pH\ 2$ and are identified by the subscript HHA. The residue of sediment left after the three previous leaches was then analyzed by the total silicate method (Buckley & Cranston, 1971); these results are designated by the subscript R. They were also recalculated on the basis of the residue sample weight to provide data that indicated the concentration in the residue only; these results are identified by the subscript TR. Two independent methods were used to determine the total concentration of metals in the sediment. The sum of the three leaches plus the residual concentration gives the results shown by the subscript SUM. An independent analysis by a single total decomposition of the sample is designated by the subscript T. These latter two results will differ by a small amount that reflects the experimental error.

Other chemical analyses used in this study include determination of total carbon (TC), and organic carbon (OC) as determined with a Leco carbon analyzer (Fitzgerald et al., 1989). $CaCO_3$ was calculated from the difference between the

TC and OC. Determination of dissolved SiO_2, SO_4, and NH_4 in the pore water were included to provide a chemical index of the extent of chemical diagenesis and reduction in the pore water associated with these sediments. A more direct measure of the redox potential of the sediments was obtained from a coupled platinum electrode inserted into the freshly split core sample (Fitzgerald at al., 1989). The results of these measurements are expressed as pE.

Size characteristics are described by the percentage of sand, silt, and clay as determined from the Coulter Counter results. Mud percentage is the sum of the silt and clay. The mean size of particles over the entire size range is expressed in phi units as well as micrometers. The standard deviation (S.D.) of the phi mean was included as a statistical measure of variation about the mean, but has limited significance as a measure of the sorting characteristic of the size distribution. The standard measures of the kurtosis (centre peak sharpness of the distribution) and the skewness of the size distribution have been included. In the convention used here, positive skew indicates a distribution with an extended tail toward the small particle sizes.

Sediments from the upper part of the core are slightly more silty and less chemically reduced than the lower part of the core (Table 22.14). The silt content and mean size (in micrometres) are negatively correlated with depth, whereas clay content and mean (phi) size are positively correlated with depth. There is also a strong indication that the sediment size distributions become less positively skewed toward the bottom of the core. Although direct correlations of the redox sensitive chemical indicators are not shown in Table 22.14, the SO_4 and NH_4 results from analyses of the pore water show that there is a strong trend toward more strongly reducing conditions in the lower part of the core. The concentration of SO_4 decreased from 27 mM to 11 mM from top to bottom of the core. The concentration of NH_4 at the top of the core was 0.2 mM and increased by an order of magnitude at the bottom of the core.

It is important to determine which chemical and sedimentological variables are linked to depositional conditions and which are linked to postdepositional diagenetic processes. The data compiled in Table 22.14 provide some indication as to which variable might be most reliable for making these determinations. Because the sand content is so small in most of these samples, this variable has little impact on the chemical variables; thus, no highly significant correlations have been found. By contrast, the silt and clay content makes up more than 99% of the textural classes in most of the sediments, and it can be expected that variations in either the silt or clay may have some relationship with the chemical variables. This is verified by the high degree of correlation between these textural quantities and Fe_{WA}, Mn_{WA}, Fe_{HA}, and Mn_{HA}. This relationship suggests that the finer particle sizes hold a higher proportion of this labile form of Fe and Mn. This may signify that oxyhydroxides or fine-grained carbonates are found preferentially in the clay size particles.

The mean particle size (in micrometres) of the sediments is a much more sensitive variable, as compared with the percentage of silt or clay, for detecting significant correlations with several chemical variables. In addition, there are several highly significant correlations with the residual, sum, and total quantities of Fe, Mn, and Cu. This indicates that most of the nonlabile (or detrital) forms of the metals are also found in the finer particle sizes. This suggests that most of the metal is from the dominant minerals in the finer particle size ranges, probably the phyllosilicates. There is also a significant correlation between the mean grain size (as well as the clay content) and the percentage of organic carbon. Because sediment texture becomes finer toward the bottom of the core, and the concentration of organic carbon also increases toward the bottom of the core, it is difficult to make a clear causal link between these variables. However, it is common for higher percentages of organic carbon to be associated with fine-grained sediments because of higher adsorption capacities of the finer particles, and also because oxidation of organic matter may be retarded due to less penetration of oxygen into finer-grained sediments. In this core there is clear evidence that the sediments in the lower part of the column are, or have been, subjected to relatively more reduced chemical conditions that would tend to preserve more of the organic matter.

It is noteworthy that both kurtosis and skewness correlate very well with the mean grain size to indicate that the coarser sediments have a broader central tendency in the grain size distribution and that they also have a larger positive skew. This would suggest that the type of deposition in the upper part of the sediment column is contributing a relatively coarser silt mode to a matrix of clay size sediments. This hypothesis may be in keeping with the present depositional environment in which winnowed silts from the surrounding Emerald Bank may be swept into the muds that have accumulated below wave base in Emerald Basin (Buckley, in press). Because the lower indices of skewness are associated with the higher content of fine particles, there is a strong tendency for negative correlation between skewness and a number of chemical variables associated with the fine-grained sediments. Kurtosis is relatively more influenced by the dominant size mode in the particle size distribution. In these sediments the dominant mode is in the silt size range, and this is the relatively less chemically enriched size fraction. Because of this characteristic there are fewer significant correlations between kurtosis and the chemical variables as compared with skewness and the chemical variables.

In order to group variables within this problem set, a factor analysis was performed with a varimax rotation. This resulted in factor loadings of the thirty-eight variables in three factors with eigenvalues >2.5. Variables with the highest loadings in each of the principal factors are shown in Table 22.15. Factor 1 contains seventeen variables with loadings >0.860. This factor is dominated by the variables of Fe concentration in both the labile and nonlabile form. This factor also includes high loadings for labile Mn (Mn_{WA} and Mn_{HA} = 49% of total Mn) and nonlabile Cu (94% of total Cu). Total carbon (TC) and organic carbon (OC) as well as $CaCO_3$ appear in this factor. The variability of NH_4 and depth are also loaded in this factor. The only negative loading in this factor is for the mean (micrometre) size. This combination of variables suggests a strong influence by sediments with the finer particle sizes on the bulk chemistry. Because of the combination of both the labile (25% of total Fe) and nonlabile forms

of Fe, this indicates a silicate (clay mineral) control on the concentration of this metal, as well as the possibility of precipitation of an iron oxyhydroxide or carbonate mineral phase. The possibility of the association of the labile forms of both Fe and Mn with a carbonate mineral phase is suggested by the common loadings of TC and $CaCO_3$ in this factor. The grouping of the depth variable in this factor reflects the change in lithostratigraphic character with time, as well as a possible influence from diagenetic processes as would be indicated by the loading of NH_4 in this factor. In spite of this latter influence, it is reasoned that this factor is dominated primarily by mineralogical control.

Factor 2 contains seven variables with factor loadings >0.380. This factor is dominated by nonlabile Mn variables associated with labile Cu_{WA} and pE. The rather weak negative loading of depth in this factor suggests that chemical variations have an opposite trend to those observed in factor 1. This factor appears to have a diagenetic influence that allows some of the Mn to be precipitated in a nonlabile form in the upper part of the core. This factor accounts for 16.5% of the total problem variance.

Factor 3 is entirely dominated by sediment texture, in which the positive loadings reflect the influence of the coarse-grained sediment variables. There are no associated chemical variables. This factor accounts for only 6.3% of the total problem variance.

Summary: Fine-grained muds deposited in a continental shelf basin retain chemical characteristics that reflect both the detrital mineralogy and the influences of early diagenesis. About 75% of the total Fe and 50% of the total Mn is contained in the detrital or nonlabile composition of the sediments. Labile Fe and Mn are associated with diagenetic processes that lead to the precipitation of oxyhydroxides in oxidizing environments, but these precipitates may be converted to hydrated sulfides or carbonates in reducing environments. Specialized analytical techniques are required to distinguish among the various sources of these metals.

Summary

The four case studies presented in this chapter provide some examples of the relation-

Table 22.15. *Geochemical and sedimentological variables in Emerald Basin core, factor loadings for factor matrix*

Variable Ranking	Factor 1		Factor 2		Factor 3	
1	Fe_{SUM}	0.957	Mn_R	0.978	Sand	0.498
2	Mn_{WA}	0.954	Mn_{TR}	0.973	Mud	-0.498
3	Fe_{HA}	0.950	Cu_{WA}	0.855	Mean (phi)	-0.460
4	Mn_{HA}	0.941	Mn_T	0.833	Clay	-0.451
5	Fe_T	0.941	Mn_{SUM}	0.734	SD (phi)	0.430
6	Fe_{TR}	0.922	pE	0.581	Silt	0.428
7	Fe_R	0.917	Depth	-0.382		
8	TC	0.917				
9	Cu_{TR}	0.910				
10	OC	0.906				
11	Fe_{WA}	0.900				
12	Cu_R	0.899				
13	Cu_{SUM}	0.896				
14	NH_4 pw	0.879				
15	Depth	0.875				
16	Mean (µm)	-0.869				
17	$CaCO_3$	0.866				
Percent of Total Problem Variance Accounted	56.6%		16.5%		6.3%	

ships between grain size characteristics and the chemical composition of natural sediments. The major influences on chemical variability have been demonstrated to be due to mineralogy and diagenesis. Mineralogical characteristics have been demonstrated to be an inherited property that reflects the provenance of the sediments, but is also dependent on the size distribution characteristics of the sediments. Diagenetic influences on chemical composition (including adsorption of chemicals on suspended sediments) is indicative of the transport and depositional environment, but is also significantly influenced by the textural characteristics of the sediments. Interpretation of the significance of geochemical and sedimentological variability requires an understanding of the intimate relationships between these properties.

In order to properly assess and interpret data derived from sedimentary geochemical studies, it is essential that both the chemical variables and the grain size distribution data have a high degree of sensitivity and precision capable of detecting subtle changes in the source of sediments and modifications that may be induced during or after deposition. Such requirements will place increased demands on analysts both in analytical chemistry and sedimentology.

REFERENCES

Ackerman, F. (1980). A procedure for correcting grain size effect in heavy metal analysis of estuarine and coastal sediments. *Environmental Technology Letters*, 1: 518–27.

Ackerman, F., Bergmann, M., & Schleichert, G. U. (1983). Monitoring of heavy metals in coastal and estuarine sediments – A question of grain size (<20 μm versus <60 μm). *Environmental Technology Letters*, 4: 317–28.

Baldi, F., & Bargagli, R. (1982). Chemical leaching and specific surface area measurements of marine sediments in the evaluation of mercury contamination near cinnabar deposits. *Marine Environmental Research*, 6: 69–82.

Balistrieri, L., Brewer, P. G., & Murray, J. W. (1981). Scavenging residence times of trace metals and surface chemistry of sinking particles in the deep ocean. *Deep Sea Research*, 28: 101–21.

Boyle, E. A., Edmond, J. M., & Sholkovitz, E. R. (1977). The mechanism of iron removal in estuaries. *Geochimica Cosmochimica Acta*, 41: 1313–24.

Brindley, G. W. (1961). Chlorite minerals. In: *The x–ray identification and crystal structures of clay minerals*, ed. G. Brown. London: Mineralogical Society, 242–96.

Buckley, D. E. (1972). Geochemical interaction of suspended silicates with river and marine estuarine water. *Proceedings of the 24th International Geological Congress*, Montréal, Canada, sec. 10, pp. 282–90.

(in press). Depositional environments and diagenetic alteration of sediments in Emerald Basin on the Scotian Shelf. *Continental Shelf Research*.

Buckley, D. E., & Cranston, R. E. (1971). Atomic absorption analyses of 18 elements from a single decomposition of aluminosilicate. *Chemical Geology*, 7: 273–84.

(1988). Early diagenesis in deep sea turbidites: The imprint of paleo-oxidation zones. *Geochimica et Cosmochimica Acta*, 52: 2925–39.

Buckley, D. E., Cranston, R. E., Fitzgerald, R. A., & Winters, G. V. (1989). Sedimentological and geochemical data from ESOPE Expedition MD-45, 1985: Great Meteor East, Madeira Abyssal Plain. In: *Geoscience investigations of two North Atlantic abyssal plains – The ESOPE International Expedition*, eds. R. T. E. Schuttenhelm, G. A. Auffret, D. E. Buckley, R. E. Cranston, L. E. Shephard, & A. E. Spijkstra. OECD/NEA Seabed Working Group, Commission of European Communities, Joint Research Centre, Ispra, 1: 423–534, EUR 12330 EN.

Buckley, D. E., & Loder, T. C. (1968). Particulate organic–inorganic geochemistry of a glacial fjord. In: *Clay–inorganic and organic–inorganic associations in aquatic environments*, eds. D. C. Burrell & D. W. Hood. Progress report to the U.S. Atomic Energy Commission, AEC Contract AT-(04-3)-310. University of Alaska: Institute of Marine Science, pp. 1–53.

Buckley, D.E., & Winters G. V. (1983). Geochemical transport through the Miramichi Estuary. *Canadian Journal of Fisheries and Aquatic Sciences* (Suppl.), Proceedings of the Conference on Pollution in the North Atlantic Ocean, 40: 162–82.

Cranston, R. E. (1976). Accumulation and distribution of total mercury in estuarine sediments. *Estuarine and Coastal Marine Science*, 4: 695–700.

Cranston, R. E., & Buckley, D. E. (1971). *Geochemical data for an Alaskan fjord*. Bedford Institute of Oceanography Report AOL Data Series 1971–8–D, Dartmouth, Nova Scotia, Canada.

(1972). Mercury pathways in a river and estuary. *Environmental Science and Technology*, 6: 274–8.

DeLange, G. J., Jarvis, I., Middelburg, J. J., & Kuijpers, A. (1989). Geochemical characteristics and provenance of Late Quaternary sediments from the Madeira and Southern Nares Abyssal Plains. In: *Geoscience investigations of two North Atlantic abyssal plains – The ESOPE International Expedition*, eds. R. T. E. Schuttenhelm, G. A. Auffret, D. E. Buckley, R. E. Cranston, L. E. Shephard, & A. E. Spijkstra. OECD/NEA Seabed Working Group, Commission of European Communities, Joint Research Centre, Ispra, 2: 785–851, EUR 12331 EN.

Eisma, D. (1986). Flocculation and de-flocculation of suspended matter in estuaries. *Netherlands Journal of Sea Research*, 20: 183–99.

Fitzgerald, R. A., Winters, G. V., & Buckley, D. E. (1987). Evaluation of a sequential leach procedure for the determination of metal partitioning in deep sea sediments. Geological Survey of Canada, Open File Report no. 1701.

Fitzgerald, R. A., Winters, G. V., Buckley, D. E., & LeBlanc, K. W. G. (1989). Geochemical data from analyses of sediments and pore water obtained from piston cores taken from Bedford Basin, La Have Basin, Emerald Basin and the slope of the Southern Scotian Shelf, HUDSON Cruise 88-010. Geological Survey of Canada, Open File Report no. 1984.

Forstner, U., & Salomons, W. (1980). Trace metal analysis on polluted sediments, Part 1: Assessment of sources and intensities. *Environmental Technology Letters*, 1: 494–505.

Fulghum, J. E., Bryan, S. R., Linton, R. W., Bauer, C. F., & Griffis, D. P. (1988). Discrimination between adsorption and coprecipitation in aquatic particle standards by surface analysis techniques: Lead distribution in calcium carbonates. *Environmental Science and Technology*, 22: 463–7.

Gipp, M. R., & Piper, D. J. W. (1989). Chronology of Late Wisconsinan glaciation, Emerald Basin, Scotian Shelf. *Canadian Journal of Earth Sciences*, 26: 333–5.

Grim, R. E. (1968). *Clay mineralogy*, 2nd ed. New York: McGraw–Hill.

Horowitz, A. J., & Elrick, K. A. (1987). The relation of stream sediment surface area, grain size and composition to trace element chemistry. *Applied Geochemistry*, 2: 437–51.

Kepkay, P. E. (1985). Kinetics of microbial manganese oxidation and trace metal binding in sediments: Results from an in situ dialysis technique. *Limnology and Oceanography*, 30: 713–26.

Li, Y. H. (1981). Ultimate removal mechanisms of elements from the ocean. *Geochimica et Cosmochimica Acta*, 45: 1659–64.

Loder, T. C. (1971). Significance of dissolved and particulate organic carbon data as a chemical parameter in sub-polar marine and estuarine waters. Ph.D. diss., University of Alaska, College, Alaska.

Loring, D. H. (1988). Trace metal geochemistry of Gulf of St. Lawrence Sediments. In: *Chemical Oceanography in the Gulf of St. Lawrence*, ed. P. M. Strain. *Canadian Bulletin of Fisheries and Aquatic Sciences*, 220: 99–122.

McCave, I. N. (1984). Size spectra and aggregation of suspended particles in the deep ocean. *Deep Sea Research*, 31: 329–52.

(1988). Deposition of ungraded muds from high-density non-turbulent turbidity currents. *Nature*, 333: 250–2.

Oliver, B. (1973). Heavy metal levels of Ottawa and Rideau River sediments. *Environmental Science and Technology*, 7: 135–7.

Piper, D. J. W. (ed.) (1988). *Cruise Report, HUDSON 88–010*. Geological Survey of Canada.

Rashid, M. A., Buckley, D. E., & Robertson, K. R. (1972). Interactions of a marine humic acid with clay minerals and a natural sediment. *Geoderma*, 8: 11–27.

Sholkovitz, E. R. (1978). The flocculation of dissolved Fe, Mn, Al, Cu, Ni, Co, and Cd during estuarine mixing. *Earth and Planetary Science Letters*, 41: 77–86.

Stumm, W., & Morgan, J. J. (1970). *Aquatic chemistry*. New York: Wiley–Interscience.

Tessier, A., Campbell, P., & Bisson, M. (1982). Particulate trace metal speciation in stream sediments and relationships with grain sizes: Implications for geochemical exploration. *Journal of Geochemical Exploration*, 16: 77–104.

Tessier, A., Rapin, R., & Carigan, R. (1985). Trace metals in oxic lake sediments: Possible adsorption onto iron hydroxides. *Geochimica et Cosmochimica Acta*, 49: 183–94.

Wangersky, P. J. (1986). Biological control of trace metal residence time and speciation: A review and synthesis. *Marine Chemistry*, 18: 269–97.

Weaver, P. P. E., Buckley, D. E., & Kuijpers, A. (1989). Geological investigations of ESOPE cores from the Madeira Abyssal Plain. In: *Geoscience investigations of two North Atlantic abyssal plains – The ESOPE International Expedition*, eds. R. T. E. Schuttenhelm, G. A. Auffret, D. E. Buckley, R. E. Cranston, L. E. Shephard, & A. E. Spijkstra. OECD/NEA Seabed Working Group, Commission of European Communities, Joint Research Centre, Ispra, 1: 535–55, EUR 12330 EN.

Weaver, P. P. E., & Kuijpers, A. (1983). Climatic control of turbidite deposition during the last 200,000 years on the Madeira Abyssal Plain. *Nature*, 306: 360–3.

23 Grain size in oceanography

KATE KRANCK AND T. G. MILLIGAN

Introduction

A significant proportion of all geological and biological processes in the sea is concerned with the behaviour and fate of particles. Living and nonliving matter exist, move about, and interact in suspension and along fluid boundaries. Most particles have no independent powers of locomotion but are dependent on the dynamic forces of waves and current to counteract their negative buoyancy. As a result, their dimensions are controlled and constrained by the physical environment. Understanding the nonfluid component of the oceans depends on research on the physical properties that control the dynamic behaviour of particles.

The principal physical properties of particles are density, shape, and size. Size varies by at least three orders of magnitude: from clay grains and bacteria, the smallest particles in the sea, to sand grains and diatoms, the largest passively suspended, single grains in the sea. In contrast, rock-forming silicate minerals and calcium carbonate grains – the principal constituents of inorganic sediment – vary in density from ~2.3–2.7 g/cm^3. Organic matter has a density range of <1 to ~1.2 g/cm^3. Thus effective density varies by less than an order of magnitude. Of even smaller relative importance are differences in dynamic properties resulting from variation in shape. According to Komar & Reimers (1978), shape may change the settling rate of fine-grained inorganic sediment at most by a factor of 2. Even the very extreme diversity of shapes exhibited by phytoplankton does not alter the basic dependence of their settling rate on particle size. Smayda (1970) measured the settling rate of a variety of phytoplankton and found up to an order of magnitude in the scatter of settling rates of similarly sized but different species. The combined pattern of all the forms, however, showed a general power-law (square)

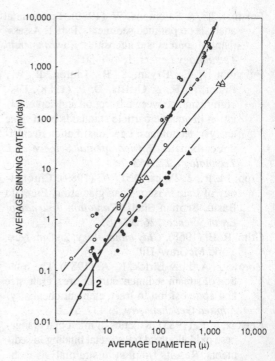

Figure 23.1. Sinking rates of phytoplankton, showing a general square relationship between rate and diameter (heavy line). Symbols give measured settling rates for different species and stages of phytoplankton. (Redrawn from Smayda, 1970.)

dependency between settling rate and diameter in accordance with Stokes's Law (Fig. 23.1).

Size is thus the single most important physical property of oceanic particles. A sediment sample may contain grains that vary in size from less than one-fifth (well-sorted sand) to five orders of magnitude (various diamicts). This chapter discusses methods and observations used in measuring the size distribution of marine particles.

Methods

Analysis

In oceanographic and environmental studies the grain size distribution of both suspended and bottom sediment are relevant. If the two are to be compared, the sample preparation methods must be chosen so as to ensure that the same properties are measured in each case. Analysis of suspended sediment presents special problems of sample preparation. For most analytical methods the sample concentration is too low for direct analysis, and the sample must be concentrated by, say, filtration or centrifugation. All natural suspended sediment is somewhat flocculated. The flocs are partly deflocculated when removed from the ambient dynamic environment so that a distribution intermediate between the natural in situ distribution and a deflocculated distribution is created. The very large marine snow or macroflocs (>200 μm) common in seawater are impossible to handle without some disruption (Alldredge & Silver, 1988; Kranck & Milligan, 1988). If analysis is performed directly on a water sample by, say, settling or Coulter counting, the results are biased by the degree of floc breakup that occurs during sample handling and analysis. Recently several techniques for analysis of in situ size distributions of suspended sediment have been described (Bale & Morris, 1987; Sternberg et al., 1988).

Preanalysis handling and treatment of samples is of crucial importance in size analysis and significantly affects the results by controlling the state of the sediment being analyzed (inorganic vs. total; natural vs. disaggregated particulate material). Most workers in the past have analyzed the single disaggregated mineral grain component of a sample. This usually requires destruction of organic matter by oxidation, disaggregation by sonification or other methods, and subsequent verification using a high-power microscope to ensure that only single mineral fragments remain.

Data presentation

Reasons for measuring grain size of sediments are twofold: descriptive and interpretative. In oceanography, grain size analysis may be utilized to characterize either the sediment itself or the water mass in which it occurs. In the latter case the relationship of the sediment distribution to some physical oceanographic variables is the important factor, and any visual presentation techniques illustrating this may be chosen. For example, in a study of plankton transport in Petpeswick Inlet, Nova Scotia (Kranck, 1980a), particle size spectra were measured using a Coulter Counter in untreated water samples from stations along the length of the inlet at two states of the tide (Fig. 23.2). The resulting size distributions consisted of a sharp peak due to phytoplankton of varying prominence superim-

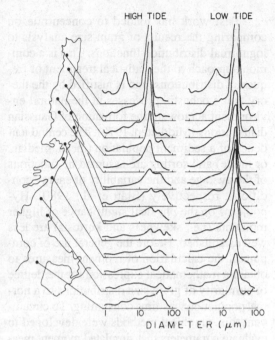

Figure 23.2. Example of the use of particle size to trace movement of tidal flow along the length of Petpeswick Inlet, N.S., a small coastal inlet. The sharp peaks in the semilog-plotted size frequency distribution spectra are due to phytoplankton cells produced near the head of the inlet and advected every tidal cycle to the entrance (Kranck, 1980a).

posed on a background of unsorted organic detritus and sediment. The change in the spectra with time reflects the generation of phytoplankton-rich water in the productive innermost part of the bay and its transport to the entrance each tidal cycle. In this study seawater was analyzed simply to identify and trace water masses rather than to measure size, obviating the need to consider aggregate breakup.

For interpretation of dynamic sedimentary processes, a plotting technique that quantitatively emphasizes the significantly indicative features of the distribution is needed. The variety of pictorial representations of size data in use at the present time reflects the many models that have been proposed for deriving descriptive variables and interpreting sediments. The merit of each plotting method is largely a function of the interpretative model applied to the data, but the variety of methods and scales used greatly hinders comparison of the large number of data sets analyzed in earlier studies.

Past work has tended to concentrate on comparing the results of grain size analysis to lognormal distribution functions. This is a common approach to the statistical treatment of frequency distributions, and is justified by the theory that stochastic processes in the natural environment act toward the formation of Gaussian distributions (Middleton, 1970). The central tendency of a sediment population (mean, median, or mode) and sorting are assumed be functions of the average and the variability of energy conditions, respectively (Blatt et al., 1980). Hypotheses on the physical significance of higher moments (i.e., skewness and kurtosis) are less well established. Before the general use of computers, the calculation of moment measures to obtain a mathematical expression for the degree of similarity of grain size distributions to a normal population was time consuming. To circumvent this, graphical methods were developed to evaluate parameters that simulated moment measures (Trask, 1930; Inman, 1952; Folk & Ward, 1957). This popularized the use of cumulative curves from which the required percentiles could easily be extracted.

Subsequently the use of log probability plotting paper, which causes a lognormal distribution to plot as a straight line (Harding, 1949), was adopted. This plotting technique became especially popular when simple normal or Gaussian distributions could not be fitted to size spectra. It was suggested that sediment size distributions could be subdivided into several truncated or overlapping populations, each one said to result from a different mode of transport (bedload, suspended load, washload) (Middleton, 1976; Visher, 1969). Individual sediment samples have been seen as combinations of as many as half a dozen individual lognormal populations (Van Andel, 1973). Doubts about whether mixture of lognormal populations do form straight line segments (e.g., Syvitski, 1977) have diminished the popularity of this form of data handling. Log-arithmetic cumulative curves have remained popular, however, partly due to habit and partly because of the ease of extracting percentile values – in particular, the fiftieth percentile or median size.

A distribution type satisfactorily explained in terms of basic physical processes is the un-

sorted sediments produced by weathering of crystalline rocks. Bennet (1936) described the generation of so-called Rosin–Rammler distributions through the probability of fracture of any grain in any direction as compared to fracture anywhere else. The resulting distribution can be described by an equation that, over much of the size range, consists of a power-law distribution with an exponent close to zero. This reflects the unsorted character of untransported weathering materials (Kranck & Milligan, 1985). A special type of plotting paper has also been constructed for linearizing this type of distribution (Kittleman, 1963).

A popular method of illustrating grain size distributions has always been the use of frequency histograms. The proportion of the total distribution occurring in a given size class is shown as a pillar straddling the size range or – in the case of smoothed frequency histograms, also referred to as frequency spectra – as a line joining distribution values (Krumbien & Pettijohn, 1938). Recently log–log plots, a specialized form of frequency spectra, have become popular. Among the earliest and most persuasive proponents of this plotting technique were Bagnold and coworkers (Bagnold & Barndorff-Nielsen, 1980; Barndorff-Nielsen 1977), who maintain that sediment size distributions in fact do not form Gaussian curves and are best characterized and plotted as log–log frequency distributions. They showed that for well-sorted sands, such as eolian deposits, the extremes of the distributions change exponentially with the logarithm of the size, although they did not have a physical explanation for the observation. Empirical studies of suspended sediment have led to a recognition of the prevalence of power-law distributions in natural particle distributions (Bader, 1970; McCave, 1984). The significance of power-law distributions was also highlighted by theoretical models for flocculation (see Hunt, 1980).

At present, log–log frequency spectra are one of the most popular plotting techniques, especially in chemistry-related fields (see, e.g., Farley & Morel, 1986). Advantages of this plotting form, beside the fact that power-law distributions plot as straight lines, are that similar relative variations in different samples form sim-

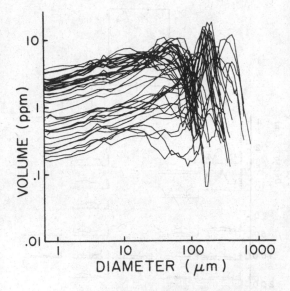

Figure 23.3. Grain size spectra of sediment samples representative of the bottom of Miramichi Bay, N.S. Plotted as log–log frequency distributions, the floc-settled components (finer than ~10 μm) with similar size characteristics form graph segments with similar slopes. (From Kranck & Milligan, 1989.)

ilar curve segments irrespective of the shape of the rest of the distribution (Fig. 23.3). Furthermore, the shape of the curve is not affected by the units used: The distributions will have the same shape whether, say, given as volume/ volume concentrations (ppm) or normalized to percent of total distribution. This is especially useful in the comparison of suspended- and bottom-sediment results. Log–log plots greatly facilitate the recognition of recurring patterns, especially if the ratio between cycle lengths on the x- and y-axes are always kept constant (of equal length is a convenient and logical choice). This plotting technique frequently brings out unexpected or not easily demonstrated relationships (see, e.g., Syvitski et al., 1985). It may be expected that with the increasing use of computers and the resulting ease of testing of different plotting techniques, log–log plotting of size distributions will become even more popular.

Observations

Fine-grained particulate matter and sediments are most relevant to marine geological processes. Most of the ocean basins are covered with fine muddy sediments. In the water column, suspended particulate matter consists of microscopic sediment grains or plankton either singly or flocculated into aggregates of various sizes and shapes; this matter is the transport vehicle of most chemical contaminants. Sediments such as coarse sands and gravels are less abundant and present special problems of sampling. Only extremely large, rugged grabs samplers will penetrate the hard bottom formed by gravel and boulders, and an adequately large sample would fill a cargo hold (Kellerhals & Bray, 1971).

Suspended particulate matter

In situ size. The in situ particle size is the size configuration in which sediment and organic matter are transported. Especially in the sea, where the high salinity promotes flocculation, this size distribution is important for understanding dynamic particle transport processes. Diver and submersible observations and underwater photography have established that large visible particles are ubiquitous in both the open ocean and coastal waters. Determination of the size distributions of these macroflocs has been rudimentary and hampered by the fact that no data are available for the fine end of the size range. Available evidence, however, suggests that the bulk of the material occurs at the coarse end of the distribution range in sizes well over 100 μm. Eisma (1986) measured 200 particles in one photograph from the Ems Estuary and found over 50% of the distribution by volume occurred between 300 and 1,000 μm, although his numbers do not allow estimation of relative amounts above and below this range. Bale & Morris (1987) also suggested that suspended sediment occurs as large uniform sized particles. Sternberg et al. (1988) digitized photographs from a Benthos Plankton Camera to analyze well-sorted distributions with modal sizes between 100 and 500 μm in San Francisco Bay, and subsequent work using the same technique has confirmed similar distribution types from other coastal areas. In the deep sea, despite intensive interest in the role of marine snow in biological processes and chemical fluxes (Alldredge & Silver, 1988), information on the complete size spectra is even scarcer than for coastal waters. Most of the observations have taken the form of numbers of visible particles per volume of water without sufficient precision for construction of size distribution spectra. The very irregular shapes of many of the open water particles complicates the task. Loosely aggregated organic matter frequently forms stringers, veils, and irregular bodies for which the standard geological concept of equivalent diameter is difficult to determine and probably irrelevant. Consensus exists on the importance of large particles, but presently the lack of suitable instrumentation makes it impossible to estimate the relative partitioning of the total particulate population between visible macroflocs and the sub-100-μm invisible sizes.

Disaggregated size. Relatively few studies of the inorganic deflocculated size distributions of suspended sediment have been carried out, but sufficient data exist to demonstrate that the carefully dispersed inorganic mineral grains from a suspension have size distributions resembling those of bottom sediments (see below). Coarse, well-sorted sands are seldom found in suspension, but the pattern of a modal hump that becomes less prominent with decreasing modal size and a poorly sorted tail of finer sediment is common in many sedimentary environments (Fig. 23.3) (Kranck, 1975, 1979, 1981). Differences exist within the water column. The surface waters of well-stratified water columns are characterized by unsorted sediment. Near the bottom and in well-mixed water columns dominated by sediment resuspended from the bottom, a prominent maximum usually occurs near the coarse end with lower concentrations toward the fine sizes (Kranck, 1979). In some waters the distribution is slightly bimodal, with a poorly developed second maximum in the fine sizes (Bay of Fundy; Kranck, unpublished).

These observations are based on limited data mainly from Canadian coastal waters. Prior to the recognition of the ephemeral character of particle size distributions in seawater and the disruption of flocs and aggregates during sampling, many workers analyzed the particulate matter in raw seawater samples. Although these results are difficult to interpret, consistent features of the recorded distributions probably re-

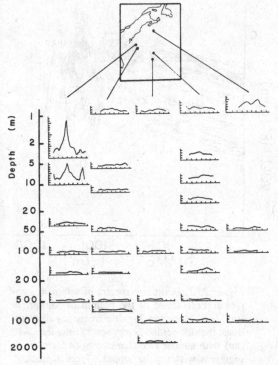

Figure 23.4. Size distribution of suspended particulate matter from various depths in the western North Atlantic. Size spectra show volume concentration vs. log diameter based on Coulter Counter analysis of untreated seawater. (From Sheldon et al., 1972.)

flect either the in situ or the disaggregated size configurations. Sheldon et al. (1972) measured particle spectra from oceanic waters and found equal volume concentrations in all size classes except near the surface, where plankton produced more irregular size spectra (Fig. 23.4). McCave (1983) also recorded unsorted particulates in Western Atlantic surface and midwaters, whereas in bottom waters size distributions were better sorted, with a mode around 5 μm.

In general it appears that, below the euphotic zone, flat, featureless spectra are the rule – except near the bottom where resuspended bottom sediment causes prominent maxima in the particle size distribution. The contrast between the apparently well-sorted, in situ particulate matter in coastal waters and its poor sorting when analyzed partly or wholly disaggregated indicates that sediment forms well-sorted flocs composed of poorly sorted constituent particles. This agrees with the laboratory observations of

Migniot (1968), who recorded a more than three orders of magnitude greater variation in the settling velocities for sediments when unflocculated than when flocculated (see also van Leussen, 1988). No in situ size spectra have been reported from the open ocean, but it is likely that here too the unsorted distributions may partly represent well-flocculated and well-sorted macroaggregates that do not survive sampling.

Bottom sediment

When flocs and aggregates come into contact with the sediment bed, they tend to lose their integrity. Through studies of microfabrics of muddy sediments it is sometimes possible to distinguish relict aggregates or "domains" indicative of relict flocs (Bennett & Hulbert, 1986). Although this may be a clue to engineering properties or depositional conditions, only a few attempts to measure the original in situ size distribution have been attempted. Pryor & Vanwie (1971) and Johnson & Stallard (1989) measured the size of sediment before and after disaggregating sand-sized aggregates, and both studies confirmed that the size distribution of flocs is well compared to the size distribution of the constituent mineral grains.

Analysis of the grain size distribution of bottom sediment after sample treatment to oxidize organic matter and agitation to break up aggregates is likely the single most studied topic in sedimentology. Countless samples have been analyzed, plotted, and studied for specific purposes, but the disparate forms of reporting make generalizations difficult, and returns in terms of practical benefits have been seen as disappointing. As summarized in a recent textbook (Collinson & Thompson, 1982):

The various measures used to characterize grain-size populations (e.g., median, mean, sorting, skewness, etc.) have been reviewed at length in many standard texts on sedimentary petrography. These measures can often be useful in describing a sediment, but despite much effort, as yet, they tell us remarkably little about processes and environments of deposition.

Certain features do, however, show up in most size analyses of detrital marine sediment suites (e.g., Krumbien & Aberdeen, 1937; Inman & Chamberlain, 1955; Swift et al., 1969;

Cita & Colombo, 1970; De Meis & Amador, 1974; Jipa, 1974; Schwarz et al., 1975; Gonthier et al., 1984; Halfman & Johnson, 1984; Andrews, 1985; Kuel et al., 1986). Sand-sized and finer sediments are usually unimodal and negatively skewed, so the modal size occurs near the coarse end of the distribution. Concentrations decrease sharply at the coarse end but tail off much more slowly toward the fine end, and some sediment is always present in all the fine-grain size classes (even in sands, although in quantities so low that it is seldom analyzed; Kranck, unpublished). With the decrease in maximum size, the excess in sediment at the coarse end decreases, so that very fine sediment usually has no pronounced mode. This pattern is reflected in the degree of sorting, which generally decreases with decreasing modal grain size, reflecting the more even partitioning of material between size classes, the abrupt falloff at the coarse end, and the lack of data at the fine end. Figure 23.5 shows examples of sediment size distributions of the major sedimentary facies of the Amazon Shelf, based on analysis using a SediGraph and plotted as semilog frequency distribution plots (Kuel et al., 1986). Facies rich in sand have lower percentages of mud. The mud fractions are characterized by about the same amounts of material in all size classes. With decreasing sand content the mud "tail" of unsorted fines becomes more prominent. When samples from a limited geographical area are plotted as log–log plots, it may be seen that the fines form similar slopes (usually close to zero, i.e., almost flat). For example, samples from throughout Miramichi Inlet, a large tidal estuary in the Gulf of St. Lawrence, show the bottom sediment ranged from fine mud to clean sands (Kranck & Milligan, 1989). The mud fraction consists of similar relative percentages of all sizes as shown by the similarity of the slopes. Even the sands containing only a small percentage of mud exhibit this same slope at the fine end of their size spectra. When grain size analyses are plotted as cumulative curves, the combination of a relatively coarse modal hump and a tail of fine material produces a sharp bend in the resulting curves. This can be seen in the curves published by Gonthier et al. (1984) describing contourite facies on the continental slope in

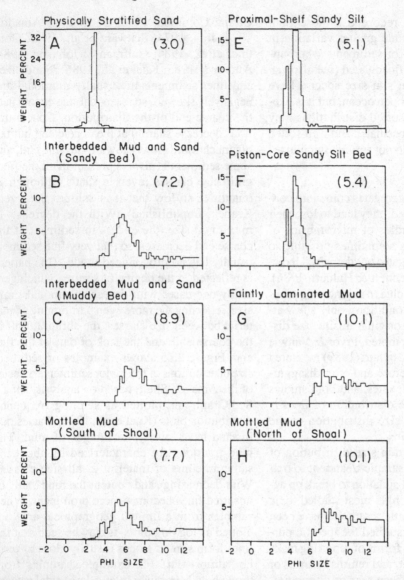

Figure 23.5. Frequency distribution spectra of examples of typical sediment facies on the Amazon Shelf. (From Kuel et al., 1986.)

the Gulf of Cadiz (Fig. 23.6). It is most pronounced in the silts and sands. The mottled silts and muds and homogeneous muds contain smaller amounts of the coarser modal fraction, and curves for these facies are relatively straight.

Floc-grain-settling model

The differentiation of sediment into sorted and unsorted (or mixed) distributions is largely a result of the existence of two modes of particle settling. Suspended particles consist either of single, discrete mineral grains or of flocs composed of many smaller constituent grains. Although the size distributions of flocs apparently are relatively well sorted, the size distributions of the constituent grains making up the flocs are generally unsorted – a reflection of the source materials from which they are formed. Unsorted "diamict"-type size distributions characterize untransported terrestrial weathering products (Kranck & Milligan, 1985). The second principal source of marine particles, plankton production, also produces distributions with equal amounts in equal logarithmic size classes (Shel-

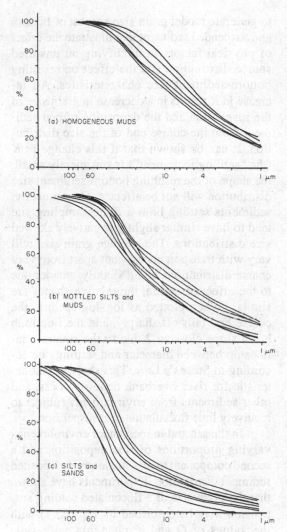

Figure 23.6. Cumulative curves of sediments from the Gulf of Cadiz. (Redrawn from Gonthier et al., 1984.)

don et al., 1972). If it is assumed that a finite site on the bottom receives only particles with a narrow range of settling velocities, the shape of the size spectra of the resulting bottom sediment will principally be a reflection of the relative amounts of flocculated and unflocculated sediment: Sediments, composed mainly of sand- and silt-sized grains, are generally well sorted with only low levels of unsorted fines. Fine muds, on the other hand, are largely composed of floc-settled material whereby the unsorted parent distribution was protected from normal hydrodynamic sorting.

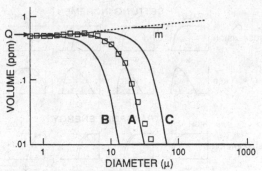

Figure 23.7. Example of grain size spectra of suspended-sediment sample (San Francisco Bay) illustrating the basic suspended-sediment equation (23.1) and the significance of the variables Q, m, and K in the equation. Squares show size distribution data obtained from Coulter Counter analysis. Curve A shows Eq. (23.1) fitted to these data using least-squares regression analysis. B and C are size spectra computed using Eq. (23.1) with the values of Q and m the same as evaluated for the sample data, but with different values of K (2 and 0.5 × the K value of A, respectively).

The floc-grain-settling process has been modeled by Kranck & Milligan (1985) and Kranck (1986a,b, 1987). The basic element of this model is the deposition of different sized grains from a turbulent unsorted source suspension at rates proportional to the exponential of their settling rates. This results in a suspended sediment size distribution of the general form

$$C = QD^m \exp(-K\alpha D^2) \qquad (23.1)$$

This equation describes the volume or weight concentration C of sediment in each size class, as defined by its midpoint diameter D, in terms of three variables: Q, m, and K (Fig. 23.7). Q positions the distributions on the x-axis; m is the exponent of a power-law function whose value usually depends on the source material and for most terrains varies in the range ± 0.3. Sediment in small regional rivers usually diverges from zero, but sediment in major world rivers frequently has a value for m close to zero (Kranck & Milligan, 1983). K is dependent on dynamic flow factors. When a steady current maintains an equilibrium between sediment diffusion, away from the bed and downward settling (as described by Rouse [1938]), the suspended sediment at a given relative depth will

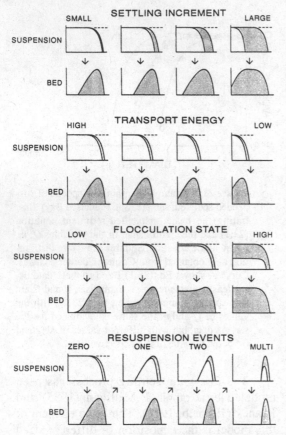

Figure 23.8. Effects of modification of the shape of an initially unsorted source sediment (- - -) by physical factors. Size spectra shown as log volume (or weight) vs. log diameter. The x-axis represents ~1–1000 μm; the y-axis is nondimensional to show maximum portions of the curves. Shaded areas represents the material settled from suspension to form the bottom sediment. For discussion of different factors, see text.

have a *K* value related to the shear velocity. If the current is reduced, settling will occur with time until a new equilibrium is reached. The resulting size distribution of the settled sediment will be defined by a higher *K* value; that is, the distribution falls off at finer grain sizes related either to current speed and depth above the bottom, or to time since the start of settling. The variable α converts grain diameter *D* to grain settling rate using Stokes's Law and is constant for any given suspension.

In Figure 23.8, Eq. (23.1) has been used

to generate model grain size spectra of bottom and suspended sediment that illustrate the effect of physical factors in modifying an unsorted source distribution and the effect on resulting bottom-sediment size characteristics. An increase in *K* results in a decrease in grain size of the suspension and the deposition of bed sediment from the coarse end of the size distribution. It can be shown that if this change in *K* (the "settling increment") is sufficiently small, the shape of the resulting bottom-sediment size distribution will not be affected. Consequently, sediments settling onto a finite sampling site tend to have similar slightly negatively skewed size distributions. The average grain size will vary with transport energy, but apart from very coarse distributions being slightly broader due to the effect of inertia, the relative shapes are similar. When plotted as log–log spectra, the coarse end falls off sharply, and the fine limb has a slope close to 2 due to the square relationship between diameter and settling rate according to Stokes's Law. This shape is characteristic for river overbank deposits, loess, and other sediments from environments subject to relatively little flocculation or resuspension.

In the sea and in most other environments, varying proportions of floc deposition add a second component to this single-grain-deposited sediment distribution. Experiments have shown that the grain size of a flocculated settling suspension may be described by Eq. (23.1) with the values of *Q* and *K* changing with time (Kranck, 1986b). Lost from suspension and deposited on the bottom will be single-grain, unimodal material, along with a floc component forming a more or less prominent tail of unsorted fines, depending on the grain size and the flocculation state of the parent suspension. This tail settles – with the same settling rate as the modal portion – in the form of well-sorted flocs, each one of which is composed of a representative portion of every size in the suspension up to some maximum size (Kranck, 1980b).

Differences in the relative proportions of grain- and floc-settled material from one sample to another is a principal reason for variability in the shape of fine-grained size distributions. The effect of this mixing is demonstrated in Figure

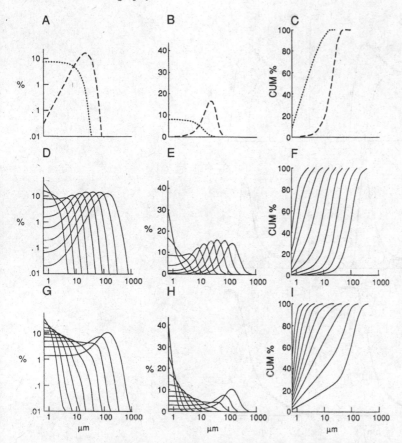

Figure 23.9. Computer-generated grain size distribution curves that illustrate the formation of typical marine sediments (D–I) from different sizes and relative mixtures of grain- (A–C, - - -) and floc- (A–C, ⋯) settled sediment. Grain size distributions are normalized to 100% and plotted using three different plotting techniques: log–log frequency (A, D, G), semilog frequency (B, E, H) and cumulative curves (C, F, I). Note that with decreasing grain size the proportion of grain relative to floc material decreases with size, reflecting the greater tendency of fine grains to flocculate.

23.9 using computer-generated model size distributions and three different plotting techniques. Different sized, single-grain-settled sediment has been mixed with corresponding sized floc-settled sediment. A–C show typical pure single grain and floc distributions; D–F show size spectra of mixtures of the two, with single-grain-settled sediment dominating; and G and I show typical size spectra for highly floc-dominated conditions. The similarity in shapes of the model spectra in Figure 23.9 to the natural sediment distributions shown in Figures 23.3, 23.5, and 23.6 is apparent.

In many environments bottom sediment is prone to resuspension; its grain size distribution then becomes the size distribution of the source

suspension when the sediment is again redeposited. The effect is to improve the sorting of the redeposited sediment (Fig. 23.8). Each resuspension event or sedimentation "round" theoretically increases the value of the slope of the fine-grained limb of the new distribution by 2 (when plotted as log–log spectra). Very well-sorted sands (e.g., beach and dune sands) are produced by repeated resuspension and settling action. It is noteworthy that each single additional round has relatively little effect on well-sorted "multiple-round" sediment.

Figure 23.10 shows equations for the floc-grain-settling model fitted to measured grain size distributions using standard linear regression techniques. Each fitted curve is defined by

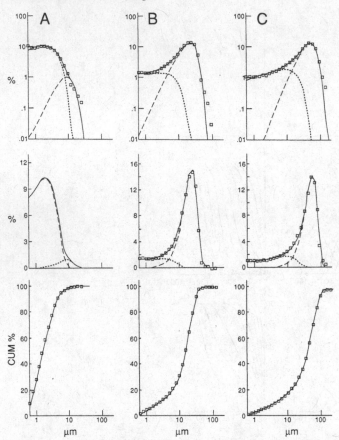

Figure 23.10. Grain size distributions from analysis (squares) of three bottom-sediment samples plotted as log–log and log–arithmetic frequency spectra and cumulative curves, along with the results of calculated distributions from model fitting of the data. - - -, grain-settled sediment; ···, floc-settled sediment; —, sum of both. Samples: A, red clay from the N.W. Pacific; B, deltaic mud from Mackenzie Delta, N.W.T.; C, tidal flat silt from Minas Basin, N.S.

constants, whose values are related to source material and depositional conditions. The excellent agreement between the model and observations illustrates the very precise reflection of sediment size characteristics to the physics of their formational environment. This was also verified by estimating K values for each sample in Figure 23.3 and plotting them against an average current velocity derived for each station location from a two-dimensional mathematical model of Miramichi Bay (Fig. 23.11).

Summary

A review of past studies of grain size of natural sediments has documented differences in grain size spectral shapes among natural sediments. Despite the difficulties in comparing sample analyses presented using a variety of reporting forms, a systematic pattern is discernible in sediments transported principally in suspension. Past studies have demonstrated that fine-grained sediments are characteristically asymmetrical, with the highest relative concentrations of material in the coarse size and lower, usually progressively decreasing or similar concentrations being found in the finer grain sizes. The relative proportions of the coarser modal hump, compared to the tail formed by the finer sizes, decrease with average grain size of a sample. Only some relatively coarse or very fine samples appear symmetrical; the former due to the tail of fines being too insignificant to warrant analysis, and the later due to the modal portion blending in with the fine tail. It is suggest-

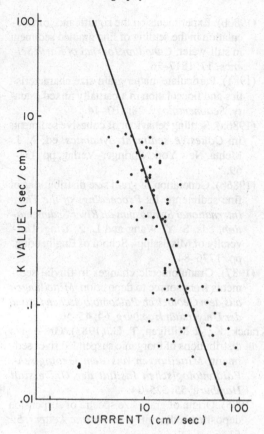

Figure 23.11. Relationship between *K* values of bottom sediment and average current speed in Miramichi Bay, N.B., based on samples shown in Figure 23.3. (From Kranck & Milligan, 1989.)

ed that this pattern is the result of the mixing of well- and poorly sorted grain size populations. A model is presented describing the sorted mode as originating from grain-by-grain settling and the poorly sorted sediment as settling of flocs that contained a representative proportion of all sizes from an unsorted source suspension. Parameterization of size distributions using equations generated from the model allows precise description of size distributions in terms of a small number of variables.

REFERENCES

Alldredge, A. L. & Silver, M. V. (1988). Characteristics, dynamics and significance of marine snow. *Progress in Oceanography*, 20: 41–82.

Andrews, J. T. (1985). Grain-size characteristics of Quaternary sediments, Baffin Island region. In: *Quaternary Environments*, ed. J. T. Andrews. Boston: Allen & Unwin, pp. 124–53.

Bader, H. (1970). The hyperbolic distribution of particle sizes. *Journal of Geophysical Research*, 15: 2831–5.

Bagnold, R. A., & Barndorff-Nielsen, O. E. (1980). The pattern of natural size distributions. *Sedimentology*, 27: 199–207.

Bale, A. J., & Morris, A. W. (1987). In situ measurements of particle size in estuarial waters. *Estuarine, Coastal and Shelf Science*, 24: 253–63.

Barndorff–Nielsen, O. (1977). Exponentially decreasing distributions for the logarithm of particle size. *Procedures of the Royal Society of London, A.* 352: 401–19.

Bennet, J. G. (1936). "Broken coal." *Journal of Industrial Fuel*, 19: 22–9.

Bennett, R. H., & Hulbert, M. H. (1986). *Clay microstructure*. Boston: International Human Resources Development Corporation, 161 pp.

Blatt, H., Middleton, G., & Murray, R. (1980). *Origin of sedimentary rocks*. Englewood, N.J.: Prentice–Hall, 782 pp.

Cita, M. B., & Colombo, L. (1970). Sedimentation in the latest Messinian at Capo Rosello (Sicily). *Sedimentology*, 26: 495–522.

Collinson, J. D., & Thompson, D. B. (1982). *Sedimentary structures*. London: Allen & Unwin, 194 pp.

De Meis, M. R. M., & Amador, E. S. (1974) Note on weathered arkosic beds. *Journal of Sedimentary Petrology*, 44: 727–37.

Eisma, D. (1986). Flocculation and de-flocculation of suspended matter in estuaries. *Netherlands Journal of Sea Research*, 20: 183–99.

Farley, K. J., & Morel, M. M. (1986). Role of coagulation in sedimentation, *Environmental Science and Technology*, 20: 187–95.

Folk, L. R., & Ward, W. C. (1957). Brazos River Bar: A study of the significance of grain size parameters. *Journal of Sedimentary Petrology*, 27: 3–26.

Gonthier, E. G., Faugerres, J.-C., & Stow, D. A. V. (1984). Contourites facies of the Faro Drift, Gulf of Cadiz. In *Fine-grained sediments: Deep water processes and facies*, eds. D. A. V. Stow & D. J. W. Piper. Oxford: Geological Society/ Blackwell Scientific Publications, pp. 272–92.

Halfman, J. D., & Johnson, T. S. (1984). The sedimentary textures of contourites in Lake Superior: Contourites facies of the Faro Drift, Gulf of Cadiz. In *Fine-grained sediments: Deep water processes and facies,* eds. D. A. V. Stow & D. J. W. Piper. Oxford: Geological Society/ Blackwell Scientific Publications, pp. 293–307.

Harding, J. P. (1949). The use of probability paper for the graphical analysis of polymodal frequency distributions. *Journal of Marine Biology,* 28: 141–53.

Hunt, J. R. (1980). Prediction of oceanic particle size distribution from coagulation and sedimentation mechanisms. In: *Particulates in water: Advances in Chemistry vol. 189,* eds. C. M. Kavanaugh & J. O. Leckie. Washington, D.C.: American Chemical Society, pp. 243–57.

Inman, D. L. (1952). Measures for describing the distribution of sediments. *Journal of Sedimentary Petrology,* 22: 1–9.

Inman, D. L., & Chamberlain, T. K. (1955). Particle-size distribution in near shore sediments. *Society of Economic Paleontologists and Mineralogists, Special Publ.* 3: 107–29.

Jipa, D. C. (1974). Graded bedding in the recent Black Sea turbidites: A textural approach. In: *The Black Sea: Geology, chemistry and biology,* eds. E. T. Degens and D. A. Ross. Tulsa, Okla.: American Association of Petroleum Geologists, Memoir no. 20, pp. 317–37.

Johnson, M. J., & Stallard, R. F. (1989). Physiographic controls on the composition of sediments derived from volcanic and sedimentary terrains on Barro Islands, Panama. *Journal of Sedimentary Petrology,* 5: 768–81.

Kellerhals, R., & Bray, D. I. (1971). Sampling procedures for coarse sediments. *ASCE Proceedings, Journal of Hydraulics Division,* 97: 1165–79.

Kittleman, L. R., Jr. (1963). Application of Rosin's distribution in size frequency analysis of clastic rocks. *Journal of Sedimentary Petrology,* 34: 483–502.

Komar, P. D., & Reimers, C. E. (1978). Grain shape effects on settling rates. *Journal of Geology,* 98: 193–209.

Kranck, K. (1975). Sediment deposition from flocculated suspensions. *Sedimentology,* 22: 111–23.

— (1979). Dynamics and distribution of suspended matter in the St. Lawrence Estuary. *Naturaliste Canadian,* 106: 163–73.

— (1980a). Variability of particulate matter in a small coastal inlet. *Canadian Journal of Fisheries and Aquatic Sciences,* 37: 1209–15.

— (1980b). Experiments on the significance of flocculation in the settling of fine grained sediment in still-water. *Canadian Journal of Earth Science,* 17: 1517–26.

— (1981). Particulate matter grain-size characteristics and flocculation in a partially mixed estuary. *Sedimentology,* 28: 107–14.

— (1986a). Settling behaviour of cohesive sediment. In: *Cohesive sediment dynamics,* ed. A. J. Mehta. New York: Springer–Verlag, pp. 151–69.

— (1986b). Generation of grain size distributions of fine sediments. In: *Proceedings of the Third International Symposium on River Sedimentation,* eds. S. Y. Wang and L. Z. Ding. University of Mississippi, School of Engineering, pp. 1776–84.

— (1987). Granulometric changes in fluvial sediments from source to deposition. *Mitteilungen aus dem Geologisch-Paläontologischen Institut der Universität Hamburg,* 64: 45–56.

Kranck, K., & Milligan, T. G. (1983). Grain size distributions of inorganic suspended river sediment. *Mitteilungen aus dem Geologisch-Paläontologischen Institut der Universität Hamburg,* 55: 525–34.

— (1985). Origin of grain size spectra of suspension deposited sediment. *Geo-Marine Letters,* 5: 61–6.

— (1988). Macroflocs from diatoms: In-situ photography of particles in Bedford Basin, Nova Scotia. *Marine Ecology – Progress Series,* 44: 183–9.

— (1989). Effects of a major dredging program on the sedimentary environment of Miramichi Bay, New Brunswick. *Canadian Technical Report of Hydrography and Ocean Sciences,* no. 112. Fisheries and Oceans, Canada.

Krumbien, W. C., & Aberdeen, E. (1937). The sediments of Barataria Bay. *Journal of Sedimentary Petrology,* 7: 3–17.

Krumbien, W. C., & Pettijohn, F. J. (1938). *Manual of sedimentary petrology.* New York: Appleton–Century–Crofts, 549 pp.

Kuel, S. A., Nittrouer, C. A., & DeMaster, D. J. (1986). Distribution of sedimentary structures in the Amazon subaqueous delta. *Continental Shelf Research,* 6: 311–36.

McCave, I. N. (1983). Particle size spectra, behaviour, and origin of nepheloid layers over the Nova Scotian Continental Rise. *Journal of Geophysical Research,* 88: 7647–66.

— (1984). Size spectra and aggregation of suspended particles in the deep sea. *Deep-Sea Research*, 31: 329–52.

Middleton, G. V. (1970). The generation of log-normal size frequency distributions in sediments. In *Topics in mathematical geology*, eds. M. A. Romanova and V. Sarmand. New York: Consultants Bureau, pp. 34–42.

(1976). Hydraulic interpretation of sand size distributions. *Journal of Geology* 84: 405–26.

Migniot, C. (1968). Action des courants, de la houle et du vent sur les sediments. *La Houille Blanche*, 1: 9–47.

Pryor, W. A., & Vanwie, W. A. (1971). The "sawdust sand": An Eocene sediment of floccule origin. *Journal of Sedimentary Petrology*, 41: 763–9.

Rouse, H. (1938). Experiments on the mechanics of sediment suspension. *Proc. 5th International Congress of Applied Mechanics*, pp. 550–4.

Schwarz, H. V., Einslie, G., & Herm, D. (1975). Quartz-sand, grazing-contoured stromatolite from coastal embayments of Mauritania, West Africa, *Sedimentology*, 22:539–61.

Sheldon, R. W., Prakash, A., & Sutcliffe, W. H. (1972). The size distribution of particles in the ocean. *Limnology and Oceanography*, 17: 327–40.

Smayda, T. J. (1970). The suspension and sinking of phytoplankton in the sea. In: *Oceanography and marine biology review*, ed. H. Barnes. London: Allen and Unwin, 8: 353–414.

Sternberg, R. W., Kranck, K. Cacchione, D. A., & Drake, D. E. (1988). Suspended sediment transport under tidal channel conditions. *Sedimentary Geology*, 57: 257–72.

Swift, D. J. P., Heron, S. D., & Dill, C. E. (1969). The Carolina Cretaceous: Petrographic reconnaissance of a graded shelf. *Journal of Sedimentary Petrology*, 39: 18–33.

Syvitski, J. P. (1977). Grain-size distribution using log-probability plots – a discussion. *Bulletin of Canadian Petroleum Geology*, 25: 683–94.

Syvitski, J. P., Asprey, K. M., Clattenburg, D. A., & Hodge, D. H. (1985). The prodelta environment of a fjord: Suspended particle dynamics. *Sedimentology*, 32: 83–107.

Trask, P. D. (1930). Mechanical analysis of sediments by centrifuge. *Economic Geology*, 25: 581–99.

Van Andel, T. H. (1973). Texture and dispersal of sediments in the Panama Basin. *Journal of Geology*, 81: 434–57.

van Leussen, W. (1988). Aggregation of particles, settling velocity of mud flocs: A review. In: *Physical processes in estuaries*, eds. J. Dronkers & W. van Leussen. Berlin: Springer–Verlag, pp. 347–403.

Visher, G. S. (1969). Grain size distributions and depositional processes. *Journal of Sedimentary Petrology*, 39: 1074–106.

24 The need for grain size analyses in marine geotechnical studies

FRANCES J. HEIN

Marine geotechnique

Marine geotechnique is the study of the physical and physicochemical properties of seafloor deposits. *Geotechnical ocean engineering* is the study of the design, analysis, and modelling techniques used in the construction of seabed facilities. Seafloor sediments are a multiphase system, mainly comprised of solid–pore water–gas (usually methane, often absent) mixtures. The response of this multiphase system to applied static and dynamic loads is usually inferred indirectly by measuring various physical, chemical, biologic, and acoustic properties of the sediments, and by applying principles of soil mechanics theory to the subsea environment. Direct in situ measurements by ocean-bottom devices and/or sampling by manned or unmanned submersibles is rarely done in deep-water (>200-m) settings because of the high cost of such endeavors.

During the 1970s geotechnical ocean engineering studies concentrated on the continental shelf in water depths of <200 m, in contrast with most of the marine geotechnique work that has been done on sediments sampled from the deep sea. Much of the interest in marine geotechnique has evolved from the Deep Sea Drilling Project (DSDP) and its successor, the Ocean Drilling Program (ODP), as well as from interest in deep-sea ocean mining, laying of transoceanic power and communications cables, and environmental impact studies of hazardous waste disposal in deep-sea sites. For slope-stability investigations involving either total or effective stress methods of analysis, the key parameters to be measured are bulk density, shear strength, excess pore pressure, and cyclic shear strength (i.e., number of cycles needed to failure; Richards, 1984). In situ testing equipment

has been deployed in water depths of less than a few hundred metres. For sites with water depths of up to thousands of meters, open-hole methods, mostly the cone or piezocone penetrometer (Shibata & Teparaksa, 1988; Moran et al., 1989), have been used. (For a review of recommendations for sampling and in situ testing, see Jamiolkowski et al. [1985]; Aas et al. [1986].)

Until recently, geologists and engineers have commonly worked independently in the assessment of the properties of seafloor sediments and stability analyses. Within the past decade, however, many teams of geologists and engineers have taken a more multidisciplinary approach, both in the mapping of seafloor deposits and in modelling the rheologic properties of seafloor sediments. Of particular interest to geologists is the application of marine geotechnical principles to quantify the physical attributes of sediments susceptible to liquefaction, creep, and mass wasting, to obtain a more quantitative (and actualistic) model for the generation of submarine sediment–gravity flows.

A number of workers have written "state-of-the-art" reviews and updates on marine geotechnique and engineering properties of marine sediments (Richards, 1967, 1984; Noorany & Gizienski, 1970; Hamilton, 1974; Inderbitzen, 1974; Silva, 1974, 1984; Richards et al., 1975; Richards & Parks, 1976; Davie et al., 1978; Almagor, 1979; Richards & Chaney, 1980; Denness, 1984; Poulos, 1988). One of the key physical properties to quantify in marine geotechnical studies is the texture and fabric (meso- and microscale) of the sediment on the seafloor. Given this information on the structure of the solid phase and of the solid–pore water–(gas) system, plus information on the stress history and in situ stress, one can reconstruct the pore geometry and predict consolidation, compression, and the response of the system to applied loads.

In this chapter, the main influence of grain size characteristics on the primary physical properties of seafloor sediments is discussed. These physical properties include the following:

1. sedimentary fabric;
2. weight–volume parameters, including

void ratio, permeability, porosity, water content, specific gravity, and unit weight;

3. Atterberg limits, including liquid and plastic limits and the plasticity index;

4. degree of consolidation and compressibility; and

5. shear strength.

Although it is difficult to isolate the effects of grain size characteristics on the physical and mechanical properties of seafloor sediments, in most cases grain size mainly affects the weight–volume parameters. Grain size, in combination with particle shape, packing geometry, sedimentary fabric, mineralogy, and organic carbon content, influences the Atterberg limits, the degree of consolidation, compressibility, and shear strength. For a review of these properties and their empirical relationships, see Lambe & Whitman (1969), Bowles (1970), Mitchell (1976), and Dunn et al. (1980). See also the "List of symbols."

Grain size variation in marine settings

As discussed by Hamilton (1974), the accurate prediction of sediment properties depends upon the empirical relationships between various physical properties and/or their expected variability. Given a sediment surface sample in which grain size, grain density, or specific gravity and water content is measured, porosity and density can then be computed (or measured). Experience has shown that empirical data are needed for proper assessment of sea-bottom sites. If only a dried sample is available, then the grain size data can provide predictive empirical relations with the other parameters (Hamilton, 1974). Available tables and empirical relations (Hamilton, 1974) can be used to obtain averaged laboratory properties, which can then be corrected to predict in situ values. Properties of surficial sediments in shallow water may be reasonably approximated using this approach; however, there is a paucity of data to predict accurately either the surficial distribution of sediments in the deep sea, or the gradients of the physical properties with depth beneath the seafloor.

Marine sediments have a very wide range of grain size, sorting, and composition. In general, from the continental shelf to the upper continental slope, both sediment types (lithofacies) and their properties vary widely over short distances. In deep-sea settings, from the base of continental slope to the abyssal plains, the lithofacies are less variable and form a more predictable pattern – particularly in the abyssal plains, where individual units can be correlated for tens of kilometres.

Grain size distributions and composition are dependent upon the source area as well as the transport and depositional mechanisms. Major sediment transport processes found in marine environments include local fluvial discharge, wave currents, ice rafting, aeolian transport, storm currents, littoral currents, geostrophic currents, submarine mass sediment movement (including creep, slides, slumps, debris flows, and turbidity currents), and pelagic/hemipelagic settling. Depending on the marine setting, one or more of the processes may be important. If one process dominates and the sediment source consists of well-sorted sediment, then uniform granular mixtures will result (e.g., the rapid settling of aeolian sands or silts in basins; or the transport of well-sorted beach sands to offshore marine settings). By contrast, if two or more processes are active and/or sediment sources are nonuniform, complex multimodal grain size mixtures result. Shallow marine shelf deposits tend to have complex lithofacies patterns, in comparison with deep-sea abyssal plains, which have more uniform patterns. Submarine channels and fans tend to have predictable vertical and lateral lithofacies patterns. Individual sediment beds in deep-sea deposits tend to be normally graded (they become finer-grained upsection), and these beds are arranged in sequences that are overall progressively fining or coarsening upward (see deep-sea facies models in Walker [1979]). In sites of active slumping and mass wasting, lithofacies patterns tend to be chaotic and random, with juxtaposition of significantly different sediment types (Hein, 1989a,b). This juxtaposition of variable sediment types results in anisotropy – that is, the condition of having different properties in different directions. Zones of vertical anisotropy are often potential failure surfaces in marine sediments (Syvitski & Farrow, 1989; Hein 1989a).

Mass sediment properties
Sedimentary fabric

The term "fabric" refers to the spatial orientation of the elements of which a rock is composed (American Geological Institute, 1957). As applied to sediments, fabric refers to the orientation of grains within a sample, specifically the orientation of the long (*a*) and intermediate (*b*) axes of symmetry of individual clasts. Sedimentary grain fabric depends upon many factors, including the following:

1. mode and rate of deposition;
2. grain size, shape, and roundness;
3. sorting of the sediment being deposited;
4. boundary conditions at the bed at the time of deposition, especially bed roughness factors and bed slope;
5. both physical and biological reworking; and
6. chemical changes.

For granular solids, the arrangement of coarse grains within the sediment may be dense, closely packed, and stable, with small void space – for example, polymodal, poorly sorted sands tend to be tightly packed, clast-supported, and matrix-filled, with void spaces filled mainly by the finer sand and silt. This sediment type would be fairly stable (Fig. 24.1A). Rapidly deposited, well-sorted sand would be very loosely packed and clast-supported, and have an openwork matrix and void spaces filled with water and/or gas (Fig. 24.1B). This type of sediment is quite unstable, would easily liquefy, and would completely lose strength during cyclic loading. Very poorly sorted mixtures, with dispersed clasts, are common in marine sediments, and may be deposited very rapidly as "freezing" from debris flows; or they may aggrade more slowly as rainout from icebergs or ice shelves. The resultant diamicton is a very poorly sorted, polymodal, matrix-supported mixture, with the coarse clasts dispersed within a finer-grained (generally muddy) matrix (Fig. 24.1C). The stability of this sediment type would depend upon its depositional origin and subsequent stress history.

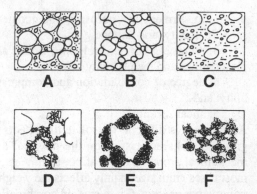

Figure 24.1. Schematic sketch of the structure of different types of sediment: (A) clast-supported, polymodal sand; (B) openwork, bimodal sand; (C) matrix-supported, multimodal, clayey sand; (D) aggregated clay showing closely packed particles connected by bridges of clay particles; (E) marine clay with large, dense aggregates of clay separated by large voids; (F) freshwater clay with small, porous aggregates of clay separated by uniform, small voids. (D–E modified from Pusch [1973].)

For cohesive sediments, the sediment fabric also depends on the organic content and chemistry of the water. In saline water, clay minerals settle out as agglomerates or flocs, with random or more ordered arrangements of individual clay particles (Fig. 24.1D) (Quigley & Thompson, 1966; Pusch 1970, 1973; Quigley, 1980; Syvitski & Murray, 1981). Thus, many marine clayey sediments have fabrics consisting of large, dense aggregates separated by large voids, which are outlined by the chains of clay particles (Fig. 24.1E) (Pusch 1970, 1973). This contrasts sharply with freshwater clays, which have small relatively porous aggregates and small voids (Fig. 24.1F). Fabric patterns may change on the seafloor, due to bioturbation, resedimentation, liquefaction, remobilization, shearing, and diagenesis of sediment (Bohlke & Bennett, 1980; Hein, 1985).

A number of workers have examined the clay microfabric patterns in deep-sea sediments (e.g., Bennett et al., 1977, 1981; Hein, 1985; O'Brien, 1970, 1987; Reynolds, 1988). Models for fabric patterns in smectite- and illite-rich submarine sediment have been proposed by Bennett et al. (1977, 1981) (Fig. 24.2). The basic building blocks of the submarine clay fabric are the "domains" (fabric elements defined by

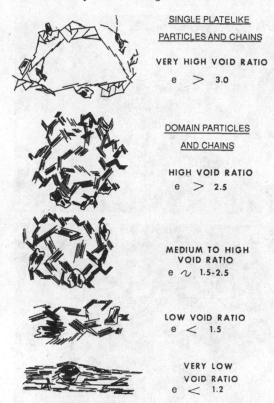

SINGLE PLATELIKE
PARTICLES AND CHAINS

VERY HIGH VOID RATIO

e > 3.0

DOMAIN PARTICLES
AND CHAINS

HIGH VOID RATIO

e > 2.5

MEDIUM TO HIGH
VOID RATIO

e $\sim$ 1.5-2.5

LOW VOID RATIO

e < 1.5

VERY LOW
VOID RATIO

e < 1.2

Figure 24.2. Clay fabric models for smectite- and illite-rich submarine sediments. (From Bennett et al., 1981.)

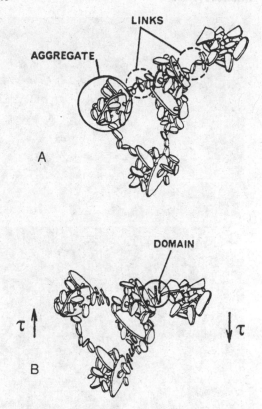

Figure 24.3. Schematic sketch of the failure processes in quick clay subjected to shear: (A) natural microstructural pattern; (B) breakdown of particle links resulting in domain formation. (From Pusch, 1970.)

structural discontinuities, within which the fabric is more or less homogeneous) and/or "single platelike" particles (top Fig. 24.2). These building blocks link together to form flocs. The void ratio of the resultant sedimentary fabric depends on the geometry and number of linkages between domain particles and chains. A very high-porosity ($n > 3$) sediment forms through the development of randomly oriented long chains (top Fig. 24.2). Low-porosity ($n < 1.2$) sediment develops when there is face-to-face particle packing resulting in a high degree of preferred particle orientation (bottom Fig. 24.2). This fabric anisotropy could result in a potential plane of weakness, with subsequent failure occurring parallel to the preferred particle orientation.

The potential of subsequent failure of submarine clays depends to a large extent on the original fabric pattern. As demonstrated experimentally by Pusch (1970) using a series of unconfined compression tests, marine clays when subject to failure deform sequentially. The links between individual particles break down successively with increasing shear stress, and form domainlike groups (Fig. 24.3A,B). With increasing shear deformation, the aggregates and domains move differentially, in turn shearing the linking particles. This results in a parallel reorientation of the linking particles between the domains and aggregates (Fig. 24.3B). This failure mechanism is thought to account for the "residual strength" phenomenon observed in some marine clays (Pusch, 1970).

Reynolds (1988) investigated the origin of clay microfabrics in marine sediments from the modern Santa Monica Basin and the ancient Los Angeles Basin, California Continental Borderland. Bioturbated sediments near the sediment–water interface have abundant bioflocs (biological agglomeration of flocs during ingestion by benthic and pelagic organisms), silt and mica

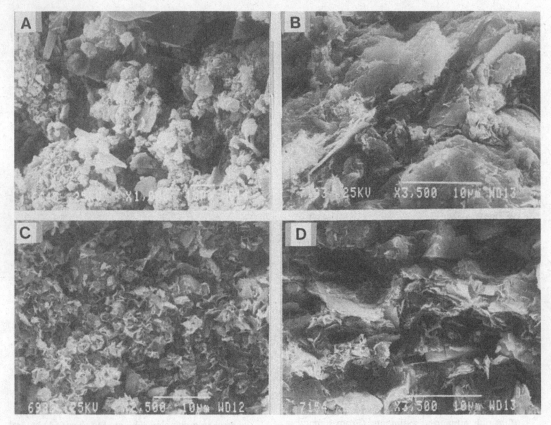

Figure 24.4. SEM micrographs showing the microstructure of submarine sediments, California Borderland. (A) Bioturbated muds, Santa Monica Basin, vertical section, 5 cm below sediment–water interface. Note agglomerations (bioflocs or pellets) 10–15 μm in diameter. Internally, clays are moderately to densely packed in random orientation. Large pores are abundant. (B) Bioturbated mudstone, Los Angeles Basin, about 300-m present-day burial depth, vertical section. Subtle packages of clay, 5–100 μm in diameter, are wedged between silt particles. Many clay faces are apparent, suggesting that many clays have a vertical orientation. Some packets are so densely welded that individual particles cannot be discerned. (C) Turbidite mud, Santa Monica Basin, vertical section, 7 cm below the sediment–water interface. Note continuously flocculated fabric of clays in dominantly edge-to-face high-angle contacts. Abundant voids are 1 μm in diameter. (D) Turbidite mudstone, Los Angeles Basin, vertical section, about 300-m present-day burial depth. Clays are densely packed in dominantly very low-angle edge-to-face and face-to-face contacts. More clay edges are visible than are faces, suggesting that most clays have a subhorizontal orientation. Clays are wrapped around silts in a "fold-belt" fashion. (From Reynolds [1988], courtesy of D. S. Gorsline.)

grains, and diatom or foraminiferal tests (Fig. 24.4A). The sedimentary fabric is characterized by abundant large voids (>5 μm). With increasing depth of burial (300 m), welded packets of clay with abundant clay faces commonly occur between the more coherent silt particles (Fig. 24.4B). These are interpreted as deformed bioflocs that were squeezed plastically between the more coherent silt grains upon compaction and burial.

Turbidite muds near the sediment–water interface show a fabric consisting of a net of small physicochemical flocs, in which there are enmeshed scattered silt grains (Fig. 24.4C). Small pores of uniform size (1 μm) are abundant. As burial depth increases, high-angle, edge-to-face clay particle contacts in the flocs change progressively to very low-angle contacts, and silt particles influence the fabric development in their immediate vicinity, where thin lamina wrap around silts (Fig. 24.4D). Individual clay particles are still discernible, and clay edges, not faces, are the dominant feature. Pelagic muds have a more heterogeneous fabric, containing abundant diatom or foraminiferal tests, fecal pellets, bioflocs, and physicochemical flocs in various proportions (Reynolds, 1988). In pelagic muds, changes in microfabric with burial depth are intermediate between turbidic and bioturbated mudstones.

As shown by Pusch (1970) and Reynolds (1988), large outsize silt grains and/or domains may profoundly affect both the original fabric pattern, as well as the postdepositional fabric developed during consolidation. The response of the sediment to subsequent shearing after initial deposition is also influenced by the occurrence of larger, outsized, rounded grains and domains, separated by finer-grained linkages of flat clay particles. The results of these clay-fabric studies clearly illustrate the need for grain size analysis along with fabric studies of marine sediments. If one were to quantify the sorting and relative proportion of larger clasts within the different sediment types, it would help predict fabric patterns with greater accuracy.

Weight–volume parameters

The weight–volume parameters include the void ratio, porosity, water content, specific gravity, and unit weight. The void ratio e is the ratio of the void volume V_v to the solid volume V_s. Porosity n is the ratio of the void volume V_v to the total volume V, and is usually expressed as a percentage:

$$n = (V_v)(V)^{-1} = (e)(1+e)^{-1} \qquad (24.1)$$

where

$$e = (V_v)(V_s)^{-1}$$

The water content w is one of the more useful relationships between the different phases of a sediment sample, and is expressed as the weight of pore water W_w divided by the weight of solid W_s in a given volume of sediment:

$$w = (W_w)(W_s)^{-1} \qquad (24.2)$$

The water content is obtained by weighing the wet sediment sample, drying it in an oven, weighing the dry sample, and calculating the ratio between the two weights. The natural water content w_n is the ratio of the weight of the pore water W_w to the total weight W. For nongas-

eous sediments, the natural water content is

$$w_n = (W_w)(W_s + W_w)^{-1} \qquad (24.3)$$

For seafloor sediments, corrections must incorporate the salinity of the seawater. Salt corrections adjust the dry-weight values of the solids based upon the salinity of the pore water (see Hamilton, 1971). Water content and natural water content are commonly used as index properties because of the relative ease in obtaining these measurements. In nongaseous sediments, the total unit weight γ of a given volume V of sediment V is

$$\gamma = (W)(V)^{-1} = (W_s + W_w)(V_s + V_w)^{-1} \qquad (24.4)$$

Hamilton (1974) presents empirical relations between mean grain diameter and density (Fig. 24.5A), and between mean grain diameter and porosity (Fig. 24.5B), for deep-sea sediment. The volume of pore space in a sediment depends upon the grain size, shape, sorting, and packing in addition to mineralogy (Hamilton, 1971). The cumulative effects of these factors results in a decrease in porosity with increasing grain size (Fig. 24.5B). In nongaseous sediments, the following general equation links bulk density of the sediment grains ρ_s, porosity n, density of the pore water ρ_w, and the saturated bulk density ρ_{sat} (Hamilton, 1974):

$$\rho_{sat} = n\,\rho_w + (1-n)\rho_s \qquad (24.5)$$

Thus, given a sediment sample in which grain size, grain density, and water content can be measured, the parameters of porosity and density can be computed. Given only a dried sample, grain size data can provide possible predictive relations with the other weight–volume parameters by using published empirical relations (Hamilton, 1974).

Atterberg limits

Atterberg limits are the water-content values that correspond with the various states of consistency in a remolded sediment. The liquid limit (LL) is the lowest water content at which a sample behaves as a liquid; the plastic limit (PL) corresponds to the lowest water content at which a sample behaves as a plastic. The difference between the plastic and liquid limit is the plasticity index (PI), and is the range of water con-

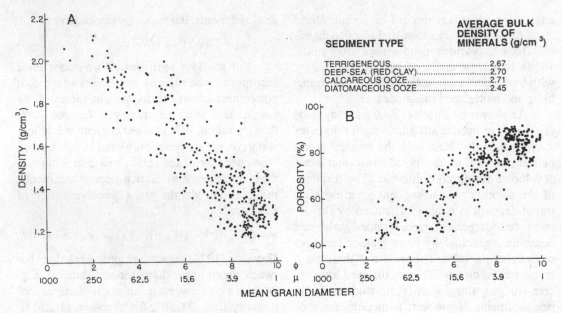

SEDIMENT TYPE	AVERAGE BULK DENSITY OF MINERALS (g/cm³)
TERRIGENEOUS	2.67
DEEP-SEA (RED CLAY)	2.70
CALCAREOUS OOZE	2.71
DIATOMACEOUS OOZE	2.45

Figure 24.5. Relationship of density (A) and porosity (B) to mean diameter of sediment grains. Table at the top lists the average bulk density of different deep-sea sediment types. (From Hamilton, 1974.)

tents over which a sediment behaves as a plastic. The liquidity index (LI) relates these Atterberg limits to the plasticity index PI by

$$LI = (w_n - PL)(PI)^{-1} \qquad (24.6)$$

where

$$PI = (LL - PL)$$

In the laboratory measurement of the Atterberg limits of seafloor samples, one should use artificial seawater (made from distilled water, at the appropriate NaCl concentration to correspond to the pore-water salinity) as opposed to distilled freshwater. Differences up to 60% in the LL and 80% in the PI can occur between salt- and distilled-water tests of Atterberg limits in marine soils (Chassefiere & Monaco, 1983) (Fig. 24.6). The liquid limit test is a function of both the grain size distribution and the mineralogy, and varies widely for different types of cohesive soil – from approximately 25 for silts up to about 600 for sodium bentonites (Skempton, 1953).

The effects of grain size on the Atterberg limits were clearly demonstrated by White 1949), and more recently by Bain (1971) for

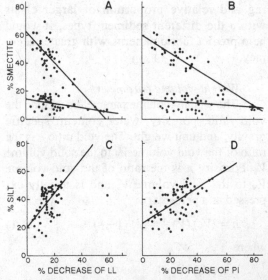

Figure 24.6. (A,B) Correlations between the variation of percentage decrease of the liquid limit (LL) and the plasticity index (PI) (between salt- and distilled-water test of Atterberg limits) and the percentage of smectite. (C,D) Correlations between the variation of percentage decrease of the liquid limit (LL) and the plasticity index (PI) (between salt- and distilled-water test of Atterberg limits) and the percentage of silt. (From Chassefiere & Monaco, 1983.)

kaolinite minerals (Fig. 24.7). Some of the kaolinites studied were virtually nonplastic, and give plasticity indices of <10; others show some plastic behavior, and plasticity indices vary ac-

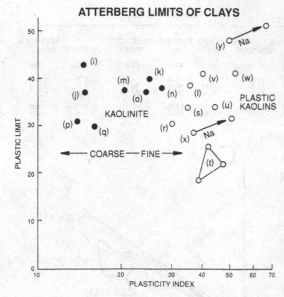

ATTERBERG LIMITS OF CLAYS

Figure 24.7. Influence of grain size on the Atterberg limits of clays: Kaolinite – Swaziland (i); china clays – Cornwall (j–l), Nigeria (m–n), Pakistan (o); differing grades of kaolin – Georgia (p–s), Gambia (t); ball clays – South Devon (u), Hong Kong (v), Dorset (w); "swamp" clay, Uganda (x); <10-μm fraction from plastic kaolin, St. Vincent (y). ○, plastic kaolins, •, kaolinite. (From Bain, 1971.)

cording to grain size. The lowest plasticity indices are from coarse-grained and well-crystallized kaolin deposits. Finer-grained and less-well-crystallized kaolin clays have higher plasticity indices.

In a typical site of mass wasting in the California Borderland, Hein & Gorsline (1981) analyzed the geotechnical and sedimentological aspects of fine-grained sediment–gravity flow deposits. Most of the samples were clayey silts, with varying amounts of sand (Fig. 24.8A). Geotechnical properties varied as follows: natural water contents, 28%–63%; LL, 94%–136%; PL, 29%–78%; and PI, 5%–58%. Using the plasticity chart, different lithofacies plot within different fields, although all samples fall below the A-line (Fig. 24.8B). The main control was interpreted as being textural, specifically the percentage of sandy and granular silts within the different lithofacies. The effect of bioturbation on the surface properties of the sediments separates sediments on the plasticity chart with regard to depth below the sediment–water inter-

face (Fig. 24.8C). This depth variation is interpreted as reflecting an alteration of the original sedimentary fabric by the burrowing activity of the organisms. Such burrowing increases porosity and provides a more open structure with enhanced porosity and permeability (Hein & Gorsline, 1981), and these sediments would be susceptible to subsequent failure – for example, under cyclic loading associated with earthquake shock.

Chassefiere & Monaco (1983), in their assessment of the use of Atterberg limits on marine soils, examined the geotechnical properties of sediments collected from a wide variety of environments within the Mediterranean Sea area. They found that sediments from different environments plotted into different zones on the plasticity chart (Fig. 24.9). This was mainly a function of the smectite content, and the percentage of silt within the samples (Figs. 24.6, 24.9) – reflecting an increase in maturity, with sediments with high silt content being more mature than those with a higher percentage of clay. Highly organic sediments plotted on the "upper limit of peats" line (Fig. 24.9).

Activity is defined as the ratio between the plasticity index and the clay size fraction percent (i.e., the percentage by weight of the dry weight of the soil that is composed of particles <2 μm in diameter). On plots of plasticity index versus percent clay fraction, the activity is the regression line (Fig. 24.10A,B). Different clays plot with different regression lines (Fig. 24.10A), with values of activity controlled more by the mineralogy (Fig. 24.10B) rather than the grain size of the material (Skempton, 1953). Much later, Grim (1962) demonstrated that there is only a general correlation between mineralogy and activity, and that particle size, solute salts, and organic material also influence the activity.

Skempton (1953) presents a series of graphs (Fig. 24.11) that illustrate the approximate relationship between void ratio e and overburden pressure σ' for clayey sediments as a function of the Atterberg limits. As can be seen here, the void ratio of high colloidal clays, with high LL and PI values, decreases very rapidly with increasing overburden pressure. By contrast, silty clays and silts with low LL and PI values show a greatly reduced regression

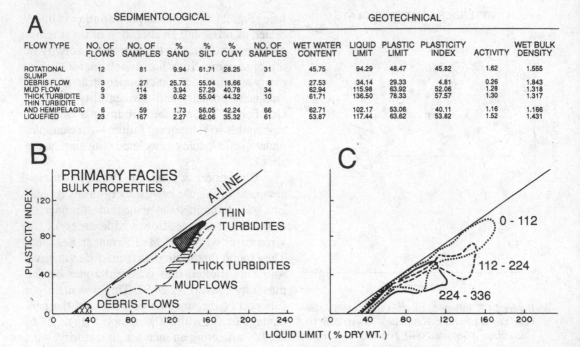

A

FLOW TYPE	SEDIMENTOLOGICAL					GEOTECHNICAL						
	NO. OF FLOWS	NO. OF SAMPLES	% SAND	% SILT	% CLAY	NO. OF SAMPLES	WET WATER CONTENT	LIQUID LIMIT	PLASTIC LIMIT	PLASTICITY INDEX	ACTIVITY	WET BULK DENSITY
ROTATIONAL SLUMP	12	81	9.94	61.71	28.25	31	45.75	94.29	48.47	45.82	1.62	1.555
DEBRIS FLOW	3	27	25.73	55.04	18.66	8	27.53	34.14	29.33	4.81	0.26	1.843
MUD FLOW	9	114	3.94	57.29	40.78	34	62.94	115.98	63.92	52.06	1.28	1.318
THICK TURBIDITE	3	28	0.62	55.04	44.32	10	61.71	136.50	78.33	57.57	1.30	1.317
THIN TURBIDITE AND HEMIPELAGIC	6	59	1.73	56.05	42.24	66	62.71	102.17	53.06	40.11	1.16	1.166
LIQUEFIED	23	167	2.27	62.06	35.32	61	53.87	117.44	63.62	53.82	1.52	1.431

Figure 24.8. (A) Sedimentological and geotechnical data of surficial (0.5–3.5-m) fine-grained mass flow deposits, California Continental Borderland. (B) Main zones on the plasticity chart occupied by the predominant facies types. (C) Main zones occupied by different depths when plotted on the plasticity chart, showing the influence of bioturbation on the Atterberg limits. Depths are in centimetres. Artificial seawater was used in measuring the Atterberg limits. (From Hein & Gorsline, 1981.)

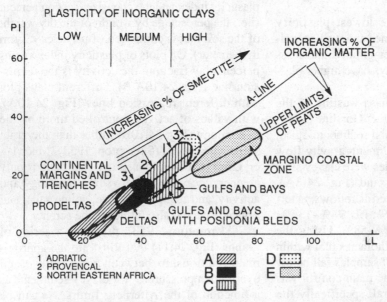

Figure 24.9. Plasticity chart showing the classification of shallow marine sediments from the Mediterranean Sea. Artificial seawater was used in measuring the Atterberg limits. (From Chassefiere & Monaco, 1983.)

curve between the void ratio and the overburden pressure. This trend reflects mainly the influence of the particle size distribution (and of the particle shape and fabric patterns within the sediment) on the development of original porosity,

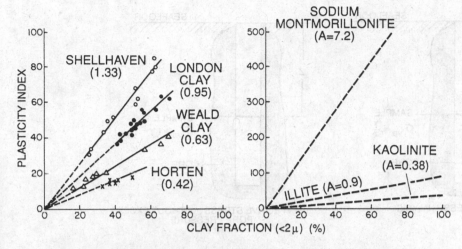

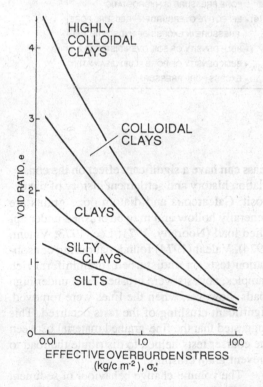

Figure 24.10. Relationship between the plasticity index and the percentage of clay. Figures in parentheses are the "activities" of the clays (i.e., the slope of the regression lines). (From Lambe & Whitman [1969]; redrawn from Skempton [1953].)

Figure 24.11. Compression curves for different sediment types, showing the relationships between the void ratio *e* and the effective overburden pressure σ′ for sediments as a function of the grain size. (From Lambe & Whitman [1969]; redrawn from Skempton [1953].)

and the effects of consolidation on porosity reduction.

Degree of consolidation and compressibility

After deposition on the seafloor, sediment consolidates, expelling pore water and collapsing the sedimentary fabric. This consolidation is generally associated with loading due to burial. Other loading mechanisms include earthquake shock, wave shock, and, in glacial environments, impact by icebergs and glacial ice advance onto the seafloor.

Using the ASTM (1988) definitions, a cohesive seafloor sediment can be classified as:
underconsolidated (not completely consolidated with respect to the existing overburden pressure);
normally consolidated (not subjected to an effective overburden pressure that exceeded the existing overburden pressure); or
overconsolidated (subjected to an effective overburden pressure that exceeded the existing overburden pressure).

In all these cases, the effective overburden pressure (effective stress) is the intergranular, average normal force per unit area (Fig. 24.12). According to Richards (1984), there may be little, if any, normally consolidated cohesive seabed sediment. Some marine sediments are apparently underconsolidated, generally due to excess pore pressure (Fig. 24.12). There have been many cases reported of apparent overconsolidation in marine sediments (e.g., Hamilton, 1974; Richards, 1984; Davie et al., 1978). The

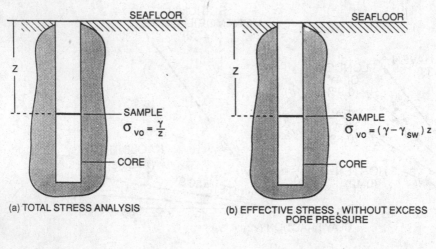

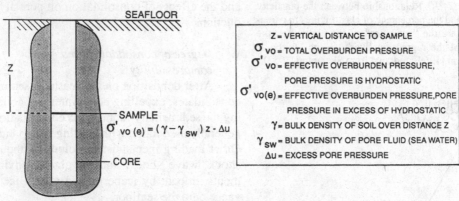

Figure 24.12. Examples showing the computation of overburden pressure or stress on a seabed sample of water-saturated soil to be collected for consolidation testing. (From Richards, 1984.)

cause of this apparent overconsolidation is not well understood, but it is clear that it represents neither removal of overburden by erosion nor the effects of dessication (Richards, 1984); rather, it may be related to strong interparticle forces or early cementation in cohesive seafloor sediments (Hamilton, 1974; Davie et al., 1978), or due to freeze–thaw conditions in marine silts of arctic regions.

The presence of fines is important in the development of overconsolidation and in the prevention of grain crushing in some marine sediments. These results suggest that the percentage of clay within a submarine sediment

mass can have a significant effect on the consolidation history and settlement history of the deposit. Calcareous and diatom ooze grains are generally hollow and may be crushed under applied load (Noorany, 1971; Lee, 1973; Valent, 1974). Valent (1974) found that during consolidation tests on undisturbed foraminiferal-rich samples, no tests were crushed even under high loads; however, when the fines were removed, significant crushing of the tests occurred. This suggested that the fine-grained material between the coarser tests helped to distribute the load to prevent grain crushing.

The volume-change behaviour of sediment is a complex of mechanical, physical, physico-chemical, and chemical changes. Although it is very difficult to quantify these relationships, empirical data show that in general the void ratio–pressure relationship is largely controlled by the grain size and plasticity of the sediment

(Fig. 24.11). As shown by Meade (1964), particle size is one of the most significant factors affecting both the void ratio and the physicochemical and physical processes of consolidation.

Shear strength

Slope failure occurs when the average shear stress T along a potential slide path is equivalent to the average shearing resistance. The strength of the sediment mass at failure (the Mohr–Coulomb criterion) is

$$T = c' + \sigma' \tan \phi' \qquad (24.7)$$

where c' is the effective intergranular or interparticle cohesion, and ϕ' is the effective angle of internal shearing resistance, in degrees. In granular sediments and normally consolidated clays, $c' = 0$; in overconsolidated submarine sediments, $c' > 0$. Some reported typical values of ϕ' for marine sediments are as follows: sand, 28°–34°; loose silt, <20°; sandy silt, 25°; calcareous ooze, 30°–42°; siliceous ooze, 36°–41°; and deep-sea clay 17.5°–37.5° (Davie et al., 1978; Silva et al., 1983). The carbonate content of deep-ocean sediments also can affect the ϕ' value: When carbonate content increased from 25% to >60%, ϕ' increased from 28° to 31° (Davie et al., 1978).

The susceptibility of a sediment mass to failure also depends upon whether there are drained or undrained conditions during shear failure (Morgenstern, 1967). Slumping under drained conditions is often associated with depositional oversteepening – for instance, on prodelta slopes or the fringes of submarine fans. Slumping under undrained conditions may occur when nonpermeable fine-grained sediments cap water-saturated, more porous, and permeable sediment.

Davie et al. (1978) summarized the downhole variation in shear strength of deep continental margin sediments obtained from 75 cores from 11 Deep Sea Drilling Project (DSDP) sites. Torvane shear strength tests were conducted on samples that fell within the clay or silty clay zones on the plasticity chart. Typical profiles of shear strength with depth are presented in Figure 24.13. All the onboard test results (Fig. 24.13A) show a general downcore increase in shear strength, with the exception of the calcar-

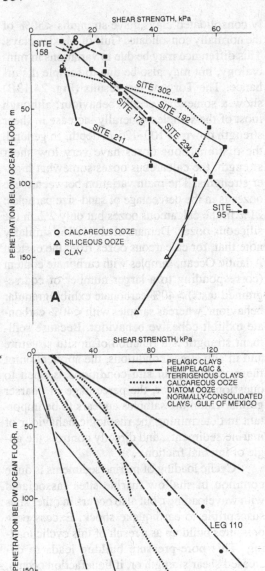

Figure 24.13. Variation of the shear strength of cohesive, deep-sea sediments with respect to depth. (A) DSDP onboard measurements. (B) Torvane strength test results from silty clays and silts recovered from DSDP sites. (From Davie et al., 1978; ODP site reports.)

eous oozes in site 294, which decrease in strength at 100 m. The DSDP clay strengths in Figure 24.13A, although obtained from normally consolidated clays, have strengths <50% of the normally consolidated Gulf of Mexico clays. This difference may be due to variations in mineralogy, but may also be due to sample disturbance. The Torvane test results (Fig. 24.13B) show a somewhat erratic behaviour, although most of the samples generally increase in shear strength down to 100–125-m depth. In general, the diatomaceous oozes have very low shear strength, the calcareous oozes somewhat higher strengths. The main variation between these oozes is in the percentage of sand-size particles: 21% in the calcareous oozes but only 2% in the siliceous oozes. Demars et al. (1976) similarly note that, for calcareous oozes from the eastern Atlantic Ocean, samples with carbonate content (corresponding to a larger number of coarse-grained tests) >40% carbonate exhibit granular behaviour, whereas samples with <40% carbonate exhibit cohesive behaviour. Because sediment strength is a function of in situ structure and in situ stress conditions, the application of these results to in situ conditions is open to question. However, the percentage of coarser grains within the samples appears to be important in determining the rheologic behaviour of marine sediments, and directly controls the angle of internal friction.

Cyclic loading of marine sediments is quite common in shallow marine sites (associated with waveloading), but also occurs in other sites susceptible to earthquake shock. Excess pore pressures build up as a result of this cyclic loading. This pore-pressure buildup leads to decreased shear strength or, if liquefaction occurs, complete loss of strength (Fig. 24.14). The susceptibility of a sediment to liquefaction is largely a function of the density, sorting, and packing arrangement of clasts in the sediment mass (Ishihara, 1985) (Fig. 24.15). Liquefaction potential due to cyclic loading increases with (1) void ratio, (2) magnitude of stress or strain, and (3) the number of stress cycles to which the material is subjected (Seed & Lee, 1967; Lee & Focht, 1976).

Liquefaction of ooze from the Mid-Atlantic Ridge under cyclic loading (associated with earthquakes) occurs for stress ratios (i.e., the ratio of shear stress to the consolidation pressure) of about 0.13 for 100 cycles, and 0.05 for

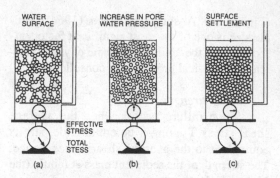

Figure 24.14. Schematic sketch illustrating the changes in sedimentary fabric for sands (a) prior to liquefaction, (b) during liquefaction, and (c) after liquefaction. (From Ishihara et al., 1985.)

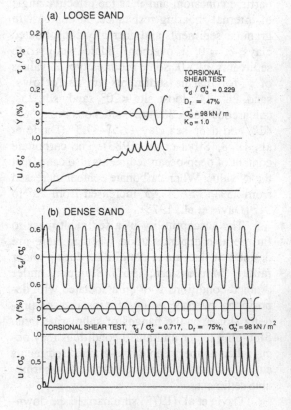

Figure 24.15. Records of cyclic torsional shear test results for (a) loose and (b) dense sand. (From Ishihara et al., 1985.)

about 5,000 cycles (Silva & Beverly, 1974). Such values are close to the lower bound of cyclic test results on sand and silty sand. The effect of cyclic loading in clays is complex and is influenced by a large number of factors (Lee & Focht, 1976). In general, the loss of strength

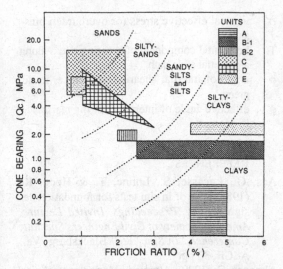

Figure 24.16. Relationships between the cone bearing (Q_c, MPa) and the friction ratio (FR, %) from piezocone penetrometer tests showing the influence of grain size on these factors. (From Moran et al., 1989.)

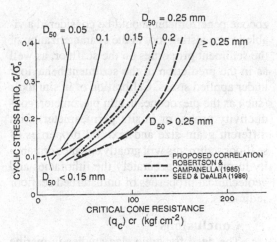

Figure 24.17. Comparisons of boundary curves for liquefaction determined by cone penetration tests (CPT) showing the influence of grain size on these factors. (From Shibata & Teparaksa, 1988.)

caused by cyclic loading for clays is less than that for loose sands. According to Lee & Focht (1976), for clays subjected to >1,000 cycles, the cyclic stress needed to cause significant strain is only 20%–50% of the static undrained strength.

Most recently workers have been using in situ methods, including the cone penetration test (CPT) (Robertson & Campanella, 1983, 1985; Shibata & Teparaksa, 1988), to assess the liquefaction potential of offshore sediments. This is largely because of the difficulty in obtaining undisturbed samples from the seafloor for subsequent testing. The piezocone penetrometer has been applied successfully by Moran et al. (1989) as a continuous, in situ testing tool to provide high-quality sedimentological data. Specifically, the ratio of cone friction to cone bearing (FR = f_c/Q_c) correlates with grain size. In plots of Q_c versus FR, sediment can be classified according to grain size and consistency (Fig. 24.16), where increasing values of FR correspond to an increase in the percentage of fines.

Shibata & Teparaksa (1988), in their evaluation of liquefaction potentials of soils using cone penetration tests, found that the normalized critical values of the CPT, which separate conditions of liquefaction from those of nonliquefaction, are a function of both the cyclic stress

ratio and the mean grain size of the sediment (Fig. 24.17). The results of their CPT-based liquefaction assessment method correlates with values proposed by Robertson & Campanella (1985) for sand and silty sand, but differs significantly from the curves proposed by Seed & De Alba (1986) for sands (Fig. 24.17). Using the results of the CPT-based liquefaction assessment method, Shibata & Teparaksa (1988) also present a chart for evaluating liquefaction potential based on the critical CPT values. Although the results of this sort of analysis have yet to be applied to deep-sea sediments, the utility of this sort of approach in predicting the liquefaction potential of offshore sediments is obvious.

Loose sand is generally very susceptible to liquefaction. Data on deep-sea oozes are generally lacking; they appear to behave like loose sand, but this behaviour needs more study. Clearly, the behaviour of sediment subjected to cyclic loading is complex, and related to many factors of the sediment fabric and the consolidation history of the sediment mass; it cannot be simply related to the grain size, although this does have a significant effect on the internal packing arrangement and fabric of the deposit. Further work must be done on the evaluation of the liquefaction potential of deep-sea sediments, and of their response to cyclic loading. The piezocone penetrometer should be considered a viable in situ testing tool in the characterization of the sediment properties on the seafloor, as well as in the prediction of the sediment behaviour under applied stress. Calibration of in situ tools (such as the piezocone, in situ piezometer, conductivity probe, or in situ permeameter) with different grain size and textural properties of seafloor sediments will greatly enhance our ability to map more accurately the lithofacies and geotechnical properties of undisturbed seafloor sediments.

Conclusions

The need for grain size studies in marine geotechnical studies is paramount in understanding the engineering properties of sediment on the seafloor. The grain size distribution either directly or indirectly (in combination with sedimentary fabric, packing, mineralogy, and organic carbon content) affects many of the mass physical properties of seafloor sediments. More detailed work is needed to understand how different facies (and depositional histories) contribute to a specific physical property or physical behaviour (e.g., consolidation, strength) of a marine sediment. Improved in situ testing of sea-bottom sediments will eventually lead to useful lithofacies and geotechnical maps, which can predict with greater accuracy the behaviour of seafloor sediments under dynamic and static loading.

LIST OF SYMBOLS

c	cohesion
e	void ratio
f_c	cone friction
FR	friction ratio
LI	liquidity index
LL	liquid limit
n	porosity
PI	plasticity index
PL	plastic limit
Q_c	cone bearing
V	volume (defined in text)
w	water content
w_n	natural water content
W	weight (defined in text)
γ	total unit weight
ρ	bulk density
σ	vertical stress (or pressure)
σ'	vertical effective stress (or overburden pressure)
T	horizontal component of shearing stress along a potential slide path
ϕ	angle of internal shearing resistance, in degrees
ϕ'	effective angle of internal shearing resistance, in degrees

REFERENCES

Aas, G., Lacasse, S., Lunne, T. & Høeg, K. (1986). Use of in situ tests for foundation design on clay. *Proceedings, Invited Lecture, American Society of Civil Engineers Specialty Conference "In Situ "86."* Blacksburg, Va.: ASCE, 30 pp.

Almagor, G. (1979). A review: Marine geotechnical studies at continental margins. *Special Committee on Oceanic Research (SCOR) Working Group 61 on Sedimentation at Continental Margins.* Canberra, Australia: SCOR, 100 pp.

American Geological Institute (1957). *Dictionary of geologic terms.* New York: Doubleday, 545 pp.

ASTM (1988). Designation D2216-80, Standard method for laboratory determination of water (moisture) content of soil, rock and soil-aggregate mixtures. In: *1988 Annual book of ASTM standards,* §4: Construction, soil and rock, building stones, geotextiles, 04.08: 262–4.

Bain, J. A. (1971). A plasticity chart as an aid to the identification and assessment of industrial clays. *Clay Minerals,* 9: 1–17.

Bennett, R. H., Bryant, W. R., & Keller, G. H. (1977). Clay fabric and geotechnical properties of selected submarine sediment cores from the Mississippi Delta. *National Oceanic and Atmospheric Administration,* Professional Paper no. 9, 86 pp.

(1981). Clay fabric of selected sediments: Fundamental properties and models. *Journal of Sedimentary Petrology,* 51: 217–32.

Bohlke, B. M., & Bennett, R. H. (1980). Mississippi prodelta crusts: A clay fabric and geotechnical analysis. *Marine Geotechnology,* 4: 55–82.

Bowles, J. E. (1970). *Engineering properties of soils and their measurement.* New York: McGraw–Hill, 187 pp.

Chassefiere, B., & Monaco, A. (1983). On the use of Atterberg limits on marine soils. *Marine Geotechnology,* 5: 153–79.

Davie, J. R., Fenske, C. W., & Serocki, S. T. (1978). Geotechnical properties of deep continental margin soils. *Marine Geotechnology*, 3: 85–119.

Demars, K. R., Nacci, V. A., Kelly, W. E., & Wang, M. C. (1976). Carbonate content: An index property for ocean sediments. *Offshore Technology Conference Proceedings*, 3: 97–106.

Denness, B. (1984). *Seabed mechanics*. London: Graham and Trotman, 206 pp.

Dunn, I. S., Anderson, L. R., & Kiefer, F. W. (1980). *Fundamentals of geotechnical analysis*. New York: John Wiley, 414 pp.

Grim, R. E. (1962). *Applied clay mineralogy*. New York: McGraw–Hill, 422 pp.

Hamilton, E. L. (1971). Prediction of in situ acoustic and elastic properties of marine sediments. *Journal of Geophysics Research*, 36: 266–84.

(1974). Prediction of deep-sea sediment properties: State-of-the-art. In: *Deep-sea sediments: Physical and mechanical properties*, ed. A. L. Inderbitzen. New York: Plenum Press, pp. 1–43.

Hein, F. J. (1985). Fine-grained slope and basin deposits, California Continental Borderland: Facies, depositional mechanisms and geotechnical properties. *Marine Geology*, 67: 237–62.

(1989a). Slope aprons and slope basins. In: *Deep marine environments: Clastic sedimentation and tectonics*, eds. K. Pickering, R. N. Hiscott, and F. J. Hein. London: Unwin–Hyman, pp. 91–108.

(1989b). Contourite drifts. In: *Deep marine environments: Clastic sedimentation and tectonics*, ed. K. Pickering, R. N. Hiscott, and F. J. Hein. London: Unwin–Hyman, pp. 219–42.

Hein, F. J., & Gorsline, D. S. (1981). Geotechnical aspects of fine-grained mass flow deposits: California Continental Borderland. *Geo-Marine Letters*, 1: 1–6.

Inderbitzen, A. L. (ed.) (1974). *Deep-sea sediments: Physical and mechanical properties*. New York: Plenum Press.

Ishihara, K. (1985). Stability of natural deposits during earthquakes. In: Publications Committee of XI ICSMFE, *Proceedings of the 11th International Conference on Soil Mechanics and Foundation Engineering, San Francisco, vol. 1, Theme lectures*. Boston: A. A. Balkema, pp. 321–77.

Jamiolkowski, M., Ladd, C. C., Germaine, J. T., & Lancellotta, R. (1985). New developments in field and laboratory testing of soils. In: Publications Committee of XI ICSMFE, *Proceedings of the 11th International Conference on Soil Mechanics and Foundation Engineering, San Francisco, vol. 1, Theme lectures*. Boston: A. A. Balkema, pp. 57–154.

Lambe, T. W., & Whitman, R. V. (1969). *Soil mechanics*. New York: John Wiley, 553 pp.

Lee, H. J. (1973). *Engineering properties of some North Pacific and Bering Sea soils*. U.S. Naval Civil Engineering Laboratory Technical Note N-1283, 27 pp.

Lee, K. L., & Focht, J. A., Jr. (1976). Strength of clay subject to cyclic loading. *Marine Geotechnology*, 1: 165–85.

Meade, R. H. (1964). *Removal of water and rearrangement of particles during compaction of clayey sediments – review*. U.S. Geological Survey Professional Paper 497–B, Washington, D.C.: USGPO.

Mitchell, J. K. (1976). *Fundamentals of soil behavior*. New York: John Wiley, 422 pp.

Moran, K., Hill, P. R., & Blasco, S. M. (1989). Interpretation of piezocone penetrometer profiles in sediment from the Mackenzie Trough, Canadian Beaufort Sea. *Journal of Sedimentary Petrology*, 59: 88–97.

Morgenstern, N. R. (1967). Submarine slumping and the initiation of turbidity currents. In: *Marine geotechnique*, ed. A. F. Richards. Urbana: University of Illinois Press, pp. 189–220.

Noorany, I. (1971). Engineering properties of calcareous submarine sediments from the Pacific. In: *Proceedings of the International Symposium on the Engineering Properties of Sea-floor Soils and Their Geophysical Identification*. Seattle: UNESCO, NSF, and University of Washington, pp. 130–9.

Noorany, I., & Gizienski, S. F. (1970). Engineering properties of submarine soils: State-of-the-art reviews. *Journal of Soil Mechanics* and *Foundation Division, Proceedings of the American Society of Civil Engineers*, 96 (no. SM5): 1735–62.

O'Brien, N. R. (1970). The fabric of shale – an electron microscope study. *Sedimentology*, 15: 229–46.

(1987). The effects of bioturbation on the fabric of shale. *Journal of Sedimentary Petrology*, 57: 449–55.

Poulos, H. G. (1988). *Marine geotechnics*, London: Unwin–Hyman, 404 pp.

Pusch, R. (1970). Microstructural changes in soft

quick clay at failure. *Canadian Geotechnical Journal*, 7: 1–7.

(1973). Influence of salinity and organic matter on the formation of clay microstructure. *Proceedings of the International Symposium on Soil Structure, Gothenburg, Sweden*, pp. 161–73.

Quigley, R. M. (1980). Geology, mineralogy and geochemistry of Canadian soft soils: A geotechnical perspective. *Canadian Geotechnical Journal*, 17: 261–85.

Quigley, R. M., & Thompson, C. D. (1966). The fabric of anisotropically consolidated sensitive marine clay. *Canadian Geotechnical Journal*, 3: 61–73.

Reynolds, S. (1988). The fabrics of deep-sea muds and mudstones: A scanning electron microscope study. Unpublished Ph.D. diss., University of Southern California, Los Angeles, 169 pp.

Richards, A. F. (ed.) (1967). *Marine geotechnique*. Urbana: University of Illinois Press, 327 pp.

(1976). Marine geotechnics of the Oslofjorden region. In: *Contributions to soil mechanics: Laurits Bjerrum memorial volume*. Oslo: Norwegian Geotechnical Institute, pp. 41–63.

(1984). Introduction: Modelling and consolidation of marine soils. In: *Seabed mechanics*, ed. B. Denness. London: Graham and Trotman, pp. 1–15.

Richards, A. F., & Chaney, R. C. (1980). Marine slope stability – a geologic approach. In: *Marine slides and other mass movements*, eds. S. Saxov and J. K. Nieuwenhuis. New York: Plenum Press, pp. 163–74.

Richards, A. F., Palmer, H. D., & Perlow, M., Jr. (1975). Review of continental shelf marine geotechnics: Distribution of soils, measurement of properties, and environmental hazards. *Marine Geotechnology*, 1: 33–67.

Richards, A. F., & Parks, J. M. (1976). Marine geotechnology. In: *The benthic boundary layer*, ed. I. N. McCave. New York: Plenum Press, pp. 157–81.

Robertson, P. K., & Campanella, R. G. (1983). Interpretation of cone penetration tests, Part I: Sand. *Canadian Geotechnical Journal*, 20: 718–33.

(1985). Liquefaction potential of sands using the CPT. *Journal of Geotechnical Engineering, American Society of Civil Engineers*, 111(3): 384–403.

Seed, H. B., & De Alba, P. M. (1986). Use of SPT and CPT tests for evaluating the liquefaction resistance of sands. In: *Proceedings of in-situ testing*. Santa Barbara: ASCE, pp. 281–302.

Seed, H. B., & Lee, K. L. (1966). Liquefaction of saturated sands during cyclic loading conditions. *Journal of the Soil Mechanics and Foundations Division, American Society of Civil Engineers*, 92 (no. SM6): 105–34.

Shibata, T., & Teparaksa, W. (1988). Evaluation of liquefaction potentials of soils using cone penetration tests. *Soils and Foundations*, 28: 49–60.

Silva, A. J. (1974). Marine geomechanics: An overview and projections. In: *Deep-sea sediments: Physical and mechanical properties:*, ed. A. L. Inderbitzen. New York: Plenum Press, pp. 45–76.

Silva, A. J., & Beverly, B. E. (1974). *Geotechnical and dynamic properties of calcareous sediments as related to the DSDP re-entry cone*. Unpublished report to the DSDP. La Jolla, Calif: Scripps Institute of Oceanography.

Silva, A. J., Moran, K., & Akers, S. A. (1983). Stress–strain–time behavior of deep sea clays. *Canadian Geotechnical Journal*, 20: 517–31.

Skempton, A. W. (1953). Soil mechanics in relation to geology. *Proceedings of the Yorkshire Geological Society*, 29: 33–62.

Syvitski, J. P. M., & Farrow, G. E. (1989). Fjord sedimentation as an analogue for small hydrocarbon bearing submarine fans. In: *Deltas: Sites and traps for petroleum accumulations*, eds. M. K. G. Whateley and K. T. Pickering. Geological Society Special Publication no. 41, pp. 21–43.

Syvitski, J. P. M., & Murray, J. W. (1981). Particle interaction in fjord suspended sediment. *Marine Geology*, 39: 215–42.

Valent, P. J. (1974). *Short-term engineering behavior of a deep-sea calcareous sediment*. U.S. Civil Engineering Laboratory Technical Note N-1334, 35 pp.

Walker, R. G. (1979). Turbidites and associated coarse clastic deposits. In: *Facies models*, ed. R. G. Walker. Ottawa: Geoscience Canada Reprint Series 1 (1st ed.), pp. 91–104.

White, W. A. (1949). Atterberg limits of clay minerals. *American Mineralogist*, 34: 508–12.

Index